Ecology and Biology of Soil

Ecology and Biology of Soil

Edited by Anthony Sherman

SYRAWOOD
PUBLISHING HOUSE
New York

Published by Syrawood Publishing House,
750 Third Avenue, 9th Floor,
New York, NY 10017, USA
www.syrawoodpublishinghouse.com

Ecology and Biology of Soil
Edited by Anthony Sherman

© 2018 Syrawood Publishing House

International Standard Book Number: 978-1-68286-500-2 (Hardback)

Cataloging-in-Publication Data

Ecology and biology of soil / edited by Anthony Sherman.
   p. cm.
Includes bibliographical references and index.
ISBN 978-1-68286-500-2
1. Soils. 2. Soil ecology. 3. Soil biology. 4. Soil science. I. Sherman, Anthony.
S591 .E26 2018
631.46--dc23

TABLE OF CONTENTS

PREFACE

It is often said that books are a boon to mankind. They document every progress and pass on the knowledge from one generation to the other. They play a crucial role in our lives. Thus I was both excited and nervous while editing this book. I was pleased by the thought of being able to make a mark but I was also nervous to do it right because the future of students depends upon it. Hence, I took a few months to research further into the discipline, revise my knowledge and also explore some more aspects. Post this process, I begun with the editing of this book.

The study of microbial activities in the soil is known as soil biology. It plays an important role in understanding the characteristics of soil. The soil biota consists of mesofauna, macrofauna, microfauna and megafauna. This book provides comprehensive insights into the fields of soil biology and ecology. A number of latest researches have been included to keep the readers up-to-date with the global concepts in this area of study. For all readers who are interested in soil biology and soil ecology the case studies included in this book will serve as an excellent guide to develop a comprehensive understanding.

I thank my publisher with all my heart for considering me worthy of this unparalleled opportunity and for showing unwavering faith in my skills. I would also like to thank the editorial team who worked closely with me at every step and contributed immensely towards the successful completion of this book. Last but not the least, I wish to thank my friends and colleagues for their support.

Editor

Carbon Stocks and Fluxes in Tropical Lowland Dipterocarp Rainforests in Sabah, Malaysian Borneo

Philippe Saner[1]*, Yen Yee Loh[2], Robert C. Ong[3], Andy Hector[1]

1 Institute of Evolutionary Biology and Environmental Studies, University of Zurich, Zurich, Switzerland, **2** School of International Tropical Forestry, Universiti Malaysia Sabah, Sabah, Malaysia, **3** Forest Research Center Institute, Sabah, Malaysia

Abstract

Deforestation in the tropics is an important source of carbon C release to the atmosphere. To provide a sound scientific base for efforts taken to reduce emissions from deforestation and degradation (REDD+) good estimates of C stocks and fluxes are important. We present components of the C balance for selectively logged lowland tropical dipterocarp rainforest in the Malua Forest Reserve of Sabah, Malaysian Borneo. Total organic C in this area was 167.9 Mg C ha^{-1}±3.8 (SD), including: Total aboveground (TAGC: 55%; 91.9 Mg C ha^{-1}±2.9 SEM) and belowground carbon in trees (TBGC: 10%; 16.5 Mg C ha^{-1}±0.5 SEM), deadwood (8%; 13.2 Mg C ha^{-1}±3.5 SEM) and soil organic matter (SOM: 24%; 39.6 Mg C ha^{-1}±0.9 SEM), understory vegetation (3%; 5.1 Mg C ha^{-1}±1.7 SEM), standing litter (<1%; 0.7 Mg C ha^{-1}±0.1 SEM) and fine root biomass (<1%; 0.9 Mg C ha^{-1}±0.1 SEM). Fluxes included litterfall, a proxy for leaf net primary productivity (4.9 Mg C ha^{-1} yr^{-1}±0.1 SEM), and soil respiration, a measure for heterotrophic ecosystem respiration (28.6 Mg C ha^{-1} yr^{-1}±1.2 SEM). The missing estimates necessary to close the C balance are wood net primary productivity and autotrophic respiration. Twenty-two years after logging TAGC stocks were 28% lower compared to unlogged forest (128 Mg C ha^{-1}±13.4 SEM); a combined weighted average mean reduction due to selective logging of −57.8 Mg C ha^{-1} (with 95% CI −75.5 to −40.2). Based on the findings we conclude that selective logging decreased the dipterocarp stock by 55–66%. Silvicultural treatments may have the potential to accelerate the recovery of dipterocarp C stocks to pre-logging levels.

Editor: Jerome Chave, Centre National de la Recherche Scientifique, France

Funding: This project was funded by the University of Zurich, with financial support from the Darwin Initiative (United Kingdom Department for Environment, Food and Rural Affairs) and is part of the Royal Society South-East Asia Rainforest Research Programme (Project No. RS243). The funders had no role in study design, data collection and analysis, decision to publish, or preparation of the manuscript.

Competing Interests: The authors have declared that no competing interests exist.

* E-mail: philippe.saner@uzh.ch

Introduction

The lowland rain forests on the island of Borneo are recognized as a living carbon density hotspot [1], with an average aboveground biomass that is roughly 60% (457.1 Mg ha^{-1}) higher than the Amazonian average of 288.6 Mg ha^{-1} [2]. Most of Borneo was covered with tropical evergreen rainforest until the 1950s [3] but an approximate annual deforestation rate of 1.7% has decreased total forest cover to 57% of the original area by 2002 [4]. The subsequent carbon C loss associated to deforestation is a representative trend for the whole Indo-Malaya region, including South Asia, Southeast Asia and Papua New Guinea, were forest cover was less than 40% of the original area by 2000 [5]. Consequently, ongoing exploitation for timber is an important source of C emissions and efforts are proposed to reduce C release to the atmosphere from deforestation and degradation (REDD+) [6]. However, there is still considerable uncertainty about C stocks [7] and fluxes [8] and their subsequent losses and accumulation rates in tropical rain forests, in particular for South East Asia [9]. An adequate understanding of the state of the remaining mixed dipterocarp forests and present C stocks and flows at the regional and local level is therefore needed [10].

An overview of recent forest carbon stocks in tropical regions is given elsewhere, where they reported a total aboveground C (TAGC) estimate of 164–196 Mg C ha^{-1} for Malaysia [7]. For the Malaysian state of Sabah (North Borneo), where this study was undertaken, logging played an important role in the tiger economy, particularly in the 1980s [11]. After decades of commercial timber exploitation and subsequent conversion of the land cover to palm oil plantations (*Elaeis guinensis*) 53% of the original forest coverage of Sabah remains [12] with less than 15% in an undisturbed primary forest state. The secondary production forests are in various stages of post-logging condition, ranging from almost pristine to selectively logged and even completely degraded (Fig. 1).

In this paper, we estimate impacts of logging on components of the C balance of an area of secondary production forest of the Malua Forest Reserve that is the location of the Sabah Biodiversity Experiment. We present it in the context of previous work made in the neighbouring primary lowland rainforest of the Danum Valley Conservation Area as well as to other forests in the region. Here we ask how selective logging has impacted C stocks and fluxes after 22 years and predict C losses for the area of study. This can later be used as a baseline to study changes in the C balance due to enrichment planting with dipterocarp seedlings and other management techniques.

Materials and Methods

Location

The selectively logged and unlogged forest study sites are located within lowland dipterocarp forest of East Sabah,

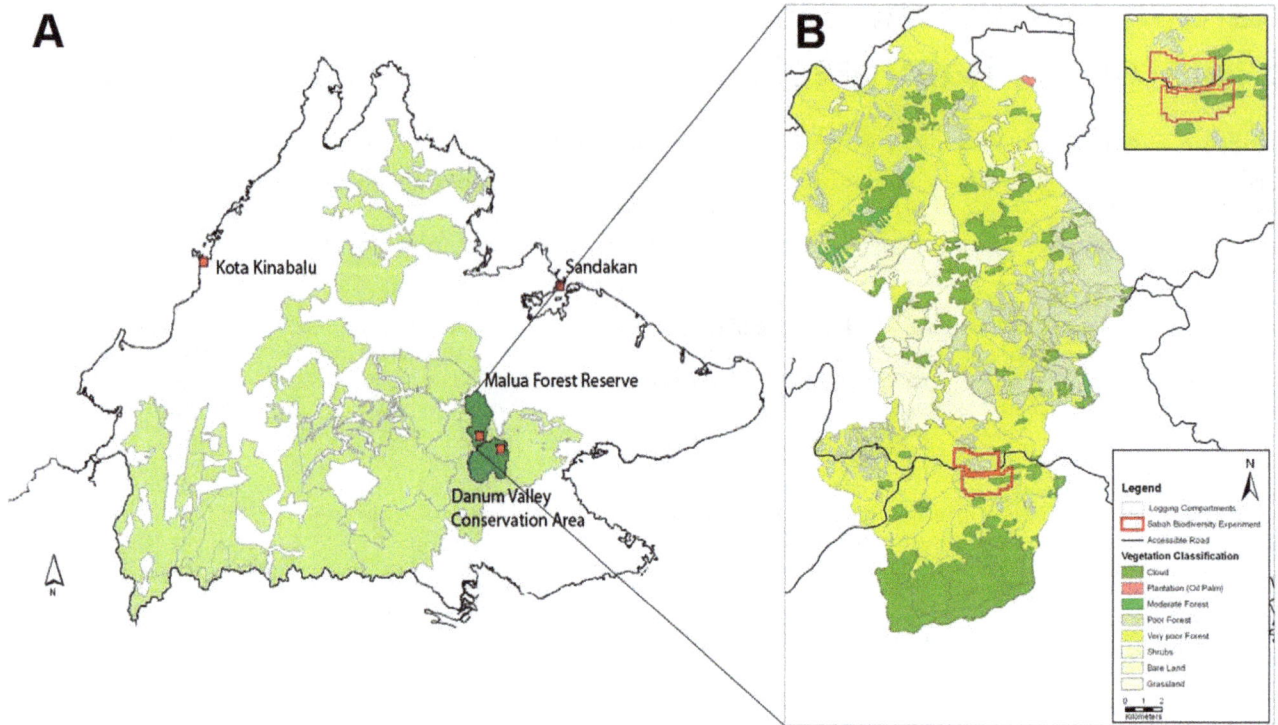

Figure 1. Map of Sabah and forest quality map of Malua Forest Reserve. (A) Allocated production forest is shown across Sabah (light green, also including Malua Forest Reserve). The study sites (marked in red) are located in the Malua Forest Reserve and the Danum Valley Conservation Area (dark green). (B) Forest quality map of the Malua Forest Reserve (visual interpretation of aerial photographs, 1:17,500). The boundary of the Sabah Biodiversity Experiment (500 ha) is outlined (red). Forest classification is based on number of trees ≥60 cm DBH derived from crown size. Cloud forest: >16 trees ha^{-1}, Moderate: 9–16 trees ha^{-1}, Poor: 5–8 trees ha^{-1}, Very Poor: 1–4 trees ha^{-1}, Shrubs/Bare Land/Grassland: none, Plantation: Oil palm monoculture.

Malaysian North Borneo (Royal Society South-East Asia Rainforest Research Programme Project No. RS243) (Fig. 1A). The region is aseasonal with an annual rainfall of >3000 mm [13]. The forest covers a concession of one million hectares which belongs to the publicly owned Sabah Foundation. Most of the area has been selectively logged twice, once in the 1980s and once within the last ten years. An area of unlogged primary forest–the Danum Valley Conservation Area–was left at the heart of the concession. Our unlogged forest site is in this area, along the main trail of the grid system to the west (W6–W9) and less than 1 km from the Danum Valley Field Centre (N05°19′21″ E117°26′26″, 220–284 m.a.s.l). A general description for the soil, geology and vegetation of Danum Valley is given elsewhere [14]. The logged forest site is part of the Sabah Biodiversity Experiment (N05°05′20″ E117°38′32″, 102 m.a.s.l.) which is located in the southern part of the Malua Forest Reserve, a 35,000 ha area of selectively logged production forest that lies immediately to the North of Danum Valley (22.6 km air distance from the Danum Valley Field Centre) (Fig. 1B). The Malua Forest Reserve was selectively logged during the early 1980s (and relogged in 2007, with the exception of the area used for this study that was established in 2003 and therefore excluded from the second round). Yearly production volume, based on data from logging coupes of the same region (Lahad Datu district), were reported as 95–157 m^3 ha^{-1} (1970–1980) and 75–134 m^3 ha^{-1} (1980–1990) (unpublished data, Danum Valley Field Centre). Our estimates of logged forest are for the background vegetation of the Sabah Biodiversity Experiment, a forest restoration project, that covers 500 ha forest that was selectively logged in 1986. The large-scale

experiment examines impacts of dipterocarp diversity on forest structure and functioning [15–18].

Site characteristics, including topography, soil characteristics and estimated pre-logging timber volume are comparable at both sites (selectively logged and unlogged) before logging (Table 1 and Table S1). Hence, we assume that changes in basal area and C stocks in the two different forest types largely indicate the effect of selective logging, rather than site specific differences in pre-logging conditions.

Table 1. Site characteristics of a pre-logging survey completed in 1983.

	Danum	Malua
Elevation	<250 m	<250 m
Topography	Slopes 15–25°	Slopes 0–20°
Parent Material	Mudstone	Mudstone
Soil Type	Orthic acrisol	Orthic acrisol
Soil Family	Tanjong Lipat	Tanjong Lipat
Soil Association	Bang	Kretam and Mentapok
Est. Pre-logging Volume	178–230 m^3 ha^{-1}	193–221 m^3 ha^{-1}
Est. Dipterocarp Volume	149–225 m^3 ha^{-1}	180–216 m^3 ha^{-1}

Site descriptions for both forest types (selectively logged and unlogged) were extracted from Yayasan Sabah Forest Management Plan 1984–2032 (unpublished data).

Components of the C balance of the logged and unlogged forests

We estimate the main C stocks, the minor C stocks and the measured C fluxes, missing to close the C balance are wood net primary productivity (NPP) and autotrophic respiration. The components were measured on site or estimated from existing biomass regression models and biomass partitioning ratios using methods given below. A less extensive survey was made for the unlogged primary forest of Danum Valley using identical regression models to estimate total aboveground tree biomass. Stocks and fluxes measured in selectively logged forest were compared to existing published literature from the unlogged primary forest at Danum Valley.

Aboveground tree basal diameter

The tree inventory was taken in July 2008 in the selectively logged forest of the Sabah Biodiversity Experiment (areas excluded from enrichment planting were used for the inventory) and in unlogged forest close to the Danum Valley Field Centre (Fig. 1). The size of the sampled areas in both forest types was one hectare divided into 4 replicate transect lines, each 250×10 m (100 m apart). Lines and orientation (North-South) were measured with a handheld GPS to 10 m accuracy (GPSMAP 60 CSx, Garmin, USA). All trees >10 cm diameter breast height (DBH at 130 cm) that were within five meters on each side of the transect line were tagged, measured with a DBH tape (Yamayo, Japan) and identified to species or, if unknown, to genus level (Table S2). We followed the RAINFOR field manual [19] to measure all trees in a comparable standardized way. If a tree showed large buttresses its DBH was measured just above using a ladder.

Wood density estimation

Species specific wood density estimates from the World Agroforestry Centre Wood Density Database (WAC; Table S2) were used and cross referenced with on site measurements on a subset of 18 local tree species ($n = 32$, $R^2 = 0.81$; unpublished data). For our own measurements wood density was defined as the oven-dry weight of wood divided by its wet volume [20]. Wood cores were taken with an increment borer at 130 cm height (Haglöf, Sweden) on trees that were located outside of the survey lines. Wood density values were calculated using the water-displacement method described in [21]. If the species was unknown or the wood density not available we took mean wood density of the genus as a substitute. In the few cases (<4 cases) where genus wood density was not available a mean overall wood density of 0.6 g cm^{-3} was taken.

Total aboveground tree biomass and C stocks

An established DBH-AGB regression model was used to estimate tree biomass and C stocks in logged and unlogged forest [22]. The general form of the equation is:

$$\ln(TAGB) = c + \alpha \ln(DBH) + \beta \ln(WD) \qquad (1)$$

where TAGB is in kg tree^{-1}, DBH is in cm, c (-0.744) is the intercept, α (2.188) and β are the slope coefficient of the regression for mixed species forest and WD is wood density in g cm^{-3}.

Several other published allometric models were considered for direct comparison [23–26]. As site specificity is known to be of importance for the selection of allometric models [22], all else being equal we took the DBH-AGB regression model from research on the geographically closest study area. This choice led to lowest estimation for aboveground tree biomass and is therefore

probably a conservative approach. All reported standard errors describe the variation among the 4 transect lines in each forest type (logged, unlogged) and do not incorporate error in allometric equations. Carbon content was assumed to be 50% by total biomass for trees and understory vegetation [24].

Total belowground tree biomass and C stocks

Coarse root biomass (>2 mm diameter) was estimated based on values derived from Pasoh Forest Reserve (peninsular Malaysia), where they reported a BGB/AGB biomass partitioning ratio of 0.18 [27]. Carbon content was assumed to be 50% by biomass [28].

Standing deadwood

DBH, height, degradation state (not degraded, degraded, heavily degraded) and tree structure (stem only/stem plus branches) was noted for all dead standing trees in the selectively logged forest that were within five meter on each side of the transect lines. The height of dead standing trees was estimated visually and tree volume was calculated following the methods described in [29,30]. Due to the lack of correct wood density estimates for sampled dead trees a wood density of 0.5 g cm^{-3} was assumed as reported from Venezuela [31].

Soil organic matter

Soil organic C from plant material was estimated from randomly located sites and represented large scale spatial variability. Soil core samples were taken from the middle of 13 unplanted 4 ha control plots in the selectively logged forest of the Sabah Biodiversity Experiment in June 2006. At each plot three 1 m soil pits were excavated, each located 10 m away from the middle of the plot directed either South, Northwest or Northeast. Soil cores (100 cm^3) were inserted horizontally down to 1 m depth and sub-samples taken from 11 different layers (0–5, 5–10, 10–20, 20–30, 30–40, 40–50, 50–60, 60–70, 70–80, 80–90, 90–100 cm) using standard soil corers (Eijkelkamp, Netherlands) and a rubber hammer ($n = 396$). The three replications (South, Northwest and Northeast) of each layer in each plot were pooled for subsequent soil analysis to include site variability. Damp soil samples were laid out to dry in trays in a well ventilated room until the soil sample weight became quite consistent (about 5 days to a week depending on the initial conditions of the samples). Rock and root components were separated from each sample. Weight and volume was determined using the water displacement method for rocks and using ethanol displacement, with a lower density for estimating fine root volume. Bulk density was derived from the equation:

$$BD = (MS - MR)/(VS - VR) \qquad (2)$$

where BD is bulk density (g cm^{-3}), MS is mass of the dry soil (g), MR is mass of rocks and roots (g), VS is volume of dry soil (ml) and VR is volume of rocks and roots (ml). Samples were further ground with a porcelain mortar and pestle to pass through a 2 mm sieve. C content was determined by the Walkley-Black method, a wet chemical analysis [32,33].

Understory vegetation

Six quadrats (5×5 m) along each of 4 transect lines in the selectively logged forest were randomly selected ($n = 24$). Within the quadrat saplings (<10 cm DBH and >2 m height), seedlings (<2 m height) and all woody vines were harvested. All saplings were individually measured using a common spring scale. Wet biomass for seedlings was measured once for all seedlings together. A subsample of saplings, seedlings and woody vines was dried to constant mass (7 days, 60°C) to relate wet to dry biomass (y = 0.55x, R^2 = 0.98, n = 76, intercept was set to zero). One *Eusideroxylon zwageri* (belian, IUCN status: vulnerable) sapling was not harvested due to its age and size (probably about 30 years, 5.2 cm DBH).

Standing litter

Within each quadrat 0.5×0.5 m subquadrats were established to collect all woody debris and leaf litter separately. The low estimate of this study may indicate that the method of collecting all woody debris within 24 subquadrats (0.5×0.5 m²) did not capture large downed woody debris adequately and that other proposed methods should be considered [30,34].

Fine root biomass

In each subquadrat of 0.5×0.5 m vertical soil cores (100 cm³) at the soil surface were taken (0–5 cm) using standard soil corers (Eijkelkamp, Netherlands). Fine roots (≤2 mm diameter) were extracted by washing the soil cores over a 210 μm sieve (Retsch, Germany). All collected samples were dried in a glasshouse (7 days, 60°C) before measuring their dry biomass with a precision scale.

Litterfall

Litterfall traps (1 m²) were randomly allocated along the 4 transect lines in the selectively logged forest at 130 cm height, using fine meshed plastic net (n = 40). The sampling complies with previously proposed standards, including a minimal sampling duration of 1 year, a total sampled surface of 40 m² and a trap surface of more than 0.25 m² [35,36]. Litter was collected every other week over one year (June 2006–June 2007, n = 25) and was further separated into leaves, twigs (typically <1 cm in diameter) and reproductive organs (flowers and fruits) [37]. Litterfall samples were dried in a glasshouse (7 days, 60°C) before measuring their biomass with a precision scale. One litterfall measurement was discarded because of a freshly fallen climber fruit that biased the analysis (>7 g day⁻¹). A carbon content of 42% for litterfall, standing leaf litter and woody debris was used [38].

Soil respiration rates

Soil respiration rates were estimated following the methods reported in [15]. Briefly, we took measurements using an Infrared Gas Analyzer CARBOCAP GMP343 (Vaisala, Finland) and a self made chamber [39] at forty sites along the transect lines in May and June 2007. We measured nine times (seven day time (08:00 am to noon) and two night time (08:00 pm to 04:00 am) measurements) over five minute intervals.

Statistical analysis

Our focus for the C balance is on the estimation of means and variabilities, so we present point estimates with standard deviations (SD; given as mean ± SD), standard errors (SEM; given as mean ± SEM), or confidence intervals (CI; given as mean with lower–upper bound) as appropriate. Standard deviations for the total sum of means are reported as the sum of the weighted variances, where the weighting is based on the relative contribution of the single mean and the number of replications (see Table S3).

Fixed effects estimates derived from meta-analytical methods were used for the across study comparison of forest types (selectively logged and unlogged) [40]. Mean differences (with 95% CI) were calculated with the metacont function from the *meta* package (version 1.6-1) for R 2.12.1 [41], using means, SD and replication (see Table S6). Inverse variance weighting was used for pooling the data.

Simple t-tests were performed for comparing components of the C balance between published literature estimates (mean ± SEM) and own measures (mean ± SEM). Differences were reported with the corresponding standard error of the difference (SED) (Table 2). Non-linear patterns in litterfall rates were examined by fitting a cubic polynomial regression to the log transformed response variable (see Fig. S1).

Table 2. Comparison of basal area, volume and selected stocks and fluxes related to C turnover processes.

	Unit	Logged	Unlogged	Difference	t	p
Basal area	m² ha⁻¹	25±0.8	29.9±0.7	4.9 (±1.1)	4.6	>0.01
– Dipterocarp	m² ha⁻¹	6.9±0.2	18.2±0.7	11.3 (±0.7)	15.5	>0.0001
– Pioneer	m² ha⁻¹	8.7±0.2	0.1±0.1	8.6 (±0.2)	38.5	>0.0001
Basal area SDW	m² ha⁻¹	6.0±0.3	5.6±0.5[a]	0.4 (±0.6)	0.7	ns
C components						
TAGC	Mg C ha⁻¹	91.9±2.9	128±13.4	36.1 (±13.7)	2.6	>0.05
Downed woody debris	Mg C ha⁻¹	20.6±8.8	70.6±9.5[a]	50.0 (±13.4)	3.7	>0.001
SDW	Mg C ha⁻¹	8.7±3.5	6.5±1.1[a]	2.2 (±3.7)	0.6	ns
Fine root biomass	Mg C ha⁻¹	0.95±0.1	0.85±0.02[b]	0.1 (±0.3)	0.3	ns
Flux sampling						
Fine litterfall rate	Mg C ha⁻¹ yr⁻¹	4.9±0.1	4.8±0.1[c]	0.1 (±0.1)	0.7	ns
Soil respiration rate	Mg C ha⁻¹ yr⁻¹	28.6±1.2	21.6±1.4[d]	7.0 (±1.9)	3.7	>0.001

Means ± SEM from selectively logged forest (Sabah Biodiversity Experiment) and unlogged forest sites (Danum Valley, except for soil respiration reported from Lambir Hills, Sarawak, North Borneo). Difference (± SED) is reported with the corresponding t-value and level of significance. SDW: Standing deadwood, TAGC: Total aboveground carbon.
[a]Gale 2000.
[b]Green et al. 2005.
[c]Burghouts et al. 1992.
[d]Katayama et al. 2009.

Results

Based on our sampling of the Sabah Biodiversity Experiment plots, we estimated the total C stock in selectively logged areas of the Malua Forest Reserve as 167.9 Mg C ha^{-1}±3.8 (SD). Total C can be split roughly 2/3 to 1/3 with aboveground stocks contributing 65% and belowground stocks 35%. We first present four main C stocks that contribute 97% to the C balance and then three minor C stocks (3%). Then we report the two C fluxes litterfall and heterotrophic soil respiration.

C stocks in the Sabah Biodiversity Experiment area of the Malua Forest Reserve

Major C stocks included total aboveground (TAGC: 55%; 91.9 Mg C ha^{-1}±2.9 SEM) and belowground biomass in trees (TBGC: 10%; 16.5 Mg C ha^{-1}±0.5 SEM), deadwood (DW: 8%; 13.2 Mg C ha^{-1}±3.5 SEM) and soil organic matter (SOM: 24%; 39.6 Mg C ha^{-1}±0.9 SEM). Minor stocks included understory vegetation (UV: 3%; 5.1 Mg C ha^{-1}±1.7 SEM), standing litter (SL: <1%; 0.7 Mg C ha^{-1}±0.1 SEM) and fine root biomass (FRB: <1%; 0.9 Mg C ha^{-1}±0.1 SEM) (Fig. 2).

Comparison of logged and unlogged forest: basal area, total aboveground tree biomass and C stocks

Total basal area of trees >10 cm DBH was significantly higher in unlogged primary forest at Danum Valley compared to the selectively logged forest of the Sabah Biodiversity Experiment in Malua (Fig. 3B and Table 2, Table S4 and S5). The observed difference was due to tree size rather than tree density, which was similar between both types (417 trees ha^{-1} for selectively logged versus 410 trees ha^{-1} for unlogged forest). To examine the impacts of logging, we separated total basal area into dipterocarp (the dominant component of unlogged primary forests and main target of logging) versus pioneer species, in our case principally *Macaranga pearsonii* and *Macaranga gigantea*, *Neolamarkia cadamba*, *Octomeles sumatrana* and *Duabanga moluccana*. Dipterocarps dominated the basal area in the unlogged forest (61%) but comprised only 28% of total basal area in the selectively logged forest. In contrast, pioneer species accounted for 35% of total basal area in selectively logged forest but less than 0.2% in the unlogged primary forest (Fig. 3B and Table 2).

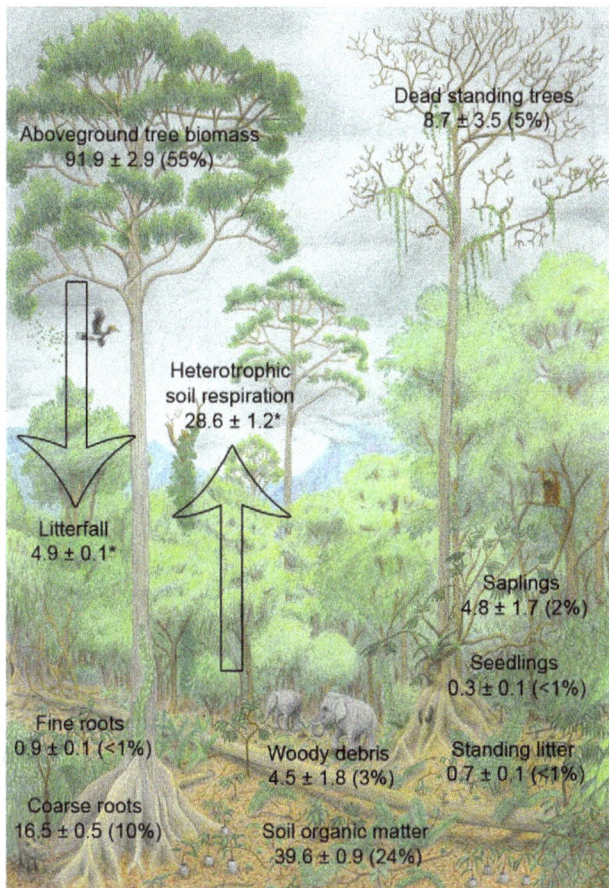

Figure 2. Pre-enrichment planting C balance for the Sabah Biodiversity Experiment. The area is located in selectively logged forest of the Malua Forest Reserve. All components are given as mean ± SEM Mg C ha^{-1}, except for C fluxes (*), which are reported as Mg C ha^{-1} yr^{-1}. Values in parenthesis indicate the percentage of single C stocks to total organic C (167.9 Mg C ha^{-1}).

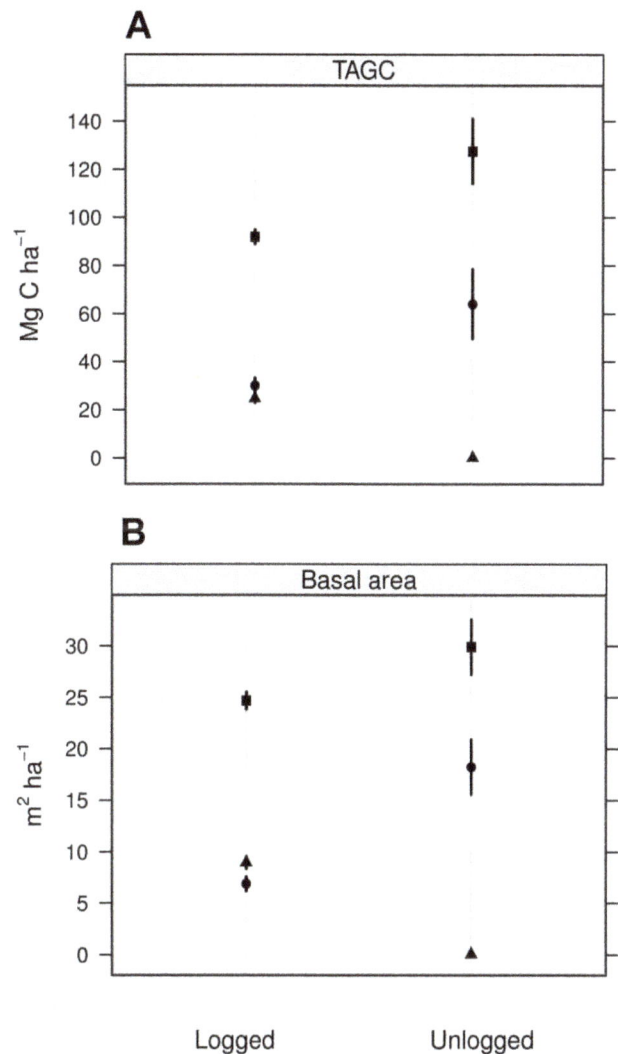

Figure 3. Total aboveground C stocks (TAGC) and basal area in selectively logged an unlogged forest. All species (square), dipterocarps (circle) and pioneer species only (triangle) (mean ± SEM; n = 4). Decreases in dipterocarp basal area but not in carbon stocks were compensated by the pioneers.

A combined weighted average for the four independent studies (Fig. 4B and Table S6) showed that tree aboveground biomass stocks (TAGB) were significantly lower ($z = 6.4$, trials $= 4$, $p < 0.0001$) in selectively logged compared to unlogged primary forest (mean difference -115.7 Mg ha^{-1}; with 95% CI -151.0 to -80.4). For the present study the mean difference (-71.3 Mg ha^{-1}; with 95% CI -158.6 to -16.1) was in large part due to the bigger canopy trees (>90 cm DBH) that were missing in selectively logged forest (DBH range selectively logged forest: 10.0–84.3 cm; primary unlogged forest 10.1–170.3 cm). With a pre-logging dipterocarp timber volume of 180–216 m^3 ha^{-1} and a mean wood density of 0.6 g cm^{-3} the total extractable wood was 108–130 Mg ha^{-1}. Based on the observed mean difference (71.3 Mg ha^{-1}) this indicates that 55–66% of the dipterocarp stock was exploited during selective logging.

Assuming that half of the biomass is stored C (see [8] for discussion), the across-study comparison predicts a C loss of -57.9 Mg C ha^{-1} (with 95% CI -75.5 to -40.2) for selectively logged forest in the region and a loss of -35.7 Mg C ha^{-1} (with 95% CI -79.3 to -8.1) for the Sabah Biodiversity Experiment in

particular. This corresponds to 28% (with 95% CI 62% to 6%) net aboveground C loss 22 years after logging.

Belowground biomass

Belowground coarse roots were assumed to contribute 10% to the total C balance based on literature values [27] and were not confirmed by measurement on site.

Deadwood

In total a mean density of 48.6 ± 10.2 (SEM) dead stems ha^{-1} was counted. Basal area and C stocks in dead standing trees were not found to significantly differ compared to unlogged primary forest at Danum Valley (Table 2).

Soil organic matter

Soil C content in organic matter was relatively low across all sites (39.6 Mg C ha^{-1} ± 0.9 SEM). Highest concentrations were found in the top 0.3 m (22.0 Mg C ha^{-1}), in the sub-soil (0.3–1 m depth) figures decreased to 17.6 Mg C ha^{-1}. Mean soil bulk density (1.1 g cm^{-3} ± 0.1 SEM) was lower in the top soil (<0.1 m)

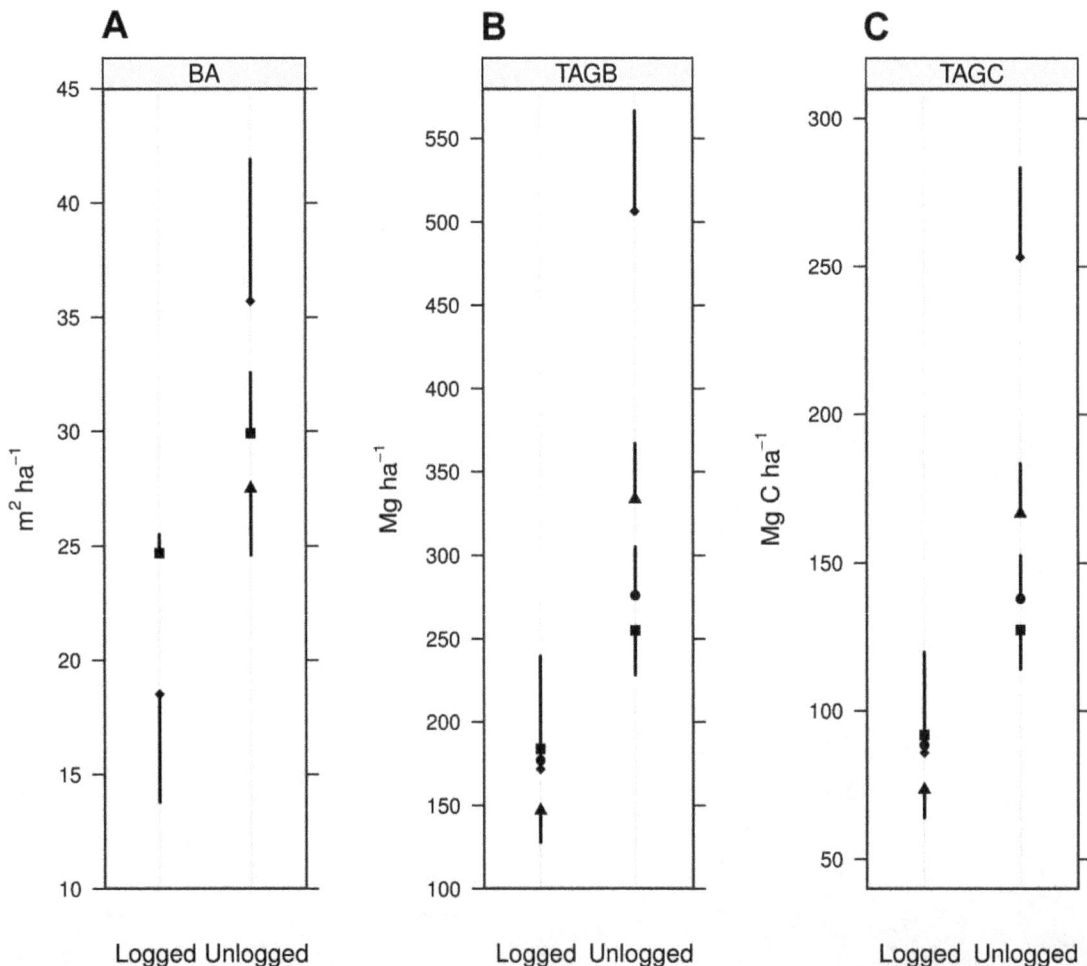

Figure 4. Across study comparison of basal area, aboveground biomass and C. Mean (plus or minus one SEM) are shown for basal area (m^2 ha^{-1}), total aboveground biomass (TAGB Mg ha^{-1}), and total aboveground carbon (TAGC Mg C ha^{-1}) assessed at Danum Valley Conservation Area and nearby selectively logged areas including the Sabah Biodiversity Experiment. Berry 2010 (circle): selectively logged measurements were taken 18 years after harvest; Pinard 1996 (triangle) and Tangki 2008 (diamond): 20 years after harvest; this study (quadrat): 22 years after harvest. All unlogged measurements are from Danum Valley Convervation Area except Pinard 1996 which were taken on unlogged Ulu Segama forest. Note that Berry 2010 (logged and unlogged) and Pinard 1996 (logged) did not specify basal area.

of selectively logged forest (0.9 g cm^{-3}±0.1 SEM) compared to 0.1 to 1 m depth (1.1 g cm^{-3}±0.1 SEM) (Fig. 5 and Table S7).

Understory vegetation

The contribution of sapling and seedling biomass to the total C balance was minor. Nevertheless, the density and diversity of the natural occuring dipterocarp saplings and seedlings are of interest for the forest rehabilitation efforts [18]. Total basal area of all dipterocarp saplings and seedlings was 0.46 m^2 ha^{-1}. A mean dipterocarp sapling and seedling density of 800±240 (SEM) ha^{-1} was measured for selectively logged forest. Dipterocarp saplings and seedlings of ten species were identified in a total area of 0.06 ha.

Standing litter

Standing litter was also a minor contributor to the total C balance (Fig. 2). In particular downed woody debris was reported to be much higher in unlogged forest compared to our estimates from selectively logged forest (Table 2). A more extensive survey

could lead to higher estimates for the contribution of woody debris to total C stocks.

Fine root biomass

Despite the fact, that we took fine roots down to 5 cm soil depth only, our estimations were not found to significantly differ from more extensive surveys to 15 cm soil depth [42] (Table 2).

Litterfall

Mean fine litterfall rate, including leaves, small twigs (<1 cm DBH) and reproductive organs was 4.9 Mg C ha^{-1} yr^{-1}±0.1 (SEM). Considering leaf litter only, the total amount was 3.2 Mg C ha^{-1} yr^{-1}±0.2 (SEM). Litterfall varied significantly between collection dates (n = 25) (Fig. 4). Peaks in litterfall occurred during the months of August 2006 (0.5 Mg C ha^{-1} month^{-1}±0.1 SEM) and May 2007 (0.6 Mg ha^{-1} month^{-1}±0.1 SEM). Lowest litterfall rates occurred in September 2006 (0.3 Mg ha^{-1} month^{-1}±0.1 SEM) and February 2007 (0.2 Mg ha^{-1} month^{-1}±0.1 SEM). There was no apparent

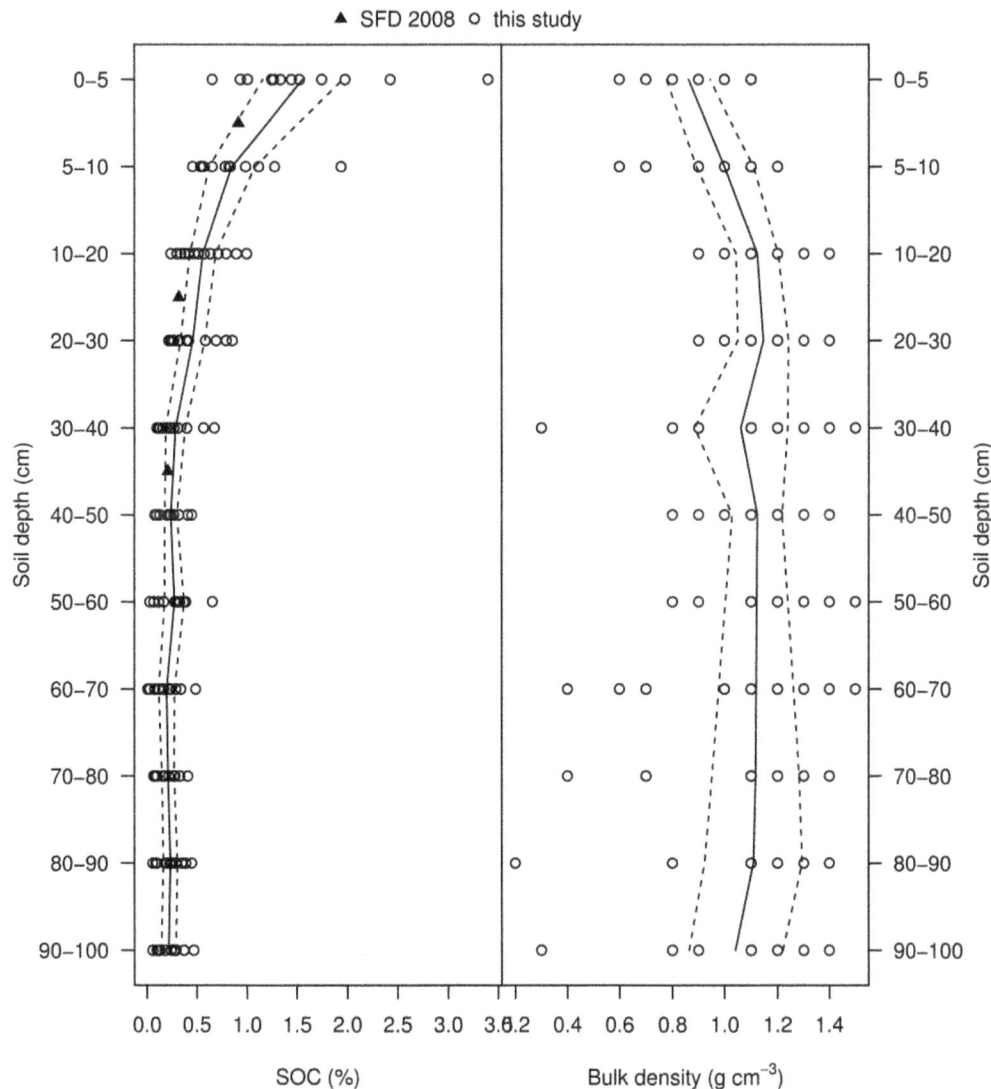

Figure 5. Soil C depth profile. Measured soil organic C (SOC) and bulk density with depth for the Sabah Biodiversity Experiment (Saner 2009). Overlayed points are soil organic C typical of the Kretam soil association (SFD 2008), which is found at Malua Forest Reserve. The line is a smoothed loess curve of mean SOC (±95% CI).

seasonal trend in litterfall rates (Fig. 6) related to the wetter months from November to March and during June and July [14]. Litterfall rates from this study were not significantly different from reported estimates of 15 year old selectively logged forest [38] and unlogged forest (Table 2).

Soil respiration rates

Overall soil respiration rate, measured over two months (May and June 2007) and extrapolated for one year were significantly higher than reported values from Pasoh Forest Reserve in West Peninsular Malaysia [43] or more recent estimates from Lambir, Sarawak [44] (Table 2). Future more extensive studies are likely to report lower estimates for yearly rates of soil respiration rates for this region.

Discussion

The main finding of this study is that the biomass stock of the Sabah Biodiversity Experiment is 28% less than that of pristine forest even 22 years after logging. We estimate that 55–66% of the pre-logging dipterocarp stock was exploited by selective logging, resulting

in significant C losses. Approximately 20% of the basal area in selectively logged forest was later occupied by fast growing pioneer trees, which contributed substantially less to the C stock. We also found that the density but not the diversity of the dipterocarp sapling and seedling stock was high [18]. In combination, these observations indicate that lowland dipterocarp rainforest that is disturbed by selective logging will lose desirable and specialized dipterocarp species, but that overall levels of diversity may be maintained by competitive release of early successional species following disturbance.

Recovery of exploitation from conventional logging may take half a century as reported by findings from French Guiana [45]. However, restoration and management practices that increase dipterocarp recruitment and basal area have the potential to increase C stocks of selectively logged forest and also may accelerate the return to pre-logging levels.

C stock calculation and potential biases

A total C stock of 167.9 Mg C ha^{-1}±3.8 (SD) was measured for selectively logged forest in this study. This is substantially lower than reported elsewhere [18,46], most likely because of the

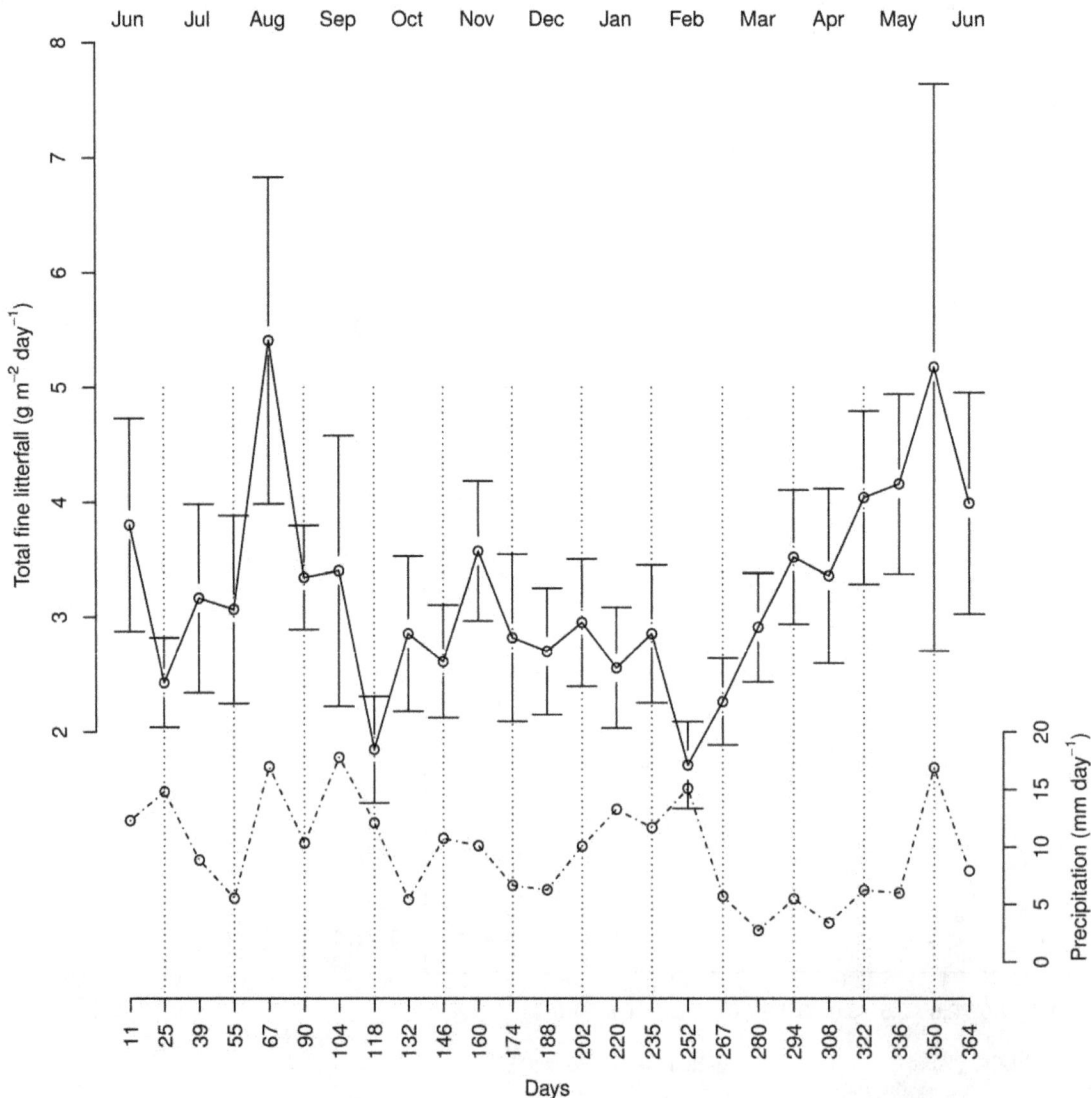

Figure 6. Monthly estimates of litterfall. Time series of total fine litterfall (solid line; mean±95% CI) and precipitation (dashed line; mean). Days indicate litter collections (n = 25) in 2006/07. Mean litterfall and rainfall were calculated over the fourteen days prior to collection.

different methodology used to calculate aboveground biomass. Estimates of the Ulu Segama Forest Reserve, adjacent to the area under study report 261 Mg C ha^{-1}, including total biomass of 200 Mg C ha$^{-1}\pm$10 SEM, woody debris and litter (28 Mg C ha^{-1}) and soil organic matter (33 Mg C ha^{-1}) for selectively logged forest [46]. Several assumptions are likely to induce large biases when estimating C stocks, including: site comparability, plot size, DBH measurement, selection of the DBH-biomass regression model, the belowground estimation of biomass and the overall C content in trees, and the C content in soil organic matter. In particular the three major components aboveground tree biomass, coarse root biomass and soil organic matter covered 89% of organic C stocks and should be estimated with greater precision.

Site comparability

Here we assumed that the Malua Forest Reserve and the primary forest of Danum Valley Conservation Area were comparable before logging to infer the impacts of selective logging. Similarity of the two areas (prior to selective logging) is supported by several lines of evidence. First, the edaphic conditions are comparable including elevation, topography, parent material and soil type (Table 1 and Table S1). Yayasan Sabah (the Sabah Foundation) estimated similar volumes of extractable timber for the two sites at Danum and Malua (Table 1). Therefore, while we cannot be sure that the differences reported here are entirely due to selective logging previous reports suggest the areas were very initially similar and suggest that the differences can be largely attributed to timber extraction. The difference is, not surprisingly most likely due to the missing large trees (>90 cm DBH), and in particular the missing large dipterocarps in selectively logged forest.

Plot size

Standing TAGB and standing deadwood was sampled based on a total sampling effort of 1 ha in both forest types. Based on analysis from La Selva (Costa Rica) [47] and Barro Colorado Island (Panama) [48] we assume that the associated sampling error is high. We therefore included estimates from the region to present an across-study loss in C stocks of −57.8 Mg C ha^{-1} (with 95% CI −75.5 to −40.2) for selectively logged forest in the region.

DBH measurement and DBH-AGB regression model

Major differences in reported TAGC stocks across studies can be also accounted to the DBH measurement and the selection of the DBH-AGB regression model. Here we measured DBH above buttresses which may substantially underestimate basal area of large trees by ignoring the area occupied at the base [47]. However, basal area estimations are comparable to those published previously in selectively logged and unlogged forest [49,50].

The quality of the allometric estimates may have led to the largest error propagation, in particular since it was applied to both, selectively logged and unlogged forest. In contrast to other studies [18,46,49,51–52] we used a more recent DBH-AGB regression model as a standard methodology [22], which resulted in lower biomass C stocks (Fig. 4). For the study site the use of the more recent methodology resulted in 12% less difference between selectively logged and unlogged forest sites compared to an earlier estimation based on older methodologies (28% instead of 40%) [18]. Previous allometric equations [23–26] estimated the aboveground tree biomass at 357–517 Mg ha^{-1}. Hence, we assume that the presented estimate is conservative.

Belowground estimation of biomass

The estimation of root biomass, which is supposedly the second largest live store of C in a tropical forest, was calculated with a BGB/AGB ratio of 0.18 [27]. This is only slightly higher compared to the one used in previous studies of the region (0.17) [46].

Overall C content in trees and soil organic matter

Another potential source for bias are general assumptions about the C content. The assumption that half of the aboveground and belowground biomass is C was challenged by recent results from Barro Colorado Island (Panama). They report C content from 41.9–51.6% among species and a mean C content of 47.4$\pm$2.5% SD. Based on their measured mean C content the biomass C stocks in the present study would decrease by 7.5 Mg C ha^{-1} (6%) [53].

Similar limitations apply to the derivation of C content in soil organic matter. It was measured down to 1 m soil depth and is based on the Walkley-Black method, a wet chemical analysis that may underestimate total content [54]. Future studies may therefore report higher estimates for the region.

C fluxes and potential biases

Measured stocks and fluxes in selectively logged forest that are directly related to C turnover rates but also to forest disturbance include standing deadwood and coarse woody debris, fine root biomass, litterfall and soil respiration rates. Of these, standing deadwood, fine root biomass and litterfall were found to be not significantly different from previous reports of unlogged forest at Danum Valley. This indicates that apart from the lower C stocks in 22 years old selectively logged forest, at least some ecosystem processes could be maintained at the rate of primary forests with regard to nutrient and C turnover (Table 2). The example of fine litterfall rates, as a proxy for net primary productivity, indicates that such processes are likely to be non-linear (Fig. S1) and may also be affected by potential time lag in the response of decomposition processes. The observed peak in fine litterfall rate in drier periods could suggest that desiccation results in increased leaf abscission to reduce evapotranspiration. On the other hand, heavy rainfall appears to also result in increased litterfall. Due to the temporal limitations of the study (1 year) it is difficult to draw any firm conclusion about the correlation of seasonality in litterfall rates and rainfall, such as presented in a recent survey with >10 year data [36]. This may especially apply to dipterocarp-dominated forests, which are characterized by supra-annual patterns such as for example mast fruiting events. Temporal limitations also apply to the estimated soil respiration rate which is a measure of heterotrophic ecosystem respiration. It is derived from a limited dataset and suggests that a more extensive spatial and temporal sampling may result in lower figures, such as the estimate reported from Lambir Hills, Sarawak (North Borneo) (Table 2) [44].

In summary, more extensive research on the main C stocks and the fluxes, including wood NPP and autotrophic respiration should be undertaken to draw a more complete picture of the C balance of selectively logged forest.

Implications for forest management and restoration

Based on the reported findings we argue that silvicultural treatments may have the potential to accelerate the recovery of dipterocarp C stocks to pre-logging levels. Nabuurs and Mohren [55] estimated on the basis of model outputs that approximately 80 Mg C ha^{-1} can be stored additionally in the short term (over several decades) by enrichment line planting which is higher than our weighted average for the region (57.8 Mg C ha^{-1}) and close to the site specific estimate of the Sabah Biodiversity Experiment (71.3 Mg C ha^{-1}). Calculating with a price of 14 € per ton of CO_2 (as currently traded on the European Energy Exchange market) reveals that the C sequestration potential by enrichment planting in selectively logged forest could achieve a market value as high as 2,065–3,879 € ha^{-1} for a project duration of approximately 60 years. Compared to the profitability of converting selectively logged forest into palm oil plantations (2,817–7,075 € over a 30 year period [56]) the revenue from C credits is still lower. A higher price for C credits could be achieved if forest rehabilitation in developing and emerging countries is accepted under the trading scheme of the CDM (Clean Development Mechanism) [57]. This is currently not the case for the guideline of the Kyoto protocol but is a goal of REDD+, which may raise international funds for tropical forest restoration [6], thus helping to close the potential funding shortfall [58]. It emphasizes that forest management in the tropics should focus on the multiple benefits that forests can provide. For example, linking CO_2 offsetting with other ecosystem functions and with biodiversity could potentially allow for a comprehensive approach towards managing forests for mitigating climatic effects and conserve biodiversity and ecosystem services [10,51,52].

Supporting Information

Figure S1 Non-linear relationship between litterfall and rainfall.

Table S1 Soil characteristics of Danum Valley and Malua Forest Reserve.

Table S2 Tree species of the selectively logged forest.

Table S3 Carbon balance of the selectively logged forest.

Table S4 Overview of most important tree families and species in unlogged forest.

Table S5 Overview of most important tree families and species in selectively logged forest.

Table S6 Study survey from in and around Danum Valley Conservation Area.

Table S7 Carbon content survey in soil organic matter.

Acknowledgments

We thank: Karin Saner, Sarah Grundy, Shan Kee Lee; Glen Reynolds, Mike Bernadus Bala Ola and the research assistants from Danum Valley Field Centre, Philip Ulok and the Sabah Biodiversity Experiment RAs, the Universiti Malaysia Sabah undergraduate research group 2006 and the Sabah Forestry Department (providing the maps for Fig. 1, for field logistics and data collection); Romano Carlo (drawing of the carbon budget in Fig. 2); David Franck (wood density estimation); Jake Snaddon and two independent reviewers (comments on earlier versions of the paper); The Economic Planning Unit Sabah, Malaysia and the Danum Valley Management Committee for research permission. This is paper no. 7 of the Sabah Biodiversity Experiment.

Author Contributions

Conceived and designed the experiments: PS AH. Performed the experiments: PS YL. Analyzed the data: PS AH. Wrote the paper: PS RO AH.

References

1. Ruesch A, Gibbs HK (2008) New IPCC Tier-1 global biomass carbon map for the year 2000. Available online from the Carbon Dioxide Information Analysis Center [http://cdiac.ornl.gov]. Oak Ridge, Tennessee, USA: Oak Ridge National Laboratory.

2. Slik JWF, Shin-Ichiro A, Brearley FQ, Cannon CH, Forshed O (2010) Environmental correlates of tree biomass, basal area, wood specific gravity and stem density gradients in Borneo's tropical forests. Global Ecology and Biogeography 19: 50–60.

3. MacKinnon K, Hatta G, Halim H, Mangalik A (1996) The ecology of Kalimantan. Hong Kong: Periplus Editions Ltd. 802 p.

4. Langner A, Miettinen J, Siegert F (2007) Land cover change 2002–2005 in Borneo and the role of fire derived from MODIS imagery. Global Change Biology 13: 2329–2340.

5. Wright SJ, Muller-Landau HC (2006) The future of tropical forest species. Biotropica 38: 287–301.

6. Kettle CJ (2010) Sowing Seeds for REDD+. Science: August 3rd eLetter.

7. Saatchi SS, Harris NL, Brown S, Lefsky M, Mitchard ETA, et al. (2011) Benchmark map of forest carbon stocks in tropical regions across three continents. Proceedings of the National Academy of Sciences 108: 9899–9904.

8. Malhi Y, Aragão LEOC, Metcalfe DB, Paiva R, Quesada CA, et al. (2009) Comprehensive assessment of carbon productivity, allocation and storage in three Amazonian forests. Global Change Biology 15: 1255–1274.

9. Pan Y, Birdsey RA, Fang J, Houghton R, Kauppi PE, et al. (2011) A large and persistent carbon sink in the world's forests. Science 333: 988–993.

10. Edwards DP, Fisher B, Boyd E (2010) Protecting degraded rainforests: enhancement of forest carbon stocks under REDD+. Conservation Letters 3: 313–316.

11. Bennett EL, Nyaoi AJ, Sompu J (2000) Saving Borneo's bacon: the sustainability of hunting in Sarawak an Sabah. In: Appanah S, Khoo KC, eds. Hunting for Sustainability in Tropical Forests. New York: Columbia University Press. pp 305–324.

12. Bhagwat SA, Willis KJ (2008) Agroforestry as a solution to the oil-palm debat. Conservation Biology 22: 1368–1369.

13. Saner P (2009) Ecosystem Carbon Dynamics in Logged Forest of Malaysian Borneo PhD thesis. Zurich, Switzerland: University of Zurich.

14. Marsh CW, Greer AG (1992) Forest land-use in Sabah, Malaysia - An introduction to Danum Valley. Philosophical Transactions of the Royal Society of London Series B-Biological Sciences 335: 331–339.

15. Saner P, Lim R, Burla B, Ong RC, Scherer-Lorenzen M, et al. (2009) Reduced soil respiration in gaps in logged lowland dipterocarp forests. Forest Ecology and Management 25: 2007–2012.

16. Hautier Y, Saner P, Philipson C, Bagchi R, Ong RC, et al. (2010) Effects of seed predators of different body size on seed mortality in Bornean logged forest. PLoS ONE 5: e11651.

17. Scherer-Lorenzen M, Potvin C, Koricheva J, Schmid B, Hector A, et al. (2005) The design of experimental tree plantations for functional biodiversity research. In: Scherer-Lorenzen M, Körner C, Schulze E, eds. Forest Diversity and Function. Temperate and Boreal systems. Berlin, Germany: Springer. pp 347–376.

18. Hector A, Philipson C, Saner P, Chamagne J, Dzulkifli D, et al. (2011) The Sabah Biodiversity Experiment: A long-term test of the role of tree diversity in restoring tropical forest structure and functioning. Philosophical Transactions of the Royal Society of London Series B-Biological Sciences 366: 3303–3315.

19. Philips O, Baker T (2002) Field manual for plot establishment and remeasurement (RAINFOR). Amazon Forest Inventory Network, Sixth Framework Programme (2002–2006).

20. Fearnside PM (1997) Wood density for estimating forest biomass in Brazilian Amazonia. Forest Ecology and Management 90: 59–87.

21. Chave J (2005) Measuring wood density for tropical forest trees; A field manual for the CTFS sites. Toulouse, France.

22. Basuki TM, van Laake PE, Skidmore AK, Hussin YA (2009) Allometric equations for estimating the above-ground biomass in tropical lowland Dipterocarp forests. Forest Ecology and Management 257: 1684–1694.

23. Kato R, Tadaki Y, Ogawa F (1978) Plant biomass and growth increment studies in Pasoh forest. Malayan Nature Journal 30: 211–224.

24. Brown S (1997) Estimating biomass and biomass change of tropical forests. A primer. FAO Forestry Paper, 134. Rome, Italy: Forest Resource Assessment.

25. Ketterings QM, Coe R, van Noordwijk M, Ambagau Y, Palm CA (2001) Reducing uncertainty in the use of allometric biomass equations for predicting above-ground tree biomass in mixed secondary forests. Forest Ecology and Management 146: 199–209.

26. Chave J, Andalo C, Brown S, Cairns MA, Chambers JQ, et al. (2005) Tree allometry and improved estimation of carbon stocks and balance in tropical forests. Oecologia 145: 87–99.

27. Niiyama K, Kajimoto T, Matsuura Y, Yamashita T, Matsuo N, et al. (2010) Estimation of root biomass based on excavation of individual root systems in a primary dipterocarp forest in Pasoh Forest Reserve, Peninsular Malaysia. Journal of Tropical Ecology 26: 271–284.

28. Nepstad DC, Decarvalho CR, Davidson EA, Jipp PH, Lefebvre PA, et al. (1994) The role of deep roots in the hydrological and carbon cycles of Amazonian forests and pastures. Nature 372: 666–669.

29. Gale N (2000) The aftermath of tree death: coarse woody debris and the topography in four tropical rain forests. Canadian Journal of Forest Research-Revue Canadienne De Recherche Forestiere 30: 1489–1493.

30. Palace MM, Keller M, Asner GP, Silva JNM, Passos C (2007) Necromass in undisturbed and logged forests in the Brazilian Amazon. Forest Ecology and Management 238: 309–318.

31. Delaney M, Brown S, Lugo AE, Torres-Lezama A, Quintero NB (1998) The quantity and turnover of dead wood in permanent forest plots in six life zones of Venezuela. Biotropica 30: 2–11.

32. Nelson DW, Sommers LE (1996) Total carbon, organic carbon and organic matter. In: Page AL, Miller RH, Keeney DR, eds. Methods of Soil Analysis Part 2 Agronomy. Madison, USA: American Society of Agronomy. pp 961–1010.

33. Walkley A, Black IA (1934) An examination of Degtjareff method for determining organic carbon in soils: effect of variations in digestion conditions and of inorganic soil constituents. Soil Science 63: 251–263.

34. von Oheimb G, Westphal C, Härdtle W (2007) Diversity and spatio-temporal dynamics of dead wood in a temperate near-natural beech forest (*Fagus sylvatica*). European Journal of Forest Research 126: 359–370.

35. Proctor J (1983) Tropical forest litterfall. I. Problems of litter comparison. In: Sutton SL, Whitmore TC, Chadwick AC, eds. Tropical rain forest: ecology and management. Oxford, UK: Blackwell. pp 267–273.

36. Chave J, Navarrete D, Almeida S, Àlvarez E, Aragão LEOC, et al. (2010) Regional and seasonal patterns of litterfall in tropical South America. Biogeosciences 7: 43–55.

37. Yamashita T, Takeda H, Kirton LG (1995) Litter production and phenological patterns of *Dipterocarpus baudii* in a plantation forest. Tropics 5: 57–68.

38. Burghouts T, Ernsting G, Korthals G, Devries T (1992) Litterfall, leaf litter decomposition and litter invertebrates in primary and selectively logged dipterocarp forest in Sabah, Malaysia. Philosophical Transactions of the Royal Society of London Series B-Biological Sciences 335: 407–416.

39. Pumpanen J, Kolari P, Ilvesniemi H, Minkkinen K, Vesala T, et al. (2004) Comparison of different chamber techniques for measuring soil CO2 efflux. Agricultural and Forest Meteorology 123: 159–176.

40. Borenstein M, Hedges LV, Higgins JPT, Rothstein HR (2009) Introduction to Meta-Analysis. Chichester, West Sussex, United Kingdom: John Wiley & Sons Ltd. 421 p.

41. R Development Core Team (2010) R: A Language and Environment for Statistical Computing. Vienna, Austria: R Foundation for Statistical Computing.

42. Green IJ, Dawson LA, Proctor J, Duff EI, Elston DA (2005) Fine root dynamics in a tropical rain forest is influenced by rainfall. Plant and Soil 276: 23–32.

43. Kira T (1978) Community architecture and organic matter dynamics in tropical lowland rain forests of Southeast Asia with special reference to Pasoh Forest, West Malaysia. In: Tomlinson PB, Zimmermann MH, eds. Tropical Trees as Living Systems. New York, USA: Cambridge University Press. pp 561–590.

44. Katayama A, Kume T, Komatsu H, Ohashi M, Nakagawa M, et al. (2009) Effect of forest structure on the spatial variation in soil respiration in a Bornean tropical rainforest. Agricultural and Forest Meteorology 149: 1666–1673.

45. Blanc L, Echard M, Herault B, Bonal D, Marcon E, et al. (2009) Dynamics of aboveground carbon stocks in selectively logged tropical forest. Ecological Applications 19: 1397–1404.

46. Pinard MA (1995) Carbon retention by reduced-impact logging. PhD thesis. Gainesville, USA: University of Florida.

47. Clark DB, Clark DA (2000) Landscape-scale variation in forest structure and biomass in a tropical rain forest. Forest Ecology and Management 137: 185–198.

48. Chave J, Condit R, Aguilar S, Hernandez A, Lao S, et al. (2003) Error propagation and scaling for tropical forest biomass estimates. Philosophical Transactions of the Royal Society of London Series B-Biological Sciences 359: 409–420.

49. Tangki H, Chappell NA (2008) Biomass variation across selectively logged forest within a 225-km(2) region of Borneo and its prediction by Landsat TM. Forest Ecology and Management 256: 1960–1970.

50. Pinard MA, Putz FE (1996) Retaining forest biomass by reducing logging damage. Biotropica 28: 278–295.

51. Imai N, Samejima H, Langner A, Ong RC, Kita S, et al. (2009) Co-Benefits of sustainable forest management in biodiversity conservation and carbon sequestration. PLoS ONE 4: e8267.

52. Berry NJ, Phillips OL, Lewis SL, Hill JK, Edwards DP, et al. (2010) The high value of logged tropical forests: lessons from northern Borneo. Biodiversity Conservation 19: 985–997.

53. Martin AR, Thomas SC (2011) A Reassessment of Carbon Content in Tropical Trees. PLoS ONE 6(8): e23533. doi:10.1371/journal.pone.0023533.

54. Krishan G, Srivastav SK, Kumar S, Saha SK, Dadhwal VK (2009) Quantifying the underestimation of soil organic carbon by the Walkley and Black technique - examples from Himalayan and Central Indian soils. Current Science 96: 1133–1136.

55. Nabuurs GJ, Mohren GMJ (1993) Carbon Fixation through Forestation Activities. Wageningen, Netherlands: Institute for Forestry and Nature Research (IBN-DLO). 205 p.

56. Butler RA, Koh LP, Ghazoul J (2009) REDD in the red: palm oil could undermine carbon payment schemes. Conservation Letters 2: 67–73.

57. Michaelowa A, Dutschke M (2009) Will Credits from Avoided Deforestation in Developing Countries Jeopardize the Balance of the Carbon Market? In: Palmer C, Engel S, eds. Avoided Deforestation: Prospects for Mitigating Climate Change. Oxon, UK: Routledge. pp 130–148.

58. Fisher B, Edwards DP, Giam X, Wilcove DS (2011) The high costs of conserving Southeast Asia's lowland rainforests. Frontiers in Ecology and the Environment 9(6): 329–334.

2

Soil Respiration in Tibetan Alpine Grasslands: Belowground Biomass and Soil Moisture, but Not Soil Temperature, Best Explain the Large-Scale Patterns

Yan Geng[1,2], Yonghui Wang[1], Kuo Yang[1], Shaopeng Wang[1], Hui Zeng[1,3], Frank Baumann[4], Peter Kuehn[4], Thomas Scholten[4], Jin-Sheng He[1,2]*

1 Department of Ecology, College of Urban and Environmental Sciences, and Key Laboratory for Earth Surface Processes of the Ministry of Education, Peking University, Beijing, China, 2 Key Laboratory of Adaptation and Evolution of Plateau Biota, Northwest Institute of Plateau Biology, Chinese Academy of Sciences, Xining, China, 3 Shenzhen Key Laboratory of Circular Economy, Shenzhen Graduate School, Peking University, Shenzhen, China, 4 Department of Geoscience, Physical Geography and Soil Science, University of Tuebingen, Tuebingen, Germany

Abstract

The Tibetan Plateau is an essential area to study the potential feedback effects of soils to climate change due to the rapid rise in its air temperature in the past several decades and the large amounts of soil organic carbon (SOC) stocks, particularly in the permafrost. Yet it is one of the most under-investigated regions in soil respiration (Rs) studies. Here, Rs rates were measured at 42 sites in alpine grasslands (including alpine steppes and meadows) along a transect across the Tibetan Plateau during the peak growing season of 2006 and 2007 in order to test whether: (1) belowground biomass (BGB) is most closely related to spatial variation in Rs due to high root biomass density, and (2) soil temperature significantly influences spatial pattern of Rs owing to metabolic limitation from the low temperature in cold, high-altitude ecosystems. The average daily mean Rs of the alpine grasslands at peak growing season was 3.92 μmol CO_2 m^{-2} s^{-1}, ranging from 0.39 to 12.88 μmol CO_2 m^{-2} s^{-1}, with average daily mean Rs of 2.01 and 5.49 μmol CO_2 m^{-2} s^{-1} for steppes and meadows, respectively. By regression tree analysis, BGB, aboveground biomass (AGB), SOC, soil moisture (SM), and vegetation type were selected out of 15 variables examined, as the factors influencing large-scale variation in Rs. With a structural equation modelling approach, we found only BGB and SM had direct effects on Rs, while other factors indirectly affecting Rs through BGB or SM. Most (80%) of the variation in Rs could be attributed to the difference in BGB among sites. BGB and SM together accounted for the majority (82%) of spatial patterns of Rs. Our results only support the first hypothesis, suggesting that models incorporating BGB and SM can improve Rs estimation at regional scale.

Editor: Ben Bond-Lamberty, DOE Pacific Northwest National Laboratory, United States of America

Funding: This study was supported by the "Program of One Hundred Talented People" of the Chinese Academy of Sciences (Grant No. KSCX2-YW-Z-0806), the National Natural Science Foundation of China (Grant No. 31025005 and 31021001), National Program on Key Basic Research Project (Grant No. 2010CB950602), and the "Strategic Priority Research Program" of the Chinese Academy of Sciences (Grant No. XDA05050304). The funders had no role in study design, data collection and analysis, decision to publish, or preparation of the manuscript.

Competing Interests: The authors have declared that no competing interests exist.

* E-mail: jshe@nwipb.cas.cn

Introduction

Soil respiration (Rs) is the major pathway for carbon (C) exiting terrestrial ecosystems and plays a central role in global carbon cycles [1–3]. Because soil is the largest carbon pool in terrestrial ecosystems, containing more than 1500 Pg C (1 PG = 10^{15} g) [4–6], small change in the rate of Rs may have a profound impact on atmospheric CO_2 concentration, exerting positive feedbacks to global warming [2,7–9]. Therefore, it is important to understand and be able to predict how Rs responds to environmental variation and climate change.

Rs has been a major research theme, particularly since the beginning of 1990s [2,6,10–16]. Many studies in a variety of ecosystems have been devoted to evaluation of various influencing factors, including microbial activity [17–19], C allocation [20,21], root dynamics [22], and regulators such as temperature, soil moisture, soil texture and other climatic and soil variables [23,24]. Nevertheless, synthetic analyses of existing data show a substan-

tially huge heterogeneity in Rs, for which reason we require comprehensive datasets before being able to discuss the uncertainties that may arise owing to differences in intensity of sampling in different ecosystems [25].

It has been well documented that Rs varies greatly with time and space [25]. With the advanced equipment for high-frequency records of Rs, temperature, moisture and other variables (e.g. [26]), within-site temporal patterns of Rs can be relatively easily obtained. However, to address patterns of ecosystem C cycling at regional scale, to predict responses of Rs to future climate change based on mechanistic data, and to scale-up from specific sites to vegetation biomes, studies on Rs need to move beyond within-site variations in soil temperature and soil moisture and to incorporate differences among broad ecosystem types [6,27,28]. At regional scale, patterns of biogeochemical cycling of different ecosystem types are governed by at least five independent controls or so-called state factors, i.e. climate, parent material, topography, biota, and time [3,29]. Hence, factors closely associated with Rs within-ecosystem and among-ecosystems are not identical. However, compared with the plenty of studies on temporal variations, relatively fewer publications have explored in-depth the regional patterns of Rs and the factors revolving around Rs process (but see [30]).

The Tibetan Plateau is one of the most under-studied regions for Rs research, despite its essential role in the global C cycles. Due to rough natural conditions, only a few studies have measured Rs. Some in alpine steppe [31], some in alpine meadow [32–35], and others in cropland [36]. Alpine grassland accounts for 62% of the total area of the plateau, out of which 32% is alpine steppe, and 30% alpine meadow [37]. Alpine grassland is of special interest because of the high C density [38,39] and potential feedbacks to climate warming [40]. We previously estimated that SOC storage in the top one meter in these alpine grasslands was 7.4 Pg C, with an average density of 6.5 kg m^{-2} [39]. Moreover, the Tibetan Plateau is the largest high-altitude and low-latitude permafrost area on the earth, with over 50% of its total surface in permafrost [41,42]. The observed rapid rises in air temperature [43], degradation of the permafrost and the associated changes in soil hydrology in the last several decades [42,44,45] will seriously impact the C cycles [34,46]. The high-altitude ecosystems, low-latitude permafrost, unique vegetation composition and physio-logical adaptation to the extreme environments, as well as the relatively low intensity of human disturbance motivated us to focus on carbon cycle and the effects of global climate change on natural ecosystems of the Tibetan Plateau.

The primary objective of this study is to investigate large-scale spatial patterns of Rs and to examine their responses to naturally occurring environmental gradients in order to identify factors most closely associated with Rs in such extreme environments. We hypothesized that:

(1) Belowground biomass is most related to large-scale variations in Rs, because alpine grasslands have a high root biomass density [47]. As a result autotrophs will contribute a large proportion of the total respiratory CO_2 efflux.

(2) Soil temperature is another important influential factor for alpine grassland Rs. This is because low growing-season temperature is a limiting factor for physiological processes in high-altitude grassland ecosystems [48,49]. Therefore, it is predicted that Rs increases with increasing soil temperature.

These two hypothesis were tested in a transect study across alpine grasslands on the Tibetan Plateau. The measurement of Rs in this vast, remote, high-altitude area complements the existing data and help to estimate the global C flux from soils.

Materials and Methods

Ethics Statement

No specific permits were required for the described field studies in the Tibetan Plateau. The research sites are not privately-owned or protected in any way and field studies did not involve endangered or protected species.

Study sites

This study was conducted during two expeditions in late July and early August of 2006 and 2007, in collaboration with University of Tuebingen, Germany. Out of the 51 sites, 42 were selected for soil respiration measurements along a transect which stretches from latitudes of 30.31 to 37.69°N and longitudes of 90.80 to 101.48°E, and elevations from 2925 to 5105 m a.s.l. (Table 1, Fig. 1). Mean annual air temperature (MAT) and mean

annual precipitation (MAP) range from −5.75 to 2.57°C and 218 to 604 mm yr^{-1}, respectively. The vegetation represents alpine grassland, including the two main ecosystem types, alpine meadow and alpine steppe [49,50]. Out of the 42 sites, 23 were alpine meadows and 19 alpine steppes. Alpine meadows are dominated by perennial tussock grasses such as *Kobresia pygmaea* and *K. tibetica*, while alpine steppes are dominated by short and dense tussock grasses such as *Stipa purpurea*; both ecosystem types have extensive distributions. The sites were selected by visual inspection of the vegetation, aiming to sample sites subject to minimal grazing and other anthropogenic disturbances.

Field measurements

At each site, we conducted (1) measurement of plant biomass after surveying the entire plant community, (2) collections of soil samples at three depths (0–5, 5–10, and 10–20 cm) using soil corer, followed by volumetric samples at equal depths for bulk density and gravimetric water content determinations, (3) on-site extraction of soil mineralized N (N_{min}) consisting of nitrate (NO_3-N) and ammonium (NH_4-N), and (4) measurement of soil respiration rates.

Plant biomass measurement. We harvested aboveground biomass (AGB) in three plots (1×1 m^2) and belowground biomass (BGB) in three soil pits (0.5×0.5 m^2) described in Yang et al. [47]. Biomass samples were dried using a custom-built portable oven in the field, and oven-dried at 60°C to a constant weight, and weighed to the nearest 0.01 g upon returning to the laboratory.

Soil property measurement. Soil sampling procedures, soil bulk density (SBD), soil total N (STN) and SOC measurements have been detailed elsewhere [39]. On-site extraction of N_{min} was carried out using a custom-designed equipment which could perform on-site extraction without any disturbances. In brief, 10 g of homogenized soil was extracted with 50 ml 1 mol KCl for 60 minutes immediately after sampling, filtered through Whatman No. 42 cellulose filter paper into 100 ml PE-vials, and conserved by acidification with 3 ml hydrochloric acid (HCl, 30%) [38].

Soil respiration measurement. At each site, seven PVC soil collars (10 cm inside diameter and 5 cm in height) were installed 2–3 cm into the soil along a straight line at one-meter intervals. Rs (CO_2 efflux) was measured with a Li-6400 infrared gas analyzer equipped with the 6400-09 soil flux chamber (Li-Cor Inc, Lincoln, NE, USA). The protocol recommended by LiCor (LI-6400-09 manual) was changed to five observations of 10 µmol mol^{-1} (for steppes) and 30 µmol mol^{-1} (for meadows) per measurement. Typically, soil respiration rates were measured 3–4 times during 4–5 hours from 10:00 to 16:00 (Beijing Standard Time) when soil respiration peaked. To obtain the diurnal pattern, we also measured the complete diurnal variation of soil respiration at nine sites (Fig. 2). We then calculated the ratios of instant Rs from 10:00 to 16:00 to the daily mean Rs for the nine sites. Using these ratios, we calculated daily mean Rs of non-diurnal sites according to similarity in community composition and closeness in distance. On average, diurnal courses of soil respiration were measured every four to five sites. Soil temperature at 10 cm was monitored simultaneously with soil respiration measurement using the attached soil temperature probe. Air temperature was measured with the temperature probe of Li-6400 infrared gas analyzer.

Laboratory analysis

Dried soil samples were grounded using a ball mill (NM200, Retsch, Germany). Total C and N concentrations were determined on 5–6 mg aliquot of the homogenously grounded material for each sample using an elemental analyzer (2400 II CHNS/O

Table 1. Description of 42 sites where soil respiration measurements were taken.

Site	Latitude	Longitude	Altitude (m)	MAT (°C)	GST (°C)	MAP (mm yr^{-1})	GSP (mm yr^{-1})	Rs (μmol m^{-2} s^{-1})	Vegetation
QZ01	36.37	101.48	3454	−1.83	7.33	466	326	4.35	Meadow
QZ02	35.80	101.30	3302	0.03	8.98	475	328	5.27	Meadow
QZ03	35.78	101.17	3263	0.39	9.37	466	322	3.26	Steppe
QZ04	35.58	101.08	3416	−0.37	8.50	488	336	2.87	Steppe
QZ06	35.41	100.97	3517	−0.79	7.99	501	346	4.07	Meadow
QZ07	34.24	100.25	4282	−4.23	3.96	604	414	5.09	Meadow
QZ08	33.96	99.88	4053	−2.11	6.06	580	395	4.08	Meadow
QZ11	33.94	99.83	4156	−2.77	5.38	589	402	5.15	Meadow
QZ13	34.06	99.40	4231	−3.22	5.05	568	389	12.9	Meadow
QZ14	34.92	98.21	4267	−3.96	4.96	464	326	5.06	Meadow
QZ15	34.89	98.23	4224	−3.63	5.27	462	325	2.14	Steppe
QZ17	34.28	97.88	4667	−5.74	2.84	522	364	9.32	Meadow
QZ18	33.32	96.28	4506	−2.89	5.48	482	333	6.00	Meadow
QZ19	34.01	95.80	4201	−1.60	7.15	390	274	3.58	Steppe
QZ22	34.06	97.60	4700	−5.55	2.97	523	365	2.46	Meadow
QZ23	35.29	99.01	4217	−4.48	4.47	478	336	1.19	Steppe
QZ24	36.01	100.25	3109	1.63	10.92	393	274	1.59	Steppe
QZ25	36.17	100.51	2925	2.57	11.93	380	264	0.89	Steppe
QZ26	36.36	100.74	3233	0.08	9.43	409	287	2.19	Steppe
QZ27	36.44	101.09	3486	−1.94	7.32	446	314	5.36	Meadow
QZ28	36.95	100.86	3130	−0.01	9.62	372	265	2.52	Steppe
QZ29	37.26	99.98	3215	−0.55	9.39	319	233	1.78	Steppe
QZ30	37.28	98.99	3437	−1.61	8.48	290	216	4.04	Steppe
QZ31	35.74	94.25	4222	−3.14	6.70	218	170	2.17	Steppe
QZ32	35.52	93.74	4564	−5.01	4.80	238	185	0.39	Steppe
QZ33	35.17	93.04	4682	−5.41	4.31	234	182	0.75	Steppe
QZ34	34.72	92.89	4801	−5.75	3.76	348	249	4.23	Meadow
QZ35	33.99	92.35	4654	−4.22	4.94	336	248	1.24	Steppe
QZ36	32.18	91.72	4903	−4.18	4.12	473	327	2.79	Meadow
QZ38	31.45	92.02	4494	−0.25	7.94	480	341	1.12	Meadow
QZ40	31.77	92.62	4605	−2.05	5.89	523	361	3.03	Meadow
QZ41	31.69	92.41	4596	−1.92	6.00	511	355	3.99	Meadow
QZ42	30.94	91.66	4756	−2.76	5.45	539	371	1.90	Steppe
QZ43	30.56	91.45	4506	−0.53	7.32	507	359	5.94	Meadow
QZ44	30.31	90.80	4324	1.23	8.81	442	326	1.99	Steppe
QZ45	32.58	91.86	5105	−5.75	2.77	488	331	4.59	Meadow
QZ46	34.37	92.61	4656	−4.56	4.78	327	241	1.78	Steppe
QZ47	36.78	99.67	3391	−1.00	8.72	348	251	8.25	Meadow
QZ48	37.61	101.31	3196	−1.53	7.74	363	309	7.26	Meadow
QZ49	37.61	101.31	3196	−2.12	7.19	364	311	10.6	Meadow
QZ50	37.69	101.28	3268	−1.89	7.67	313	270	5.32	Meadow
QZ51	37.28	98.99	3437	−1.61	8.48	290	216	1.99	Steppe

MAT, Mean annual temperature; GST, growing season temperature; MAP, mean annual precipitation; GSP, growing season precipitation; Rs, daily mean soil respiration rate.

Elemental Analyzer, Perkin-Elmer, Boston, MA, USA) with a combustion temperature of 950°C and a reduction temperature of 640°C. Soil inorganic carbon (SIC) was measured volumetrically using an inorganic carbon analyzer (Calcimeter 08.53, Eijkelk-amp, Netherland). Thus SOC was calculated as the difference between STC and SIC. Soil pH was determined in both 0.01 M CaCl$_2$ and bi-distilled H$_2$O potentiometrically, but only those of water solution were used in the current study. The KCl-

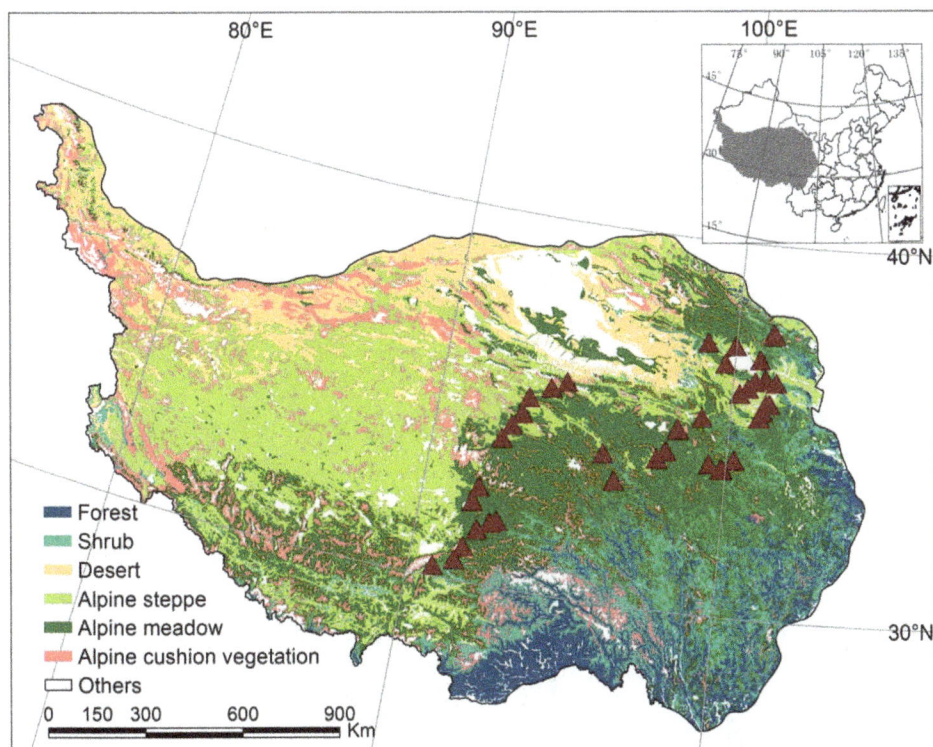

Figure 1. Vegetation map of the sampling sites, selected from the Vegetation Map of China [80]. Triangles represent sampling sites.

extractions for N_{min}-analysis were measured photometrically using a Continuous Flow Analyzer (SAN Plus, Skalar, Netherlands). Soil moisture (SM) was determined gravimetrically by taking the skeleton content into account.

Climate data and statistical analysis

At each site, we installed temperature data loggers (Hobo U12, Onset Computer Corporation, Pocasset, MA) in July 2006 to measure soil temperature (−10 cm) at 1 h interval. We revisited these sites in July or August in 2007, 2008 and 2009 to download the recorded temperature data. Based on those measurements mean annual soil temperature (MAST) of each site was calculated. The climate data used in this study were calculated based on linear models using latitude, longitude, and altitude as variables from 55-year averaged temperature and precipitation records (1951–2005) at 680 well-distributed climate stations across China [48,51,52]

The variables to explain the spatial variation of soil respiration consist of (1) soil properties, measured by SOC, SM, MAST, soil C/N ratio, SBD, soil acidity (pH), soil texture (sand content, clay content), and N_{min}, (2) average climate, encompassing growing season temperature (GST), growing season precipitation (GSP), and (3) plant community characteristics, including vegetation type (VT, meadow or steppe), AGB and BGB (Table 2). We used regression tree analysis [53], as implemented in the SAS statistical software package version 8.01 [54], to screen important variables influencing soil respiration, as tree-based modeling is an exploratory data analytic technique for summarizing multivariable and uncovering its structure in large datasets [55]. We selected F test's p-value as splitting criterion, and set observations required for a split search at 5. Our sample size (42 sites) doesn't allow us to do cross validation, but when we set the F test significant level at 0.20, the tree developed was adequate in complexity (depth) and explanation (R^2). From the relative importance in the regression

tree which was calculated as the cumulative variance reduction at each split for a particular independent variable, five variables with the importance values greater than 0.4 were screened out, i.e. AGB, BGB, VT, SOC, and SWC (Table 2).

To address how these variables affect soil respiration both directly and indirectly is challenging because variables measured in field are cross-correlated [11,14,28]. Structural equation modelling (SEM) [56–58] has been used in recent studies to explicitly evaluate the causal relationships among multiple interacting variables (e.g. [59–61]). SEM aims to account for the roles of multiple variables in a single analysis, providing mechanisms behind the overall patterns by partitioning direct from indirect effects that act through other components of the system.We used SEM here to partition the total effect of variables on soil respiration into direct effects and indirect effects. A path model was developed to relate soil respiration to AGB, BGB, VT, SOC, and SWC, based on theoretical knowledge of the major factors associated with soil respiration at ecosystem level [3]. The model was fitted using EQS 6.1 for Windows [62].

As the results of SEM are dependent on correctly specifying theoretical causal relationships between variables prior to analysis [56,58], the initial theoretical model was modified to improve the fit between model and data. The final model was strong: Bentler's comparative fit index (CFI) = 0.96, Bentler-Bonett normed fit index (NFI) = 0.95. Furthermore, R-squares for Rs, AGB, BGB are very high in the path model.

Results

Overall soil respiration

Across 42 sites, the daily mean Rs of alpine grassland at peak growing season was 3.92 µmol CO_2 m^{-2} s^{-1}, and ranged from 0.39 to 12.88 µmol CO_2 m^{-2} s^{-1} (Table 1), with a coefficient of variation (CV) of 69.1%. The daily mean Rs of steppes was

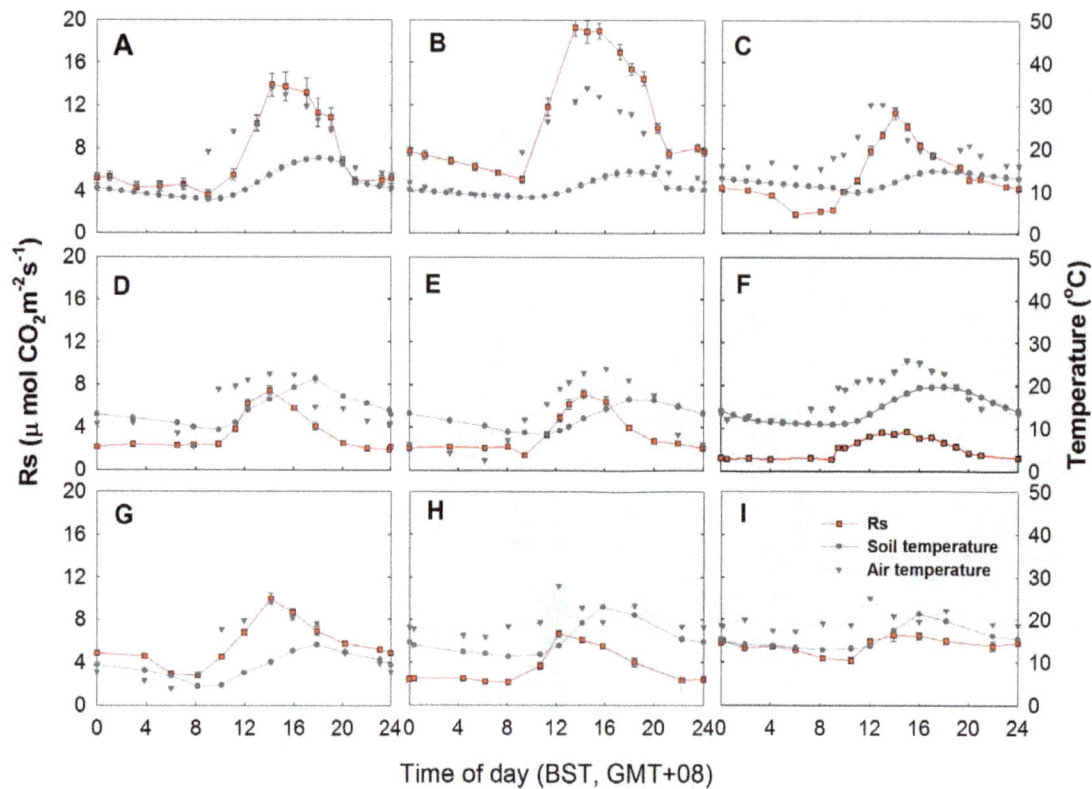

Figure 2. Diurnal changes of soil respiration rate, soil temperature and air temperature. Complete diurnal courses of soil respiration were measured for seven alpine meadows and two alpine steppes on the Tibetan Plateau. Vertical bars indicate the standard error of the measurement mean (n = 5–7) for each time. (A), Haibei, *Kobresia* and *Festuca* mixed meadow; (B), Haibei, *Kobresia tibetica* meadow; (C), Haibei, *Kobresia pygmaea* meadow; (D) Naqu, *Kobresia pygmaea* meadow; (E) Naqu, *Kobresia tibetica* meadow; (F) Tianjun, *Stipa purpurea* steppe; (G) Fenghuoshan, *Kobresia pygmaea* meadow; (H) Qumalai, *Kobresia pygmaea* meadow; (I) Qumalai, *Festuca* steppe.

2.01 μmol CO_2 m^{-2} s^{-1} (ranged from 0.39 to 4.04), while Rs of meadows, 5.49 μmol CO_2 m^{-2} s^{-1} (ranged from 1.12 to 12.88), was approximately two and half times that of the steppes.

Although the meadows had a significantly higher Rs than steppes, their CV were similar, being 48.9 and 47.1% for meadow and steppe, respectively.

Table 2. Variables included in the regression tree analysis and their importance value.

Variable	n	Mean	SD	Range	Importance in regression tree
Soil organic carbon (SOC, %)	42	5.25	4.79	0.339–19.4	1.0000
Aboveground biomass (AGB, g m^{-2})	42	119	100	29.9–530	0.8997
Belowground biomass (BGB, g m^{-2})	42	1816	1957	202–9393	0.8889
Vegetation type (VT)	42	-	-	-	0.4577
Soil moisture (SM, v/v, %)	42	38.3	50.2	0.44–220	0.4383
Growing season temperature (GST, °C)	42	6.67	2.25	2.77–11.93	0.1719
Mean annual soil temperature (MAST, °C)	42	17.0	5.53	−1.12–8.14	0.1621
Growing season precipitation (GSP, mm yr^{-1})	42	306	61.5	170–414	0.0000
Soil temperature (ST, °C)	42	17.0	5.53	6.30–31.55	0.0000
Soil C/N ratio (C/N, g g^{-1})	39	12.1	2.85	7.97–20.1	0.0000
Soil bulk density (SBD, g cm^{-3})	38	0.94	0.32	0.31–1.65	0.0000
pH	38	7.3	0.52	6.0–8.1	0.0000
Sand content (%)	37	42.3	18.4	20.0–80.0	0.0000
Clay content (%)	37	7.60	6.59	3.0–24	0.0000
Available nitrogen (mmol l^{-1})	37	0.080	0.046	0.026–0.218	0.0000

Table 3. Soil respiration, community biomass, soil properties, and climatic variables in different ecosystem types.

Variable	Kobresia pygmaea meadow			Kobresia tibetica meadow			Stipa spp. steppe			Mixed-specie meadow			Others		
	n	Mean	SD	n	Mean	SD	n	Mean	SD	n	Mean	SD	n	Mean	SD
Rs (μmol m⁻² s⁻¹)	11	4.36	1.52	4	9.34	3.49	12	2.18	0.95	3	6.93	1.52	12	2.68	1.49
SOC (%)	11	6.23	3.61	4	11.51	2.54	12	1.96	1.07	3	8.82	6.50	12	4.66	5.68
AGB (g m⁻²)	11	107	65	4	285	188	12	78	40	3	253	114	12	84	41
BGB (g m⁻²)	11	2390	1261	4	5852	2682	12	528	272	3	3299	2051	12	862	619
SM (g water g⁻¹ dry soil)	11	32.2	20.4	4	125.1	26.3	12	7.0	4.2	3	45.4	37.7	12	44.6	68.3
GST (°C)	11	6.28	1.46	4	4.46	2.10	12	7.61	2.20	3	8.48	0.66	12	6.38	2.64
Soil MAT (°C)	11	3.37	1.61	4	0.91	1.88	12	3.80	2.36	3	3.66	0.31	12	2.72	3.09
GSP (mm yr⁻¹)	11	351	42	4	349	35	12	257	55	3	296	40	12	301	59
C/N (g g⁻¹)	10	13.75	3.20	3	15.19	0.77	12	10.36	1.16	2	11.79	1.56	12	11.81	3.10
SBD (g cm⁻³)	10	0.79	0.20	3	0.44	0.01	12	1.14	0.24	2	0.74	0.20	12	1.00	0.35
pH	8	6.9	0.59	3	6.8	0.32	12	7.7	0.26	3	7.4	0.09	12	7.3	0.53
Sand content (%)	10	34.6	6.0	3	31.0	4.5	12	55.7	23.7	2	37.5	9.2	10	36.1	13.5
Clay content (%)	10	8.2	6.0	3	4.0	3.8	12	5.1	3.7	2	12.5	12.0	10	9.4	8.7
Available N (mmol l⁻¹)	10	0.11	0.05	3	0.12	0.03	11	0.05	0.02	2	0.15	0.02	11	0.06	0.04

Rs, daily mean soil respiration rate; SOC, soil organic carbon content; AGB, above-ground biomass; BGB, below-ground biomass; SM, soil moisture; GST, growing season temperature; MAT, Mean annual temperature; GSP, growing season precipitation; SBD, soil bulk density; n, number of sampling sites; SD, standard deviation.

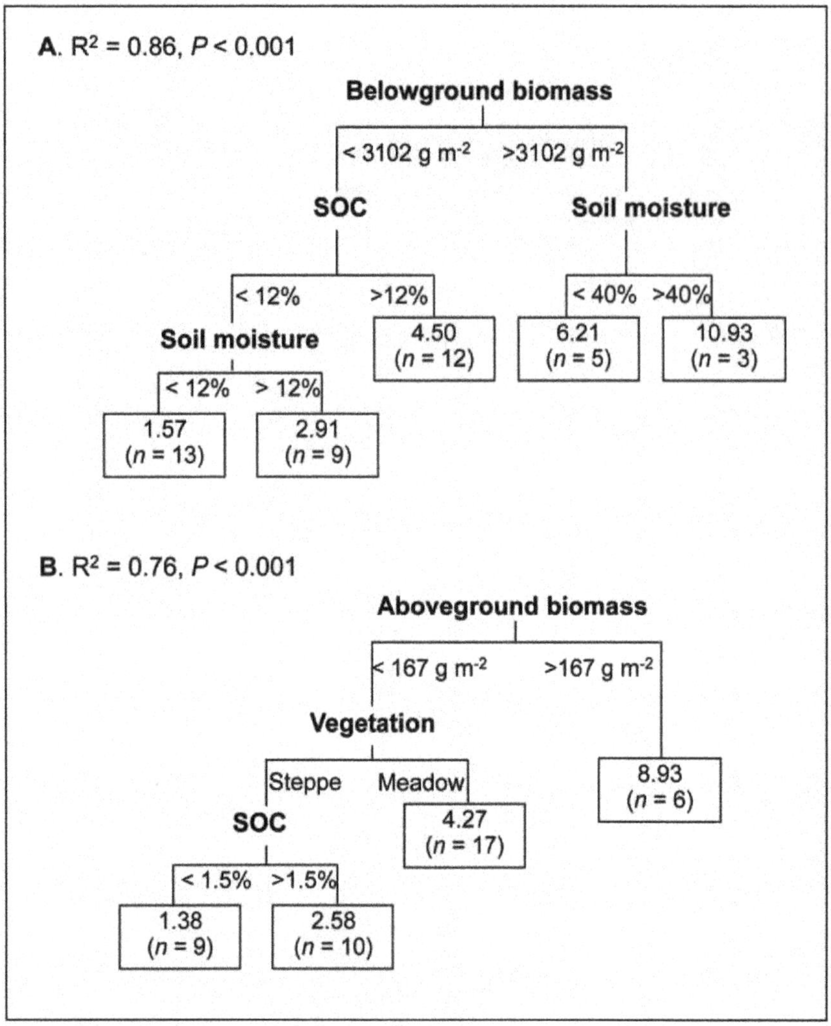

Figure 3. Regression tree showing generalized relationships between daily mean soil respiration rate and environmental variables. Relationships between soil respiration rate and belowground biomass, soil organic carbon content (SOC) and soil moisture (A), aboveground biomass, vegetation type, and SOC (B). Branches are labelled with criteria used to segregate data. Values in terminal nodes represent mean soil respiration rate of sites grouped within the cluster. The tree explained 86% (A) and 76% (B) of the variance in soil respiration rate, which is significantly more than a random tree ($P<0.001$). n = number of plots in the category.

Large diurnal variations in Rs were observed, although the diurnal patterns were generally similar for meadow and steppe (Fig. 2), both exhibiting the highest Rs during the time from 12:00 to 14:00 BST. Rs and their climatic, community and soil properties for the important ecosystem types, such as *Kobresia pygmaea* meadow, *K. tibetica* meadow, species-rich meadow (mixed-species meadow), and *Stipa* spp. steppe are lised in Table 3. *K. tibetica* meadow had the highest Rs, while *Stipa* steppe had the lowest Rs.

Factors associated with spatial variations in soil respiration

Based on regression analysis, five variables with an importance value greater than 0.4383 were selected (Table 2), and thus were included in the development of the structural equation models. Other variables had negligible or no impact on soil respiration.

When all five variables were entered into the model, a tree with AGB, vegetation type, and SOC as explanatory variables

was developed (Fig. 3B), while BGB and SM were excluded from the model because of the close correlations between BGB and AGB, and SM and vegetation type. When BGB and SM were entered into the model, another tree was developed (Fig. 3A). Both trees are significantly more than a random tree ($P<0.001$), explaining 86% (Fig. 3A) or 76% (Fig. 3B) of the variance in Rs rate.

These analyses indicated that BGB, SOC, SM, AGB, and vegetation types are biotic and abiotic factors that are most closely associated with large-scale variations in soil respiration. For the first tree (Fig. 3A), in the areas with BGB>3102 g m^{-2}, only SM had a statistically significant influence on soil respiration rate; while in the areas with BGB<3102 g m^{-2}, both SOC and SM had a detectable effect. For the second tree (Fig. 3B), when AGB>167 g m^{-2}, soil respiration rate was not significantly affected by vegetation type or SOC; by contrast, when AGB<167 g m^{-2}, soil respiration rate was influenced by both vegetation type and SOC.

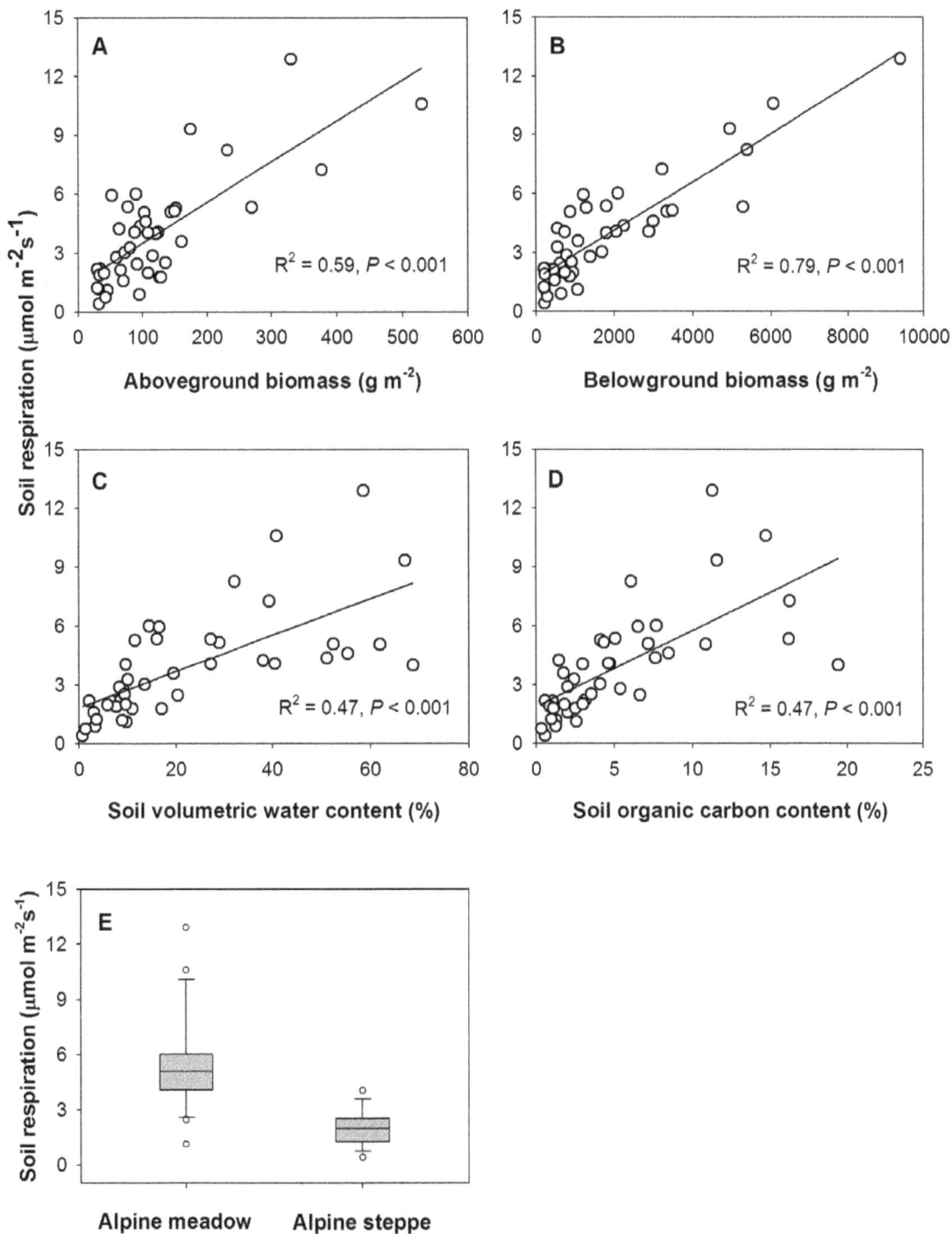

Figure 4. Scatterplots and box plot for daily mean soil respiration rate versus biotic and abiotic factors. Relationships between soil respiration rate and aboveground biomass (A), belowground biomass (B), soil moisture (C), soil organic carbon content (D), and vegetation type (E).

Structural equation modelling to explain variations in soil respiration

From the scatter plots and the box plot (Fig. 4), each of the selected variables such as AGB, BGB, SM, SOC and vegetation type was closely related to Rs. However, because these five variables were intercorrelated, these apparent relationships combined both direct and indirect correlations. Thus, we further used SEM to explicitly evaluate the causal relationships among these interacting variables.

The final SEM explained 82.1% of the variation in Rs (Fig. 5). Direct, indirect and total effects of the variables are summarized in Table 4. Increasing BGB and SM were strongly associated with increases in Rs, indicating that Rs could be well-predicted from these two variables ($R^2 = 0.82$). Even though there were significant bivariate relationships between AGB, SOC and Rs, they only had strong indirect positive effects on RS. Vegetation type had only an indirect effect on Rs (0.379) through its direct effect on BGB and its indirect effect on SOC and AGB. The rank of total effects, in

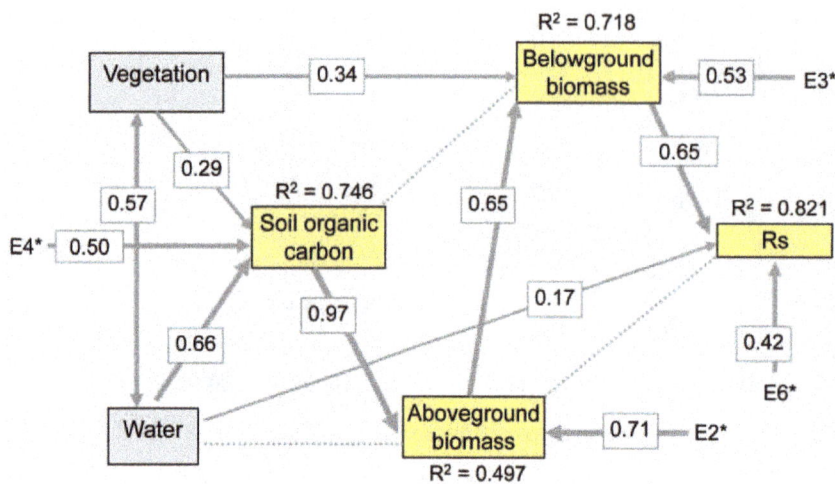

Figure 5. Final structural equation model for soil respiration. Non-significant paths are showed in dashed lines. The thickness of the solid arrows reflects the magnitude of the standardized SEM coefficients. Standardized coefficients are listed on each significant path.

decreasing order, was: BGB, AGB, SOC, vegetation type, and SM (Table 4).

It is also evident that, from the SEM (Fig. 5), BGB can well be predicted from vegetation type and AGB, explaining 71.8% of the variation. Moreover, SOC explained about 50% of the variations in AGB.

Discussion

One common feature of natural grasslands is the climate, usually characterized by periodic droughts [63]. For a specific

region, it may also be associated with basic parameters such as soil characteristics, frequent fires, grazing pressure and human activities. Chinese grasslands are generally distributed in three different regions: temperate grassland on the Inner Mongolian Plateau, alpine grassland on the Tibetan Plateau, and mountain grassland in the Xinjiang mountain areas [64]. Tibetan alpine grasslands, which are associated with cold climate of the high altitudes [49], differ from tropical and temperate grasslands. Yet, they are poorly documented in C cycles. Our survey on the large-scale patterns of Rs was preliminary, but the trend and relationships were clear.

Table 4. Total direct and indirect effects in the structural model. Effects were calculated using standardized path coefficients.

Variable	Direct effect	Indirect effect	Total
Rs			
Belowground biomass	0.654	-	0.654
Aboveground biomass	0.191 ns	0.427	0.618
Soil organic carbon	-	0.586	0.586
Vegetation type	-	0.397	0.397
Soil moisture	0.165	0.175 ns	0.335
Belowground biomass			
Aboveground biomass	0.652	-	0.652
Soil organic carbon	−0.021 ns	0.634	0.613
Vegetation type	0.345	0.179	0.524
Soil moisture	-	0.175 ns	0.175 ns
Aboveground biomass			
Soil organic carbon	0.971	-	0.971
Vegetation type	-	0.283	0.283
Soil moisture	−0.355 ns	0.644	0.29
Soil organic carbon			
Vegetation type	0.292	-	0.292
Soil moisture	0.663	-	0.663

Nonsignificant effects are indicated by "ns".

Magnitude of soil respiration of alpine grasslands

Large differences were observed between Rs from two vegetation types, alpine meadow and alpine steppe, being about two and half times greater in the alpine meadows. The daily mean Rs rates measured in alpine meadows (5.49 µmol CO_2 m^{-2} s^{-1} by daily average) are similar to previously reported results. For example, Cao et al. [32] reported that during peak growing season (Mid-July or August), daily Rs was 4.4 and 3.2 µmol CO_2 m^{-2} s^{-1} for light and heavy grazed meadows on the north-eastern edge of the Plateau. Li and Sun [33] reported a range of Rs from 0.93 to 8.02 µmol CO_2 m^{-2} s^{-1} during growing season in their recently published results. However, the only study from the alpine steppe by Zhang et al. [31], with a daily mean Rs rate of 0.38 µmol CO_2 m^{-2} s^{-1} at peak growing season, and an annual mean soil respiration rate of 0.248 µmol CO_2 m^{-2} s^{-1} using a closed static chamber-gas chromatograph method in a *Stipa purpurea* and *Carex moocroftii* community, was at the lower end of our measurement. The Rs rates of alpine steppe from this study (2.01 µmol CO_2 m^{-2} s^{-1} by daily average) are similar to the temperate steppe on the Inner Mongolia Plateau [23,65–69].

Consequently, the question arises: why is there such a difference in Rs rates between the two main grassland types, alpine meadow and alpine steppe? We suggest biological differences in standing biomass and productivity as well as physical differences in soil water availability were the major factors affecting Rs. On average, AGB (proxy of aboveground productivity) and BGB of the typical *Kobresia* meadows were much greater than the typical *Stipa* steppe (Table 3). Furthermore, SM of alpine meadow was also much higher than *Stipa* steppe. These high BGB and SM in alpine meadows significantly increased Rs rate.

The alpine grasslands of the Tibetan Plateau are sometimes called alpine tundra, despite their different species composition and environmental conditions compared to arctic tundra. Nevertheless, Tibetan alpine grassland and arctic tundra share some common features, such as large below ground standing biomass (averaging 1658 g m^{-2} for arctic tundra in Alaska [70], and 1816 g m^{-2} on the Tibetan Plateau in current study), relatively large soil C density [39,40], relatively high soil moisture (particularly in alpine meadow), and influences of permafrost. These characteristics mean that they are more responsive to global warming than other ecosystems, because their soils have the potential to release significant amounts of carbon-based greenhouse gases [46,71,72].

Factors associated with the large-scale patterns of soil respiration

Our analysis showed that among biotic and abiotic factors, BGB and SM together well explained the spatial patterns of peak growing-season Rs, accounting for 82% of the variation among 42 sampling sites. The important role of SM for Rs is in good accordance with results from other studies on soil nitrogen and carbon contents across the Tibetan Plateau [39]. Most of the variation in Rs could be attributed to the difference in BGB among sites (80%), with a small proportion further explained by SM (2%, SM entered after BGB in general linear models, because BGB and SM covaried). Thus, the results support our first hypothesis that BGB is most closely associated with the large-scale variations in Rs. This finding implies that autotrophic Rs (including plant roots and closely associated organisms) contributes a large proportion to total Rs, or/and autotrophic Rs is strongly related to heterotrophic Rs in these alpine grassland ecosystems.

A few studies with data compilation have addressed the general patterns of Rs across biomes. For example, on a global scale, Raich & Schlesinger [6] found Rs is positively correlated with MAT and MAP, as well as a close correlation between mean annual net primary productivity (NPP) of different vegetation biomes and their mean annual Rs. Bond-Lamberty & Thomson [25] built a global database of Rs from 3379 records spanning publication years 1963–2008, and found MAT, MAP and leaf area index together explained approximately 41% of the observed variability in annual Rs. Across the northern hemisphere, Hibbard et al. [28] found Rs and soil temperature are closely correlated for the deciduous and mixed forests, but not for non-forest biomes. These across-biome patterns of Rs are generally controlled by climate and NPP. Furthermore, Mahecha et al. [73] approximated the sensitivity of terrestrial ecosystem respiration to MAT across 60 sites worldwide, and offers substantial evidence for a general temperature sensitivities of soil respiration. Within the grassland biome, aboveground net primary productivity (ANPP, approximation to AGB of peak growing season as in this study) was shown to be positively correlated with Rs rate [12]. Craine et al. [27] also reported in Minnesota grasslands that both AGB and BGB are positively correlated with Rs. These previous studies in grasslands are consistent with the current results, since we observed a positive correlation between AGB, BGB and Rs as well. The novel part of our study is that we found only BGB and SM had direct effects on Rs at regional scale, with other factors indirectly affecting Rs through BGB or SM. It is also evident that factors most closely associated with Rs within-biome and across biomes are different.

In contrast, intra-annual variation in Rs at individual sites are mainly explained by soil temperature and soil moisture, but not by ANPP or AGB [74]. Temporal variations of Rs have been well simulated by using the continuous records of temperature and moisture [75]. Our measurements, across altitudes from 2925 to 5105 m and mean soil temperature (-10 cm) of midday (10:00 to 16:00 BST) from 6.3 to 31.6°C (the highest soil temperature of 31.6°C was recorded in an alpine steppe at 2925 m) during the field measurement exhibited that soil temperature did not have a strong effect on Rs across study sites. For example, *Kobresia tibetica* meadow on permafrost with a soil temperature of 6.3°C still had a daily mean Rs rate as high as 5.1 μmol CO$_2$ m^{-2} s^{-1}. Our results from the Tibetan grassland do not support the second hypothesis that Rs increases with increasing soil temperature in alpine grassland, but support the argument by Hibbard et al. [28] that within-site robust relationships with temperature and/or moisture are not adequate to characterize soil CO$_2$ effluxes across space, because for regional variation BGB is the most important factor.

Separating direct and indirect factors influencing soil respiration

In the present study, we used regression tree analysis [53] and SEM [56–58] as new approaches to conduct variable selection, to identify direct and indirect factors, and to determine the extent to which these factors may constrain Rs. To our knowledge, the efficiency of these approaches has not been evaluated empirically in soil respiration research.

Traditionally, stepwise selection and linear regression are used to identify and rank the limiting factors in Rs studies. However, when performing stepwise selection, closely covariated parameters cannot be selected simultaneously in the final model, because the explanatory power would not increase when a closely related variable is included. In our case, when BGB retain in the model, AGB will not be selected due to their close correlation. As a matter of fact, AGB has a strong indirect effect through BGB on Rs. This problem can be solved by a regression tree analysis which has the advantage to rank the limiting factors based on their importance [55].

Field studies examining ecosystem responses to climatic and other environmental changes typically use naturally occurring climatic gradients. However, some studies have realized the limitations of correlation method in analyzing factors influencing Rs [12,76]. For example, Rs rates vary significantly among major plant biomes, suggesting that vegetation type influences the rate of soil respiration. Nevertheless, the correlations among climatic factors, vegetation distributions, and Rs make cause-effect arguments difficult [12]. Burke et al. [76] raised the issue that there are inherent problems with utilizing simple statistical relationships of spatial variability as a foundation for understanding ecosystem function, because complex covariance along the gradient occurs across large spatial scales, leading to the problem that actual and apparent controlling factors may be confounded. Without field experiments, which are difficult to conduct across numerous sites, and without simulation of ecological processes, which need to be based on mechanistic data, SEM is one option. The quantitative procedure in the current study showed that the direct factors influencing Rs at large-scale were BGB and SM, AGB, SOC and vegetation type only had indirect influences despite their significant correlations with Rs. This holistic approach is appropriate in across-site comparisons of ecosystem structure and function.

Limitations of the current study

In the present study the soil PVC collars were installed only one hour before measurement due to the low accessibility of most sites, while the placement of collars are at least 24 hour prior to measurement in most Rs studies. Althouth the insertion of collars may cause unrealistic readings of soil CO_2 efflux because of the high fluxes after colloar installation, fluxes stabilize after 10–30 min [77,78]. In addition, our measurement of Rs followed the same procedure throughout our survey. Therefore the error introduced by soil disturbance could be treated as a systematic error which is weak.

Complete diurnal courses were obtained at nine sites, whereas for most of our sites soil respiration were measured 3–4 times during 4–5 hours when Rs peaked. We acknowledge that soil respiration is a dynamic process that may not be well represented by a few replicated measurements during several hours of a day. However, we found average midday Rs rates of the nine sites were well correlated with their daily mean Rs. Furthermore, we calculate daily mean Rs of each site by extrapolating the nine diurnal courses to all 42 sites according to community composition and closeness in distance. This extrapolation might add uncertainty to the estimates of daily mean Rs. Nevertheless, sites of similar vegetation composition and closest in distance generally share comparable features of geology, climate, soil and vegetation, which in combination are the major determinants of soil respiration.

The main objective of this study is to investigate the large-scale regional patterns of Rs in the Tibetan Plateau. Rs of 42 sites were measured during peak growing season of late July and early August. Measurements over a time span of one month may lead to problems as spatial variation of Rs could interfere with temporal changes. However, a four-year observation on soil CO_2 efflux in Haibei Alpine Grassland Research Station of Northwest Institute of Plateau Biology, the Chinese Academy of Sciences (3200 m a.s.l.) revealed that Rs values peak and stabilize in late July and early August (unpublished data by YHW and JSH). This phenomenon was observed in north America as well [79]. Therefore, compared with the large variation of Rs across the plateau, the temporal interference should be minor.

Conclusions and implications

Our understanding of the controls and magnitudes of regional Rs is limited by the uncertainties due to spatial heterogeneity of

vegetation across regional environmental gradients. In the current study, we moved beyond within-site differences in soil temperature and moisture to incorporate differences among broad ecosystem types (e.g. biomes). We can conclude with certainty that BGB is the factor most closely associated with Rs rate at regional scale for the grassland ecosystems, suggesting that in future we could develop models for Rs from plant standing biomass, which has a much larger database with wider biogeographic coverage, particularly in remote areas, such as the Tibetan Plateau. We acknowledge that only Rs rates during peak growing season were measured in the current study. Therefore, intensive measurements should be taken on a few sites across environmental gradients to develop more precise prediction models for annual Rs. Our results also have the implication that if we take Rs rates at peak growing season as a parameter of ecosystem metabolic activity, then compared with the plant physiology at individual level, ecosystem metabolism is not so much influenced by temperature itself. Furthermore, our results imply that a shift from alpine meadow to steppe due to changes of soil hydrological properties as a consequence of permafrost degradation will significantly alter Rs.

Acknowledgments

The authors are grateful to members of the Peking University expedition team, particularly Tong Shen, Wenhong Ma, Cunzhu Liang, Liang Wang, Yi Wu, Shanmin Mou and Shanxue Qi for assistance with field measurement, to Bernhard Schmid of University of Zurich, Switzerland, for statistical advice, and to Jingyun Fang and Dan Flynn for helpful comments.

Author Contributions

Conceived and designed the experiments: JSH. Performed the experiments: YG YHW KY FB PK. Analyzed the data: YG SW JSH. Contributed reagents/materials/analysis tools: JSH FB HZ TS. Wrote the paper: JSH YG.

References

1. Schlesinger WH (1997) Biogeochemistry: an analysis of global change. San Diego: Academic Press. 588 p.
2. Schlesinger WH, Andrews JA (2000) Soil respiration and the global carbon cycle. Biogeochemistry 48: 7–20.
3. Chapin FS, III, Matson PA, Mooney H (2002) Principles of terrestrial ecosystem ecology. New York: Springer-Verlag. 529 p.
4. Amundson R (2001) The carbon budget in soils. Annu Rev Earth Pl Sc 29: 535–562.
5. Eswaran H, van der Berg E, Reich P (1993) Organic carbon in soils of the world. Soil Sci Soc Am J 57: 192–194.
6. Raich JW, Schlesinger WH (1992) The global carbon dioxide flux in soil respiration and its relationship to vegetation and climate. Tellus B 44: 81–99.
7. Davidson EA, Janssens IA (2006) Temperature sensitivity of soil carbon decomposition and feedbacks to climate change. Nature 440: 165–173.
8. Luo YQ (2007) Terrestrial carbon-cycle feedback to climate warming. Annu Rev Ecol Evol S 38: 683–712.
9. Melillo JM, Steudler PA, Aber JD, Newkirk K, Lux H, et al. (2002) Soil warming and carbon-cycle feedbacks to the climate system. Science 298: 2173–2176.
10. Bond-Lamberty B, Thomson A (2010) Temperature-associated increases in the global soil respiration record. Nature 464: 579–582.
11. Luo YQ, Zhou XH (2006) Soil respiration and the environment. San Diego: Academic Press. 319 p.
12. Raich JW, Tufekcioglu A (2000) Vegetation and soil respiration: Correlations and controls. Biogeochemistry 48: 71–90.
13. Rustad LE, Campbell JL, Marion GM, Norby RJ, Mitchell MJ, et al. (2001) A meta-analysis of the response of soil respiration, net nitrogen mineralization, and aboveground plant growth to experimental ecosystem warming. Oecologia 126: 543–562.
14. Ryan MG, Law BE (2005) Interpreting, measuring, and modeling soil respiration. Biogeochemistry 73: 3–27.
15. Subke J-A, Inglima I, Cotrufo MF (2006) Trends and methodological impacts in soil CO_2 efflux partitioning: A metaanalytical review. Global Change Biol 12: 1–23.
16. Xu M, Qi Y (2001) Spatial and seasonal variations of $Q_{(10)}$ determined by soil respiration measurements at a Sierra Nevadan forest. Global Biogeochem Cy 15: 687–696.
17. Allison SD, Czimczik CI, Treseder KK (2008) Microbial activity and soil respiration under nitrogen addition in Alaskan boreal forest. Global Change Biol 14: 1156–1168.
18. Fierer N, Colman BP, Schimel JP, Jackson RB (2006) Predicting the temperature dependence of microbial respiration in soil: A continental-scale analysis. Global Biogeochem Cy 20: GB3026.
19. Kutsch WL, Persson T, Schrumpf M, Moyano FE, Mund M, et al. (2010) Heterotrophic soil respiration and soil carbon dynamics in the deciduous Hainich forest obtained by three approaches. Biogeochemistry 100: 167–183.
20. Högberg P, Nordgren A, Buchmann N, Taylor AFS, Ekblad A, et al. (2001) Large-scale forest girdling shows that current photosynthesis drives soil respiration. Nature 411: 789–792.
21. Wan S, Luo Y (2003) Substrate regulation of soil respiration in a tallgrass prairie: Results of a clipping and shading experiment. Global Biogeochem Cy 17: 1054.
22. Misson L, Gershenson A, Tang J, McKay M, Cheng W, et al. (2006) Influences of canopyphotosynthesis and summer rain pulses on root dynamics and soil respiration ina young ponderosa pine forest. Tree Physiol 26: 833–844.
23. Chen Q, Wang Q, Han X, Wan S, Li L (2010) Temporal and spatial variability and controls of soil respiration in a temperate steppe in northern China. Global Biogeochem Cy 24: 1–10.
24. Wan S, Norby RJ, Ledford J, Weltzin JF (2007) Responses of soil respiration to elevated CO_2, air warming, and changing soil water availability in a model old-field grassland. Global Change Biol 13: 2411–2424.
25. Bond-Lamberty B, Thomson A (2010) A global database of soil respiration data. Biogeosciences 7: 1915–1926.

26. Savage K, Davidson EA, Richardson AD (2008) A conceptual and practical approach to data quality and analysis procedures for high-frequency soil respiration measurements. Funct Ecol 22: 1000–1007.

27. Craine JM, Tilman D, Wedin D, Reich P, Tjoelkker M, et al. (2002) Functional traits, productivity and effects on nitrogen cycling of 33 grassland species. Funct Ecol 16: 563–574.

28. Hibbard KA, Law BE, Reichstein M, Sulzman J (2005) An analysis of soil respiration across northern hemisphere temperate ecosystems. Biogeochemistry 73: 29–70.

29. Jenny H (1941) Factors of soil formation. New York: McGraw-Hill. 229 p.

30. McCulley RL, Burke IC, Nelson JA, Lauenroth WK, Knapp AK, et al. (2005) Regional patterns in carbon cycling across the Great Plains of North America. Ecosystems 8: 106–121.

31. Zhang XZ, Shi PL, Liu YF, Ouyang H (2005) Experimental study on soil CO_2 emission in the alpine grassland ecosystem on Tibetan Plateau. Sci China Ser D 48: 218–224.

32. Cao G, Tang Y, Mo W, Wang Y, Li Y, et al. (2004) Grazing intensity alters soil respiration in an alpine meadow on the Tibetan plateau. Soil Biol Biochem 36: 237–243.

33. Li G, Sun S (2011) Plant clipping may cause overestimation of soil respiration in a Tibetan alpine meadow, Southwest China. Ecol Res 26: 497–504.

34. Lin X, Zhang Z, Wang S, Hu Y, Xu G, et al. (2011) Response of ecosystem respiration to warming and grazing during the growing seasons in the alpine meadow on the Tibetan Plateau. Agric For Meteor 151: 792–802.

35. Wang JF, Wang GX, Hu HC, Wu QB (2010) The influence of degradation of the swamp and alpine meadows on CH_4 and CO_2 fluxes on the Qinghai-Tibetan Plateau. Environ Earth Sci 60: 537–548.

36. Shi PL, Zhang XZ, Zhong ZM, Ouyang H (2006) Diurnal and seasonal variability of soil CO_2 efflux in a cropland ecosystem on the Tibetan Plateau. Agric For Meteor 137: 220–233.

37. Hou XY (1982) Vegetation map of the People's Republic of China (1:4M). Beijing: Chinese Map Publisher.

38. Baumann F, He J-S, Schmidt K, Kühn P, Scholten T (2009) Pedogenesis, permafrost, soil temperature and soil moisture as controlling factors for soil nitrogen and carbon contents across the Qinghai-Tibetan Plateau. Global Change Biol 15: 3001–3017.

39. Yang YH, Fang JY, Tang YH, Ji CJ, Zheng CY, et al. (2008) Storage, patterns and controls of soil organic carbon in the Tibetan grasslands. Global Change Biol 14: 1592–1599.

40. Zimov SA, Schuur EAG, Chapin FS, III (2006) Permafrost and the global carbon budget. Science 312: 1612–1613.

41. Cheng GD (2005) Permafrost studies in the Qinghai-Tibet Plateau for road construction. J Cold Reg Eng 19: 19–29.

42. Nan ZT, Li SX, Cheng GD (2005) Prediction of permafrost distribution on the Qinghai-Tibet Plateau in the next 50 and 100 years. Sci China Ser D 48: 797–804.

43. Wu S, Yin Y, Zheng D, Yang Q (2005) Climate change in the Tibetan Plateau during the last three decades. Acta Geogr Sin 60: 3–11.

44. Zhao L, Ping CL, Yang DQ, Cheng GD, Ding YJ, et al. (2004) Changes of climate and seasonally frozen ground over the past 30 years in Qinghai-Xizang (Tibetan) Plateau, China. Global Planet Change 43: 19–31.

45. Böhner J, Lehmkuhl F (2005) Environmental change modeling for Central and High Asia:Pleistocene, present and future scenarios. Boreas 34: 220–231.

46. Cheng GD, Wu TH (2007) Responses of permafrost to climate change and their environmental significance, Qinghai-Tibet Plateau. J Geophys Res Earth Surface 112.

47. Yang YH, Fang JY, Ji CJ, Han WX (2009) Above- and belowground biomass allocation in Tibetan grasslands. J Veg Sci 20: 177–184.

48. He JS, Wang ZH, Wang XP, Schmid B, Zuo WY, et al. (2006) A test of the generality of leaf trait relationships on the Tibetan Plateau. New Phytol 170: 835–848.

49. Zhang J, Wang JT, Chen W, Li B, Zhao K (1988) Vegetation of Xizang (Tibet). Beijing: Science Press. 589 p.

50. Wang JT (1988) The steppes and deserts of the Xizang Plateau (Tibet). Plant Ecol 75: 135–142.

51. Fang JY, Piao SL, Tang ZY, Peng CH, Ji W (2001) Interannual variability in net primary production and precipitation. Science 293: 1723.

52. He JS, Wang X, Flynn DFB, Wang L, Schmid B, et al. (2009) Taxonomic, phylogenetic, and environmental trade-offs between leaf productivity and persistence. Ecology 90: 2779–2791.

53. Breiman L, Friedman J, Olshen R, Stone C (1984) Classification and regression trees. Belmont: Wadsworth International Group. 358 p.

54. SASInstitute (1999) SAS/STAT User's guide, Version 8.01 (On-line Docs). Cary, NC: SAS Institute.

55. De'ath G, Fabricius KE (2000) Classification and regression trees: a powerful yet simple technique for ecological data analysis. Ecology 81: 3178–3192.

56. Grace JB (2006) Structural equation modeling and natural systems. Cambridge: Cambridge University Press. 365 p.

57. Grace JB, Pugesek BH (1977) A structural equation model of plant species richness and its application to a coastal wetland. Am Nat 149: 436–460.

58. Shipley B (2002) Cause and correlation in biology: A user's guide to path analysis, structural equations and causal inference. Cambridge: Cambridge University Press. 317 p.

59. Grace JB, Keeley JE (2006) A structural equation model analysis of post-fire plant diversity in California shrublands. Ecol Appl 16: 503–514.

60. Lamb EG (2008) Direct and indirect control of grassland community structure by litter, resources, and biomass. Ecology 89: 216–225.

61. Shipley B, Lechowicz MJ, Wright IJ, Reich PB (2006) Fundamental trade-offs generating the worldwide leaf economics spectrum. Ecology 87: 535–541.

62. Bentler PM (2006) EQS 6 Structural equations program manual. Encino, CA: Multivariate Software, Inc.

63. Ripley EA (1992) Grassland climate. In: Coupland RT, ed. Natural grasslands: Introduction and western hemisphere. Amsterdam: Elsevier. pp 151–182.

64. Wu ZY (1980) Vegetation of China. Beijing: Science Press. 1382 p.

65. Dong YS, Qi YC, Liu JY, Geng YB, Domroes M, et al. (2005) Variation characteristics of soil respiration fluxes in four types of grassland communities under different precipitation intensity. Chin Sci Bull 50: 583–591.

66. Li LH, Wang QB, Bai YF, Zhou GS, Xing XR (2000) Soil respiration of a *Leymus chinensis* grassland stand in the Xilin River Basin as affected by over-grazing and climate. Acta Phytoecol Sin 24: 680–686.

67. Qi Y, Dong Y, Domroes M, Geng Y, Liu L, et al. (2006) Comparison of CO_2 effluxes and their driving factors between two temperate steppes in Inner Mongolia, China. Adv Atmos Sci 23: 726–736.

68. Wang G, Du R, Kong Q, Lu D (2004) Experimental study on soil respiration of temperate grassland in China. Chin Sci Bull 49: 642–646.

69. Yan L, Chen S, Huang J, Lin G (2010) Differential responses of auto- and heterotrophic soil respiration to water and nitrogen addition in a semiarid temperate steppe. Global Change Biol 16: 2345–2357.

70. Dennis JG (1977) Distribution patterns of belowground standing crop in arctic tundra at Barrow, Alaska. Arct Alp Res 9: 113–127.

71. Billings WD, Luken JO, Mortensen DA, Peterson KM (1982) Arctic tundra: A source or sink for atmospheric carbon dioxide in a changing climate? Oecologia 53: 7–11.

72. Nobrega S, Grogan P (2008) Landscape and ecosystem-level controls on net carbon dioxide exchange along a natural moisture gradient in Canadian low arctic tundra. Ecosystems 11: 377–396.

73. Mahecha MD, Reichstein M, Carvalhais N, Lasslop G, Lange H, et al. (2010) Global convergence in the temperature sensitivity of respiration at ecosystem level. Science 329: 838–840.

74. Dornbush ME, Raich JW (2006) Soil temperature, not aboveground plant productivity, best predicts intra-annual variations of soil respiration in central Iowa grassland. Ecosystems 9: 909–920.

75. Raich JW, Potter CS, Bhagawati D (2002) Interannual variability in global soil respiration, 1980–94. Global Change Biol 8: 800–812.

76. Burke IC, Lauenroth WK, Parton WJ (1997) Regional and temporal variation in net primary production and nitrogen mineralization in grasslands. Ecology 78: 1330–1340.

77. Davidson EA, Savage K, Verchot LV, Navarro R (2002) Minimizing artifacts and biases in chamber-based measurements of soil respiration. Agric For Meteor 113: 21–37.

78. Norman JM, Kucharik CJ, Gower ST, Baldocchi DD, Crill PM, et al. (1997) A comparison of six methods for measuring soil-surface carbon dioxide fluxes. J Geophys Res 102: 28771–28778.

79. Zhou X, Wan S, Luo Y (2007) Source components and interannual variability of soil CO_2 efflux under experimental warming and clipping in a grassland ecosystem. Global Change Biol 13: 761–775.

80. Editorial Board of Vegetation Map of China (2001) Vegetation Atlas of China (1:1,000,000). Beijing: Science Press. 260 p.

3

Comparison of Soil Respiration in Typical Conventional and New Alternative Cereal Cropping Systems on the North China Plain

Bing Gao[1], Xiaotang Ju[1]*, Fang Su[1], Fengbin Gao[1], Qingsen Cao[1], Oene Oenema[2], Peter Christie[1], Xinping Chen[1], Fusuo Zhang[1]

1 College of Resources and Environmental Sciences, China Agricultural University, Beijing, China, **2** Wageningen University and Research Center, Alterra, Wageningen, The Netherlands

Abstract

We monitored soil respiration (Rs), soil temperature (T) and volumetric water content (VWC%) over four years in one typical conventional and four alternative cropping systems to understand Rs in different cropping systems with their respective management practices and environmental conditions. The control was conventional double-cropping system (winter wheat and summer maize in one year - Con.W/M). Four alternative cropping systems were designed with optimum water and N management, i.e. optimized winter wheat and summer maize (Opt.W/M), three harvests every two years (first year, winter wheat and summer maize or soybean; second year, fallow then spring maize - W/M-M and W/S-M), and single spring maize per year (M). Our results show that Rs responded mainly to the seasonal variation in T but was also greatly affected by straw return, root growth and soil moisture changes under different cropping systems. The mean seasonal CO_2 emissions in Con.W/M were 16.8 and 15.1 Mg CO_2 ha^{-1} for summer maize and winter wheat, respectively, without straw return. They increased significantly by 26 and 35% in Opt.W/M, respectively, with straw return. Under the new alternative cropping systems with straw return, W/M-M showed similar Rs to Opt.W/M, but total CO_2 emissions of W/S-M decreased sharply relative to Opt.W/M when soybean was planted to replace summer maize. Total CO_2 emissions expressed as the complete rotation cycles of W/S-M, Con.W/M and M treatments were not significantly different. Seasonal CO_2 emissions were significantly correlated with the sum of carbon inputs of straw return from the previous season and the aboveground biomass in the current season, which explained 60% of seasonal CO_2 emissions. T and VWC% explained up to 65% of Rs using the exponential-power and double exponential models, and the impacts of tillage and straw return must therefore be considered for accurate modeling of Rs in this geographical region.

Editor: Xiujun Wang, University of Maryland, United States of America

Funding: This work was funded by the National Natural Science Foundation of China (41230856, 31172033), the Special Fund for the Agricultural Profession (201103039), the '973' Project (2009CB118606) and the Innovation Group Grant of the National Natural Science Foundation of China (31121062). The funders had no role in study design, data collection and analysis, decision to publish, or preparation of the manuscript.] into [This work was funded by the '973' Project (2012CB417105, 2009CB118606), the National Natural Science Foundation of China (41230856, 31172033), the Special Fund for the Agricultural Profession (201103039), China Postdoctoral Science Foundation (2013M 530778), and the Innovation Group Grant of the National Natural Science Foundation of China (31121062). The funders had no role in study design, data collection and analysis, decision to publish, or preparation of the manuscript.

Competing Interests: The authors have declared that no competing interests exist.

* E-mail: juxt@cau.edu.cn

Introduction

Soils provide a very large sink of carbon (C) in terrestrial ecosystems with C reserves of about 1500 Pg C (1 Pg = 10^{15} g) and make a major contribution to the global carbon equilibrium [1]. Slight changes in soil C might therefore lead to significant changes in the concentration of CO_2 in the atmosphere. Soil respiration is the main terrestrial source of C return to the atmosphere with a flux reaching 98±12 Pg C in 2008 and increasing at a rate of 0.1 Pg C y^{-1} from 1989 [2]. Agricultural soils play a very important role in the global C cycle [3,4] and account for 11% of global anthropogenic CO_2 emissions [5]. It is therefore important to minimize soil respiration and retain more C sequestered in agricultural soils.

Soil respiration comprises mainly autotrophic respiration by plant roots and heterotrophic respiration of plant residues, root

litter and exudates, and soil organic matter by soil microorganisms [6,7]. Its magnitude is affected mainly by soil and climatic conditions [8] such as soil temperature and moisture [1,9,10], vegetation characteristics and management practices [11–14]. Soil respiration therefore shows high spatial and temporal variation [1]. Understanding this variation in different cropping systems in specific region will make a large contribution to the efficient management of C flow in agricultural ecosystems.

Soil respiration in cropland is greatly affected by tillage practices and straw management, with the greatest increase occurring immediately after tillage operations, and cumulative soil CO_2 emissions can be lowered significantly by reducing the intensity of tillage [15,16]. Daily CO_2 fluxes can differ significantly at some sampling dates between conventional moldboard plow tillage and no tillage in continuous corn [17]. Soil CO_2 emission can be enhanced in the short term after crop residues are returned to the field [15,18] but this practice may build up the soil organic carbon (SOC) pools in the long term and may therefore be regarded as a more sustainable way of managing SOC compared to straw burning or other uses for straw [19]. Differences in fertilizer N rates had no significant effect on the CO_2 exchange rates in the same crop rotation and CO_2 fluxes did not differ with crop rotation under no till practices [16]. In addition, crop species and/or other management practices affect soil CO_2 emission as a result of their influence on soil biological and biochemical properties [14,20].

Soil temperature and moisture are two of the most important environmental factors controlling soil respiration [1,21,22]. Soil temperature is significantly positively correlated with soil respiration using linear [7], exponential [1,7], improved Arrhenius [8], power and quadratic [9] and Q_{10} [10] models in different regions. Soil moisture is also a key factor controlling soil respiration, especially in arid or semiarid regions where it can be more important than temperature and become the dominant factor [11]. This shows that when one factor linking soil temperature and moisture is in a higher or lower range, the other might become a major factor controlling soil respiration [13,23,24]. The respiration rate will be limited when soil volumetric water content (VWC%) drops below a threshold of 15% [20]. Soil CO_2 emission increased significantly with increasing temperature up to 40°C, with emissions reduced at the lowest and highest soil moisture contents [20]. Therefore, the single-factor models cannot describe soil CO_2 emission well because they neglect the impacts of interactions between factors. The multiple polynomial models considering both soil temperature and moisture result in a much better description of CO_2 ($r^2 = 0.70$–0.78, $P<0.0001$) emissions than using temperature ($r^2 = 0.27$–0.54, $P<0.01$) or moisture ($r^2 = 0.29$–0.45, $P<0.01$) alone [20].

China has broad climatic regimes and the different ecosystems depend on regional climatic conditions [25]. The North China Plain (NCP) is a major agricultural region. The soil type is Fluvo-aquic soil and the climate is sub-humid temperate monsoon with abundant solar radiation but with cold and dry conditions in winter and spring and warm and wet weather in summer. Evapotranspiration is intense and the spring drought is an important feature. Winter wheat-summer maize is the typical double cropping system and current farming practice involves application of 300 kg N ha^{-1} yr^{-1} for winter wheat and 250 kg N ha^{-1} yr^{-1} for summer maize with a ratio of basal to topdressing applications of 1:1 and 1:1.5, respectively [26,27]. The soil is rotary tilled to 20 cm depth after maize straw removal for sowing winter wheat, and maize is sown directly after removing the wheat straw. Generally, wheat is irrigated three to four times and maize once or twice depending on precipitation. The amount of irrigation water ranges from 60 to 100 mm on each occasion [26]. About 30–60% of N input could be saved without sacrificing yields while significantly reducing environmental risk by adopting optimum N management in the winter wheat-summer maize system as shown by our earlier study [28]. However, over-exploitation of groundwater has become the main factor restricting sustainable agricultural development [29]. There is therefore concern to explore new alternative cropping systems for sustainable use of groundwater and optimum N fertilization to reduce pollution. Winter wheat–summer maize–spring maize with three harvests over two years and a single spring maize system have shown great potential to reduce water use and N use and can achieve balanced use of groundwater [26], and this cropping system may serve as a new alternative system for efficient resource use and sustainable development. However, it is still unclear how these changes will affect soil respiration in the study region.

Low frequency of measurement, lack of data at some growth stages, and failure to consider the interactive effects of soil moisture and temperature on soil respiration may lead to failure to describe the characteristics of soil respiration in this region [18,30,31]. There are indications that the correlation between soil respiration and soil temperature to 5 cm depth is 0.51 but the study that produced this result involved measurement only 21 times over one year [30]. Meng et al. [31] found that soil respiration had a higher correlation with soil temperature to 5 cm depth using the exponential model through weekly measurements of soil respiration under the typical double-cropping system over a whole year. Soil temperature at 5 cm depth explained 63–74% of soil respiration using the exponential model except during the winter, and the application of crop residues had significant positive impacts on soil respiration [18]. The management of N and water, crop residues and tillage practices will change significantly after conversion to new alternative cropping systems [26], an effect closely related to soil respiration. However, no quantitative information is yet available regarding soil respiration in new alternative cropping systems in this region.

In the present study we have compared soil respiration characteristics in different cropping systems with their respective management practices and environmental variables and we explore the factors affecting these differences. We have also analyzed the effects of straw return on variation in seasonal CO_2 emissions on the North China Plain.

Materials and Methods

Site description

A long-term field experiment was set up in October 2007 at Quzhou experimental station (36.87°N, 115.02°E) of China Agricultural University in Hebei province. The site is a sub-humid temperate monsoon area at an altitude of 40 m. The annual mean temperature is 13.2°C. Annual mean precipitation was 494 mm from 1980 to 2010 with a range of 213–840 mm, and 68% of precipitation falls from June to September [26]. The typical double-cropping system is a winter wheat and summer maize rotation which accounts for >80% of agricultural fields in Quzhou county. The soil type is Fluvo-aquic soil and the bulk density of the top 30 cm of the soil profile is 1.37 g cm^{-3}, soil pH is 7.72 (1:2.5, soil:water), SOC content 7.31 g kg^{-1}, total N 0.7 g kg^{-1}, Olsen-P 4.8 mg kg^{-1} and available K 72.7 mg kg^{-1}. Fig. 1 shows the daily mean air temperatures and precipitation during the measurement period (also see Table S1).

Field experiment treatments and management

A completely randomized design was employed with five treatments and four replicates. Each plot is 1800 m^2 (30×60 m). The control is conventional winter wheat and summer maize based on local farming practice (Con.W/M). Four new alternative cropping systems were designed with high-yielding varieties (using optimum planting density and crop management) and optimum water and N fertilizer management compared with conventional practice. They are: optimized two harvests in one year (winter wheat and summer maize - Opt.W/M), three harvests within two years (first year, winter wheat and summer maize or winter wheat and summer soybean; second year fallow then spring maize - W/M-M and W/S-M) and single spring maize per year (M).

Nitrogen input and irrigation for Con.W/M were described in the Introduction above. The basal fertilizer for wheat was surface broadcast before rotary tillage to 20 cm depth after removal of maize straw from the soil and topdressing was broadcast at shooting for wheat followed by irrigation, with both fertilizer

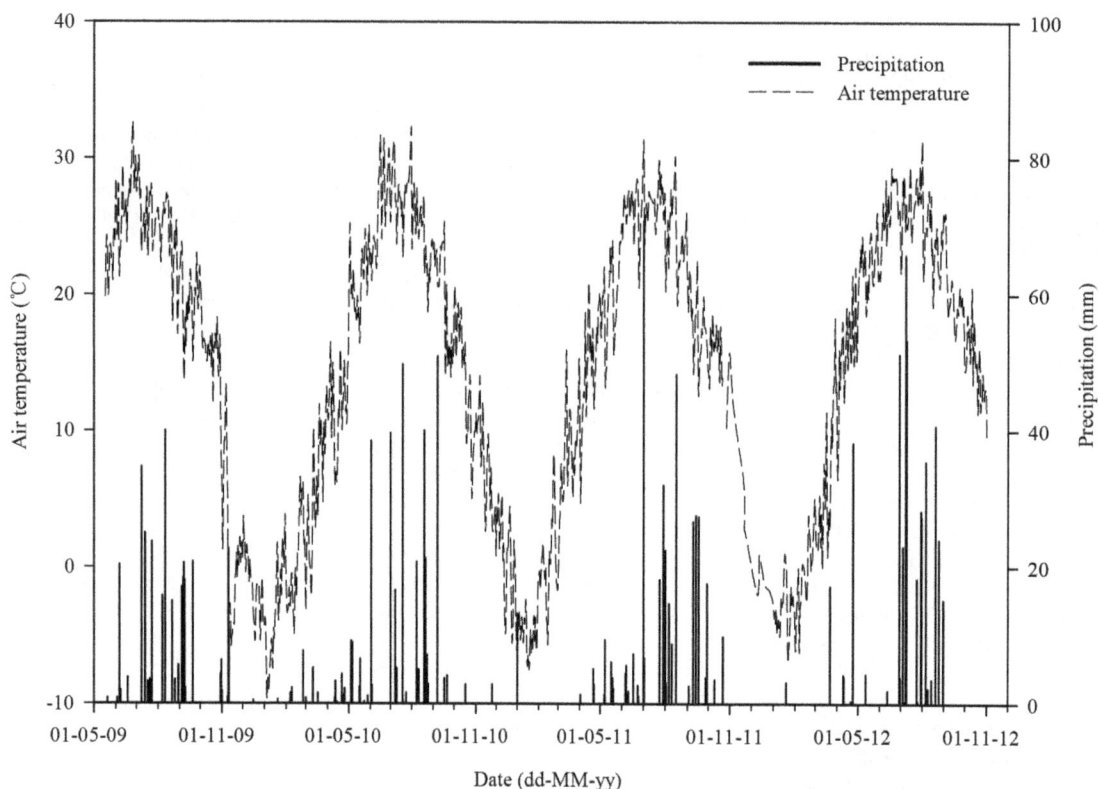

Figure 1. Daily mean air temperature (°C) and precipitation (mm) during the field experiment.

applications at 150 kg N ha^{-1} in the form of urea. The basal application for summer maize comprised 45 kg N ha^{-1} applied to the soil as 15-15-15 compound fertilizer with a seed drill after removing the wheat straw from the soil, and 55 kg N ha^{-1} surface broadcast as urea followed by irrigation and the topdressing 150 kg N ha^{-1} was applied at the ten-leaf stage of summer maize in the form of urea. In the other systems optimized N management was devised according to the N target values minus the soil nitrate-N content in the root zone before side-dressing as described by Cui et al. [27]. For summer maize 45 kg N ha^{-1} was applied as a basal dressing in the same way as for Con.W/M and 80 and 60 kg N ha^{-1} were side-dressed using a soil cover of 0–5 cm after band application at the six- and ten-leaf stages of summer maize, respectively. No other N fertilizer except 45 kg N ha^{-1} was applied as a basal application for soybean as for Con.W/M. Irrigation times and rates were determined by testing the soil water content before the critical growing seasons as described by Meng et al. [26]. The details of nitrogen input and irrigation rate over the whole study are shown in Table 1. Wheat straw was mulched after chopping into 5–10 cm pieces and summer maize or soybean was sown directly. Summer and spring maize and soybean residues were also chopped into 5–10 cm pieces and mechanically ploughed into the top 30 cm of the soil after maize and soybean were harvested and then winter wheat was sown if there was no fallow the following season. The soil was rotavated to 20 cm depth before sowing spring maize.

We measured soil respiration in each plot of the experiment from May 2009 to October 2012. Crops present in the different treatments during gas measurement are shown in Fig. 2.

Soil respiration measurement

Soil respiration was representatively determined in every plot using an automatic soil CO_2 flux system (LI-COR LI-8100, Lincoln, NE). Measurements were carried out daily for 10 days after fertilization events and 3–5 days after irrigation or precipitation events (>10 mm) depending on the size of gas fluxes; for the remaining periods emissions were measured twice per week and once a week when the soil was frozen. Two bases were used in each plot, one on a row and the other in the middle of the row during the maize and soybean seasons. Each base was a PVC tube with an inner diameter of 20 cm and a height of 13 cm inserted 9 cm into the soil for measurement and was removed only before sowing. Soil respiration was measured directly by LI-8100 in units of μmol CO_2 m^{-2} s^{-1} in the field between 08:30 and 11:00 am. Soil respiration is presented as the mean values of four replicated measurements on four different plots. The seasonal amounts of CO_2 emissions were sequentially linearly determined from the emissions between every two adjacent intervals of the measurements.

Auxiliary measurements

Soil temperature to 5 cm depth was measured directly by Li-8100 through a temperature sensing probe during the measurement time. Soil moisture at 0–5 cm is expressed as volumetric water content (VWC%) and was measured directly by Li-8100 through an ECH$_2$O type of EC-10 soil water sensing probe (Decagon Devices, Inc, Pullman, WA). We also measured the top 20 cm depth SOC content in each plot of this field experiment after summer maize harvest in 2011 using the method described by Huang et al. [19]. The daily mean air temperatures and precipitation data during the field experiment were obtained from

Table 1. Nitrogen fertilizer rates and irrigation rates throughout the study period.

Year	N application rate (kg N ha⁻¹)					Irrigation rate (mm)				
	Con.W/M[1]	Opt.W/M	W/M-M	W/S-M	M	Con.W/M	Opt.W/M	W/M-M	W/S-M	M
2009	W[2] 300	W 263	F -[3]	F -	F -	W 250	W 215	F -	F -	F -
	M_1 250	M_1 185	M_2 135	M_2 210	M_2 95	M_1 60	M_1 60	M_2 125	M_2 135	M_2 125
2010	W 300	W 100	W 140	W 140	F -	W 180	W 120	W 120	W 120	F -
	M_1 250	M_1 185	M_1 185	S 45	M_2 150	M_1 60	M_1 60	M_1 60	S 60	M_2 110
2011	W 300	W 139	F -	F -	F -	W 240	W 275	F -	F -	F -
	M_1 250	M_1 185	M_2 162	M_2 178	M_2 150	M_1 70	M_1 70	M_2 60	M_2 60	M_2 60
2012	W 300	W 140	W 162	W 158	F -	W 180	W 160	W 160	W 170	F -
	M_1 250	M_1 185	M_1 185	S 45	M_2 266	M_1 90	M_1 90	M_1 90	S 90	M_2 120
Total	2200	1382	969	776	661	1130	1050	615	635	415

[1]Con.W/M, Opt.W/M, W/M-M, W/S-M and M represent conventional and optimized winter wheat–summer maize, winter wheat–summer maize–spring maize, winter wheat–summer soybean–spring maize and spring maize treatment, respectively.
[2]W, M_1, M_2, S and F represent winter wheat, summer maize, spring maize, summer soybean and fallow.
[3]Denotes no data in the fallow season.

an automatic weather station located 50 m from our experimental site as shown in Fig. 1. Soil respiration and environmental variable data from the present study are presented in the Supplementary Data (Table S1).

Correlations between soil respiration and soil temperature and moisture

The compound factor models of soil respiration with soil temperature and moisture (equations 1–4) were employed as follows:

$$Rs = a + bTs + cWs \qquad (1)$$

$$Rs = aTs^b Ws^c \qquad (2)$$

$$Rs = ae^{bTs} Ws^c \qquad (3)$$

$$Rs = ae^{bTs + cWs} \qquad (4)$$

We established the four compound factor models above among soil respiration (Rs), soil temperature (Ts) and VWC(%) (Ws) using the measured fluxes from May 2009 to May 2012, and compared MAE (the mean absolute error), ME (model efficiency, the ratio of difference in measured and predicted flux in total variation in measured flux, expressed as significant correlation coefficient from −1 to 1), d (the percentage of mean square error and potential error, expressed as significant correlation from 0 to 1) [32,33], RMSE (root mean square error, reflecting the degree of dispersion of one variable), MSEₛ (systematic error) and MSEᵤ (random error) [34] among the models. We comprehensively evaluated the model performances by the sizes of MAE, ME, d, RMSE, MSEₛ and MSEᵤ, and the value of MSEₛ/(MSEᵤ+MSEᵤ). In general, MSEᵤ is close to RMSE in a well fitting model.

Statistical analysis

The primary data were processed using Microsoft Excel 2003 spreadsheets. Total CO_2 emissions in the different treatments were tested by analysis of variance and mean values were compared using SAS statistical software (Version 9.2; SAS Institute, Inc., Cary, NC) to calculate least significant difference (LSD) at the 5% level. Compound factor regression analysis among soil respiration, T and VWC% were performed using Sigmaplot 12.0 (Systat Software Inc., Erkrath, Germany).

Results

Characteristics of soil respiration in the different cropping systems

Over a complete rotation cycle soil respiration gradually increased from March, reached a maximum in July and gradually decreased from August to November, and then remained at the lowest values during winter, in a pattern similar to soil temperature (Figs. 2 and 3A). The mean soil respiration values were 3.35, 4.55, 4.03, 3.35 and 3.25 µmol CO_2 m⁻² s⁻¹ for Con.W/M, Opt.W/M, W/M-M, W/S-M and M throughout the study period, with ranges of 0.02–12.4, 0.26–14.9, 0.31–12.1, 0.34–11.3 and 0.30–11.2 µmol CO_2 m⁻² s⁻¹, respectively. Three peaks per year occurred in the typical double-cropping system, at the shooting stage of winter wheat, six-leaf of summer maize and the period after winter wheat sowing, the first two peaks caused by rapid crop growth and the last by the return of summer maize straw combined with soil tillage. Soil respiration of Opt.W/M was higher than of Con.W/M at the six-leaf stage of summer maize in the middle of July and the period after winter wheat sowing. The maximum peaks of soil respiration in Con.W/M were 8.2, 7.7, 7.8, 12.4 and 4.9, 2.8, 3.6, 2.9 µmol CO_2 m⁻² s⁻¹ during these two periods for four growing seasons, respectively, and they increased to 10.8, 9.6, 10.1, 14.9 and 7.6, 10.3, 6.5, 11.2 µmol CO_2 m⁻² s⁻¹ in Opt.W/M during the corresponding periods. Under the new alternative cropping systems one peak disappeared in the fallow season (season with no winter wheat planted). The highest value of soil respiration was around 7.0 µmol CO_2 m⁻² s⁻¹ in the spring maize season under the new alternative cropping systems, but it increased to more than 10.0 µmol CO_2 m⁻² s⁻¹ for summer maize in Opt.W/M at the corresponding time (Fig. 2).

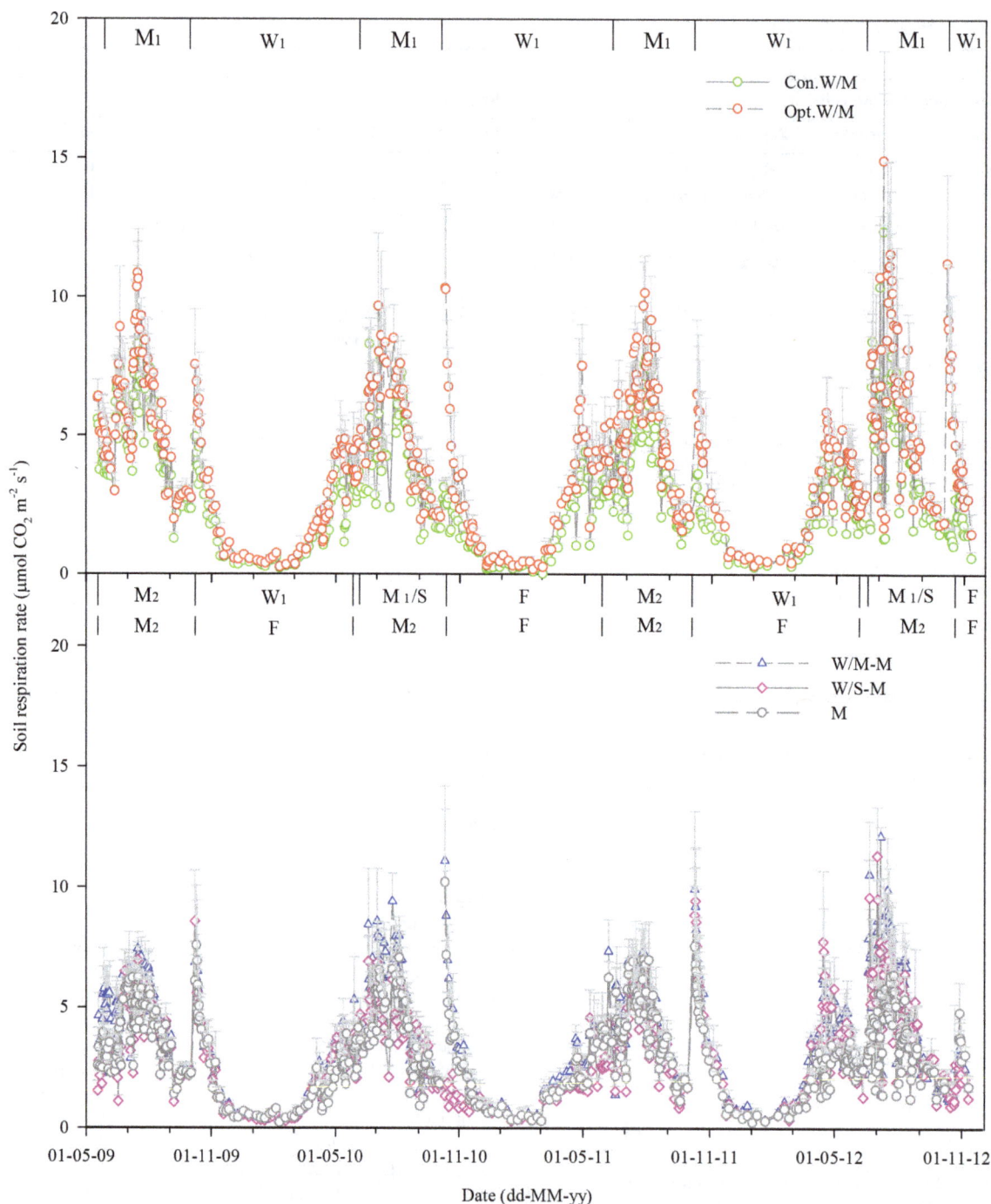

Figure 2. CO₂ emissions of different cropping systems. Con.W/M, Opt.W/M, W/M-M, W/S-M and M represent conventional winter wheat–summer maize in one year, optimized winter wheat–summer maize in one year, winter wheat–summer maize (or summer soybean) –spring maize three harvests in two years and single spring maize system in one year; W, M₁, M₂, S and F represent winter wheat, summer maize, spring maize, soybean and fallow.

Soil respiration was very low even after summer soybean stover return to the field in W/S-M in mid-November 2010 when the soil temperature in the top 5 cm ranged from −2.3 to +4.7°C within a month of soil tillage. A similar phenomenon occurred at the end of October 2012 due to the late spring maize and summer soybean harvests and the soil was tilled when soil temperature to 5 cm depth was around 10°C, and the peaks of W/M-M, W/S-M and M were only one third of the values of those at the corresponding times in other years. In addition, soil respiration showed large between-year change, so that peaks of soil respiration occurred after irrigation at shooting of winter wheat in other years, but not in winter wheat in 2010.

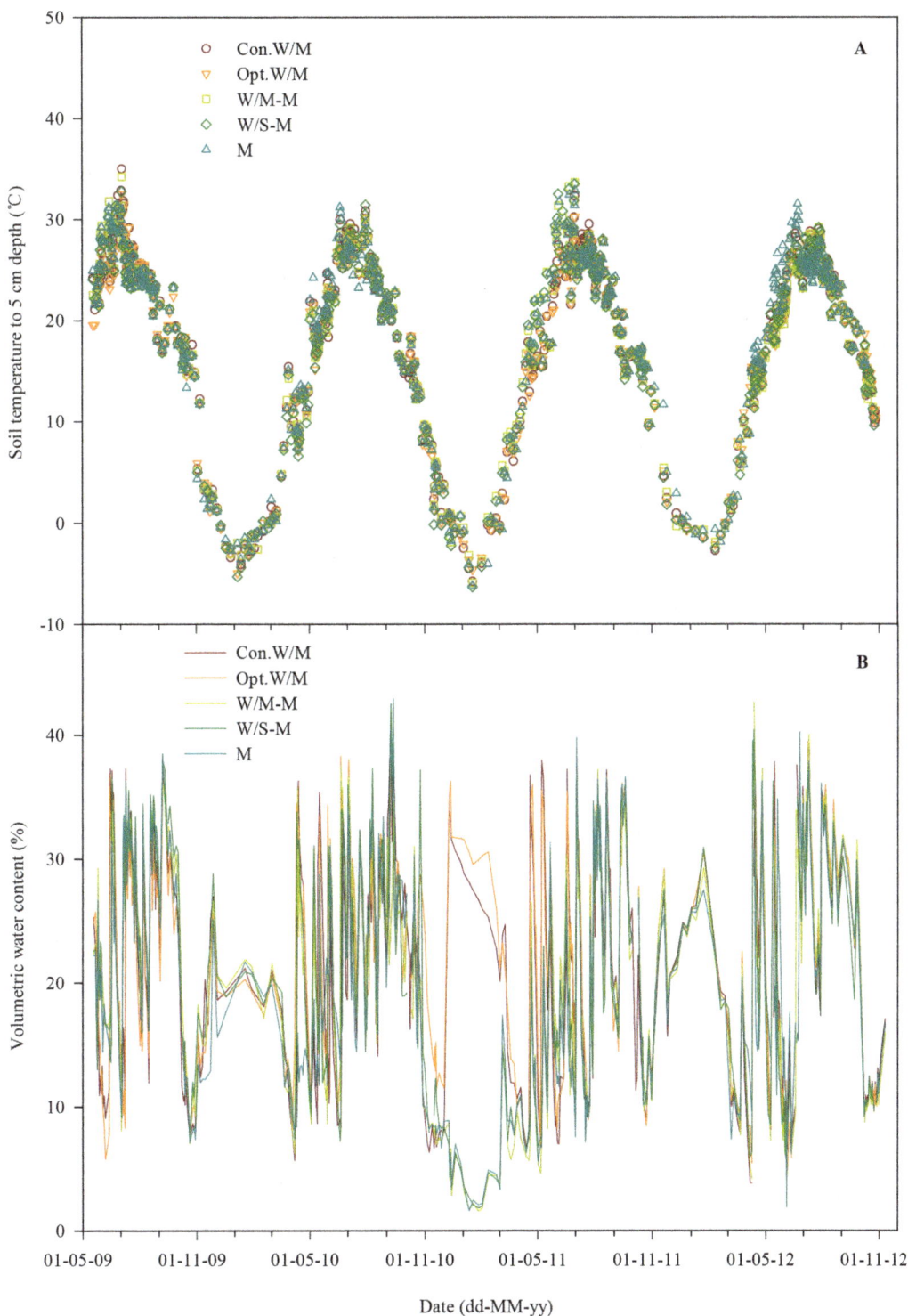

Figure 3. Dynamics of (A) soil temperature and (B) soil VWC% to 5 cm depth.

Total CO_2 emissions in each cropping season and each rotation cycle

Total CO_2 emissions in each cropping season and each rotation cycle were system dependent (Table 2). The mean seasonal total CO_2 emissions of Con.W/M were 16.8 and 15.1 Mg CO_2 ha^{-1} for summer maize and winter wheat, respectively. They increased significantly by 26 and 35% in Opt.W/M in the corresponding season. Under the new alternative cropping systems W/M-M showed similar results to Opt.W/M, and the seasonal total CO_2 emission of W/M-M was significantly higher than the corresponding season of Con.W/M except spring maize in 2009. However, W/S-M showed no significant difference from Con.W/M in each cropping season and the total CO_2 emissions in the fallow season and spring maize of W/S-M were clearly affected by summer soybean planting. Total CO_2 emission of M in each cropping season also showed no clear difference from Con.W/M except spring maize in 2011. In order to compare the impacts of cropping systems on CO_2 emissions of each rotation cycle we calculated the total CO_2 emissions during the period 2011–2012, which included two rotation cycles of Con.W/M, Opt.W/M, and M and a completely rotation cycle of W/M-M and W/S-M. Total CO_2 emission of Con.W/M was 61.9 Mg CO_2 ha^{-1}, and increased significantly by 37 and 29% in Opt.W/M and W/M-M treatment, respectively. The total CO_2 emission of W/M-M was not significantly different from Opt.W/M when there was only one season of winter wheat in two years, but total CO_2 emission of W/S-M decreased sharply in contrast to Opt.W/M when summer soybean was planted to replace summer maize of W/M-M because soil respiration was reduced significantly in the following fallow and spring maize seasons after the low biomass of soybean straw was returned to the soil. Total CO_2 emissions expressed as one complete rotation cycle of W/S-M, Con.W/M and M treatments were not significantly different (Table 2).

Soil respiration as affected by C input in each growth season

The measured soil respiration rates in this study consisted mainly of autotrophic respiration by crop roots in the current season, heterotrophic respiration of root litter and exudates in the current season, and heterotrophic respiration of crop straw return to the soil from the previous season and soil organic matter. As Fig. 2 and Table 2 show, the characteristics and total seasonal cumulative CO_2 emissions were greatly affected by straw return and crop growth status. To further explain soil respiration driven by C input in each growing season, we analyzed the correlation of seasonal cumulative CO_2 emissions with: (1) current-season aboveground biomass only; (2) the sum of C input of straw return from the previous season and the aboveground biomass in the current season. The relationship is improved significantly by inclusion of straw inputs (Fig. 4, equation A) compared to current-season aboveground biomass only (Fig. 4, equation B). Carbon input of straw return from the previous season and the aboveground biomass in the current season explains up to 60% of seasonal cumulative CO_2 emissions, much higher than that of 27% with current-season aboveground biomass only, which demonstrates that straw C inputs from the previous season can significantly affect soil respiration.

Correlation between soil respiration and soil temperature and VWC% to 5 cm depth

Large changes in soil respiration followed the variation in temperature over a complete year (Figs. 2 and 3A). Soil temperature explained 45% of soil respiration using the quadratic model (Fig. 5, equation A). In addition, soil moisture exerted some impacts on soil respiration under our climatic conditions such as inhibition within a short period after irrigation at shooting and grain filling stages of winter wheat and then a sharp increase which was derived from the effects of drying and wetting cycles. Soil respiration showed significant correlations with soil VWC% using the linear ($R_s = 2.7535 + 0.0447V$, $R^2 = 0.04$, n = 2282) and power ($R_s = 1.7708V^{0.2488}$, $R^2 = 0.06$, n = 2282) models at $P < 0.001$. However, soil VWC% explained only 4–6% of soil respiration. We further examined the combined effects of soil temperature and VWC% using four compound models, namely the linear, power, exponential-power and double exponential models (Table 3). The results indicate that the R^2 values combining temperature and VWC% are significantly higher than using the quadratic model only considering soil temperature. Soil temperature and VWC% explained up to 65% of soil respiration using the exponential-power and double exponential models.

The exponential-power and double exponential models (Table 3) gave significant improvements compared to the linear and power models. We again compared MAE, ME and d, RMSE, MSE_s and MSE_u among the four models and comprehensively evaluated the model performances by the sizes of these indicators and the value of $MSE_s/(MSE_u + MSE_u)$. The exponential-power model was much better for description of soil respiration in response to soil temperature and VWC% (both in the top 5 cm) in our study because it had lower MAE, RMSE, MSE_u, and higher

Table 2. Total CO_2 emissions in each cropping season and each rotation cycle (Mg CO_2 ha^{-1}).

Year	Con.W/M		Opt.W/M	W/M-M		W/S-M		M	
	Crop	CO₂	CO₂	Crop	CO₂	Crop	CO₂	Crop	CO₂
2009	M₁	19.1±0.8bc¹	22.0±1.5a	M₂	21.4±1.9ab	M₂	17.5±1.3c	M₂	17.4±1.4c
2010	W	15.5±1.2bc	17.2±0.6ab	W	17.8±1.2a	W	17.0±0.9abc	F	14.9±0.5c
	M₁	16.0±0.9b	21.4±1.7a	M₁	20.6±1.8a	S	15.7±0.9b	M₂	18.6±2.5ab
2011	W	13.4±0.7bc	22.4±0.4a	F	19.6±0.4a	F	11.1±1.6c	F	15.9±1.6b
	M₁	15.5±1.4b	20.4±1.9a	M₂	19.6±2.1a	M₂	16.0±1.1b	M₂	18.7±1.2a
2012	W	16.5±0.9bc	21.7±3.5a	W	21.3±2.4a	W	19.3±1.6ab	F	15.0±1.8c
	M₁	16.4±0.6c	20.4±1.9a	M₁	19.4±1.4ab	S	17.0±2.0bc	M₂	18.0±0.7abc
2009-2012 Mean	M₁	16.8	21.1	M₁	20.0	S	16.4	-	
	W	15.1	20.4	W	19.6	W	18.2	-	
	-	-	-	M₂	20.5	M₂	16.8	M₂	18.2
	-	-	-	F	19.6	F	11.1	F	15.3
2011-2012	2 W-M₁²	61.9±1.3b	84.9±8.2a	F-M₂-W-M₁	79.8±5.6a	F-M₂-W-S	64.1±2.7b	2 F-M₂	67.1±2.0b

¹The same letter in the same line denotes no significant difference in different cropping systems by LSD at $P < 0.05$.
²2 W-M₁, F-M₂-W-M₁ (or S) and 2 F-M₂ represent two winter wheat-summer maize rotation cycles, fallow-spring maize-winter wheat-summer maize (or summer soybean) rotation cycle and two fallow-spring maize rotation cycles.

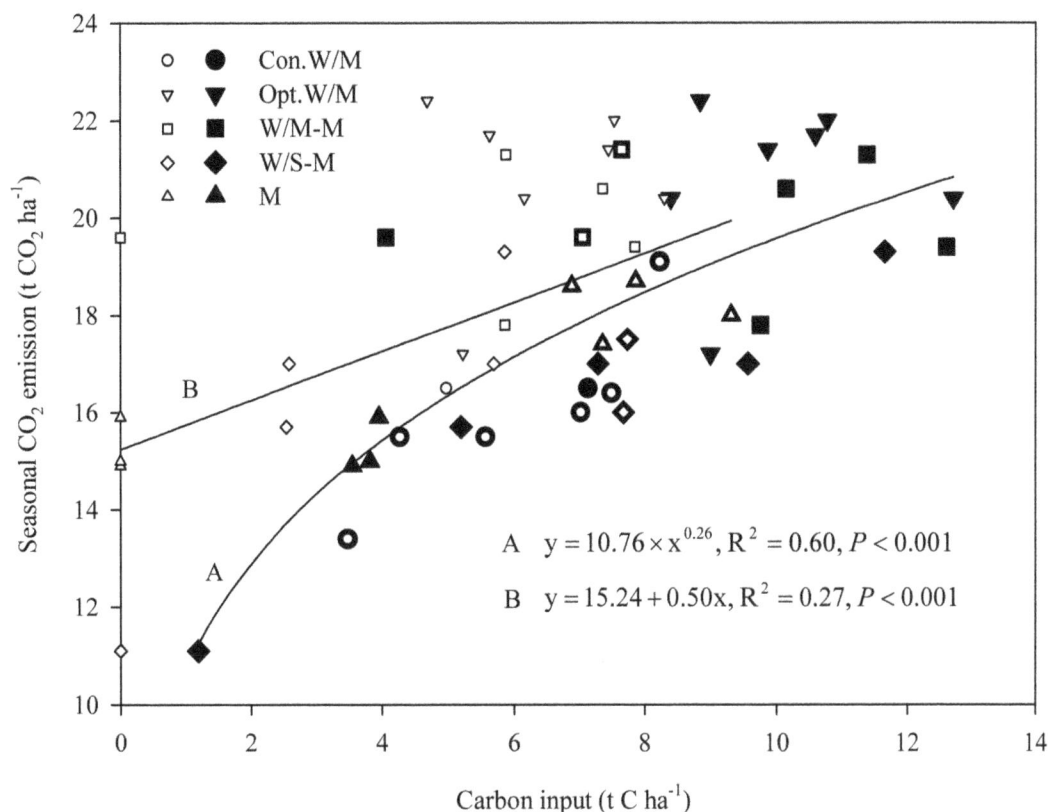

Figure 4. Correlation between seasonal CO_2 emission and carbon input. Carbon input was calculated from current-season aboveground biomass only (A); and calculated from straw return of the previous season and the aboveground biomass in the current season (B); the abbreviations of the treatment are shown in the footnotes in Fig. 2.

ME than the double exponential model and the values of $MSE_s/(MSE_s+MSE_u)$ were similar using both models.

Discussion

Soil respiration in croplands is affected mainly by soil properties, cropping system (which is related to crop species), tillage and straw management, water and nutrient management, and environmental variables (soil temperature, moisture etc.) [1,6,20,35]. There is temporal variation within the same cropping system and spatial variation among different cropping systems [16,17,31]. Changes in soil respiration in our sub-humid temperate monsoon region are largely affected by the seasonal variation in temperature, which is in line with most previous reports [30,31,36]. However, soil respiration responded little to soil temperature as shown in Fig. 5, equation A using the quadratic model because some data points did not fit the model with the impacts of soil tillage before the wheat crop was sown. The R^2 value improved by 18%, and up to 53% when the data within one month after tilling were excluded (Fig. 5, equation B). Moreover, we found that soil respiration tended to follow the variation in temperature from August to the following March when the data after tillage were excluded (Figs. 2 and 3A). Soil temperature explained 74% of soil respiration when only the data from August to March were included (Fig. 6, equation A). Therefore, the impacts of tillage must be considered for modeling soil respiration on the NCP.

The short decline in soil respiration after irrigation might be attributable to blocked diffusion of CO_2 with high moisture and limited oxygen concentrations in the soil matrix [37], and the

flushes afterwards may be due to the stimulation of decomposition of plant residues [21], root litter and exudates or autotrophic respiration of rapid root growth, which taken together induced the effects of drying and wetting cycles. Soil respiration would be limited when soil moisture was too high or too low and the maximum range is usually close to field water holding capacity [38]. The disappearance of respiration flushes was due to the low soil temperatures within a week after irrigation at the shooting stage of wheat in 2010 relative to other years (6–10°C in 2010 vs 12–21°C in 2011 and 12–19°C in 2012) (Fig. 3A). Soil moisture was not the key driving factor over the whole study period but did affect soil respiration slightly at particular stages and therefore only explained a very small proportion of the variation in soil respiration in our study area.

Numerous studies have reported that soil respiration is significantly affected by tillage practices combined with straw management [14–16]. Total soil respiration was significantly higher in Opt.W/M than Con.W/M as the latter soil was rotary tilled to 20 cm depth after maize straw removal and Opt.W/M was ploughed into the top 30 cm of the soil after maize straw return to the soil, soil respiration increased sharply after soil disturbance by tillage operations possibly because increased soil aeration accelerated the decomposition rate of crop residues which was associated with higher microbial activity [14,15,39]. However, the impacts of maize straw return and tillage were lowered by delaying tillage until the soil temperature to a depth of 5 cm reached 10°C or lower.

Although seasonal cumulative CO_2 emission in Opt.W/M and W/M-M increased significantly relative to Con.W/M as a result of

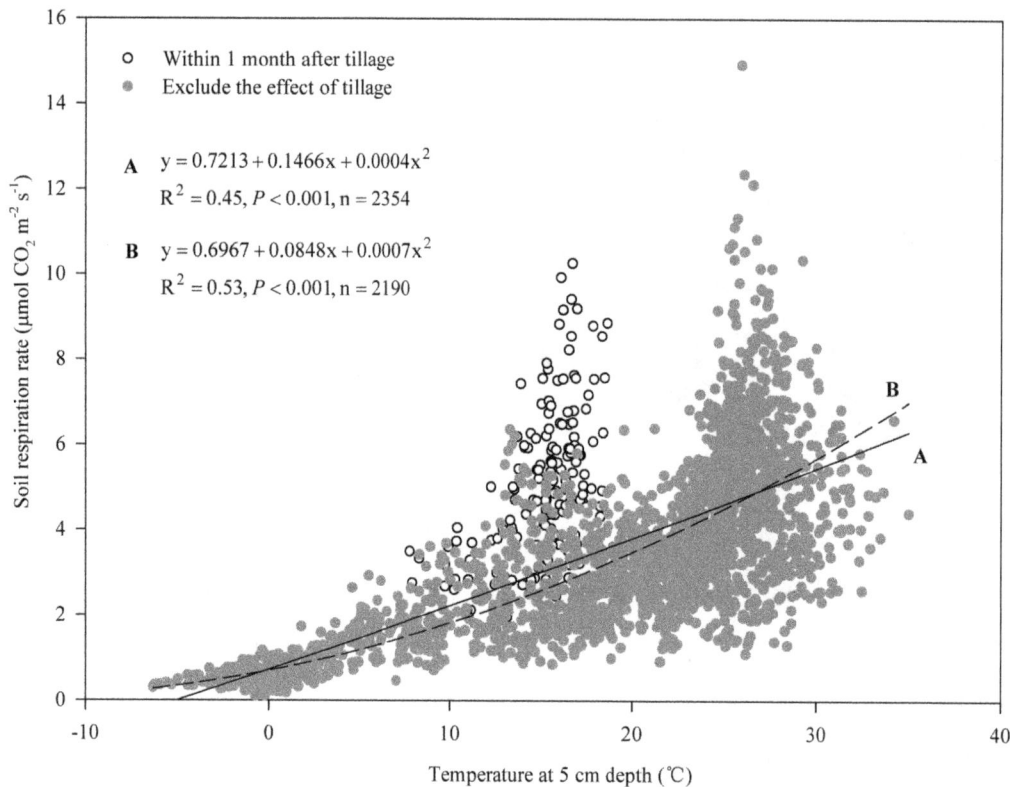

Figure 5. Impacts of soil tillage combined with straw return on soil respiration. Correlation between soil respiration and soil temperature at 5 cm depth over the whole year (equation A) and the correlation between soil respiration and soil temperature at 5 cm depth excluding the data within one month of tillage (equation B).

straw return [14,15], this practice also increases the SOC content over the long term [16,17,40]. The SOC content in the top 20 cm of the soil profile in straw return treatments increased by 3.9–16.5% relative to the straw removal treatments in winter wheat–summer maize double-cropping systems on the NCP, with a mean increase in rate of 0.04 to 1.44 t C ha^{-1} y^{-1} over a six-year period as shown by Huang et al. [19]. We also measured SOC to a depth of 20 cm in the present field experiment after summer maize harvest in 2011 and all values increased to 8.07, 8.71, 7.93 and 7.52 g kg^{-1} in Con.W/M, Opt.W/M, W/M-M and W/S-M, respectively, with the sole exception of a slight decrease to 7.18 g kg^{-1} in M (from 7.31 g kg^{-1} at the start of the field experiment in 2007). Although there was no crop straw return, Con.W/M also showed a clear increment relative to the initial value in line with Huang et al. [19], and this may have been due to the large amounts of crop roots and rhizo-deposited carbon. Con.W/M

showed a greater increase in SOC than W/M-M, W/S-M and M, possibly due to the lower intensity of tillage in Con.W/M than in W/M-M, W/S-M and M. Our results show that soil respiration responded mainly to the seasonal variation in soil temperature but was also greatly affected by straw return, root growth and soil moisture changes under the different cropping systems.

Supporting Information

Table S1 Measured soil respiration fluxes, soil temperature and soil volumetric water content to 5 cm depth, daily mean air temperature and precipitation in the field experiments from 18th May 2009 to 11th November 2012.

Table 3. Correlation between soil respiration and soil temperature and VWC(%) to 5 cm depth.

Model	Fitting equation	n	R^2	MAE	ME	RMSE	MSE$_s$	MSE$_u$	d
Linear	$R_s^1 = 0.7712 + 0.1581T - 0.0030V$	1905	0.47*[2]	1.11	0.48	1.48	1.28	1.03	0.80
Power (T>0)	$R_s = 0.6291T^{0.5929}V^{-0.0096}$	1811	0.56*	1.17	0.36	1.57	2.13	0.59	0.70
Exponential-power	$R_s = 0.9347e^{0.069T}V^{-0.0464}$	1905	0.65*	1.14	0.38	1.61	1.41	1.67	0.80
Double exponential	$R_s = 0.8924e^{0.0693T-0.0045V}$	1905	0.65*	1.20	0.31	1.69	1.36	1.81	0.78

[1]R_s, T and V represent soil respiration, soil temperature and VWC% to 5 cm depth, respectively.
[2]* represents highly significant correlation at $P<0.001$.

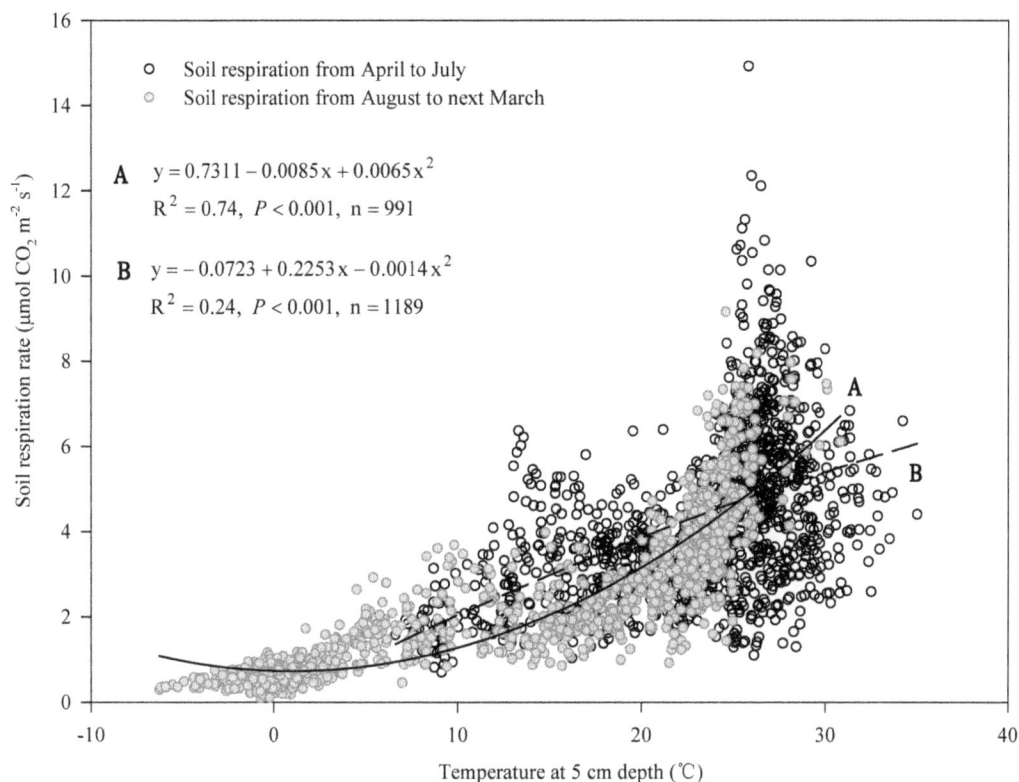

Figure 6. Correlation between soil respiration and soil temperature at 5 cm depth. Equations A and B represent the correlations between soil respiration and soil temperature at 5 cm depth from August to March after removing the impacts of tillage and from April to July, respectively.

Author Contributions

Conceived and designed the experiments: XJ XC FZ. Performed the experiments: BG FS FG QC. Analyzed the data: BG XJ. Contributed reagents/materials/analysis tools: XJ. Wrote the paper: BG XJ. Given some suggestion and modified the language for the manuscript OO PC.

References

1. Davidson EA, Belk E, Boone RD (1998) Soil water content and temperature as independent or confounded factors controlling soil respiration in temperate mixed hardwood forest. Global Change Biol 4: 217–227.
2. Bond-Lamberty B, Thomson A (2010) Temperature-associated increases in the global soil respiration record. Nature 464: 579–582.
3. Robertson GP, Paul EA, Harwood RR (2000) Greenhouse gases in intensive agriculture: contributions of individual gases to the radiative forcing of the atmosphere. Science 289: 1922–1926.
4. Mahecha MD, Reichstein M, Carvalhais N, Lasslop G, Lange H, et al. (2010) Global convergence in the temperature sensitivity of respiration at ecosystem level. Science 329: 838–840.
5. Grace J, Rayment M (2000) Respiration in the balance. Nature 404: 819–820.
6. Raich JW, Schlesinger WH (1992) The global carbon dioxide flux in soil respiration and its relationship to vegetation and climate. Tellus B 44: 81–99.
7. Melling L, Hatano R, Goh KJ (2005) Soil CO_2 flux from three ecosystems in tropical peatland of Sarawak, Malaysia. Tellus B 57: 1–11.
8. Lloyd J, Taylor JA (1994) On the temperature dependence of soil respiration. Funct Ecol 8: 315–323.
9. Fang C, Moncrieff JB (2001) The dependence of soil efflux on temperature. Soil Biol Biochem 33: 155–165.
10. Li HJ (2008) Studies on soil respiration and its relations to environmental factors in different ecosystems. Shanxi: PhD thesis, Shanxi University (in Chinese).
11. Wang WJ, Dalal RC, Moody PW, Smith CJ (2003) Relationships of soil respiration to microbial biomass, substrate availability and clay content. Soil Biol Biochem 35: 273–284.
12. Carlisle EA, Steenwerth KL, Smart DR (2006) Effects of land use on soil respiration: Conversion of Oak woodlands to vineyards. J Environ Qual 35: 1396–1404.
13. Rey A, Pegoraro E, Tedeschi V, De Parri I, Jarvis PG, et al. (2002) Annual variation in soil respiration and its components in a coppice oak forest in central Italy. Global Change Biol 8: 851–866.
14. Bavin TK, Griffis TJ, Baker JM, Venterea RT (2009) Impact of reduced tillage and cover cropping on the greenhouse gas budget of a maize/soybean rotation ecosystem. Agric Ecosyst Environ 134: 234–242.
15. Al-Kaisi MM, Yin XH (2005) Tillage and crop residue effects on soil carbon and carbon dioxide emission in corn-soybean rotations. J Environ Qual 34: 437–445.
16. Mosier AR, Halvorson AD, Reule CA, Liu XJ (2006) Net global warming potential and greenhouse gas intensity in irrigated cropping systems in Northeastern Colorado. J Environ Qual 35: 1584–1598.
17. Alluvione F, Halvorson AD, Del Grosso SJ (2009) Nitrogen, tillage, and crop rotation effects on carbon dioxide and methane fluxes from irrigated cropping systems. J Environ Qual 38: 2023–2033.
18. Zhang QZ, Wu WL, Wang MX, Zhou ZR, Chen SF (2005) The effects of crop residue amendment and N rate on soil respiration. Acta Ecologica Sinica 25: 2883–2887 (in Chinese with English abstract).
19. Huang T, Gao B, Christie P, Ju XT (2013) Net global warming potential and greenhouse gas intensity in a double cropping cereal rotation as affected by nitrogen and straw management. Biogeosciences Discuss 10: 13191–13229.
20. Iqbal J, Hu RG, Lin S, Ahamadou B, Feng ML (2009) Carbon dioxide emissions from Ultisol under different land uses in mid-subtropical China. Geoderma 152: 63–73.
21. Xu LK, Baldocchi DD, Tang JW (2004) How soil moisture, rain pulses, and growth alter the response of ecosystem respiration to temperature. Global Biogeochem Cycl 18: GB4002.
22. Conant RT, Dalla-Bett P, Klopatek CC, Klopatek JM (2004) Controls on soil respiration in semiarid soils. Soil Biol Biochem 36: 945–951.
23. Howard DM, Howard PJA (1993) Relationships between CO_2 evolution, moisture content, and temperature for a range of soil types. Soil Biol Biochem 25: 1537–1546.
24. Lellei-Kovács E, Kovács-Láng E, Botta-Dukát Z, Kalapos T, Emmett B, et al. (2011) Thresholds and interactive effects of soil moisture on the temperature response of soil respiration. Eur J Soil Biol 47: 247–255.
25. Fang JY, Chen AP, Peng CH, Zhao SQ, Ci LJ (2001) Changes in forest biomass carbon storage in China between 1949 and 1998. Science 292: 2320–2322.

26. Meng QF, Sun QP, Chen XP, Cui ZL, Yue SC, et al. (2012) Alternative cropping systems for sustainable water and nitrogen use in the North China Plain. Agric Ecosyst Environ 146: 93–102.

27. Cui ZL, Chen XP, Zhang FS (2010) Current Nitrogen management status and measures to improve the intensive wheat-maize system in China. Ambio 39: 376–384.

28. Ju XT, Xing GX, Chen XP, Zhang SL, Zhang LJ, et al. (2009) Reducing environmental risk by improving N management in intensive Chinese agricultural systems. Proc Natl Acad Sci USA 106: 3041–3046.

29. Hu C, Delgado JA, Zhang X, Ma L (2005) Assessment of groundwater use by wheat (*Triticum aestivum* L.) in the Luancheng Xian Region and potential implications for water conservation in the Northwestern North China Plain. J Soil Water Conserv 60: 80–88.

30. Niu LA, Hao JM, Zhang BZ, Niu XS, Lu ZY (2009) Soil respiration and carbon balance in farmland ecosystems on North China Plains. Ecol Environ Sci 18: 1054–1060 (in Chinese with English abstract).

31. Meng FQ, Guan GH, Zhang QZ, Si YJ, Qu B, et al. (2006) Seasonal variation in soil respiration under different long-term cultivation practices on high yield farm land in the North China Plain. Acta Scientiae Circumstantiae 26: 992–999 (in Chinese with English abstract).

32. Guo RY, Nendel C, Rahn C, Jiang CG, Chen Q (2010) Tracking nitrogen losses in a greenhouse crop rotation experiment in North China using the EU-Rotate_N simulation model. Environ Pollut 158: 2218–2229.

33. Willmott CJ (1982) Some comments on the evaluation of model performance. Am Meteorological Soc 63: 1309–1313.

34. Xiao M (2006) Effect of long-term fertilization on soil carbon and nitrogen storages and dynamics of nitrogen mineralization. Beijing: China Agricultural University (in Chinese).

35. Lohila A, Aurela M, Regina K, Laurila T (2003) Soil and total ecosystem respiration in agricultural fields: effect of soil and crop type. Plant Soil 251: 303–317.

36. Han GX, Zhou GS, Xu ZZ, Yang Y, Liu JL, et al. (2007) Soil temperature and biotic factors drive the seasonal variation of soil respiration in a maize (*Zea mays* L.) agricultural ecosystem. Plant Soil 291: 15–26.

37. Gaumont-Guay D, Black TA, Griffis TJ, Barr AG, Jassal RS, et al. (2006) Interpreting the dependence of soil respiration on soil temperature and water content in a boreal aspen stand. Agric For Mete 140: 220–235.

38. Davidson EA, Verchot LV, Cattanio JH, Ackerman IL, Carvalho JEM (2000) Effects of soil water on soil respiration in forests and cattle pastures of eastern Amazonia. Biogeochemistry 48: 53–69.

39. Jackson LE, Calderon FJ, Steenwerth KL, Scow KM, Rolston DE (2003) Responses of soil microbial processes and community structure to tillage events and implications for soil quality. Geoderma 114: 305–317.

40. Huang Y, Sun WJ (2006) Changes in topsoil organic carbon of croplands in mainland China over the last two decades, Chinese Sci Bull 51: 1785–1803.

4

Light and Heavy Fractions of Soil Organic Matter in Response to Climate Warming and Increased Precipitation in a Temperate Steppe

Bing Song[1,2], Shuli Niu[1]*, Zhe Zhang[1,2], Haijun Yang[1,2], Linghao Li[1], Shiqiang Wan[3]

1 State Key Laboratory of Vegetation and Environmental Change, Institute of Botany, Chinese Academy of Sciences, Xiangshan, Beijing, China, **2** Graduate School of Chinese Academy of Sciences, Yuquanlu, Beijing, China, **3** Key Laboratory of Plant Stress Biology, College of Life Sciences, Henan University, Kaifeng, Henan, China

Abstract

Soil is one of the most important carbon (C) and nitrogen (N) pools and plays a crucial role in ecosystem C and N cycling. Climate change profoundly affects soil C and N storage via changing C and N inputs and outputs. However, the influences of climate warming and changing precipitation regime on labile and recalcitrant fractions of soil organic C and N remain unclear. Here, we investigated soil labile and recalcitrant C and N under 6 years' treatments of experimental warming and increased precipitation in a temperate steppe in Northern China. We measured soil light fraction C (LFC) and N (LFN), microbial biomass C (MBC) and N (MBN), dissolved organic C (DOC) and heavy fraction C (HFC) and N (HFN). The results showed that increased precipitation significantly stimulated soil LFC and LFN by 16.1% and 18.5%, respectively, and increased LFC:HFC ratio and LFN:HFN ratio, suggesting that increased precipitation transferred more soil organic carbon into the quick-decayed carbon pool. Experimental warming reduced soil labile C (LFC, MBC, and DOC). In contrast, soil heavy fraction C and N, and total C and N were not significantly impacted by increased precipitation or warming. Soil labile C significantly correlated with gross ecosystem productivity, ecosystem respiration and soil respiration, but not with soil moisture and temperature, suggesting that biotic processes rather than abiotic factors determine variations in soil labile C. Our results indicate that certain soil carbon fraction is sensitive to climate change in the temperate steppe, which may in turn impact ecosystem carbon fluxes in response and feedback to climate change.

Editor: Jack Anthony Gilbert, Argonne National Laboratory, United States of America

Funding: This study was financially supported by the National Natural Science Foundation of China (31000227), Chinese Academy of Sciences (KSCX-YW-Z-1022), Key Project of National Natural Science Foundation of China (31130008), Distinguished Young Scientist program (30925009), and State Key Laboratory of Vegetation and Environmental Change. The funders had no role in study design, data collection and analysis, decision to publish, or preparation of the manuscript.

Competing Interests: The authors have declared that no competing interests exist.

* E-mail: niushuli@ibcas.ac.cn

Introduction

As atmospheric CO_2 concentrations are rising, global temperature has increased and will continue to increase in the future [1]. Simultaneously, changes in global and regional precipitation regimes are expected [2,3]. The climate change could profoundly affect ecosystem carbon (C) and nitrogen (N) cycles, with consequent increase or decrease in soil C and N storage and negative or positive feedback to climate change. Global soils contain 1500 Pg ($1\,Pg = 10^{15}$ g) of organic carbon in the top soil layer to the depth of 1 m, which is more than the amounts of C in the atmosphere and vegetation combined [4,5]. Even slight change in the amount of soil C may dramatically influence atmospheric CO_2 concentration [6]. The stocks of soil C result from the balance between carbon inputs and outputs. Therefore, any factors impacting carbon inputs (net primary productivity) and outputs (dominated by soil respiration) could change the quantity of organic carbon in soils [7].

Soils contain thousands of different organic-C compounds which have mean residence times ranging from years to millennia [7,8]. Soil organic carbon (SOC) is usually divided into labile C with a small size and rapid turnover and recalcitrant C with a large

size but slow turnover. Generally, it is difficult to detect significant changes in soil total C and N contents because of their high background values and great heterogeneity of soils [9]. However, separating soil carbon into different physical or chemical components and then examining their individual responses to climate change is a useful way to detect signals in soil carbon changes [7]. For example, soil labile C was documented to be more sensitive to alterations in moisture, temperature and plant species [10] in comparison with recalcitrant C. So, identifying the fractions of soil organic matter (SOM) into different pools and quantitatively analyzing these pools changes are critical for better understanding C and N dynamics and their responses to climate change.

Previous studies have documented that SOC contents would decrease greatly with global warming [11,12,13], and increase [5,14] or stay constant [15] with precipitation change, but how labile and recalcitrant soil organic C respond to climate change remains unclear. The key issue whether soil labile C and recalcitrant C respond differently to temperature change is still in debate [16,17,18]. In particular, there is lack of field experimental evidence on the responses of soil labile and recalcitrant C to climate change [19].

Grassland soil represents an important global C reservoir, and stores as much as 20% of global soil C [5]. Thus, the response of grassland soil C to climate change will be an important component of the global soil C feedback to climate change. Here we designed a field experiment manipulating temperature and precipitation changes in a semiarid steppe, which has been conducted for 6 years since April 2005, to examine the influence of climate change on labile and recalcitrant fractions of soil carbon and nitrogen. Previous studies in this field experiment have shown that the gross ecosystem productivity (GEP) and soil respiration (SR) of this steppe ecosystem were both decreased by warming and enhanced by increased precipitation [20,21]. However, the reduction in GEP by warming was greater than that in SR, and the stimulation in GEP was greater than that in SR under increased precipitation. Thus, we hypothesize that soil carbon will change due to shifts of the balance between carbon gains and carbon losses. The specific objectives of this study were to evaluate (1) how labile and recalcitrant fractions of soil carbon and nitrogen respond to climate warming and increased precipitation, and (2) what factors or processes determine the variations in soil carbon fractions in the context of climate change.

Materials and Methods

Study site

The study site is located in a semiarid temperate steppe in Duolun County (42°02′N, 116°17′E, 1324 m a.s.l.), Inner Mongolia, China. Mean annual precipitation is 382.3 mm with approximately 90% occurring in the growing season from May to October. Mean annual temperature is 2.1°C with monthly mean temperature ranging from −17.5°C in January to 18.9°C in July. Dominant species in this grassland are *Stipa krylovii* Roshev., *Artemisia frigida* Willd, *Potentilla acaulis* L., *Agropyron cristatum* (L.) Gaertn, *Cleistogenes squarrosa* (Trin.) Keng and *Allium bidentatum* Fisch. ex Prokh. [20]. The soil in the study site is a chestnut soil according to the Chinese classification or Haplic Calcisol according to the FAO classification, with 62.75±0.04% sand, 20.30±0.01% silt and 16.95±0.01% clay. Soil bulk density and pH are 1.31±0.02 g cm^{-3} and 7.34, respectively.

Experimental design

The experiment has received the permit for the field study from the land owner, Institute of Botany, the Chinese Academy of Sciences. The experiment used a nested design, with increased precipitation manipulated at the plot level and warming manipulated at the subplot level. There were three blocks with a 44×28 m area. In each block, there were two 10×15 m plots. One plot was assigned as the increased precipitation treatment and the other as the control. Six sprinklers were evenly arranged into two rows in each of the increased precipitation treatment plots. In July and August, 15 mm of water was added weekly to the increased precipitation treatment plots. Thus, a total of 120 mm precipitation was supplied each year.

Within each 10×15 m plot, four 3×4 m subplots with two warmed subplots and two control subplots were arranged randomly. In the warmed subplot, a 1.65×0.15 m MSR-2420 infrared radiant heater (Kalglo Electronics Inc., Bethlehem, PA, USA) that was suspended 2.5 m above the ground had heated the subplot continuously since April 28, 2005. A previous study by Niu et al. has documented that experimental warming elevated soil temperature at 10 cm depth by 1.17°C [22]. In the control subplot, a "dummy" heater with the same shape and size as the infrared radiator was suspended at the same height to simulate the shading effect of the heater. Therefore, the experimental design

consisted of 24 subplots with six replicates for four treatments (control, warming, increased precipitation, and warming plus increased precipitation).

Soil sampling and measurements

Soil samples were collected from the topsoil (0–10 cm) of all the 24 subplots on August 29, 2010. Two soil cores (6 cm in diameter and 10 cm in depth) were taken from each subplot, and then completely mixed to one fresh sample. Each soil sample was divided into two parts after sieving by a 2 mm mesh and removing any visible plant materials. One part of each sample was stored in iceboxes and transported to the laboratory for microbial analysis, and the other part was air-dried for chemical analysis.

Soil microbial biomass C (MBC) and N (MBN) were determined using the chloroform fumigation-extraction method [23]. Briefly, fresh soil samples were adjusted to 60–70% of field water-holding capacity and incubated for 1 week at 25°C. After that parts of each moist soil sample (30 g) were fumigated for 24 h by ethanol-free $CHCl_3$. The remainders (30 g) were used as non-fumigated controls. Both the fumigated and non-fumigated samples were extracted with 75 ml of 0.5 M K_2SO_4 for 30 min on a shaker. The extracts were filtered through 0.45 μm filters and determined for extracted C by potassium dichromate-bitriol oxidation method [23] and N by Kjeldahl digestion [24]. MBC and MBN were calculated from the differences between extracted C and N contents in the fumigated and non-fumigated samples using conversion factors of 0.38 [25] and 0.45 [24], respectively. And the extracted C in non-fumigated samples was considered as soil dissolved organic C (DOC).

Soil total C and N were measured by a CHNOS elemental analyzer (vario El III, Elementar Analysensysteme GmbH, Hanau, Germany) after the air-dried samples were ground finely.

In this study, we separated soil labile and recalcitrant fractions of SOM by density fractionation which is one of physical fractionation methods used widely [26]. The light fractions with low density (<1.7 g cm^{-3}) are partly decayed plant and animal products, while heavy fractions with high density (>1.7 g cm^{-3}) referred to humic substance which are generally mineral associated [27,28]. Specifically, 15 g air-dried soil was placed in a centrifuge tube and added 50 ml of NaI solution with a density of 1.7 g cm^{-3}. The tubes were shaken on a shaker for 30 min, and then centrifuged at 3000 rpm for 10 min. The floating light fraction was sucked on a fiberglass filter in a Büchner funnel. This process was repeated twice in order to separate the light and heavy fractions totally. After that, the material remaining at the bottom of the tube (the heavy fraction) was added 50 ml of deionized water, shaken and centrifuged for three times to wash. The light fraction was washed with 50 ml of 0.01 M $CaCl_2$ and then 50 ml of deionized water. Both the light fraction and heavy fraction were dried at 60°C for 48 h, weighed and ground to determine the C and N contents using a CHNOS elemental analyzer (vario El III, Elementar Analysensysteme GmbH, Hanau, Germany).

Statistical analysis

Three-way ANOVA for a blocked nested design was used to test the effects of block, warming and increased precipitation on all measured variables. Linear regression analyses were used to evaluate relationships between soil labile fractions (light fraction C and N, microbial biomass C and N, and dissolved organic C) and gross ecosystem productivity (GEP), ecosystem respiration (ER) and soil respiration (SR). The value of GEP, ER or SR was the yearly mean value from 2005 to 2009. The effects were considered to be significantly different if $p < 0.05$. All statistical analyses were conducted with SAS V.8.1 software (SAS Institute Inc., Cary, NC, USA).

Results

Total C and N in soil

Soil total C content (TC) was 18.48 ± 1.82 g C kg^{-1} dry soil and total N content (TN) was 1.88 ± 0.14 g N kg^{-1} dry soil (Fig. 1a). Neither warming nor increased precipitation had significant effects on TC or TN after six years of treatments (Table 1, Fig. 1a). The interactive effects of warming and increased precipitation on TC and TN were also not statistically significant (Table 1).

Light and heavy fraction C and N

Heavy fraction C (HFC) and N (HFN) accounted for 84.3% and 89.2% of TC and TN, respectively, in the control plots (Fig. 1). Increased precipitation significantly increased light fraction C (LFC) and N (LFN) by 16.1% and 18.5%, respectively (Table 1, Fig. 1b). The overall warming effects were marginally significant on LFC ($p = 0.07$, Table 1) but insignificant on LFN ($p = 0.16$, Table 1). For example, LFC changed from 3.03 ± 0.32 g C kg^{-1} dry soil in the control plots to 2.66 ± 0.20 g C kg^{-1} dry soil in the warmed plots with ambient precipitation. The interactions between warming and increased precipitation had no significant impacts on LFC and LFN. Neither HFC nor HFN were changed by warming or increased precipitation (Table 1, Fig. 1c).

The ratio of LFC to HFC (LFC:HFC) was significantly enhanced from 0.18 in the control plots to 0.23 in the increased precipitation plots across both warmed and unwarmed plots. Similarly, ratio of LFN to HFN (LFN:HFN) was enhanced from 0.12 in the control plots to 0.15 in the increased precipitation plots (Table 1, Fig. 1d).

Soil C to N ratios

Total soil C:N ratio (TC:TN) was 9.74 ± 0.29 in the control plots (Fig. 2). Warming significantly decreased TC:TN ratio from 9.90 ± 0.10 in the control plots to 9.40 ± 0.14 in the warmed plots across both ambient and increased precipitation treatments. C:N ratio of heavy fraction (HFC:HFN) was also decreased from 8.96 ± 0.13 to 8.54 ± 0.13 by warming (Table 2, Fig. 2). Nevertheless, warming did not significantly change C:N ratio of light fraction (LFC:LFN) which was 13.81 ± 0.20 in the control plots (Table 2, Fig. 2). Increased precipitation or its interaction with experimental warming had no impacts on any of these variables (Table 2).

Soil microbial biomass C and N and dissolved organic C

The main effects of warming significantly reduced soil microbial biomass C (MBC) by 12.6% (Table 2, Fig. 3). However, there were no effects of increased precipitation or its interaction with warming on MBC. Soil microbial biomass N (MBN) was not changed by any treatments (Table 2, Fig. 3).

Soil dissolved organic C was decreased from 38.08 ± 5.75 mg kg^{-1} in the control plots to 27.78 ± 2.92 mg kg^{-1} in the warmed plots across the ambient and increased precipitation treatments (Table 2, Fig. 3). Neither increased precipitation nor its interaction with warming had significant effects on DOC (Table 2, Fig. 3).

Relationships between carbon fluxes and soil C or N fractions

Across all the 24 subplots, LFC showed a positive relationship with the yearly mean values of GEP ($R^2 = 0.26$, $p = 0.01$; Fig. 4a), ER ($R^2 = 0.39$, $p < 0.01$; Fig. 5a) or SR ($R^2 = 0.42$, $p < 0.01$; Fig. 5b). Similarly, LFN showed a positive linear correlation with GEP

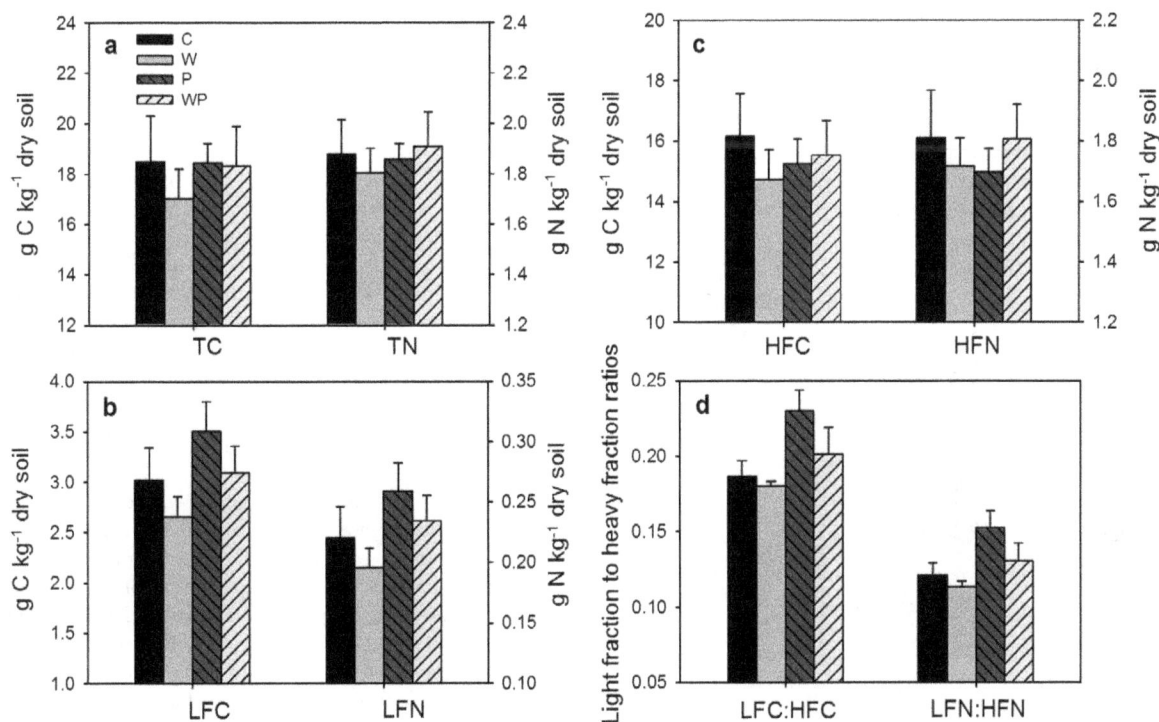

Figure 1. Effects of warming and increased precipitation on soil total C and N (TC, TN) (a), light fraction C and N (LFC, LFN) (b), heavy fraction C and N (HFC, HFN) (c), LFC:HFC ratio and LFN:HFN ratio (d) (means ± SE). C, control; W, warming; P, increased precipitation; WP, warming plus increased precipitation.

Table 1. Results (F and p values) of three-way ANOVA on the effects of block (B), warming (W), increased precipitation (P) and their interaction on measured soil variables.

		TC		TN		LFC		LFN		HFC		HFN		LFC:HFC		LFN:HFN	
	df	F	p	F	p	F	p	F	p	F	p	F	p	F	p	F	p
B	2	60.31	**<0.001**	56.13	**<0.001**	7.94	**0.006**	7.44	**0.008**	74.00	**<0.001**	77.54	**<0.001**	0.58	0.573	0.87	0.442
P	1	1.74	0.211	1.23	0.289	5.37	**0.039**	5.51	**0.037**	0.03	0.875	0.11	0.741	5.81	**0.033**	6.39	**0.027**
W	1	2.78	0.121	0.11	0.745	3.93	0.071	2.23	0.161	2.51	0.139	0.05	0.830	1.78	0.207	2.39	0.148
P×W	1	2.05	0.178	2.80	0.120	0.01	0.917	0.00	0.996	5.48	**0.037**	7.67	**0.017**	0.65	0.435	0.52	0.484

TC, soil total C; TN, soil total N; LFC, light fraction C; LFN, light fraction N; HFC, heavy fraction C; HFN, heavy faction N.

($R^2 = 0.28$, $p<0.01$; Fig. 4b), ER ($R^2 = 0.42$, $p<0.01$; Fig. 5c) or SR ($R^2 = 0.44$, $p<0.01$; Fig. 5d). Moreover, MBC and DOC, also showed positive linear correlations with GEP, ER and SR, except that DOC had no significant correlation with GEP ($p>0.05$; Fig. 4, Fig. 6). No significant relationship of soil light or heavy fraction C or N was found with soil temperature or moisture across the 24 subplots ($p>0.05$).

Discussion

Positive effects of increased precipitation on soil light fraction

Although density fractionation has some uncertainties, such as black carbon issue [26,29] and potential deficiency in operation [29], it has been widely used for more than 50 years and is well documented to be an effective way for assessing light and heavy pools of SOM that are differently sensitive to environmental changes [29]. As predicted, increased precipitation enhanced soil light fraction C and N in this study (Fig. 1b). The light fraction is a short-term reservoir of plant nutrients and the primary fraction for soil carbon formation, and serves as a readily decomposable substrate for soil microorganisms [10,30]. Its size is a balance between residue inputs and decomposition [31]. Increased precipitation could stimulate plant growth, leading to more carbon inputs to soil. On the other hand, increased precipitation may directly enhance soil microbial activities and accelerate soil carbon

decomposition, inducing more carbon losses from soil. Although increased precipitation could indirectly suppress plant growth and soil microbial activities via reducing soil temperature, in this temperate steppe where water availability plays a dominant role, the positive effects of increased precipitation were much stronger than the indirectly negative effects via reducing soil temperature [21]. In this study, light fraction showed a positive linear correlation with gross ecosystem production (GEP) (Fig. 4) and soil respiration (SR) (Fig. 5). Though both GEP and SR were enhanced by increased precipitation, the stimulation of GEP was greater than that of SR [20,21]. In addition, plant root production was also improved by increased precipitation [32]. These imply that increased precipitation has resulted in greater substrate inputs to soil than carbon outputs from soil, leading to the positive effects of increased precipitation on soil light fraction C and N.

Soil microbial biomass carbon (MBC) and dissolved organic carbon (DOC) are vital components of ecosystem carbon cycling, which have relatively rapid turnover rate and sever as a source or a sink of labile nutrients. In our study, MBC and DOC were not changed by increased precipitation (Fig. 3), which is consistent with some previous studies [33], but not in accordance with some others [21,34,35,36]. Since microbial biomass and activity are sensitive to changes in soil microenvironment [37,38], the responses of MBC to increased precipitation can be rapid but short-lived [39]. As we sampled soil in late August when water addition treatment was over, the response of MBC to increased precipitation would not be detected. Another possible reason is that soil temperature in the increased precipitation plots (18.60°C) was markedly lower than that in the control plots (22.65°C) in late August, 2010. Lower soil temperature would constrain microbial activity and growth, which partly compensate the directly positive effect of increased precipitation, leading to little change in MBC.

Negative effects of experimental warming on soil light fraction

The finding that soil light fraction C was decreased by experimental warming is in accordance with a previous study, in which soil labile C and N were reduced by warming in two forest ecosystems [40]. However, another experiment conducted in a tallgrass prairie showed that experimental warming increased soil labile C and N contents [19]. The discrepancy between our result and the above-mentioned result could be explained by the different controlling factors of C fluxes in different ecosystems. In the tallgrass prairie ecosystem where water is not as limited as in our arid ecosystem, experimental warming could directly stimulate plant growth and microbial activity. The enhancement of above- and below-ground biomass by warming was greater than the stimulation of soil respiration [41], so labile C and N fractions

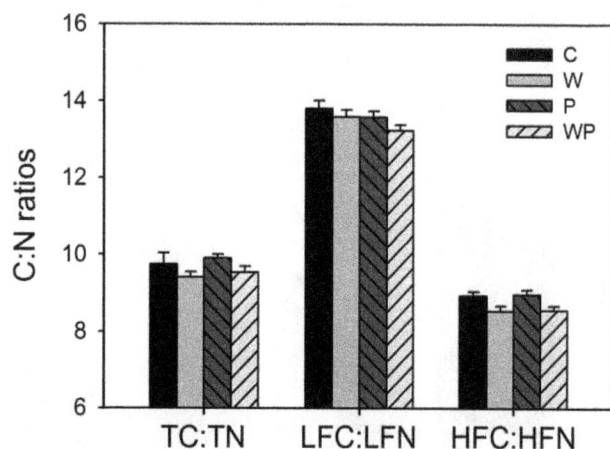

Figure 2. Effects of warming and increased precipitation on ratios of soil C:N (TC:TN), light fraction C:N (LFC:LFN) and heavy fraction C:N (HFC:HFN) (mean ± SE). See Fig. 1 for abbreviations.

Table 2. Results (F and p values) of three-way ANOVA on the effects of block (B), warming (W), increased precipitation (P) and their interaction on soil C:N ratios, MBC, MBN and DOC.

	df	TC:TN		LFC:LFN		HFC:HFN		MBC		MBN		DOC	
		F	p	F	p	F	p	F	p	F	p	F	p
B	2	18.15	**<0.001**	1.12	0.359	6.80	**0.011**	0.33	0.725	0.63	0.549	7.83	**0.007**
P	1	1.92	0.191	2.69	0.127	0.11	0.751	1.60	0.230	4.68	0.051	0.05	0.835
W	1	11.12	**0.006**	2.62	0.131	16.67	**0.002**	5.81	**0.033**	1.76	0.210	6.27	**0.028**
P×W	1	0.01	0.917	0.13	0.723	0.00	0.947	0.19	0.667	0.18	0.676	3.01	0.109

MBC, soil microbial biomass C; MBN, soil microbial biomass N; DOC, soil dissolved organic C. See Table 1 for abbreviations of TC, TN, LFC, LFN, HFC, and HFN.

were increased as a result of higher substrate inputs in the tallgrass prairie. However, in the semiarid ecosystem where water availability is the predominate limiting factor [20], warming can exacerbate the dry condition. The negatively indirect warming effect by reducing soil moisture is much stronger than the positively direct effect of improving temperature on ecosystem C fluxes. Previous studies conducted in the same experiment have showed that GEP, ecosystem respiration (ER) [20], SR [21], and plant root production [32] were all reduced by warming. Because light fraction C or N depends on both GEP (Fig. 4) and ER (Fig. 5), more reductions in GEP than those in ER and SR [21] leads to the decrease of light fraction C in soil. There were similar impacts on soil microbial biomass C (MBC) and soil dissolved organic C (DOC). The positive linear relationships between soil MBC and DOC with GEP, ER and SR (Fig. 4, Fig. 6) suggest that the decreases in MBC and DOC are partly due to the decrease in substrate inputs under warming (Fig. 4). This is consistent with a previous study which documented that DOC was positively related to the amount of organic matter inputs [42]. So, we conclude that greater reductions in C gains relative to C losses under climate warming decreased soil light fraction C, MBC and DOC.

Warming decreased total soil C:N ratio and the C:N ratio of heavy fraction (Fig. 2), which suggests that soil heavy fraction C has the potential to be decomposed more under warming than in control. Previous studies have documented that soil microbial community structure will change under warming and that the microorganisms preferring more recalcitrant carbon could establish as temperature increases [43]. This means that soil heavy fraction carbon could be preferentially respired by microbes under warming. The decreases of total soil C:N ratio and heavy fraction

C:N ratio under experimental warming implies that, as global temperature increases, soil heavy fraction C which constitutes the majority of soil carbon may potentially induce increasing C emissions from soil to the atmosphere.

Although it is assumed that abiotic factors associated with climate change may interact to affect ecosystem carbon cycling, there were no significant interactive effects between warming and increased precipitation on soil light fractions of C and N. This is probably due to that 30% increase of precipitation is not enough to alleviate water limitation and to change the negative warming effects. The insignificant interactions between warming and increased precipitation were also reported in previous studies on soil respiration [44,45] and above-ground biomass production [46].

Changes in soil total and heavy fraction C and N

Because of the large pool size, significant change in soil total C content in response to climate change is usually difficult to detect in a short time. In the present study, we did not detect significant changes in soil total C or N contents even after 6 years' treatment of experimental warming or increased precipitation (Fig. 1a). Soil heavy fraction C and N, which account for approximately 85% of total C or N, were not affected by either experimental warming or increased precipitation (Fig. 1c). The results are consistent with previous studies which found that soil mineral C did not change after 13 years of increased rainfall [47] and that soil recalcitrant C were not influenced by experimental warming [19,40]. These indicate that soil heavy fraction C is relative stable to climate change [16,48].

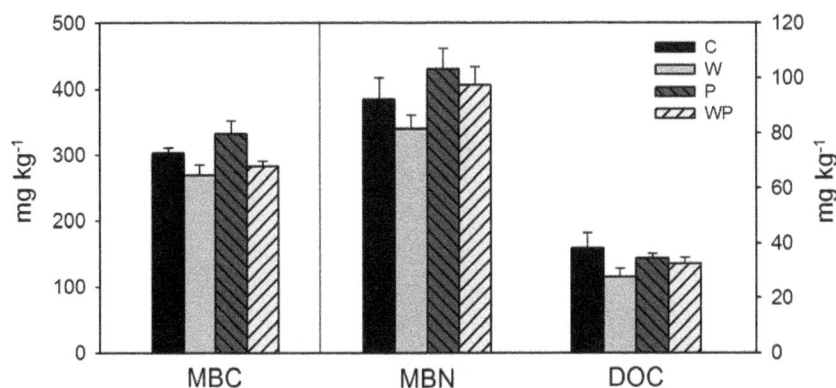

Figure 3. Effects of warming and increased precipitation on soil microbial biomass C and N (MBC, MBN) and soil dissolved organic C (DOC) (means ± SE). See Fig. 1 for abbreviations.

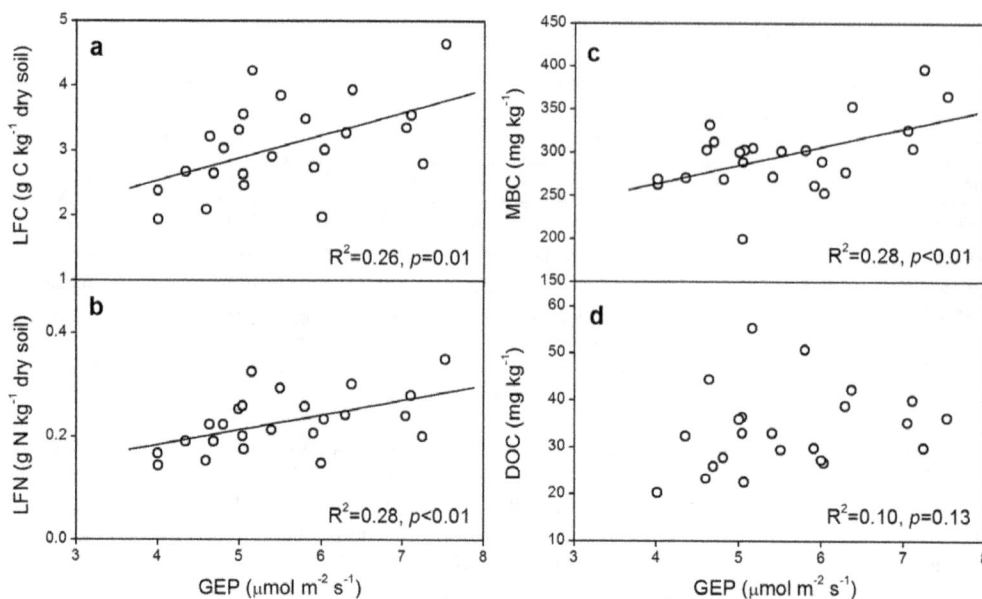

Figure 4. Linear correlations between GEP and light fraction C (a) or N (b), MBC (c) and DOC (d) across all the 24 subplots. GEP, gross ecosystem productivity, whose values were the yearly mean values from 2005 to 2009.

In conclusion, although soil total carbon and heavy fraction carbon were not affected by increased precipitation or warming, soil light fraction carbon was significantly increased by water addition and decreased by experimental warming after 6 years of treatments in a semiarid temperate steppe. The changes in soil labile C and N were primarily due to the different responses of carbon uptake and release rather than the changes in environ-mental conditions under treatments. The sensitive responses of soil light fraction C and N, microbial biomass C, and dissolved organic C to climate change indicate that climate warming and increased precipitation may impact carbon cycling by changing certain fractions of soil organic matter. Models should take into account of the fractions of soil organic matter to more accurately predict ecosystem's response and feedback to climate change.

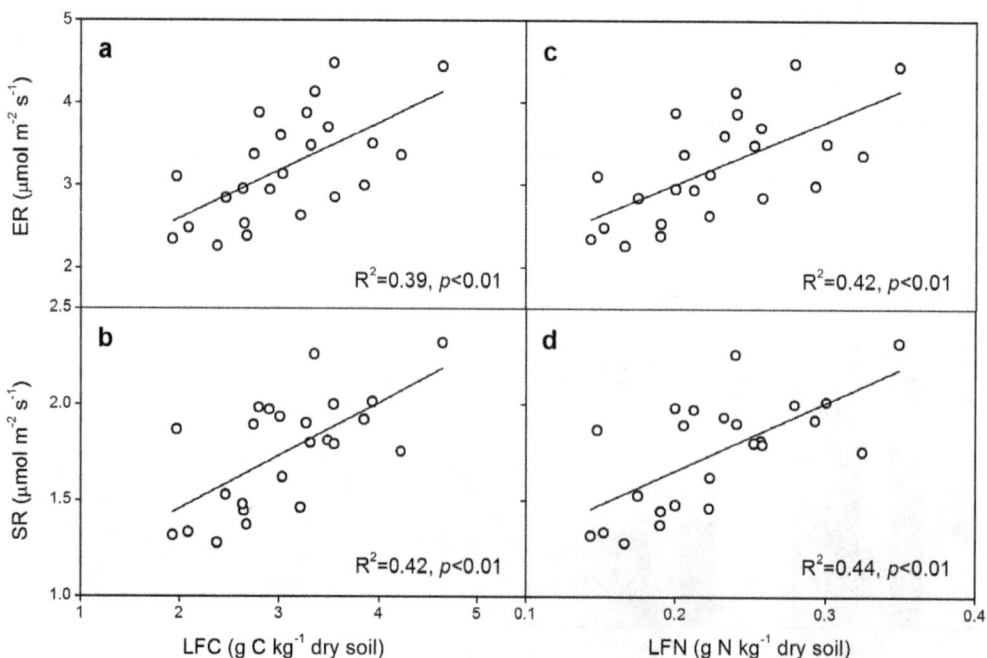

Figure 5. Linear correlations between carbon flux (ER or SR) and light fraction C (a, b) or N (c, d) across all the 24 subplots. ER, ecosystem respiration; SR, soil respiration. The values of ER and SR were the yearly mean values from 2005 to 2009.

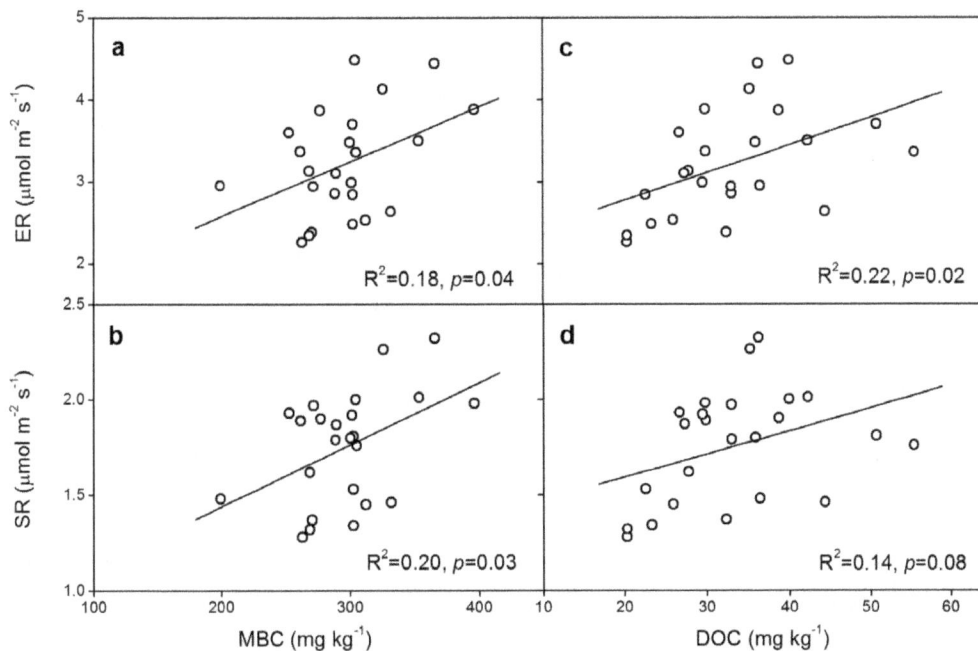

Figure 6. Linear correlations between carbon flux (ER or SR) and MBC (a, b) or DOC (c, d) across all the 24 subplots. ER, ecosystem respiration; SR, soil respiration; MBC, microbial biomass C; DOC, soil dissolved organic C. The values of ER and SR were the yearly mean values from 2005 to 2009.

Acknowledgments

The study was conducted as part of a comprehensive research project (Global Change Multifactor Experiment – Duolun) sponsored by Institute of Botany, Chinese Academy of Sciences. The authors thank Naili Zhang, Changhui Wang, Weixing Liu and Wendong Zhang for their help in instrument support and laboratory analysis. We thank the staff of Duolun Restoration Ecology Experimentation and Demonstration Station.

Author Contributions

Conceived and designed the experiments: BS SN LL SW. Performed the experiments: BS SN ZZ HY. Analyzed the data: BS. Wrote the paper: BS SN.

References

1. IPCC (2007) Climatic Change 2007: The Physical Science Basis. Cambridge, UK: Cambridge University Press.

2. Min SK, Zhang XB, Zwiers FW, Hegerl GC (2011) Human contribution to more-intense precipitation extremes. Nature 470: 376–379.

3. Dore MHI (2005) Climate change and changes in global precipitation patterns: What do we know? Environment International 31: 1167–1181.

4. Schlesinger WH (1997) Biogeochemistry: An Analysis of Global Change. San Diego, USA: Academic Press.

5. Jobbagy EG, Jackson RB (2000) The vertical distribution of soil organic carbon and its relation to climate and vegetation. Ecological Applications 10: 423–436.

6. Raich JW, Potter CS (1995) Global Patterns of Carbon-Dioxide Emissions from Soils. Global Biogeochemical Cycles 9: 23–36.

7. Davidson EA, Janssens IA (2006) Temperature sensitivity of soil carbon decomposition and feedbacks to climate change. Nature 440: 165–173.

8. Schmidt MWI, Torn MS, Abiven S, Dittmar T, Guggenberger G, et al. (2011) Persistence of soil organic matter as an ecosystem property. Nature 478: 49–56.

9. Goidts E, van Wesemael B (2007) Regional assessment of soil organic carbon changes under agriculture in Southern Belgium (1955–2005). Geoderma 141: 341–354.

10. Neff JC, Townsend AR, Gleixner G, Lehman SJ, Turnbull J, et al. (2002) Variable effects of nitrogen additions on the stability and turnover of soil carbon. Nature 419: 915–917.

11. Kirschbaum MUF (1995) The Temperature-Dependence of Soil Organic-Matter Decomposition, and the Effect of Global Warming on Soil Organic-C Storage. Soil Biology & Biochemistry 27: 753–760.

12. Kirschbaum MUF (2000) Will changes in soil organic carbon act as a positive or negative feedback on global warming? Biogeochemistry 48: 21–51.

13. Lal R (2004) Soil carbon sequestration to mitigate climate change. Geoderma 123: 1–22.

14. Zhou G, Wang Y, Wang S (2002) Responses of grassland ecosystems to precipitation and land use along the Northeast China Transect. Journal of Vegetation Science 13: 361–368.

15. Zhou XH, Talley M, Luo YQ (2009) Biomass, Litter, and Soil Respiration Along a Precipitation Gradient in Southern Great Plains, USA. Ecosystems 12: 1369–1380.

16. Giardina CP, Ryan MG (2000) Evidence that decomposition rates of organic carbon in mineral soil do not vary with temperature. Nature 404: 858–861.

17. Fang CM, Smith P, Moncrieff JB, Smith JU (2005) Similar response of labile and resistant soil organic matter pools to changes in temperature. Nature 433: 57–59.

18. Knorr W, Prentice IC, House JI, Holland EA (2005) Long-term sensitivity of soil carbon turnover to warming. Nature 433: 298–301.

19. Belay-Tedla A, Zhou XH, Su B, Wan SQ, Luo YQ (2009) Labile, recalcitrant, and microbial carbon and nitrogen pools of a tallgrass prairie soil in the US Great Plains subjected to experimental warming and clipping. Soil Biology & Biochemistry 41: 110–116.

20. Niu SL, Wu MY, Han Y, Xia JY, Li LH, et al. (2008) Water-mediated responses of ecosystem carbon fluxes to climatic change in a temperate steppe. New Phytologist 177: 209–219.

21. Liu WX, Zhang Z, Wan SQ (2009) Predominant role of water in regulating soil and microbial respiration and their responses to climate change in a semiarid grassland. Global Change Biology 15: 184–195.

22. Niu SL, Xing XR, Zhang Z, Xia JY, Zhou XH, et al. (2011) Water-use efficiency in response to climate change: from leaf to ecosystem in a temperate steppe. Global Change Biology 17: 1073–1082.

23. Vance ED, Brookes PC, Jenkinson DS (1987) An Extraction Method for Measuring Soil Microbial Biomass-C. Soil Biology & Biochemistry 19: 703–707.

24. Brookes PC, Landman A, Pruden G, Jenkinson DS (1985) Chloroform Fumigation and the Release of Soil-Nitrogen - a Rapid Direct Extraction Method to Measure Microbial Biomass Nitrogen in Soil. Soil Biology & Biochemistry 17: 837–842.

25. Ocio JA, Brookes PC (1990) Soil Microbial Biomass Measurements in Sieved and Unsieved Soil. Soil Biology & Biochemistry 22: 999–1000.

26. von Lutzowa M, Kogel-Knabner I, Ekschmittb K, Flessa H, Guggenberger G, et al. (2007) SOM fractionation methods: Relevance to functional pools and to stabilization mechanisms. Soil Biology & Biochemistry 39: 2183–2207.

27. Six J, Elliott ET, Paustian K, Doran JW (1998) Aggregation and soil organic matter accumulation in cultivated and native grassland soils. Soil Science Society of America Journal 62: 1367–1377.

28. Aanderud ZT, Richards JH, Svejcar T, James JJ (2010) A Shift in Seasonal Rainfall Reduces Soil Organic Carbon Storage in a Cold Desert. Ecosystems 13: 673–682.

29. Crow SE, Swanston CW, Lajtha K, Brooks JR, Keirstead H (2007) Density fractionation of forest soils: methodological questions and interpretation of incubation results and turnover time in an ecosystem context. Biogeochemistry 85: 69–90.

30. Gregorich EG, Carter MR, Angers DA, Monreal CM, Ellert BH (1994) Towards a Minimum Data Set to Assess Soil Organic-Matter Quality in Agricultural Soils. Canadian Journal of Soil Science 74: 367–385.

31. Gregorich EG, Janzen HH (1996) Storage of soil carbon in the light fraction and macroorganic matter. In: Carter MR, Stewart BA, eds. Structure and Organic Matter Storage in Agricultural Soils. Boca Raton, USA: Lewis Publishers. pp 167–190.

32. Bai WM, Wan SQ, Niu SL, Liu WX, Chen QS, et al. (2010) Increased temperature and precipitation interact to affect root production, mortality, and turnover in a temperate steppe: implications for ecosystem C cycling. Global Change Biology 16: 1306–1316.

33. Landesman WJ, Dighton J (2010) Response of soil microbial communities and the production of plant-available nitrogen to a two-year rainfall manipulation in the New Jersey Pinelands. Soil Biology & Biochemistry 42: 1751–1758.

34. Bell C, McIntyre N, Cox S, Tissue D, Zak J (2008) Soil microbial responses to temporal variations of moisture and temperature in a Chihuahuan Desert Grassland. Microbial Ecology 56: 153–167.

35. Sponseller RA (2007) Precipitation pulses and soil CO_2 flux in a Sonoran Desert ecosystem. Global Change Biology 13: 426–436.

36. Yan LM, Chen SP, Huang JH, Lin GH (2011) Water regulated effects of photosynthetic substrate supply on soil respiration in a semiarid steppe. Global Change Biology 17: 1990–2001.

37. Sparling GP (1992) Ratio of Microbial Biomass Carbon to Soil Organic-Carbon as a Sensitive Indicator of Changes in Soil Organic-Matter. Australian Journal of Soil Research 30: 195–207.

38. Skopp J, Jawson MD, Doran JW (1990) Steady-State Aerobic Microbial Activity as a Function of Soil-Water Content. Soil Science Society of America Journal 54: 1619–1625.

39. Norton U, Mosier AR, Morgan JA, Derner JD, Ingram LJ, et al. (2008) Moisture pulses, trace gas emissions and soil C and N in cheatgrass and native grass-dominated sagebrush-steppe in Wyoming, USA. Soil Biology & Biochemistry 40: 1421–1431.

40. Liu Q, Xu ZF, Wan CA, Xiong P, Tang Z, et al. (2010) Initial responses of soil CO_2 efflux and C, N pools to experimental warming in two contrasting forest ecosystems, Eastern Tibetan Plateau, China. Plant and Soil 336: 183–195.

41. Luo YQ, Sherry R, Zhou XH, Wan SQ (2009) Terrestrial carbon-cycle feedback to climate warming: experimental evidence on plant regulation and impacts of biofuel feedstock harvest. Global Change Biology Bioenergy 1: 62–74.

42. Kalbitz K, Solinger S, Park JH, Michalzik B, Matzner E (2000) Controls on the dynamics of dissolved organic matter in soils: A review. Soil Science 165: 277–304.

43. Richter A, Biasi C, Rusalimova O, Meyer H, Kaiser C, et al. (2005) Temperature-dependent shift from labile to recalcitrant carbon sources of arctic heterotrophs. Rapid Communications in Mass Spectrometry 19: 1401–1408.

44. Zhou XH, Sherry RA, An Y, Wallace LL, Luo YQ (2006) Main and interactive effects of warming, clipping, and doubled precipitation on soil CO_2 efflux in a grassland ecosystem. Global Biogeochemical Cycles 20: GB1003.

45. Wan SQ, Norby RJ, Ledford J, Weltzin JF (2007) Responses of soil respiration to elevated CO_2, air warming, and changing soil water availability in a model old-field grassland. Global Change Biology 13: 2411–2424.

46. Sherry RA, Weng E, Arnone III JA, Johnson DW, Schimel DS, et al. (2008) Lagged effects of experimental warming and doubled precipitation on annual and seasonal aboveground biomass production in a tallgrass prairie. Global Change Biology 14: 2923–2936.

47. Froberg M, Hanson PJ, Todd DE, Johnson DW (2008) Evaluation of effects of sustained decadal precipitation manipulations on soil carbon stocks. Biogeochemistry 89: 151–161.

48. Liski J, Ilvesniemi H, Makela A, Westman CJ (1999) CO_2 emissions from soil in response to climatic warming are overestimated - The decomposition of old soil organic matter is tolerant of temperature. Ambio 28: 171–174.

5

Impact of Environmental Factors and Biological Soil Crust Types on Soil Respiration in a Desert Ecosystem

Wei Feng, Yuqing Zhang*, Xin Jia, Bin Wu*, Tianshan Zha, Shugao Qin, Ben Wang, Chenxi Shao, Jiabin Liu, Keyu Fa

Yanchi Research Station, College of Soil and Water Conservation, Beijing Forestry University, Beijing, China

Abstract

The responses of soil respiration to environmental conditions have been studied extensively in various ecosystems. However, little is known about the impacts of temperature and moisture on soils respiration under biological soil crusts. In this study, CO_2 efflux from biologically-crusted soils was measured continuously with an automated chamber system in Ningxia, northwest China, from June to October 2012. The highest soil respiration was observed in lichen-crusted soil (0.93 ± 0.43 $\mu mol\ m^{-2}\ s^{-1}$) and the lowest values in algae-crusted soil (0.73 ± 0.31 $\mu mol\ m^{-2}\ s^{-1}$). Over the diurnal scale, soil respiration was highest in the morning whereas soil temperature was highest in the midday, which resulted in diurnal hysteresis between the two variables. In addition, the lag time between soil respiration and soil temperature was negatively correlated with the soil volumetric water content and was reduced as soil water content increased. Over the seasonal scale, daily mean nighttime soil respiration was positively correlated with soil temperature when moisture exceeded 0.075 and 0.085 $m^3\ m^{-3}$ in lichen- and moss-crusted soil, respectively. However, moisture did not affect on soil respiration in algae-crusted soil during the study period. Daily mean nighttime soil respiration normalized by soil temperature increased with water content in lichen- and moss-crusted soil. Our results indicated that different types of biological soil crusts could affect response of soil respiration to environmental factors. There is a need to consider the spatial distribution of different types of biological soil crusts and their relative contributions to the total C budgets at the ecosystem or landscape level.

Editor: Xiujun Wang, University of Maryland, United States of America

Funding: This research was fund by the National Key Technology Research and Development Program of China for 12th Five-year Plan (2012BAD16B02) and the National Natural Science Foundation of China (31170666). The funders had no role in study design, data collection and analysis, decision to publish, or preparation of the manuscript.

Competing Interests: The authors have declared that no competing interests exist.

* Email: zhangyqbjfu@gmail.com (YZ); wubin@bjfu.edu.cn (BW)

Introduction

Soil respiration (R_s) accounts for the second largest carbon flux between terrestrial ecosystems and atmosphere, after gross primary productivity. Physical (e.g., soil temperature, moisture) and biological factors (e.g., microbial community) affecting Rs should be taken into consideration in order to accurately estimate global carbon balance [1]. However, we have limited knowledge on the biophysical controls of R_s in dryland ecosystems. Drylands cover 41–47% of the terrestrial surface [2]. Biological soil crusts (BSCs) as a biological factor commonly cover 70% of the inter-canopy earth in dryland and are found in all ecosystems around the world [3]. BSCs consist of algae, lichen, moss, fungi, cyanobacteria, and bacteria and cover the top few millimeters of the soil surface [3,4]. However, knowledge about the role of BSCs as a modulator of R_s is still lacking [5–7]. It is important to study the effects of environmental factors, such as temperature and moisture, on R_s under BSCs. This knowledge can reduce bias in ecosystem-level estimation of R_s and can help us predict how climate changes will affect CO_2 flux in desert ecosystems.

BSCs are an integral part of the soil system in arid regions worldwide [4]. R_s studies in relation to BSCs have drawn much attention in the past decade [4]. In the Gurbantunggute desert, the mean R_s of cyanobacteria/lichen-crusted soil is significantly higher than that of bare land after 15 mm rainfall [8]. In Kalahari sand, the CO_2 flux of cyanobacteria-crusted soil is lower than that of disturbed crusted soil [6]. In the Iberian Peninsula, lichen-crusted soils are the main contributor to R_s [9]. In the Mu Us desert, R_s does not differ between BSC-dominated areas and bare land [10]. However, the limited knowledge about the role of BSCs as a modulator of R_s on C cycle merely focused on particular species or communities. Although those have provided valuable insights on the effects of BSCs on C fluxes, in-situ data remain rare and we have incomplete understanding of the impact of different types of BSCs on R_s.

Soil temperature (T_s) and soil water content (VWC) are the key environmental factors responsible for variation in R_s [11]. T_s is the major control of R_s through its influence on the kinetics of microbial decomposition, root respiration, and the diffusion of enzymes and substrate [12]. VWC controls the decomposition of soil organic matter, root respiration, and microbial actively [3,4,12,13]. T_s and VWC were been predicted to increase at global scales in the following decades [2]. In order to assess the impact of the changing climate on ecosystem C flux, quantification of the effects of T_s and VWC on R_s is needed. Recent studies have shown that diurnal variations in R_s are usually highly correlated with temperature of the surface soil layers [14,15]. However, a few studies have reported a hysteresis effect and a decoupling between R_s and soil surface temperature during drought conditions in boreal forests [16], tropical forests [17], Mediterranean ecosystems [18], and desert ecosystems [19]. Low water content may increase the degree of hysteresis between R_s and T_s [17,18,19] or, in some cases, may reduce it [20]. At the seasonal scale, R_s is also highly correlated with changes in T_s when water content is not limited [19,21,22]. Strong inhibition of R_s has often been observed when soil water content is low [23]. All those are mainly focused on shrub soils or bare-land soils. However, our ability to capture the effects of environmental factors on R_s in biologically-crusted soil is still lacking.

Understanding of how biologically-crusted soil types and environmental factors influence R_s in a desert ecosystem, we measured R_s in algae-, lichen-, and moss-crusted soil in the Mu Us Desert, northwestern China. The specific objectives of this study were: (1) to examine and compare the temporal variability of R_s in three crusted soils; (2) to determine seasonal and diurnal patterns of R_s; and (3) to assess the contributions of the three crusted soils to the amount of C released by R_s at the ecosystem level.

Materials and Methods

2.1 Ethics Statement

The study site is owned by Beijing Forestry University. The field work did not involve any endangered or protected species, and did not involve destructive sampling. Specific permits were required for the described study.

2.2 Site description

The research was conducted at the Yanchi Research Station (37°04′ to 38°10′ N and 106°30′ to 107°41′ E, 1550 m a.s.l.), Ningxia, northwest China. The area is located in the mid-temperate zone and characterized by a semiarid continental monsoon climate. The mean annual temperature is 8.1°C, the mean annual rainfall is 292 mm, 62% of which falls between July and September. The mean annual potential evaporation is 2100–2500 mm. All meteorological data were provided by the meteorological station of Yanchi County and represent 51 year averages (1954–2004). The vegetation in the area is dominated by *Artemisia ordosica*. The soil surface of inter-canopy is commonly covered by algae, lichen, and moss crusts, which are mainly composed of *Microcoleus vaginatus*, *Oscillatoria chlorine*, *Collema tenax*, and *Byumargenteum*, respectively [10,24]. The physical and chemical characteristics of the three crusted soils are shown in Table 1. The soil of the area is aripsamment with 1.61 g cm^{-3} in soil bulk density.

2.3 Soil respiration measurements

Continuous measurements of soil surface CO_2 efflux (R_s) were made in an open area at *Artemisia ordosica* shrub land with intact algae, lichen and moss crusts between June and October in 2012. An automated soil respiration system (Model LI 8100A fitted with a LI-8150 multiplexer, LI-COR, Nebraska, USA) was used to measure R_s. Three permanent PVC collars (20.3 cm in diameter, 10 cm in height, inserted ~7 cm) were separately installed in intact algae-, lichen- and moss-crusted soil in March 2012, three months before the start of measurements. A permanent opaque chamber (model LI-104, LI-COR, Nebraska, USA) was set on each collar. The measurement time for each chamber was 3 min and 15 s, including a 30 s pre-purge, a 45 s post-purge, and a 2 min observation period. Hourly T_s and VWC at 5-cm depth were measured near the chamber using an 8150-203 temperature sensor and an EC_{H2O} soil moisture sensor (Li-COR, Nebraska USA), respectively. During observation, any plants re-growing within collars were manually removed. Rainfall was measured near the chamber by a manual rain gauge and a tipping-bucket rain gauge (model TE525MM, Campbell Scientific, UT, USA). Half-hourly incident photosynthetically active radiation (PAR) was measured using a quantum sensor (PAR-LITE, Kipp & Zonen, The Netherlands) near the chambers.

2.4 Data treatment and analysis

The CO_2 efflux values greater than 15 μmol m^2 s^{-1} or less than -1 μmol m^2 s^{-1} were considered abnormal and removed from the dataset. Instrument failure, sensor calibration, and poor-quality measurements together resulted in the loss of 4% to 5.4% of the values for three chambers from June to October 2012 (Fig. 1).

To avoid including the impacts of photosynthesis and Birch effects on the seasonal responses of R_s to T_s and VWC, certain observations were removed from the dataset. (1) Daytime (photosynthetically active radiation, PAR >5 μmol m^{-2} s^{-1}) CO_2 efflux values were removed to ensure that no photosynthesis effects were included. (2) Measurements recorded immediately (within 30 min) after a rain event were excluded because they were potentially affected by the rewetting of the upper soil layers, which could stimulate respiration [25,26]. The daily mean nighttime value (R_s, T_s, and VWC) was computed as the average of the hourly values when PAR was below 5 μmol m^{-2} s^{-1}. Daily mean nighttime values were used to examine the seasonal responses of R_s to T_s and VWC. The seasonal relationships between R_s and T_s were estimated using four common models: Exponential (Q_{10}), Arrhenius, Quadratic, and Logistic (see Table 2). The four models were fitted separately for each crusted soil. Root mean square error (RMSE) and the coefficient of determination (R^2) were used to evaluate model performance. Temperature-normalized daily mean nighttime R_s (R_{sN}), calculated as the ratio of the observed nighttime R_s to the value predicted by the Q_{10} model, was used to analyze the seasonal dependence of daily mean nighttime R_s on VWC. Three bivariate models with T_s and VWC as independent variables were developed to show the combined effect of both variables (Table 3).

To ensure that the measurements of diurnal responses of R_s to T_s and VWC were not affected by photosynthesis, CO_2 flux measurements taken within two days after a significant rain event (>10 mm) were removed from the dataset. Field observation

Table 1. Physical and chemical characteristics of BSC layer in the study sites [41,42].

Soil type	SOC (%)	TNC (%)	SBD (g·cm^{-3})	pH	Particle content (<0.05 mm)(%)
Algae-crusted soil	0.34±0.13	0.02±0.01	1.69±0.10	8.81±1.40	6.16±1.14
Lichen-crusted soil	1.33±0.09	0.07±0.01	1.60±0.03	8.62±1.10	8.43±1.41
Moss-crusted soil	2.14±0.19	0.10±0.02	1.70±0.45	7.84±1.60	11.07±0.81

SOC: soil organic carbon; TNC: total nitrogen content; SBD: soil bulk density.

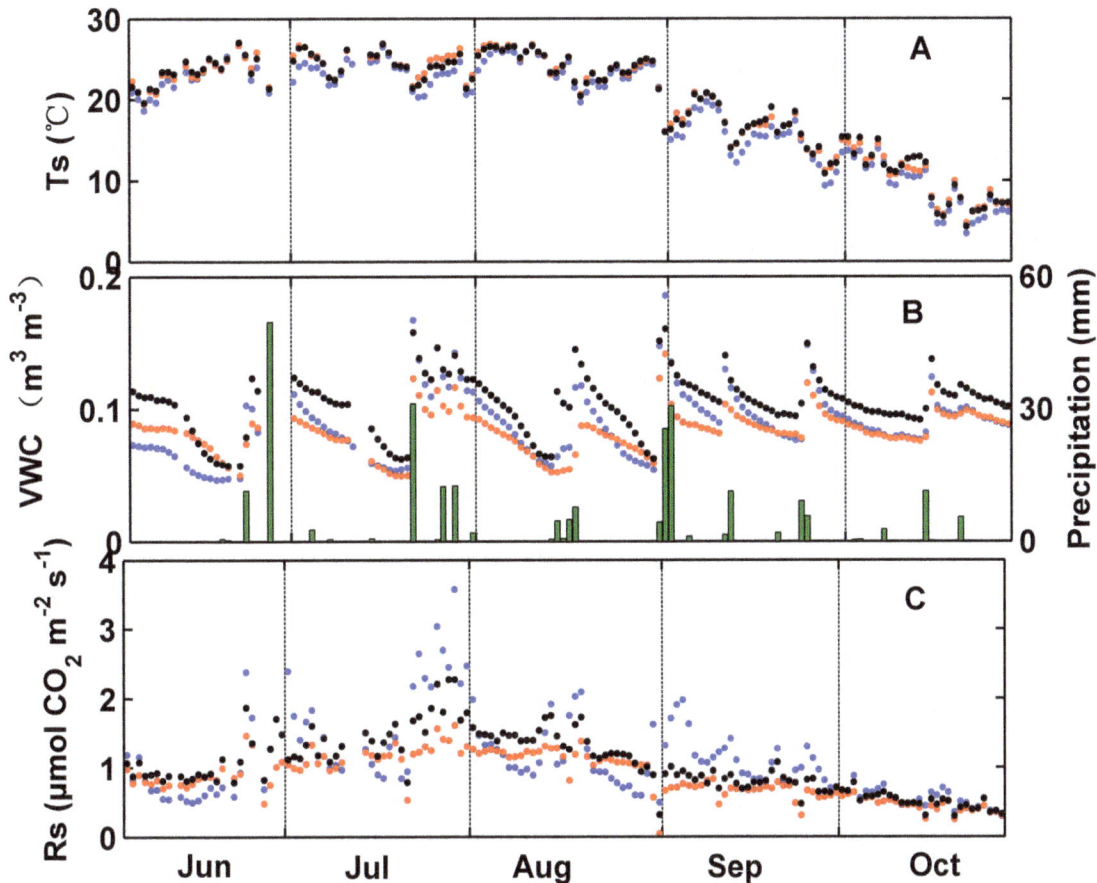

Figure 1. Daily mean of soil respiration (R_s), soil temperature (T_s), and soil volumetric water content (VWC) in soil crusted with algae (red), lichen (black), and moss (blue).

revealed that the water content of BSCs layers decreased to the water compensation point of photosynthesis within two days after the last significant rain event (>10 mm) in all three crusted soils [24,27]. The mean diurnal courses of R_s, T_s, and VWC were computed for each month by averaging the hourly means for each time of day. Cross-correlation analysis was used to detect hysteresis between R_s and T_s at the diurnal scale. Correlation analysis was used to evaluate the relationship between R_s and T_s (Table 4). All analyses were processed in Matlab 7.11.1 (R2010b, the Mathworks Inc., Natick, MA, USA).

To examine whether daily mean nighttime R_s, T_s, and VWC differed among biologically-crusted soils, we used a two-way (biologically-crusted soil types and time) ANOVA, with repeated measures of one of the factors (time). The environmental factors show relatively small variation within three days. Thus, we selected consecutive three-day periods as the three replication for statistical requirements. When significant biologically-crusted soils effects were found ($P<0.05$), the Tukey HSD post hoc test was employed to evaluate differences between biologically-crusted soil types. Prior to these analyses, data were tested for assumptions of normality and homogeneity of variances and were log-transformed when necessary. All the ANOVA analyses were performed using the SPSS 15.0 statistical software (SPSS Inc., Chicago, Illinois, USA).

Results

3.1. Hysteresis between R_s and T_s

Over the course of the diurnal period, R_s (μmol m^{-2} s^{-1}) reached its minimum at 6:00 and peaked at around 10:00–11:00 (Fig. 2), and T_s arrived at its minimum at 7:00–8:00 and peaked at 16:00 in the three crusted soils. The diurnal variation of R_s was out of phase with T_s, causing hysteresis between R_s and T_s. The maximum mean lag time between R_s and T_s was 5 h in June in moss-crusted soil, and the minimum mean lag time was 1 h in August in lichen-crusted soil, with R_s peaking earlier than T_s (Table 4). The degree of hysteresis was small in lichen-crusted soil, and large in moss-crusted soil (Table 4). The lag time between R_s and T_s was negatively and linearly correlated with VWC in crusted soil (Fig. 3). The lag time was reduced as VWC increased. The r values, derived from the data set with synchronized R_s and T_s, were higher than that without synchronization (Table 4).

3.2. Seasonal variation in R_s, T_s, and VWC

Similar changes in daily mean T_s, VWC, and CO_2 flux (including both daytime and nighttime data) were detected in algae-, lichen-, and moss-crusted soils (Fig. 1). Daily mean T_s was high from June to August, after which it gradually declined (Fig. 1A). No differences were observed in the daily mean nighttime T_s between algae- (18.15±5.61°C, mean ± standard deviation, SD) and lichen-crusted soil (18.14±7.13°C). However, daily mean nighttime T_s in moss-crusted soil (17.45±5.56°C) was

Table 2. Parameters and statistics for the analysis of the dependence of daily mean nighttime R_s (µmol m^{-2} s^{-1}) on daily mean nighttime T_s (°C) at 5-cm depth when daily mean nighttime VWC (m^3 m^{-3}) was above and below 0.075 m^3 m^{-3} in algae-and moss-crusted soil, and 0.085 m^3 m^{-3} in lichen-crusted soil.

Soil Type	Model	VWC >0.075 m^3 m^{-3}					VWC <0.075 m^3 m^{-3}				
		a	b	c	Adj.R^2	RMSE	a	b	c	Adj.R^2	RMSE
Algae-crusted soil	Q$_{10}$	0.38	2.01		**0.82**	0.1254	0.55	1.52		0.57	0.1482
	Quadratic	0.0014	−0.002	0.25	**0.82**	0.1262	−0.01	0.68	−7.67	0.52	0.1511
	Logistic	32.2	0.07	71.79	**0.82**	0.1262	1.10	0.51	19.28	0.57	0.1513
	Arrhenius	0.38	0.0005		**0.82**	0.1255	0.54	0.00031		0.57	0.1482
Moss-crusted soil	Q$_{10}$	0.55	1.97		**0.53**	0.377	0.32	1.81		0.08	0.2446
	Quadratic	0.00053	0.06	−0.01	**0.53**	0.3708	0.01	−0.45	5.158	0.10	0.2419
	Logistic	1.52	0.19	13.58	**0.53**	0.3639	3.60	0.026	79.02	0.0006	0.2545
	Arrhenius	0.55	0.00047		**0.53**	0.3759	0.32	0.00043		0.076	0.244

Type	Model	VWC>0.085 m^3 m^{-3}					VWC<0.085 m^3 m^{-3}				
		a	b	c	Adj.R^2	RMSE	a	b	c	Adj.R^2	RMSE
Lichen-crusted soil	Q$_{10}$	0.46	2.13		**0.74**	0.2196	0.43	2.00		0.062	0.2849
	Quadratic	0.002	0.007	0.22	**0.74**	0.2198	−0.07	3.30	−39.6	0.12	0.2759
	Logistic	3.31	0.10	27.95	**0.74**	0.2198	1.26	1.33	21.65	0.092	0.2803
	Arrhenius	0.46	0.00053		**0.74**	0.2191	0.41	0.00051		0.063	0.2848

Q_{10}: $R_s = ab^{(T-10)/10}$; Arrhenius: $R_s = a\exp(b/283.15\ 8.314)(1-283.15/T_s)$; Quadratic: $R_s = a\cdot T_s^2 + b\cdot T_s + c$; Logistic: $R_s = a/(1+\exp(b(c-T_s))$; Q_{10}: relative increase in R_s for a 10°C increase in T_s; Adj.R^2 is the adjusted coefficient of determination; RMSE is the root-mean-square error; a, b, and c are fitted parameters; values in bold indicate best fits according to Adj.R^2 and RMSE.

Table 3. Parameters, statistics, and predicted values from temperature-only and bivariate models of soil respiration on the basis of daily mean values.

Soil Type	Model	a	b	c	d	Adj.R^2	RMSE	Predicted R_s (g C m^{-2})
Algae-crusted soil	Q_{10}	0.38	1.98			0.82	0.1293	123.22
	Q_{10}-power	0.53	2.03	0.13		0.82	0.1268	127.46
	Q_{10}-linear	1.76	1.99	0.13	0.21	0.82	0.1262	126.65
	Q_{10}-hyperbolic	2.042	−0.13	3.83	0.015	**0.83**	**0.1268**	**126.21**
Lichen-crusted soil	Q_{10}	0.48	1.98			0.68	0.2393	155.92
	Q_{10}-power	1.34	2.20	0.53		0.74	0.2151	132.43
	Q_{10}-linear	1.19	2.07	0.68	0.32	0.72	0.7129	158.84
	Q_{10}-hyperbolic	2.167	−0.62	6.86	0.036	**0.76**	**0.2066**	**165.39**
Moss-crusted soil	Q_{10}	0.56	1.54			0.22	0.4127	130.43
	Q_{10}-power	12. 61	2.07	1.36		0.67	0.2699	146.92
	Q_{10}-linear	1.43	1.82	1.72	0.19	0.43	0.3522	134.57
	Q_{10}-hyperbolic	2.08	−0.51	9.31	0.013	**0.67**	**0.2691**	**147.08**

Q_{10}-power: $R_s = a \cdot b^{((T-10)/10)}$ VWC. Q_{10}-linear: $R_s = a \cdot b^{((T-10)/10)}$ (c $VWC+d$). Q_{10}-hyperbolic: $R_s = a \cdot b^{((T-10)/10)} \cdot (b+c \cdot VWC+d/VWC)$; $Adj.R^2$ is the adjusted coefficient of determination; RMSE is the root-mean-square error; a, b, and c are fitted parameters; values in bold indicate best fits according to $Adj.R^2$ and RMSE.

Table 4. Correlation and hysteresis analysis of monthly diurnal courses of soil respiration (R_s) and soil temperature (T_s) at 5-cm depth.

Soil Type		Jun		Jul		Aug		Sep		Oct
		r	P	r	P	r	P	r	P	r	P
		Non-Synchronized									
T_s-R_s	Algae-crusted soil	0.231	0.279	0.521**	0.009	0.441*	0.031	0.571**	0.001	0.438*	0.032
	Lichen-crusted soil	0.435*	0.034	0.728**	0.001	0.591**	0.002	0.625**	0.01	0.405*	0.05
	Moss-crusted soil	−0.210	0.362	0.531**	0.008	0.208	0.331	0.668**	0.001	0.212	0.781
		Synchronized									
T_s-R_s	Algae-crusted soil	0.844	0.001	0.943	0.001	0.914	0.001	0.962	0.001	0.971	0.001
	Lichen-crusted soil	0.701	0.001	0.875	0.001	0.658	0.001	0.952	0.001	0.917	0.001
	Moss-crusted soil	0.932	0.001	0.903	0.001	0.851	0.001	0.966	0.001	0.781	0.0001
Lag time	Algae-crusted soil	2		3		2		3		4	
	Lichen-crusted soil	3		2		1		3		3	
	Moss-crusted soil	5		3		3		3		4	

r is the Pearson correlation coefficient; P is the significance level.

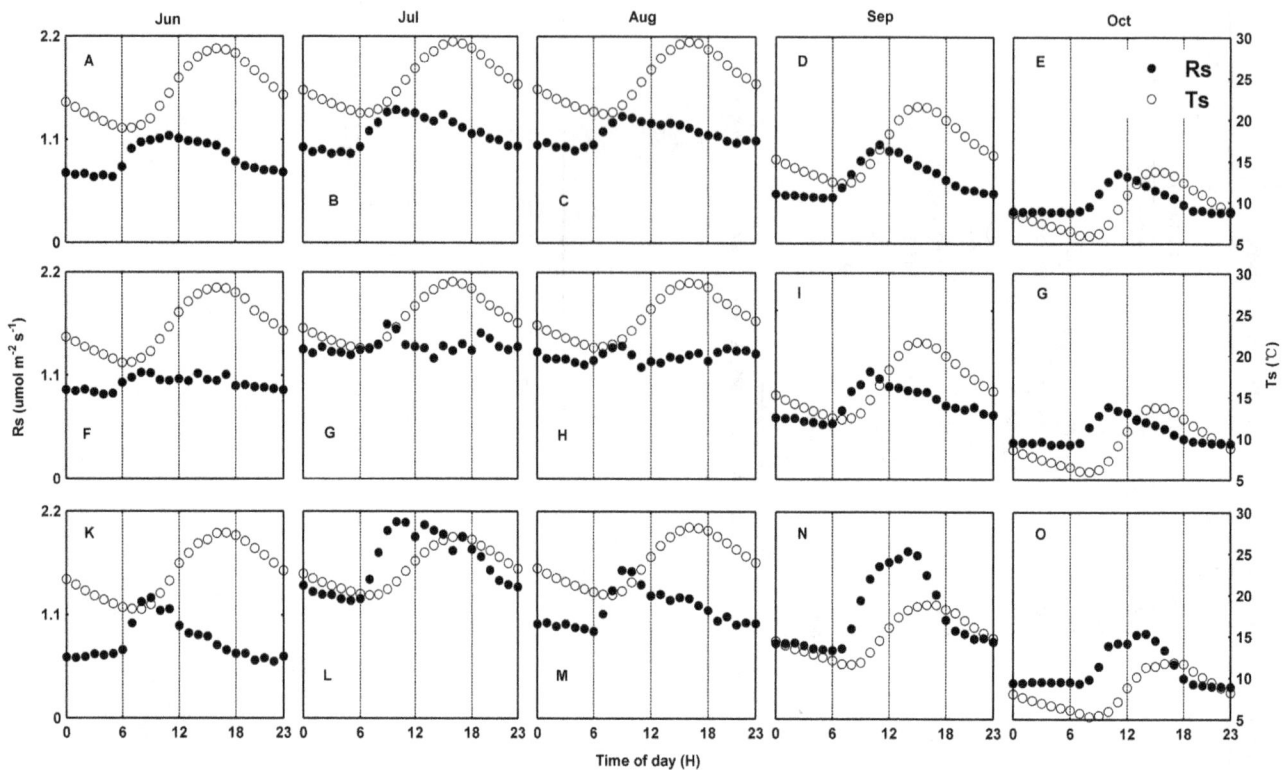

Figure 2. Monthly diurnal courses of soil respiration (R_s) and soil temperature (T_s) in soil crusted with algae (A-E), lichen (F-J), and moss (K-O). Each point is the monthly mean for a particular time of day.

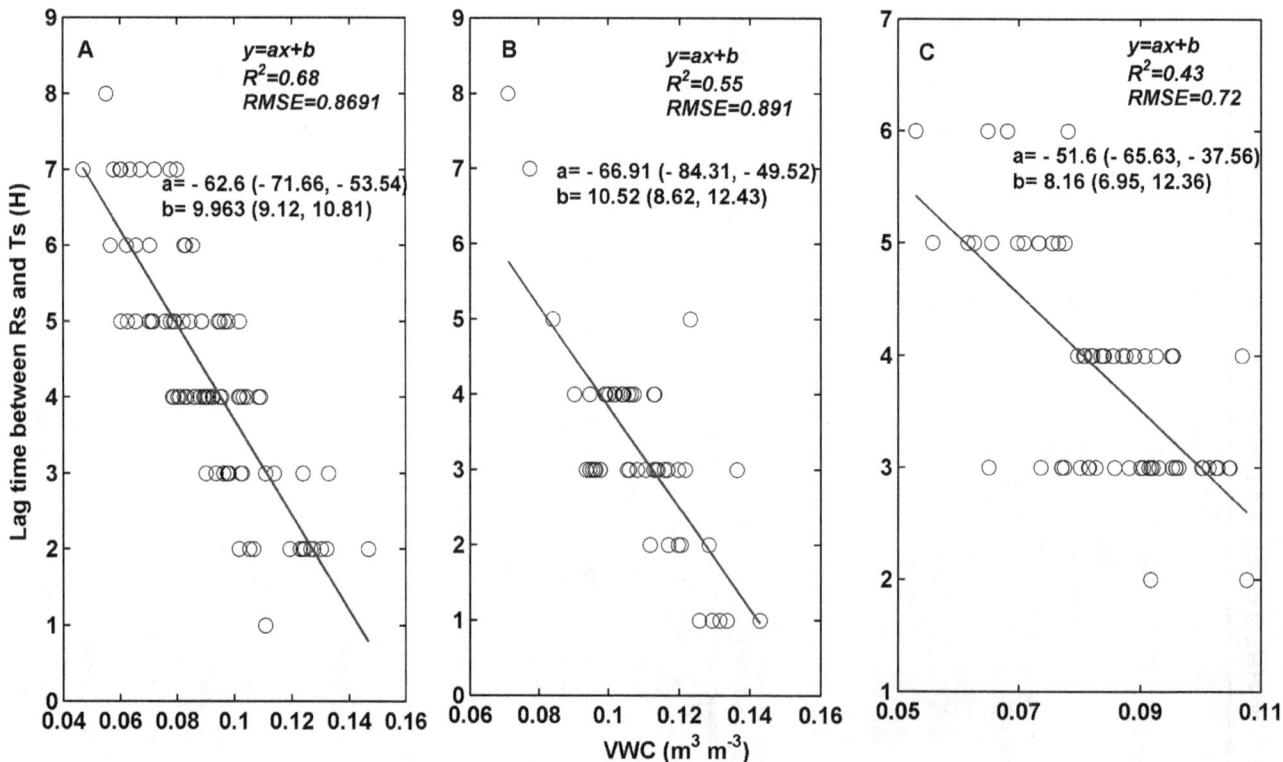

Figure 3. Lag time between soil respiration (R_s) and soil temperature (T_s) over diurnal courses, in relation to soil volumetric water content (*VWC*) in soil crusted with moss (A), lichen (B), and algae (C). The solid line is fitted using linear regression.

significantly lower than that in algae- and lichen-crusted soil ($df = 2$, $F = 11.92$, $P = 0.013$). Daily mean VWC sharply increased after each precipitation pulse (Fig. 1B). Daily mean nighttime VWC ranged from 0.049 to 0.14 m^3 m^{-3}, 0.057 to 0.16 m^3 m^{-3}, and 0.046 to 0.19 m^3 m^{-3} in algae-, lichen-, and moss-crusted soil, respectively. Daily mean nighttime VWC was significantly higher in lichen-crusted soil (0.104±0.026 m^3 m^{-3}) than in algae- and moss-crusted soils (0.083±0.015 m^3 m^{-3} and 0.089±0.026 m^{-3}, respectively) ($df = 2$, $F = 251.91$, $P<0.001$). Daily mean CO_2 flux varied markedly following the changes in T_s and VWC, especially after a rain pulse. Daily mean CO_2 flux peaked in late July and then generally declined following the decrease in T_s (Fig. 1C). The limiting effect of VWC on CO_2 flux was clear as CO_2 flux reached its highest value in a quick, sharp response to each rain event and then decreased to pre-rain values (Fig. 1B, C). Daily mean nighttime R_s was significantly different in three crusted soils ($df = 2$, $F = 56.69$, $P<0.001$) with the highest values in lichen-crusted soil (0.93±0.43 µmol m^{-2} s^{-1}) and lowest values in algae-crusted soil (0.73±0.31 µmol m^{-2} s^{-1}).

Daily mean nighttime R_s was positively related to T_s when VWC was higher than 0.075 m^3 m^{-3} in moss-crusted soil and 0.085 m^3 m^{-3} in lichen-crusted soil (Fig. 4). There were no differences among the four temperature-response models examined (Table 2). T_s at the 5-cm depth explained 82%, 74%, and 51% of the seasonal variation of daily mean nighttime R_s when VWC was not a limiting factor in algae-, lichen-, and moss-crusted soil, respectively (Table 2). In algae-crusted soil, however, R_s was controlled by T_s below the VWC threshold value (Table 2). As no differences were observed among the temperature-response models, the remainder of the analysis was performed using the Q_{10} model. Over the study period, daily mean nighttime R_s normalized using the Q_{10} model with T_s at 5 cm depth (R_{sN}) increased with VWC, except in algae-crusted soil (Fig. 5).

The seasonal sensitivity of R_s to T_s (parameter b from the Q_{10} model in Table 2) were 2.01, 2.13, and 1.97 in algae-, lichen-, and moss-crusted soil, respectively. The long-term basal respiration rate at 10°C (R_{s10}, parameter a from the Q_{10} model in Table 2) for these same soils was 0.38, 0.46, and 0.55 µol m^{-2} s^{-1}.

The bivariate model Q_{10}-hyperbolic with T_s and VWC as independent variables produced higher R^2 and lower RMSE values than the other models in lichen- and moss-crusted soil (Table 3). There was no significant difference observed between the temperature-only and the bivariate model in algae-crusted soil (Table 3), and the estimated total C release calculated with the Q_{10} model and gap-filled T_s was 123.22 g C m^{-2} in algae-crusted soil (Table 3). The estimated total R_s, as computed using the Q_{10}-hyperbolic model and gap-filled T_s and VWC, was 165.39 and 147.08 g C m^{-2} over the study period in lichen- and moss-crusted soils, respectively. Lichen-crusted soil was the main contributor to this flux among crusted soils during the study period.

Discussion

4.1. Interactive effects of T_s and VWC on R_s

Over the course of the diurnal cycle, there was a significant hysteresis between R_s and T_s (Table 4, Fig. 2). Diurnal hysteresis has been observed in many other ecosystems [16–19,28,29] and is affected by many physical and biological processes, such as mismatch between the depth of temperature measurement and the depth of CO_2 production, photosynthetic carbon supply for diurnal R_s [30], wind-induced pressure pumping [31], and different responses of autotrophic and heterotrophic respiration to environmental factors [20]. We observed that the lag time between R_s and T_s was negatively related to VWC in the three crusted soils, which is consistent with the finding from the Mu Us desert [19]. The increased lag time at low VWC in crusted soils was mainly due to the decoupling of R_s from T_s when VWC is low, and which indicate the sensitivity of root and microbial activity to soil moisture. The timing of the diurnal R_s peak is highly sensitive to VWC, with progressively earlier peaks as the VWC reduces. At

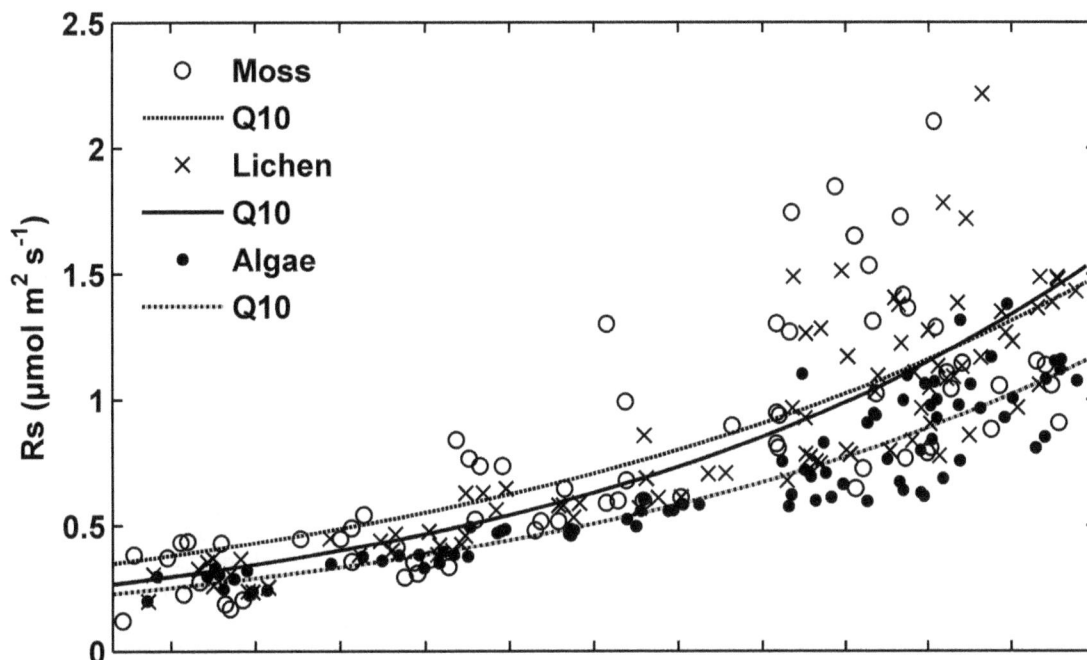

Figure 4. Relationships between daily mean nighttime soil respiration (R_s) and soil temperature (T_s) in algae-, lichen-, and moss-crusted soil.

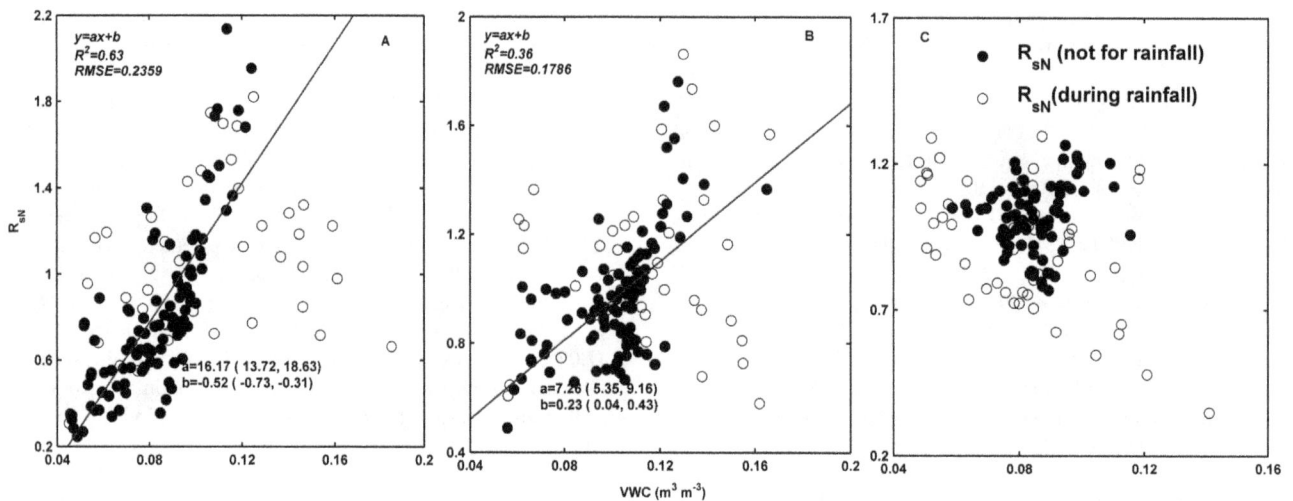

Figure 5. Relationship between daily temperature-normalized mean nighttime soil respiration (R_{sN}) and soil volumetric water content (VWC) at 5-cm depth in moss- (A), lichen- (B), and algae- (C) crusted soil, respectively. R_{sN} is the ration of the observed soil respiration (R_s) value to the value predicted by the Q_{10} function. The solid line is fitted using linear regression.

low VWC, R_s peaks in the early morning due to root and microbial activity may strongly increased with condensation water, resulting to significant hysteresis between T_s and R_s (Fig. 2, Table 4) [19].

The seasonal changes in daily mean nighttime R_s were mainly controlled by T_s (Table 2, Fig. 4). The four temperature-only models performed well with the same R^2. T_s explained 74% and 53% of the variation in R_s when VWC was above 0.085 and 0.075 m^3 m^{-3} in lichen- and moss-crusted soil, respectively, but it was uncorrelated with R_s when VWC fell below those thresholds (Table 2). Our observations are in line with those of previous studies in many other ecosystems [8,21,28,32]. Wang et al. [19] reported that T_s explained 76% of the variation in R_s for VWC values above 0.08 m^3 m^{-3}, but it was uncorrelated with R_s when VWC fell below 0.08 m^3 m^{-3}. Castillo-Monroy et al. [9] found that R_s was controlled by T_s when soil moisture was higher than 11% in microsites dominated by BSCs. Below this level, R_s was driven by soil moisture alone. The decreased R_s under low VWC was limited by reduced microbial contact with the available substrate, dormancy and/or death of microorganisms, and substrate supply, which was affected by reduced photosynthesis and drying out of the litter in the surface layer [15,33].

R_{sN} increased with VWC and did not show a threshold value in moss- and lichen-crusted soils during the seasonal cycle. Our observation contrasts with the results of previous studies that found a distinct VWC threshold [16]. The difference mainly resulted from low VWC (0.04–0.16 m^3 m^{-3}) and high soil porosity did not limit CO_2 transport out of soil and CO_2 production due to a lack of O_2.

4.2. Differences in R_s among biologically-crusted soil types

Daily mean nighttime R_s was significantly different in three types of crusted soils (algae-, lichen- and moss-crusted soil) ($df = 2$, $F = 56.69$, $P < 0.001$) with the highest values in lichen-crusted soil and lowest values in algae-crusted soil. This result contrasts with those of other studies in desert ecosystems. Su et al.'s [8] study of Gurbantunggute Desert reported no differences in carbon flux between moss- and lichen/cyanobacteria-crusted soil. The differences in the present study can be explained by the following aspects. It is possible that the lowest R_s in algae-crusted soil

resulted from the differences in soil fertility induced by BSCs, total N was significantly lower in algae-crusted soil (0.17±0.09 g kg^{-1}) than in lichen- (0.23±0.08 g kg^{-1}) and moss-crusted soil (0.28±0.13 g kg^{-1}) (unpublished data). In addition, the assemblage of microbial and microfaunal organisms varied in the three crusted soils [10,24,34–36]. The observation of the highest values occurred in lichen-crusted soil was in line with the result conducted in dry condition in the Mu Us desert. The highest values in lichen-crusted soil is mainly due to highest water content and total porosity of lichen layer [10]. T_s was significantly lower in moss-crusted soil than in algae- and lichen-crusted soil (Fig. 1). This result is attributed to the darkening of the surface by cyanobacteria and lichens, resulting in greater absorption of solar radiation and a higher surface temperature [39]. VWC in lichen-crusted soil was consistently significantly higher than in moss- and algae-crusted soils (Fig. 1). The difference may be attributed to higher dew deposition (soil moisture input by dewfall can be an important mechanism in dryland environment) and water infiltration in lichen-crusted soil than in moss- and algae-crusted soil [37].

The lag time between R_s and T_s differed depending on the type of crusted soil, suggesting that the response of species in biologically-crusted soils to VWC was different among crusted types. The timing of the diurnal R_s peak is highly sensitive to VWC, with progressively earlier peaks as the soil VWC declines [19]. Moss crusts need more VWC than lichen and algae crust to achieve metabolic activity [24]. In water stressed ecosystems, algae and lichen can utilize dew and light rainfall that moss are unable to use [24,27]. Thus the diurnal R_s in moss-crusted soil peaks earlier than algae- and lichen-crusted soils, which lead to significant hysteresis between R_s and T_s in moss-crusted soil. Hysteresis had a smaller impact on lichen-crusted soil than on algae-crusted soil. The result may be partly attributed to the higher water level in lichen- than in algae-crusted soil.

The average Q_{10} of 1.83 from three biologically-crusted soil types from June to October is at the lower end of the range of 1.28 to 4.75 from alpine, temperate, and tropical ecosystems across China [38]. The low Q_{10} value is attributed to their low levels of soil organic matter, small microbial community, and dry soil conditions [19,39,40].The Q_{10} of algae-, lichen-, and moss-crusted soil was 1.98, 1.98, and 1.54, respectively. The majority of C associated with BSCs, in the forms of microbial biomass or their secretions [31,32], is close to or at the soil surface and is directly in contact with small precipitation or dew captured by algae and lichen crusts. However, small amounts of hydration cannot directly reach the soil surface because the soil is covered with moss. The relatively small amounts of hydration in moss-crusted soils result in the lower Q_{10} [16,21,22,32].

The effects of VWC and T_s on R_s should be considered in carbon cycle models in moss- and lichen-crusted soils. However, we did not find any effect of VWC on daily mean nighttime R_s in algae-crusted soil from June to October 2012 (Tables 2, 3). This observation coincided with the result that R_{sN} was independent of VWC in algae-crusted soil. The independence of VWC from R_s in algae-crusted soil may be attributed to the low water requirement of algae for active metabolism [24,27]. Even a very small hydration event, such as water vapor and dew in the early morning, is sufficient to allow algae to achieve microbial metabolism. Further examination is needed to justify our conclusion regarding the role of VWC on algae-crusted soil due to the dew data gap. We used the Q_{10}-hyperbolic model, with T_s and VWC as independent variables, to predict changes in R_s. Using Q_{10}-hyperbolic model to predict R_s was also reported in a boreal trembling aspen stand [16].

Using temperature-only and Q_{10}-hyperbolic model, we obtained an approximate estimate of the total amount of C released at each crusted soil via soil respiration of 123.2, 165.4, and 147.1 g C m^{-2} over 5 months studied in algae-, lichen- and moss-crusted soils, respectively. Lichen-crusted soil was the main contributor to the total C released by R_s. We found that total C released by R_s in lichen-crusted soil was 2.5% higher than the mean total C released by R_s (161.4 g C m^{-2}, unpublished data) over 5 months, whereas total C released by R_s in algae- and moss-crusted soil were 23.65% and 8.87% smaller than the mean total C released by R_s, respectively. Our results show the importance of BSCs as modulators of R_s in the C release and indicate that we should

not ignore their relative contributions to the total C budgets in desert ecosystems.

Conclusions

Our study showed that R_s was significantly different in three crusted soils with highest values in lichen-crusted soil and lowest values in algae-crusted soil. Lichen-crusted soil was the main contributor to the total C released by R_s. Over the diurnal cycle, T_s exerted dominant control over R_s in the three crusted soils. There was a significant lag between T_s and R_s over the diurnal cycle, and that the lag time increased as VWC decreased. Over the seasonal scale, the response of R_s to T_s was regulated by VWC, and R_s was uncorrelated with T_s when VWC dropped below 0.075 and 0.085 m^3 m^{-3} in lichen- and moss-crusted soils, respectively. However, VWC was not a limiting factor on R_s in algae-crusted soil. Our results indicated that different types of BSCs may affect response of R_s to environmental factors. There is a need to consider the spatial distribution of different types of BSCs and their relative contributions to the total C budgets at the ecosystem or landscape level.

Acknowledgments

We thank Su Lu, Huishu Shi, Yuming Zhang, Xuewu Yang for their assistance with the field measurements and instrumentation maintenance. We are grateful to the anonymous reviewers and the Academic Editor for providing insightful comments and suggestions. We also thank language service company for their help with language revision, and valuable comments to the manuscript.

Author Contributions

Conceived and designed the experiments: WF YZ BW TZ SQ XJ CS. Performed the experiments: SQ WF BW KF. Analyzed the data: WF SQ XJ BW. Contributed reagents/materials/analysis tools: YZ BW TZ XJ BW JL KF. Wrote the paper: WF YZ BW XJ SQ. Designed the software used in analysis: XJ.

References

1. Schimel DS (1995) Terrestrial ecosystems and the carbon cycle. Glob Change Biol 1: 77–91.

2. Le Houérou HN (1996) Climate change, drought and desertification. J Arid Environ 34: 133–185.

3. Belnap J (2003a) Comparative structure of physical and biological soil crusts. In: Belnap J, Lange OL, editors. Biological soil crusts: Structure, function, and management. Berlin: Springer-Verlag. pp. 177–191.

4. Belnap J (2003) The world at your feet: desert biological soil crusts. Front Ecol Environ 1: 181–189.

5. Maestre FT, Cortina J (2003) Small-scale spatial variation in soil CO$_2$ efflux in a Mediterranean semiarid steppe. Appl Soil Ecol 23: 199–209.

6. Thomas AD, Hoon SR, Dougill AJ (2011) soil respiration at five sites along the Kalahari Transect: effects of temperature, precipitation pulse and biological soil crust cover. Geoderma 167: 284–294.

7. Thomas AD, Hoon SR (2010) Carbon dioxide fluxes from biologically-crusted Kalahari Sands after simulated wetting. J Arid Environ 74: 131–139.

8. Su YG, Wu L, Zhou ZB, Liu YB, Zhang YM (2013) Carbon flux in deserts depends on soil cover type: A case study in the Gurbantunggute desert, North China. Soil Biol Biochem 58: 332–340.

9. Castillo-Monroy AP, Maestre FT, Rey A, Soliveres S, García-Palacios P (2011) Biological soil crust microsites are the main contributor to soil respiration in a semiarid ecosystem. Ecosystems 14: 835–847.

10. Feng W, Zhang YQ, Wu B, Zha TS, Jia X, et al. (2013) Influence of disturbance on soil respiration in biologically-crusted soil during the dry season. The Scientific World J 2013.

11. Fang C, Moncrieff JB (2001) The dependence of soil CO$_2$ efflux on temperature. Soil Biol Biochem 33: 155–165.

12. Jassal RS, Black TA, Novak MD, Gaumont-Guay D, Nesic Z (2008) Effects of soil water stress on soil respiration and its temperature sensitivity in an 18-year-old temperate Douglas-fir stand. Glob Change Biol 14: 1305–1318.

13. Bouma TJ, Nielsen KL, Eissenstat DM, Lynch JP (1997) Estimating respiration of roots in soil: interactions with soil CO$_2$, soil temperature and soil water content. Plant Soil 195: 221–232.

14. Drewitt GB, Black TA, Nesic Z, Humphreys ER, Jork EM, et al. (2002) Measuring forest floor CO$_2$ fluxes in a Douglas-fir forest. Agr Forest Meteorol 110: 299–317.

15. Jassal R, Black A, Novak M, Morgenstern K, Nesic Z, et al. (2005) Relationship between soil CO$_2$ concentrations and forest-floor CO$_2$ effluxes. Agr Forest Meteorol 130: 176–192.

16. Gaumont-Guay D, Black TA, Griffis TJ, Barr AG, Jassal RS, et al. (2006) Interpreting the dependence of soil respiration on soil temperature and water content in a boreal aspen stand. Agr Forest Meteorol 140: 220–235.

17. Vargas R, Allen MF (2008) Environmental controls and the influence of vegetation type, fine roots and rhizomorphs on diel and seasonal variation in soil respiration. New Phytol 179: 460–471.

18. Tang J, Baldocchi DD, Xu L (2005) Tree photosynthesis modulates soil respiration on a diurnal time scale. Glob Change Biol 11: 1298–1304.

19. Wang B, Zha TX, Jia X, Wu B, Zhang YQ, et al. (2014) Soil moisture modifies the response of soil respiration to temperature in a desert shrub ecosystem. Biogeosciences 11: 259–268.

20. Riveros-Iregui DA, Emanuel RE, Muth DJ, McGlynn BL, Epstein HE, et al. (2007) Diurnal hysteresis between soil CO$_2$ and soil temperature is controlled by soil water content. Geophys Res Lett 34.

21. Yuste JC, Janssens IA, Carrara A, Meiresonne L, Ceulemans R (2003) Interactive effects of temperature and precipitation on soil respiration in a temperate maritime pine forest. Tree Physiol 23: 1263–1270.

22. Jassal RS, Black TA, Novak MD, Gaumont-Guay D, Nesic Z (2008) Effects of soil water stress on soil respiration and its temperature sensitivity in an 18-year-old temperate Douglas-fir stand. Glob Change Biol 14: 1305–1318.

23. Harper CW, Blair JM, Fay PA, Knapp AK, Carlisle JD (2005) Increased rainfall variability and reduced rainfall amount decreases soil CO$_2$ flux in a grassland ecosystem. Glob Change Biol 11: 322–334.

24. Feng W, Zhang YQ, Wu B, Qin SG, Lai ZR (2014) Influence of environmental factors on carbon dioxide exchange in biological soil crusts in desert areas. Arid Land Res Manage 28: 186–196.

25. Birch H (1958) The effect of soil drying on humus decomposition and nitrogen availability. Plant Soil 10: 9–31.

26. Rey A, Pegoraro E, Tedeschi V, De Parri I, Jarvis PG, et al. (2002) Annual variation in soil respiration and its components in a coppice oak forest in Central Italy. Glob Change Biol 8: 851–866.

27. Lange OL (2003) Photosynthesis of soil-crust biota as dependent on environmental factors. In: Belnap J, Lange OL, editors. Biological soil crusts: Structure, function, and management. Berlin: Springer-Verlag. pp. 217–240.

28. Vargas R, Baldocchi DD, Allen MF, Bahn M, Black TA, et al. (2010) Looking deeper into the soil: biophysical controls and seasonal lags of soil CO_2 production and efflux. Ecol Appl 20: 1569–1582.

29. Jia X, Zha TS, Wu B, Zhang YQ, Chen WJ, et al. (2013) Temperature response of soil respiration in a Chinese pine plantation: hysteresis and seasonal vs. diel Q_{10}. PLoS one 8: e57858.

30. Stoy PC, Palmroth S, Oishi AC, Siqueira MB, Juang JY, et al. (2007) Are ecosystem carbon inputs and outputs coupled at short time scales? A case study from adjacent pine and hardwood forests using impulse-response analysis. Plant Cell Environ 30: 700–710.

31. Flechard CR, Neftel A, Jocher M, Ammann C, Leifeld J, et al. (2007) Temporal changes in soil pore space CO_2 concentration and storage under permanent grassland. Agr Forest Meteorol 142: 66–84.

32. Xu M, Qi Y (2001) Soil-surface CO_2 efflux and its spatial and temporal variations in a young ponderosa pine plantation in northern California. Glob Change Biol 7: 667–677.

33. Högberg P, Nordgren A, Buchmann N, Taylor AFS, Ekblad A, et al. (2001) Large-scale forest girdling shows that current photosynthesis drives soil respiration. Nature 411: 789–792.

34. Housman DC, Yeager CM, Darby BJ, Sanford RL, Kuske CR, et al. (2007) Heterogeneity of soil nutrients and subsurface biota in a dryland ecosystem. Soil Biol Biochem 39: 2138–2149.

35. CastilloMonroy AP, Bowker MA, Maestre FT, Rodríguez-Echeverría S, Martinez I, et al. (2011) Relationships between biological soil crusts, bacterial diversity and abundance, and ecosystem functioning: Insights from a semi-arid Mediterranean environment. J Veg Sci 22: 165–174.

36. Warren S (2003) Biological soil crusts and hydrology in North American deserts. In: Belnap J, Lange OL, editors. Biological soil crusts: Structure, function, and management. Berlin: Springer-Verlag.pp. 327–337.

37. Liu LC, Li SZ, Duan ZH, Wang T, Zhang ZS, et al. (2006) Effects of microbiotic crusts on dew deposition in the restored vegetation area at Shapotou, northwest China. J Hydrol 328: 331–337.

38. Zheng ZM, Yu GR, Fu YL, Wang YS, Sun XM, et al. (2009) Temperature sensitivity of soil respiration is affected by prevailing climatic conditions and soil organic carbon content: a trans-China based case study. Soil Biol Biochem 41: 1531–1540.

39. Gershenson A, Bader NE, Cheng W (2009) Effects of substrate availability on the temperature sensitivity of soil organic matter decomposition. Glob Change Biol 15: 176–183.

40. Cable JM, Ogle K, Lucas RW, Huxman TE, Loik ME, et al. (2011) The temperature responses of soil respiration in deserts: a seven desert synthesis. Biogeochemistry 103: 71–90.

41. Gao GL, Ding GD, Wu B, Zhang YQ, Qin SG, et al. (2014) Fractal scaling of particle size distribution and relationships with topsoil properties affected by biological soil crusts. PloSone 9: e88559.

42. Bao YF, Ding GD, Wu B, Zhang YQ, Liang WJ, et al. (2013) Study on the wind-sand flow structure of windward slope in the Mu Us Desert, China. J Food Agric Environ 11: 1449–1454.

6

Microbial Growth and Carbon Use Efficiency in the Rhizosphere and Root-Free Soil

Evgenia Blagodatskaya[1,2,3*], Sergey Blagodatsky[2,4], Traute-Heidi Anderson[5], Yakov Kuzyakov[1,3]

1 Soil Science of Temperate Ecosystems, Büsgen-Institute, University of Göttingen, Göttingen, Germany, 2 Institute of Physicochemical and Biological Problems in Soil Science, Russian Academy of Sciences, Pushchino, Russia, 3 Agricultural Soil Science, Büsgen-Institute, University of Göttingen, Göttingen, Germany, 4 Institute for Plant Production and Agroecology in the Tropics and Subtropics, University of Hohenheim, Stuttgart, Germany, 5 Thünen-Institute of Climate-Smart Agriculture (vTI), Braunschweig, Germany

Abstract

Plant-microbial interactions alter C and N balance in the rhizosphere and affect the microbial carbon use efficiency (CUE)–the fundamental characteristic of microbial metabolism. Estimation of CUE in microbial hotspots with high dynamics of activity and changes of microbial physiological state from dormancy to activity is a challenge in soil microbiology. We analyzed respiratory activity, microbial DNA content and CUE by manipulation the C and nutrients availability in the soil under *Beta vulgaris*. All measurements were done in root-free and rhizosphere soil under steady-state conditions and during microbial growth induced by addition of glucose. Microorganisms in the rhizosphere and root-free soil differed in their CUE dynamics due to varying time delays between respiration burst and DNA increase. Constant CUE in an exponentially-growing microbial community in rhizosphere demonstrated the balanced growth. In contrast, the CUE in the root-free soil increased more than three times at the end of exponential growth and was 1.5 times higher than in the rhizosphere. Plants alter the dynamics of microbial CUE by balancing the catabolic and anabolic processes, which were decoupled in the root-free soil. The effects of N and C availability on CUE in rhizosphere and root-free soil are discussed.

Editor: Jeffrey L. Blanchard, University of Massachusetts, United States of America

Funding: This authors acknowledge the following: the European Commission (Marie Curie IIF program, project IIF 039907-MICROSOM) for supporting EB (http://ec.europa.eu/research/mariecurieactions/); the Alexander von Humboldt Foundation for supporting SB (http://www.humboldt-foundation.de/web/start.html); Russian Academy of Sciences (Scientific School Program 6123.2014.4, https://grants.extech.ru/grants/res/winners.php?OZ = 4&TZ = S&year = 2014); Russian Foundation for Basic Research (grant No 12-04-01170, http://www.rfbr.ru/rffi/ru/); and German Research Foundation (DFG) within project KU 1184/13-1/2 (http://www.dfg.de/en/). The funders had no role in study, design, data collection and analysis, decision to manuscript, or preparation of the manuscript.

Competing Interests: The authors have declared that no competing interests exist.

* E-mail: janeblag@mail.ru

Introduction

Analysis of microbial carbon use efficiency (CUE) and microbial turnover rates are critical for accounting of C balance in soil with the goal of correct estimation of C sequestration potential as well as for modelling the turnover of soil C and CO_2 fluxes [1–3]. The efficiency of microbial growth on a carbonaceous substrate coming with plant residues is positively related to formation rates of soil organic carbon [4]. A magnitude and dynamics of CUE is a function of numerous physical, chemical and ecological factors, e.g. soil quality [5], microbial community composition [6], [7], substrate and nutrient availability [3], [8], etc. At that the factor specific mechanisms, which control the CUE, remain uncertain [9]. This calls for the case studies under control conditions, so that the number of influencing factors can be reduced. So, preferential objects for CUE studies are the soils similar in physico-chemical characteristics but contrasting in substrate availability: e.g. rhizosphere and root-free soil. Higher microbial abundance and diversity and faster microbial growth occur in the rhizosphere soil as compared to root-free soil [10], [11] due to the high availability of C exuded by roots [1], [12]. Contrary to this, permanent limitation by available substrates in root-free soil leads to the selection of microorganisms with slower growth rates and more efficient metabolism [13]. So, rhizosphere and root-free soil can serve as good model for an *in situ* comparison of microbial physiology and CUE in microhabitats with contrasting resource levels.

CUE has become a very popular but ambiguous term in soil science. It is often used with a broad meaning, combining the efficiency of growth and the efficiency of maintenance of soil microorganisms [3]. Here, we introduce basic terms and approaches applicable either for distinct growth or for sustaining microbial biomass.

CUE Estimation for Growing Microbial Biomass

During microbial growth, CUE is equivalent to the microbial yield coefficient (Y, g C_{mic} g^{-1} C_s), i.e. biomass-C increment per amount of substrate-C used (Eq. 1, [14]):

$$Y = -\frac{\Delta C_{mic}}{\Delta C_s} \qquad (1)$$

where ΔC_{mic} is the increase in microbial biomass-C caused by the consumption of substrate-C ΔC_s. So, for **estimation of CUE for growing microbial biomass,** we used the microbial yield coefficient (Y). In spite of wide variability of the experimental Y estimations in the range of 0.1 to 0.8 [6], [15], [16] and a maximal theoretical value of 0.62 for glucose [17], the fixed value of

Y = 0.45 is often assumed in soil studies and models [1], [18]. Considering very high variation (about 8 times) such a rough overall assumption of the average of 0.45 applied for different soils can distort the estimations and predictions of C stocks and fluxes [5], [18].

CUE Estimation under Steady-state Conditions

In the absence of microbial growth, the estimation of Y (Eq. 1) is not applicable. However, even without distinct exponential growth, the substrate can be used both for maintenance and for the very slow, "cryptic" growth [19], so that microbial biomass does not decrease in time. Under such steady-state conditions, the estimation of the efficiency of microbial metabolism by specific respiration (CO_2 produced per time and microbial biomass unit) can be used as a physiological characteristic.

The dormancy or maintenance state of microbial community reveals itself as a low respiration-to-biomass ratio which has been suggested as a physiological index of soil microbial communities [20]. The maintenance requirements are higher for microorganisms adapted to permanent input of available substrates than for microbial communities from nutrient-limited microhabitats [21]. The similar relationship is valid for growth expenses: the amount of respired CO_2 during growth is larger for microbial communities with a higher growth rate and comparatively less efficient metabolism [22]. So, we hypothesised that both in the presence and absence of an available substrate, microbial communities in rhizosphere soil will have higher specific respiration rates than those in root-free soil.

CUE Estimation during Shift from Dormancy to Active Stage

It is important to consider the CUE not only as a growth parameter (Y) and as a dormancy characteristic (maintenance coefficient), but also as the amount of CO_2 produced per biomass unit in the course of the famine-to-feast transition. How such a transition alters CUE dynamics under changing environmental conditions, i.e. from substrate-limited to substrate-rich microhabitats, remains unclear. In contrast to steady-state or growth conditions where CUE remains constant, the experimental estimation of CUE during the famine-to-feast microbial transition remains a challenge for environmental microbiology. This is because the application of standard methods (fumigation-extraction or substrate-induced respiration) is restricted for biomass assessment in growing microbial communities.

A strong positive correlation between DNA and microbial C in soil [5], [23–25] led us use the DNA content as a proxy of microbial biomass. The increase in microbial DNA content corresponds to the respiratory response during exponential microbial growth after substrate addition [24], [26]. Therefore, we used the CO_2/DNA ratio for comparison of the CUE by transition from dormant to active stage for microbial communities with contrasting growth strategies. Experimentally, the growth strategies can be evaluated by the maximal specific growth rate under unlimited conditions that is greater for r- than for K-strategists [27], [28]. So, we used two parameters of microbial metabolism: microbial maximal specific growth rates and CUE, to evaluate the relative abundance of slow- or fast-growing microorganisms in rhizosphere and root-free soil.

Nitrogen Effect on CUE

The efficiency of microbial metabolism depends strongly on nitrogen (N) availability [29]. Lower respiration due to higher efficiency of microbial C reutilisation has been observed in the absence of N limitation as compared to N-limited conditions [30]. Nitrogen addition reduces cumulative microbial respiration in soil amended with glucose [31] and plant litter [32] and increased the growth yield efficiency [18]. While the CUE decline under N limitation is commonly expected [3], it is unknown whether N availability affects equally microbial respiration and growth rates in microhabitats with contrast substrate availability, e.g. in root-free and rhizosphere soil [33]. Therefore, we compared the specific respiration and microbial growth kinetics in the root-free and in rhizosphere soil with different N fertilization rates. We expected to find more distinct effect of N availability in the rhizosphere where microbial activity and abundance are higher and N limitation may be more important as compared to root-free soil. We hypothesized that the increase of N availability improves CUE and decreases specific respiration, especially in the rhizosphere.

We analyzed the ratio between respiration and microbial DNA content 1) under steady state conditions (in unamended soil), 2) during microbial growth in soil amended with glucose, and 3) during transition from steady state conditions to growth. In addition, effect of N availability on microbial growth rate and CUE was determined. Three complementary indices were applied as indicators of the efficiency of microbial metabolism in the rhizosphere and in root-free soil: 1) the CO_2/DNA ratio further referred to as 'specific respiration rate', 2) the $\Delta CO_2/\Delta DNA$ ratio for growing biomass, and 3) CUE during microbial growth on glucose.

Materials and Methods

Soil Sampling

Soil samples were taken from the field experimental station at the Institute of Agroecology (FAL, Braunschweig, Germany). No specific permission was required as one of the co-authors (THA) had been working in the Institute of Agroecology, and soil was regularly sampled in the course of long-term field trial described elsewhere [34]. The soil is a loamy sand Haplic Cambisol (C_{org} 1.1%; N_{tot} 0.087%; pH_{CaCl2} 6.7). The plots under sugar beet (*Beta vulgaris* subsp. *rapacea* (KOCH-DÖLL, cv. Wiebke) with full and half the recommended rate of mineral N fertiliser (126 and 63 kg N ha^{-1} $year^{-1}$, respectively) were chosen for analysis of the N effects on microbial communities of root-free and rhizosphere soil. Soil was sampled during harvesting the sugar beet at a mature stage (age 4.5-month). Soil samples were taken from the 0–10 cm layer from five randomly chosen replicate microsites and then mixed. Rhizosphere soil was sampled at a distance 1–5 mm adjacent to the roots (i.e. collecting the soil aggregates falling off when shaking the root system), whereas root-free soil was taken between rows of sugar beets. Fine roots and other plant debris were carefully removed during sampling. No significant differences were detected in pH, C_t or N_t content of the rhizosphere and root-free soil. The soil was stored field-fresh in aerated polyethylene bags at 4°C for 1–2 weeks. Prior to analysis the soil was sieved (< 2 mm), moistured to 60% of WHC, and preincubated at 22°C for 24 h.

Soil Respiration and Chemical Analysis

Microbial biomass (C_{mic}) was determined by the initial rate of substrate-induced respiration after soil amendment with glucose and according to the equation of Anderson & Domsch [35]:

$$C_{mic}(\mu g \cdot g^{-1} soil) = (\mu l CO_2 \cdot g^{-1} soil \cdot h^{-1}) \cdot 40.04 \quad (2)$$

Rate of basal respiration (V_{basal}) was estimated for soil without glucose as the hourly mean of 10 h of CO_2 evolution at 22°C, after 2–3 hours diminishing of the initial CO_2 flush caused by soil disturbance during sample preparation [36]. The CO_2 emission rate (V_{CO2}) was measured hourly at 22°C using an automated infrared-gas analyser system [37].

Soil organic C and total N were analysed by dry combustion (C-IR 12, Leco, and Macro-N, Hereaus, respectively). Soil pH was measured in 0.01 M $CaCl_2$ with a soil-to-solution ratio of 1:2.

Total DNA

Quantity of double-stranded DNA was determined by direct DNA isolation from the soil with mechanic and enzymatic disruption of microbial cell walls and subsequent spectrofluorimetric detection with PicoGreen [23], [24]. For rhizosphere and root-free soil from plot fertilized with 126 kg N ha^{-1} $year^{-1}$ the dsDNA determination was done at 0, 12, 15, 20, 25 and 36 hours after addition of glucose and nutrients (as described below for respiration kinetics).

The procedure of DNA isolation involved sonication of the soil suspension in Tris-EDTA buffer (TE) at pH 8, addition of aurintricarboxylic acid (a nuclease inhibitor) and sodium dodecyl sulphate. Then two cycles of quick freeze at −80°C in Deep Freezer (ProfiMaster EPF3080/N, National Lab GmbH, Mölln, Germany) for 1 h and subsequent thaw at +65°C in water bath with thermostat (Model 1002, GFL Gesellschaft für Labortechnik mbH, Burgwedel, Germany) were performed to destroy microbial cells. Enzymatic digestion was accomplished with lysozyme and Proteinase K for 1 h at 37°C. Mechanical destruction of microbial cells was implemented by shaking with sterile acid-washed glass-beads (Sigma-Aldrich, Inc.) of three sizes (710–1180, 212–300, and <106 μm) on a Vortex homogeniser at 2000 rpm. The samples were diluted with an equal volume of TE-buffer and centrifuged for 10 min at 5500 g. Half a millilitre of the diluted supernatant (1:100) was mixed with 0.5 ml of a 1:200 dilution of PicoGreen^{TM} (Molecular Probes). After 4 min incubation, the fluorescence was measured on an SFM-25 spectrofluorimeter (Kontron, Germany) at an excitation wavelength of 480 nm and an emission wavelength of 523 nm. The dsDNA of bacteriophage lambda was used as a standard; samples for the standard curve were prepared in TE-buffer in the same way as the experimental samples.

Kinetic Parameters of Microbial Growth

Kinetics of microbial growth was determined indirectly by the rate of CO_2 emission from soil amended with glucose and mineral nutrients [38]. It has to be noted that despite substrate addition is required for the estimation of kinetic parameters (specific growth rate, active and total microbial biomass, see below), the results obtained by this approach (substrate induced growth response – SIGR) are the characteristics of the soil microbial community at the sampling instant, i.e. before substrate addition. Samples of 10 g (dry weight) soil were amended with a powder-mixture containing glucose (10 mg g^{-1}), talcum (20 mg g^{-1}) and mineral salts: $(NH_4)_2SO_4$−1.9 mg g^{-1}, K_2HPO_4−2.25 mg g^{-1} and $MgSO_4 \cdot 7H_2O$−3.8 mg g^{-1} [39]. These optimal concentrations of the substrates were selected in preliminary experiments and are sufficient for unlimited exponential growth of soil microorganisms at least during several hours needed for recording of respiration kinetics. Mineral salts were chosen considering the pH value and buffer capacity of the soil so that the pH was not changed more than 0.1 pH units. Soil samples were placed (in triplicate) in an ADC2250 24-channel Soil Respiration System (ADC Bioscientific, Herts, UK) at 22°C. Each sample was continuously aerated

(300 mL min^{-1}), and the rate of CO_2 production from each sample was measured every hour using an infrared detector and mass-flow meter [37].

Maximal specific microbial growth rate (μ_m) was determined by fitting the model parameters to the measured data on CO_2 production:

$$v(t) = A + B \cdot \exp(\mu_m \cdot t) \qquad (3)$$

where $v(t)$ - CO_2 evolution rate at time (t), A - initial rate of uncoupled (non-growth) respiration, B - initial rate of coupled (growth) respiration [19], [40]. Fitting was restricted to the initial phase of the curve, which corresponded to unlimited exponential growth [41]. Maximum values of statistic criteria: r^2, the fraction of total variation explained by the model were used for fitting optimisation. Further goodness of fit estimations were made and based on the Q value derived from χ^2 [42].

Activity status of the microbial biomass r_0 was calculated from the ratio of A:B [19]:

$$r_0 = \frac{B(1-\lambda)}{A + B(1-\lambda)} \qquad (4)$$

where λ may be accepted as a basic stoichiometric constant = 0.9 [19]. The total glucose-metabolizing microbial biomass (sustaining + growing; x_0) was calculated as following:

$$x_0 = \frac{B \cdot \lambda \cdot Y_{CO_2}}{r_0 \cdot \mu_m} \qquad (5)$$

where Y_{CO2} is yield of biomass C per unit of respired C-CO2.

The *growing microbial biomass* (x_0') was calculated using the equation:

$$x'_0 = x_0 \cdot r_0 \qquad (6)$$

More complete theoretical background and details on equations derivation were described elsewhere [28], [38], [40].

The duration of lag-period (t_{lag}) – a period characterised by stable respiration preceding microbial growth – was defined as the time from glucose addition to the time when the increasing rate of growth-associated respiration ($B* exp(\mu_m*t)$) equalled the rate of non-growth respiration (A) [43]. The lag-period was calculated using parameters of Eq. 3:

$$t_{lag} = \ln(A/B)/\mu_m \qquad (7)$$

The ratio of *CO_2 increment-to-DNA increment* ($\Delta CO_2/\Delta DNA$) was calculated as the amount of CO_2 in μg C evolved per μg of DNA increment during the same period. The amount of respired CO_2 in soil amended with glucose was corrected for basal respiration, i.e. the corresponding amount of CO_2 respired from the unamended soil during the same period was subtracted from the CO_2 increment for glucose-amended soil.

The carbon use efficiency or CUE (in the growth phase, this is equivalent to the growth yield quotient, Y, Eq.1) was calculated as biomass C increment per amount of consumed C-substrate, which is in turn equal to biomass C increment plus CO_2 evolved:

$$CUE = \Delta C_{mic}/(\Delta C_{mic} + \Delta C_{CO_2}) \qquad (8)$$

where ΔC_{mic} is the net increase in microbial biomass C ($\mu g\ C\ g^{-1}$) and ΔC_{CO2} is the net increase in cumulative respiration ($\mu g\ C\ g^{-1}$) corrected for basal respiration. Microbial C content was calculated from mean measured DNA content found in our study (11% of dry biomass), assuming that the C content in microbial biomass is 45% [5], [44].

Statistical Analyses

The means of three replicates with standard errors are presented in tables and figures. Two-way ANOVA was applied to characterise the effects of C and N availability: 1) C availability: rhizosphere versus root-free soil, and 2) N availability: half versus full N fertilisation. When significant effects were found, a multiple comparison using the Student-Newman-Keuls test ($P<0.05$) was performed. All variables passed normality and equal variance tests.

Results

Basal Respiration Rate, DNA Content and Microbial Biomass

The basal respiration rate (V_{basal}) was significantly higher in the rhizosphere as compared to root-free soil (Fig. 1a). This rhizosphere effect amounted to 66% at the half N rate while it was only 14% at the full rate of N application. The V_{basal} in root-free soil was significantly higher at the full versus half rate of N-fertilisation (Fig. 1a). In rhizosphere soil, however, N fertilisation significantly decreased basal respiration.

Microbial DNA content was higher at the full N rate than in the corresponding treatments with the half N (Fig. 1b). Higher DNA content in rhizosphere versus root-free soil (28% at the full and 21% at the half N rate) reflects a pronounced rhizosphere effect.

Microbial respiration curves during growth on glucose were clearly different between the rhizosphere and root-free soil (Fig. 2). These differences were more pronounced under N limitation (Fig. 2). Maximal specific growth rates (μ_m) were significantly higher, while the duration of the lag-period was 1.7–1.9 h shorter in the rhizosphere than in root-free soil (Table 1).

Both the total microbial biomass C and its growing fraction were always higher in the rhizosphere as compared to root-free soil (Table 1). This rhizosphere effect was most pronounced at half versus the full N rate (Table 1) and amounted to 31% and 14% of the total microbial biomass, respectively. Actively growing microbial biomass did not exceed 0.34% of total microbial C and was much more sensitive to the presence of roots as compared to total microbial biomass. So, the rhizosphere effect for growing microbial biomass was much greater than for the total microbial biomass and amounted to 45% at full N and to 83% at the half N rate (Table 1). The direct effect of N on total microbial biomass was insignificant in rhizosphere soil, while in root-free soil significantly higher microbial biomass C was observed at the full N rate.

Two-way ANOVA confirmed the strong effects of roots of *Beta vulgaris* on all microbial parameters tested (Table 2). The portion of active microbial biomass and the lag-period were affected by roots at the largest extent: more than 90% of their variation was explained by the rhizosphere effect. The direct effect of N on the specific growth rate (μ_m) and DNA was even stronger than the effect of roots (Table 2).

We conclude that significantly higher basal respiration, DNA content and total and actively growing microbial biomass were observed in the rhizosphere versus root-free soil and this effect was more pronounced under low N fertilization.

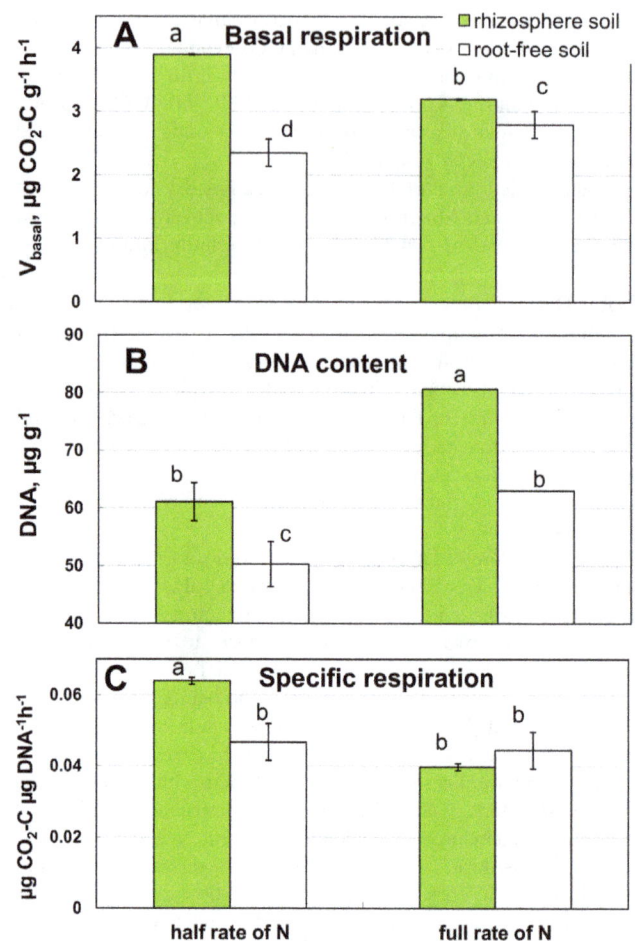

Figure 1. Respiration rate and microbial DNA in soil and rhizosphere. Basal respiration rate (a), microbial DNA content (b), and ratio of basal respiration rate (V_{basal}) to DNA content (c) of rhizosphere and root-free soil under *Beta vulgaris* at half (63 kg N ha^{-1}) and full (126 kg N ha^{-1}) rates of nitrogen fertilisation.

Respiratory Activity in Relation to DNA Content in Rhizosphere and Root-free Soil

The CO_2/DNA ratio in the non-growing microbial community varied between 0.038 and 0.064 $\mu g\ CO_2$-C μg^{-1} DNA h^{-1} (Fig. 1c). The rhizosphere effect on the CO_2/DNA ratio was significant only at the half N rate (Fig. 1c). A significant N effect was observed only in rhizosphere soil: the CO_2/DNA ratio was 64% greater at the half versus the full N rate (Fig. 1c).

Respiratory Response and Microbial DNA Dynamics during Glucose-induced Growth

According to respiratory kinetics, we defined three phases of microbial growth on glucose (Fig. 2): an initial phase corresponding to the absence of microbial growth lasting in rhizosphere soil between 0 and −10.7 h (Table 1, see lag period); followed by the phase of exponential growth to 25.5 h; and by the phase of growth retardation thereafter. In root-free soil duration of corresponding microbial growth phases was for ca. 2 h (lag-phase) and even for 4 h longer than in the rhizosphere (Table 1, Fig. 2). The DNA content in the rhizosphere significantly increased within two hours after the end of the lag-period (t_{lag} 10.3 h, Tables 1, 3). Thus, the amount of DNA in the rhizosphere soil increased almost

Table 1. Biomass and kinetic parameters of the respiratory response of microorganisms growing on glucose.

Soil	N rate	Microbial biomass C			Total cell mass	Maximal growth rate (μ_m)	Lag-period (t_{lag})
		Total	Growing	Growing			
		μg C g^{-1}	μg C g^{-1}	% of total	μg g^{-1}	h^{-1}	h
Root-free	50%	221[c]±1	0.486[c]±0.04	0.22	491[a]±2	0.250[b]±0.003	12.2[a]±0.3
Rhizosphere	50%	290[a]±20	0.888[a]±0.02	0.31	644±44	0.260[a]±0.001	10.3[b]±0.2
Root-free	100%	245[b]±14	0.637[b]±0.03	0.26	544±31	0.238[c]±0.002	12.4[a]±0.2
Rhizosphere	100%	280[a]±1	0.922[a]±0.05	0.33	622±2	0.246[b]±0.002	10.7[b]±0.5

Total cell mass was calculated assuming a C content of the microbial biomass of 45% of dry weight (Christensen et al., 1993). Small letters show significant differences within the same column (p<0.05).

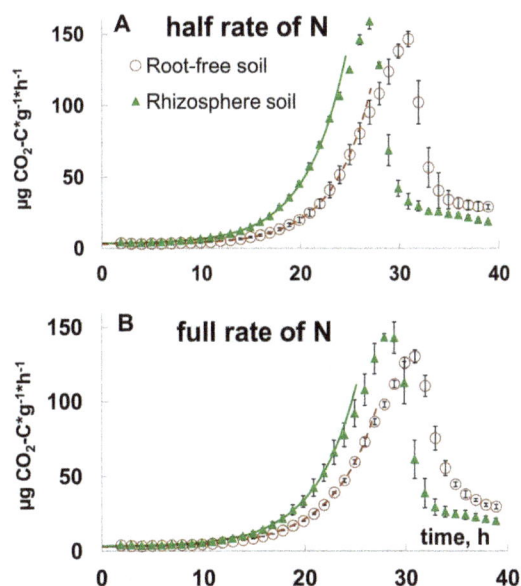

Figure 2. Dynamics of microbial respiration after glucose addition to root-free and rhizosphere soil. Glucose and nutrients induced respiration rate in root-free and rhizosphere soil under *Beta vulgaris* at half (a) and full (b) rates of N fertiliser. Experimental points and curves fitted by Eq. 3 for unlimited growth period are presented.

simultaneously with the respiration (Fig. 3a). In contrast, there were no changes in DNA content 15 hours after glucose application in root-free soil (Fig. 3b). So, contrary to the rhizosphere a time shift of at least three hours was observed between the increase of CO_2 and of DNA.

During the exponential growth, the specific rate of CO_2 emission (V_{CO2}/DNA ratio) steadily increased in both soils (Fig. 3 inserts). Despite the DNA content was significantly lower in root-free as compared to rhizosphere soil during the 35 h after glucose addition (Fig. 3), no significant differences (exception for one point at 20 h) between root-free and rhizosphere soil were found for the V_{CO2}/DNA ratio, which peaked at 25 h after glucose addition and exceeded 1 μg C μg^{-1} DNA h^{-1}. After growth retardation, the V_{CO2}/DNA ratios returned to the initial state and were close to 0.1 μg C μg^{-1} DNA h^{-1} (Fig. 3 inserts).

The quantity of CO_2 evolved per unit of newly-formed DNA ($\Delta CO_2/\Delta DNA$) from the rhizosphere soil continuously increased until the middle of the exponential growth, then stabilised until the end of incubation at 13.6±0.3 μg CO_2-C μg^{-1} DNA (Fig. 4a), indicating a proportional increase in CO_2 and DNA content. In the root-free soil however, the $\Delta CO_2/\Delta DNA$ ratio was 1.5–2 times lower than in rhizosphere during exponential growth (until 20–23 h after glucose addition) and increased only after growth retardation (Fig. 4b). The microbial respiration rate decreased in the rhizosphere after 25 h, and in the root-free soil after 30 hours (Fig. 2), but the DNA content increased for at least 10 more hours in both soils (Fig. 3,). Twice as much CO_2 was produced during exponential growth in rhizosphere versus root-free soil (Table 3), but only 8% more CO_2 was evolved from rhizosphere as compared to root-free soil during the whole incubation (36 h after glucose addition). Thus, the more efficient growth in the exponential phase (according to the $\Delta CO_2/\Delta DNA$ ratio) was counterbalanced by a less efficient metabolism after substrate exhaustion in the root-free soil.

The CUE (Eq. 8) also indicated more efficient microbial metabolism in root-free versus rhizosphere soil during exponential

Table 2. Contribution of two factors: living roots (Roots) and N fertilisation rate (N) and their interactions (Roots x N) to the variance of microbial parameters.

Factor	Basal respiration	Microbial biomass		dsDNA content	Maximal growth rate, μ_m	Lag-period
		total	active			
Roots	67.2***	86.7***	89.8***	40.7***	30.6**	95.1***
N	0.6^{ns}	1.5^{ns}	6.7**	48.1***	63.8**	1.7^{ns}
Roots x N	28.6**	8.5*	2.5*	7.5***	0.4^{ns}	0.3^{ns}
Residual	3.9	3.3	1	3.7	5.2	2.9

two-way ANOVA, % of explained variance.
***, **, * - significant effects at P<0.001, <0.01 and <0.05, respectively.
^{ns}– not significant.

growth (Table 3). At the early stage of glucose utilization and after growth retardation, however, the efficiency of microbial metabolism was lower in root-free than in rhizosphere soil. Remarkably, CUE estimated for the whole incubation period did not differ between both soils (Table 3).

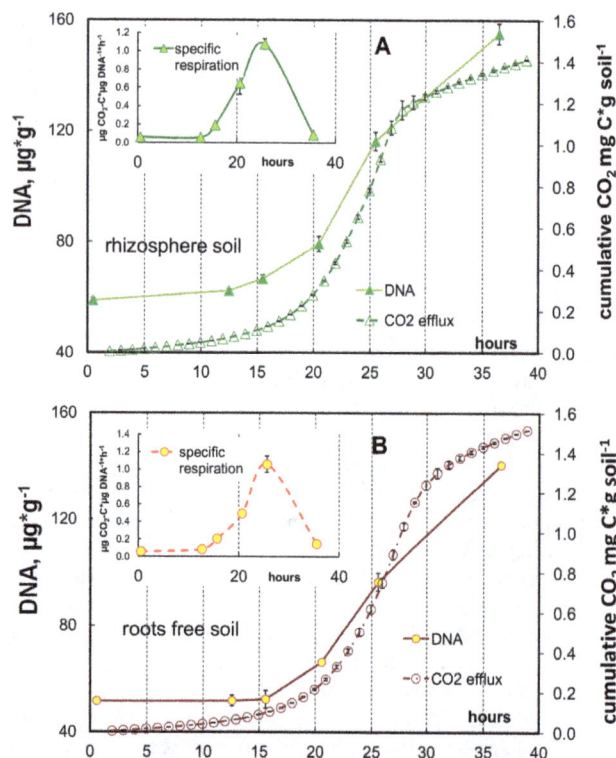

Figure 3. Microbial DNA dynamics and cumulative CO$_2$ production in root-free and rhizosphere soil. Dynamics of microbial DNA content and CO$_2$ accumulation after glucose addition in rhizosphere (a) and root-free (b) soil collected from the plot fertilized with 126 kg N ha^{-1} year^{-1}. Dynamics of specific CO$_2$ production (V$_{CO2}$-to-DNA ratio) are shown in the inserted graphs for rhizosphere and root-free soil, correspondingly.

Discussion

Microbial Biomass and DNA Content as a Basis for CUE Estimation

Assuming a C content of microbial biomass of 45% of dry weight [5], the total cell mass in soil without glucose varied from 491 to 644 μg g^{-1} soil (according to the SIR method, Eq. 2, Table 1). Therefore, the DNA content in microbial biomass amounted to 9.5–13% of dry weight which is in the upper range of the values reported for cultures extracted or isolated from soil bacteria, 5.2–13% [45] and is very close to the microbial DNA content in situ in soil (7–9%) when microbial biomass was assessed by a fumigation-extraction technique [26]. The comparison of several independent observations indicated that approximately 13% of the soil microbial biomass consisted of DNA [25]. However, the DNA content per biomass unit was not constant and decreased with increasing cell size from 13 to 5.2% [45] and was greater in non-growing than in growing bacterial cells. Therefore, the high DNA percentage in microbial biomass in our soil reflected the domination of small-sized cells in the non-growing microbial community.

Respiration and DNA Content under Steady-state and Unlimited Growth Conditions

Our results (Fig. 3, insert) confirm the findings of Marstorp & Witter [26] for a sandy loam soil from central Sweden, where CO_2/DNA ratios were lower than 0.1 μg CO$_2$-C μg^{-1} DNA h^{-1} for a non-growing microbial community. During exponential growth, however, we observed a quick increase in CO_2/DNA ratios. The CO_2/DNA ratio calculated according to Figure 1 in Marstorp & Witter [26] also increased during glucose-induced growth up to 0.5 μg CO$_2$-C μg^{-1} DNA h^{-1}. The CO_2/DNA ratio changed along with the physiological state of microorganisms and therefore, together with the metabolic quotient qCO$_2$, can be used as a valuable ecophysiological indicator reflecting the activity status of microbial biomass in soil.

A constant DNA content during the lag-period has been observed for in situ soil conditions [26]. We noticed, however, that the increase in DNA content in root-free soil began several hours after the increase in respiration, reflecting a period necessary for the activation of microbial metabolism (CO$_2$ increase) before the real growth (DNA increase) start. Such behaviour is common for K-strategists [46]. The delay between respiratory increase and DNA synthesis after the stimulation of microbial growth was much shorter in rhizosphere than in root-free soil, where no increase in DNA content was evident, even at the start of the exponential respiration increase. This was supported by the amount of active

Table 3. The amount of produced CO_2, DNA increment and carbon use efficiency (CUE) at different phases of microbial growth after glucose addition.

Period after glucose addition, (h)	Location	Phase of microbial growth	DNA increase during specified period	CO_2 accumulated during specified period	CUE, calculated according Eq.8, see details in text
			$\mu g\ g\ soil^{-1}$	$\mu g\ C\ g\ soil^{-1}$	$g\ C\ g\ C^{-1}$
0–12.5	Rhizosphere	lag-phase & initial growth	3.5±1.3	59[d]±3	0.41[a]±0.04
	Root-free soil	lag-phase	0.2±3.6	40[d]±2	0.39[a]±0.05
12.5–25.5	Rhizosphere	exponential growth	54.1±3.4	772[b]±22	0.23[b]±0.02
	Root-free soil	exponential growth	44.8±8.2	383[c]±39	0.35[a]±0.07
25.5–36.5	Rhizosphere	growth retardation growth	38.6±4.9	578[c]±26	0.22[b]±0.04
	Root-free soil	& growth retardation	43.5±8.5	877[b]±65	0.17[b]±0.06
0–36.5	Rhizosphere	all phases	96±3.8	1408[a]±1	0.23[b]±0.01
	Root-free soil	all phases	87.7±4.3	1300[a]±24	0.23[b]±0.02

Small letters show significant differences within the same column (p<0.05).

microbial biomass capable for immediate growth that was twice as large in rhizosphere as compared to root-free soil (Table 1).

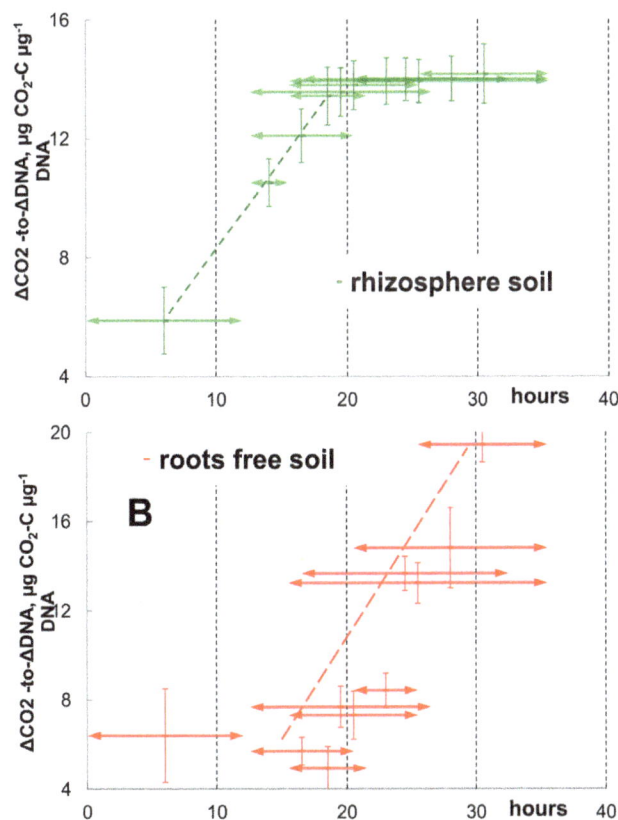

Figure 4. The ratio of CO_2 increment-to-DNA increment in rhizosphere and root-free soil. Soil was collected from the plot fertilized with 126 kg N ha^{-1} $year^{-1}$, (a) – rhizosphere soil, (b) – root-free soil. Horizontal arrows show the time period used for $\Delta CO_2/\Delta DNA$ ratio calculation. Vertical bars show standard deviations.

We demonstrated two kinds of physiological responses to glucose addition in microbial communities in rhizosphere and root-free soil. The DNA synthesis after glucose addition was more closely coupled with CO_2 production in rhizosphere soil as compared to root-free soil, where the dynamics of DNA synthesis and CO_2 production were decoupled both immediately after glucose addition and after its exhaustion. Microorganisms in the root-free soil persisted in a dormant state and reacted to increased substrate availability with a distinct delay between respiration response and DNA synthesis. In the rhizosphere, where the fraction of active microorganisms capable for immediate growth was two-fold larger than in root-free soil, the microbial community responded to glucose earlier in terms of both respiration and DNA synthesis (Figs. 2, 3).

Lag Period and Specific Growth Rates of Microorganisms in the Rhizosphere and Root-free Soil

The significantly greater μ_m values in rhizosphere as compared to root-free soil (Table 1) indicated a greater portion of fast growing microorganisms with r-strategy in the rhizosphere. Selective stimulation of some bacterial species in the rhizosphere (e. g. *Pseudomonas sp.*), [12], [47] with higher specific growth rates than most other soil bacteria [38] explains this phenomenon. The microbial community of the rhizosphere has a shorter lag-period and was ready for immediate growth on available substrate compared to the microbial community in root-free soil. According to Eq. 7, the duration of t_{lag} is dependent both on μ_m and on the fraction of actively growing microorganisms in the total microbial biomass. The negative correlation between lag-period and the amount of active biomass ($r^2 = -0.78$, p<0.12) was stronger compared to correlation between t_{lag} and μ_m ($r^2 = -0.49$, p<0.30). Thus, we conclude that the activity state of microbial biomass rather than such feature of the microorganisms as maximal specific growth rate (μ_m) is responsible for the duration of t_{lag}.

Basal Respiration as a Response to N Limitation in Rhizosphere versus Root-free Soil

The inverse response of basal respiration rate to N fertilization level in the rhizosphere and root-free soil (Fig. 1a) reflected the different strategies of microbial growth in soil microhabitats. Microorganisms with r-strategy dominating in rhizosphere soil increased basal respiration under N limitation. This resulted in highest values of specific respiration (maintenance efficiency) and consequently in lowest CUE. Contrary to that, the K-strategists prevailing in root-free soil even decreased basal respiration in low N treatment, thus, maintaining CUE similar to that in high N plot under steady-state. There were no differences in fine root development between the plots with full and half rate of N at time of soil sampling [34]; therefore we do not attribute the observed differences in V_{basal} to the variation in C input from roots to the soil [48]. Double limitation by C and N in the root-free soil at the half N rate decreased both microbial DNA content and basal respiration compared to root-free soil at the full N rate. However, specific respiration (maintenance efficiency) did not differ significantly between half and full rate of N fertilization in root-free soil (Fig. 1c) demonstrating stronger competitive abilities of K-strategists under N limitation. Therefore, both the level of metabolic activity and CUE should be considered when the N effect on soil respiration is estimated.

CUE in Rhizosphere and Root-free Soil: Dynamics and Proof of Estimates

Our study revealed the basic differences between microbial communities in rhizosphere and root-free soil in catabolic and anabolic processes traced by the dynamics of two fundamental microbial parameters: respiration activity (CO_2) and cell proliferation (DNA), which were used for estimation of CUE. Lower CUE during exponential growth of the r-selected rhizosphere community (Table 3) was confirmed by the two-fold higher $\Delta CO_2/\Delta DNA$ ratios in rhizosphere versus root-free soil (Fig. 4, 15–20 hours). This agrees with the negative correlation between growth rate and yield [22], [49]. Contrary to r-strategists, the K-strategists relatively more abundant in root-free soil do not mineralise glucose immediately, but can partly store it as an intracellular reserve during lag-phase and use it later after substrate exhaustion [38], [50], [51], thus maintaining their respiratory activity longer. Remarkably, distinct differences in CUE between rhizosphere and root-free soil observed during exponential growth were completely smoothed for CUE estimated for the whole incubation period. Thus, the same energy input caused different patterns of catabolic and anabolic processes in r- and K-selected communities resulting in similar energy output per unit of newly formed DNA in rhizosphere and root-free soils. This demonstrates that the shift in balance between catabolic and anabolic processes can serve as a tool for microbial community to maintain CUE independently of changing environment.

The CUE estimated during the exponential growth was 22% and 35% for rhizosphere and root-free soil, respectively. This is close to the range of 20–30% found for a cultured population of indigenous soil bacteria in the growth phase [45] and it is in the range of 14–51% observed for 8 agricultural soils [5]. However, much higher CUE has been obtained by other methods for in situ microbial communities growing on ^{14}C- or ^{13}C-labeled glucose (50–61%, [30]; 69–78%, [18], see for review [3]).

We used the average DNA value of 11% of total microbial biomass that was determined in soil without glucose addition [5], [45]. Considering lower DNA content in growing cells versus the cells in stationary phase [45], and that the DNA content in fungal

mycelium can be much lower than in bacterial cells [52,53] the CUE of 38% and 51% can be obtained for rhizosphere and root-free soil, respectively (based on the lowest DNA content of 5.2% of cell mass for pure cultures [45]). These CUE exactly fits to the estimates for glucose use efficiency in N-amended and in N-limited soil ($Y = 0.52$ and 0.38, respectively) using a balance calculation [30]. The scatter of CUE values found in the literature can be explained by the variation in growth conditions of microorganisms affecting also the DNA content in microbial cells. More experimental studies on the variability of DNA content in situ are needed for narrowing CUE estimates in experiments similar to ours.

Conclusions

The applied combination of approaches: analysis of the double-stranded DNA content in soil and of respiration kinetics allows quantitative distinguishing of microbial traits in the rhizosphere versus root free soil. Total microbial biomass in the rhizosphere was 14–31% higher than that in the root free soil, while the growing (active) part of microbial biomass was 45–83% higher. The higher microbial specific growth rate (μ_m) and lower CUE indicated the greater contribution of r-strategists in rhizosphere as compared with root-free soil. We partly confirmed hypotheses posed in the introduction: microbial communities in rhizosphere soil have specific respiration rate higher than microorganisms in root-free soil. This holds true under N limiting conditions but no difference was observed for fully fertilized N plot. Lower content of available N decreased microbial DNA, but increased the μ_m values. The N limitation in the rhizosphere increased microbial respiration, presumably due to lower C use efficiency confirming domination of r-selected species in rhizosphere microbial community and supporting our second hypotheses.

The $\Delta CO_2/\Delta DNA$ ratio was stable in the growing microbial community in the rhizosphere while it increased consistently in root-free soil, revealing contrasting patterns of microbial metabolism in different microhabitats. The K-strategy typical for root-free soil manifested itself by decoupling of the respiration burst after glucose addition and DNA increase, more efficient growth (high CUE) and longer persistence of respiratory activity. The r-strategy (common for rhizosphere) was exhibited as a faster and simultaneous response on substrate addition, lower growth efficiency and a shorter period of high activity following by more abrupt respiration decrease after substrate exhaustion. The CUE during exponential growth was by the factor of 1.5 higher in root-free than in rhizosphere soil indicating the necessity to consider variable Y depending on substrate availability in soil microhabitats. Further studies are necessary for the determination of the range of differences in CUE in soil microhabitats, because microbial community composition depends on multiple factors such as host plant species, soil properties, plant development stage [10], [54] and these factors will affect also the microbial physiology in rhizosphere and root-free soils.

Author Contributions

Conceived and designed the experiments: EB SB THA YK. Performed the experiments: EB SB. Analyzed the data: EB SB THA YK. Contributed reagents/materials/analysis tools: THA YK. Wrote the paper: EB SB THA YK. Obtained permission for use of soil samples: THA.

References

1. Cheng W (2009) Rhizosphere priming effect: Its functional relationships with microbial turnover, evapotranspiration, and C-N budgets. Soil Biol Biochem 41: 1795–1801.

2. Gärdenäs AI, Ågren GI, Bird JA, Clarholm M, Hallin S, et al. (2011) Knowledge gaps in soil carbon and nitrogen interactions - From molecular to global scale. Soil Biol Biochem 43: 702–717.

3. Manzoni S, Taylor P, Richter A, Porporato A, Agren GI (2013) Environmental and stoichiometric controls on microbial carbon-use efficiency in soils. New Phytol 196: 79–91.

4. Bradford M, Keiser A, Davies C, Mersmann C, Strickland M (2013) Empirical evidence that soil carbon formation from plant inputs is positively related to microbial growth. Biogeochem 113: 271–281.

5. Anderson TH, Martens R (2013) DNA determinations during growth of soil microbial biomasses. Soil Biol Biochem 57: 487–495.

6. Keiblinger KM, Hall EK, Wanek W, Szukics U, Hammerle I, et al. (2010) The effect of resource quantity and resource stoichiometry on microbial carbon-use-efficiency. FEMS Microbiol Ecol 73: 430–440.

7. Schimel J, Schaeffer SM (2012) Microbial control over carbon cycling in soil. Frontiers in Microbiology 3: 348.

8. Allison SD, Wallenstein MD, Bradford MA (2010) Soil-carbon response to warming dependent on microbial physiology. Nature Geosci 3: 336–340.

9. Paterson E (2009) Comments on the regulatory gate hypothesis and implications for C-cycling in soil. Soil Biol Biochem 41: 1352–1354.

10. Berg G, Smalla K (2009) Plant species and soil type cooperatively shape the structure and function of microbial communities in the rhizosphere. FEMS Microbiol Ecol 68: 1–13.

11. Paterson E, Midwood AJ, Millard P (2009) Through the eye of the needle: a review of isotope approaches to quantify microbial processes mediating soil carbon balance. New Phytol 184: 19–33.

12. Grayston SJ, Wang S, Campbell CD, Edwards AC (1998) Selective influence of plant species on microbial diversity in the rhizosphere. Soil Biol Biochem 30: 369–378.

13. Blagodatskaya EV, Blagodatsky SA, Anderson TH, Kuzyakov Y (2007) Priming effects in Chernozem induced by glucose and N in relation to microbial growth strategies. Applied Soil Ecology 37: 95–105.

14. Pirt SJ (1975) Principles of microbe and cell cultivation: John Wiley & Sons. 274 p.

15. Blagodatsky SA, Demyanova EG, Kobzeva EI, Kudeyarov VN (2002) Changes in the efficiency of microbial growth upon soil amendment with available substrates. Euras Soil Sci 35: 874–880.

16. Herron PM, Stark JM, Holt C, Hooker T, Cardon ZG (2009) Microbial growth efficiencies across a soil moisture gradient assessed using 13C-acetic acid vapor and ^{15}N-ammonia gas. Soil Biol Biochem 41: 1262–1269.

17. Payne JW (1970) Energy yields and growth of heterotrophs. Annual Reviews in Microbiology 24: 17–52.

18. Thiet RK, Frey SD, Six J (2006) Do growth yield efficiencies differ between soil microbial communities differing in fungal: bacterial ratios? Reality check and methodological issues. Soil Biol Biochem 38: 837–844.

19. Panikov NS, Sizova MV (1996) A kinetic method for estimating the biomass of microbial functional groups in soil. J Microbiol Meth 24: 219–230.

20. Anderson TH, Domsch KH (1985) Maintenance carbon requirements of actively metabolizing microbial populations under in situ conditions. Soil Biol Biochem 17: 197–203.

21. van Bodegom P (2007) Microbial Maintenance: A Critical Review on Its Quantification. Microbial Ecol 53: 513–523.

22. Lipson D, Monson R, Schmidt S, Weintraub M (2009) The trade-off between growth rate and yield in microbial communities and the consequences for under-snow soil respiration in a high elevation coniferous forest. Biogeochem 95: 23–35.

23. Marstorp H, Guan X, Gong P (2000) Relationship between dsDNA, chloroform labile C and ergosterol in soils of different organic matter contents and pH. Soil Biol Biochem 32: 879–882.

24. Blagodatskaya EV, Blagodatskii SA, Anderson TH (2003) Quantitative isolation of microbial DNA from different types of soils from natural and agricultural ecosystems. Mikrobiology 72: 840–846.

25. Joergensen RG, Emmerling C (2006) Methods for evaluating human impact on soil microorganisms based on their activity, biomass, and diversity in agricultural soils. J Plant Nutr Soil Sci 169: 295–309.

26. Marstorp H, Witter E (1999) Extractable dsDNA and product formation as measures of microbial growth in soil upon substrate addition. Soil Biol Biochem 31: 1443–1453.

27. Andrews JH, Harris RF (1986) r and K-selection and microbial ecology. In: Marshall KC, editor. Adv Microb Ecol. New York. 99–144.

28. Dorodnikov M, Blagodatskaya E, Blagodatsky S, Fangmeier A, Kuzyakov Y (2009) Stimulation of r- vs. K-selected microorganisms by elevated atmospheric CO2 depends on soil aggregate size. FEMS Microbiol Ecol 69.

29. del Giorgio PA, Cole JJ (1998) Bacterial growth efficiency in natural aquatic systems. Ann Rev Ecol Systematics 29: 503–541.

30. Blagodatskiy SA, Larionova AA, Yevdokimov IV (1993) Effect of mineral nitrogen on the respiration rate and growth efficiency of soil microorganisms. Euras Soil Sci 25: 85–95.

31. Blagodatsky SA, Yevdokimov IV, Larionova AA, Richter J (1998) Microbial growth in soil and nitrogen turnover: Model calibration with laboratory data. Soil Biol Biochem 30: 1757–1764.

32. Rousk J, Bååth E (2007) Fungal and bacterial growth in soil with plant materials of different C/N ratios. FEMS Microbiol Ecol 62: 258–267.

33. Kuzyakov Y, Xu X (2013) Competition between roots and microorganisms for nitrogen: mechanisms and ecological relevance. New Phytologist 198: 656–669.

34. Weigel HJ, Pacholski A, Burkart S, Helal M, Heinemeyer O, et al. (2005) Carbon turnover in a crop rotation under free air CO$_2$ enrichment (FACE). Pedosphere 15: 728–738.

35. Anderson JPE, Domsch KH (1978) A physiological method for the quantative measurement of microbial biomass in soils. Soil Biol Biochem 10: 215–221.

36. Anderson TH, Domsch KH (1986) Carbon assimilation and microbial activity in soil. Zeitschrift für Pflanzenernährung und Bodenkunde 149: 457–468.

37. Heinemeyer O (1989) Soil microbial biomass and respiration measurements: An automated technique based on infra-red gas analysis. Plant Soil: 191–195.

38. Panikov NS (1995) Microbial Growth Kinetics. London, Glasgow: Chapman and Hall. 378 p.

39. Blagodatsky SA, Blagodatskaya EV, Anderson TH, Weigel HJ (2006) Kinetics of the respiratory response of the soil and rhizosphere microbial communities in a field experiment with an elevated concentration of atmospheric CO$_2$. Euras Soil Sci 39: 290–297.

40. Blagodatsky SA, Heinemeyer O, Richter J (2000) Estimating the active and total soil microbial biomass by kinetic respiration analysis. Biol Fertil Soils 32: 73–81.

41. Wutzler T, Blagodatsky SA, Blagodatskaya E, Kuzyakov Y (2012) Soil microbial biomass and its activity estimated by kinetic respiration analysis - Statistical guidelines. Soil Biol Biochem 45: 102–112.

42. ModelMaker (1997) ModelMaker Version 3.0.3 Software. CherwellScientific Publishing Limited, Oxford.

43. Blagodatskaya EV, Blagodatsky SA, Anderson TH, Kuzyakov Y (2009) Contrasting effects of glucose, living roots and maize straw on microbial growth kinetics and substrate availability in soil. Europ J Soil Sci 60.

44. Christensen H, Bakken LR, Olsen RA (1993) Soil bacterial DNA and biovolume profiles measured by flow-cytometry. FEMS Microbiol Ecol 102: 129–140.

45. Christensen H, Olsen RA, Bakken LR (1995) Flow Cytometric Measurements of Cell Volumes and DNA Contents During Culture of Indigenous Soil Bacteria. Microbial Ecol 29: 49.

46. Panikov NS (2010) Microbial Ecology. Environmental Biotechnology. 121–191.

47. Goddard VJ, Bailey MJ, Darrah P, Lilley AK, Thompson IP (2001) Monitoring temporal and spatial variation in rhizosphere bacterial population diversity: A community approach for the improved selection of rhizosphere competent bacteria. Plant Soil 232: 181–193.

48. Gershenson A, Bader NE, Cheng W (2009) Effects of substrate availability on the temperature sensitivity of soil organic matter decomposition. Global Change Biol 15: 176–183.

49. Pfeiffer T, Schuster S, Bonhoeffer S (2001) Cooperation and competition in the evolution of ATP-producing pathways. Science (Washington D C) 292: 504–507.

50. Hill PW, Farrar JF, Jones DL (2008) Decoupling of microbial glucose uptake and mineralization in soil. Soil Biol Biochem 40: 616–624.

51. Schneckenberger K, Demin D, Stahr K, Kuzyakov Y (2008) Microbial utilization and mineralization of [14C] glucose added in six orders of concentration to soil. Soil Biology and Biochemistry 40: 1981–1988.

52. Anderson TH (2008) Assessment of DNA contents of soil fungi. Landbauforsch Volkenrode 58: 19–28.

53. Leckie SE, Prescott CE, Grayston SJ, Neufeld JD, Mohn WW (2004) Comparison of chloroform fumigation-extraction, phospholipid fatty acid, and DNA methods to determine microbial biomass in forest humus. Soil Biology and Biochemistry 36: 529–532.

54. Zachow C, Tilcher R, Berg G (2008) Sugar beet-associated bacterial and fungal communities show a high indigenous antagonistic potential against plant pathogens. Microbial Ecol 55: 119–129.

7

Sensitivity of Soil Respiration to Variability in Soil Moisture and Temperature in a Humid Tropical Forest

Tana E. Wood[1,2]*, Matteo Detto[3], Whendee L. Silver[4]

1 International Institute of Tropical Forestry, USDA Forest Service, Río Piedras, Puerto Rico, United States of America, **2** Fundación Puertorriqueña de Conservación, San Juan, Puerto Rico, United States of America, **3** Smithsonian Tropical Research Institute, Apartado Balboa, Republic of Panama, **4** Department of Environmental Science, Policy and Management, University of California, Berkeley, California, United States of America

Abstract

Precipitation and temperature are important drivers of soil respiration. The role of moisture and temperature are generally explored at seasonal or inter-annual timescales; however, significant variability also occurs on hourly to daily time-scales. We used small (1.54 m^2), throughfall exclusion shelters to evaluate the role soil moisture and temperature as temporal controls on soil CO_2 efflux from a humid tropical forest in Puerto Rico. We measured hourly soil CO_2 efflux, temperature and moisture in control and exclusion plots (n = 6) for 6-months. The variance of each time series was analyzed using orthonormal wavelet transformation and Haar-wavelet coherence. We found strong negative coherence between soil moisture and soil respiration in control plots corresponding to a two-day periodicity. Across all plots, there was a significant parabolic relationship between soil moisture and soil CO_2 efflux with peak soil respiration occurring at volumetric soil moisture of approximately 0.375 m^3/m^3. We additionally found a weak positive coherence between CO_2 and temperature at longer time-scales and a significant positive relationship between soil temperature and CO_2 efflux when the analysis was limited to the control plots. The coherence between CO_2 and both temperature and soil moisture were reduced in exclusion plots. The reduced CO_2 response to temperature in exclusion plots suggests that the positive effect of temperature on CO_2 is constrained by soil moisture availability.

Editor: Han Y.H. Chen, Lakehead University, Canada

Funding: Research support was provided by the National Oceanic and Atmospheric Administration (NOAA) Climate and Global Change Postdoctoral Fellowship, DEB-0620910 from the National Science Foundation (NSF) to the Institute of Tropical Ecosystem Studies (IEET), University of Puerto Rico, and the United States Department of Agriculture (USDA) Forest Service International Institute of Tropical Forestry as part of the Long Term Ecological Research Program. The funders had no role in study design, data collection and analysis, decision to publish, or preparation of the manuscript.

Competing Interests: The authors have declared that no competing interests exist.

* E-mail: wood.tana@gmail.com

Introduction

In an era of significant and rapid environmental change, understanding biophysical controls on soil respiration is of immense importance. Tropical forests account for approximately one third of the world's soil carbon (C) pool [1], and have the highest soil respiration rates globally [2]. Temperature and soil moisture are known to affect the production and release of carbon dioxide (CO_2) from tropical forest soils through their effects on soil redox dynamics, diffusion, root and microbial activity as well as C and nutrient availability [3,4,5,6,7,8,9,10,11,12]. While considerable research has addressed seasonal and inter-annual patterns in soil respiration in tropical forests [5,7,10,13], less is known about the role of temperature and precipitation on shorter time-scales (e.g., hours to days) [8,12].

In the tropics, mean month-to-month temperature variation is generally much smaller than that observed on shorter, diel time-scales (e.g., 2 to 4°C versus 6 to 12°C, respectively) [14]. Kinetic theory suggests that reaction rates increase with increasing temperature [15,16]. Laboratory incubations of tropical forest soils support this theory, showing increased soil respiration rates with increasing temperature when carbon (C) and nutrients are not limiting [17,18,19]. It follows that soil respiration under field conditions will also respond to short-term variation in soil temperature (i.e., hours to days).

Light and temperature tend to co-vary in tropical forest ecosystems. Soil respiration is a combination of root and heterotrophic respiration and thus changes in light availability could drive changes in soil respiration via affects on plant activity. In high latitude ecosystems, light limitation of photosynthesis reduced allocation of photosynthate to roots leading to reduced root respiration [20]. A field study in the eastern Amazon found a weak correlation between soil CO_2 efflux and temperature on a diel time-scale in an active pasture, but no correlation in neighboring old growth forest or in a degraded pasture [21]. Given the sharp drop in soil CO_2 efflux that was observed at the end of the daylight period, the authors hypothesized that the diel pattern may be related more to the response of grass metabolism to light than to a response of soil processes to soil temperature. Thus apparent relationships between diel or seasonal variation in soil CO_2 efflux with temperature may actually be due to effects of light availability on root respiration.

Tropical forests experience a wide range of variation in precipitation, at both short (hour to day) to long (seasonal and interannual) temporal scales. This variability in the timing and magnitude of precipitation events can drive changes in biophysical and biogeochemical conditions that can affect soil CO_2 effluxes in complex ways [8]. High soil water content creates a barrier at the soil-atmosphere surface, which could inhibit the diffusion of CO_2 out of the soil [8,9,22]. In humid tropical forests, the consistently moist conditions combined with finely textured clay soils and high biological demand for oxygen (O_2) can facilitate the periodic depletion of O_2 in surface soils [23,24]. Declines in soil O_2 concentrations have been found to occur within hours of even small precipitation events (~ 1 mm) [24]. Low soil O_2 availability can limit aerobic respiration, decreasing soil CO_2 effluxes [25]. However, highly weathered tropical forests are typically rich in poorly crystalline, reactive iron (Fe) minerals; declines in soil redox potential in humid tropical forest soils can drive high rates of iron (Fe) reduction and anaerobic CO_2 respiration [26]. In controlled laboratory experiments, rates of CO_2 production under anaerobic conditions were similar to rates of aerobic respiration [27]. Iron reduction can also increase soil phosphorus (P) availability by decreasing the affinity of Fe for P. Biological activity is generally assumed to be limited by P in these ecosystems [28,29], and thus alleviation of P limitation during low or fluctuating redox conditions has the potential to fuel increased soil respiration [30,31,32,33].

Moisture limitation can also reduce microbial activity and restrict microbial access to C substrates [34,35]. The associated increase in O_2 diffusion into dry soils would increase the concentration of oxidized Fe, decrease P availability through Fe-P bonding, and potentially limit CO_2 production [30,33]. Although the relationships between moisture and soil respiration are complex, theory generally predicts a parabolic relationship between soil CO_2 efflux and soil moisture with the highest soil CO_2 emissions occurring at an intermediate moisture level [7,36,37,38,39]. During periods of soil water saturation and extreme soil drying soil moisture is likely to exert a stronger control over soil CO_2 efflux than that of soil temperature [10].

In this study we investigated hourly to daily changes in soil CO_2 efflux in a relatively a-seasonal humid tropical forest in Puerto Rico to determine (1) the timescale over which CO_2 efflux varies, (2) the relationship of this variation to soil temperature and moisture, and (3) how these relationships are affected by experimental reduction in soil moisture.

Materials and Methods

Site Description

We conducted this research in the Bisley Experimental Watersheds of the Luquillo Experimental Forest in northeastern Puerto Rico (18°18′N, 65°50′W) [33]. Permission to work in this site was granted by the USDA Forest Service International Institute of Tropical Forestry. This research was conducted in collaboration with the Luquillo Long-term Ecological Research (Luq-LTER), and as such, data will be made available via the Luq-LTER database (http://luq.lternet.edu/data/datacatalog). The forest is classified as subtropical wet forest [40]. The elevation is approximately 300 m above sea level, receives an average 3500 mm of precipitation annually, and the mean annual temperature is 23°C. Mean month-to-month variation in temperature is approximately 4°C throughout the year. While precipitation is highly variable throughout the year, there is no significant dry season [41]. The soils are deep, highly weathered, clay-rich and acidic. The study site was dominated by the palm *Prestoea montana* R. Graham Nichols [33].

Experimental Design

We created an experimental drought using small (1.54 m^2) throughfall exclusion shelters that were in place from June through

August of 2008 (3 months total). This study included a total of three exclusion and three control plots. We used time-domain reflectometry (TDR, Campbell Scientific Model CS616) to estimate hourly soil moisture in all plots (0–30 cm), and measured hourly soil temperature (10 cm; Campbell Scientific, Model 108 L) in one control and one exclusion plot. Automated soil respiration chambers (Li-Cor LI-8100/8150 Multiplexer; Li-Cor Biosciences, Lincoln, NE, USA) were installed in all six plots to measure hourly changes in soil CO_2 efflux. Due to limited power access at the field site, soil CO_2 efflux was measured in a series of field campaigns conducted over a six-month period that included three months with the throughfall exclusion shelters in place and three months following shelter removal (average 8 days per campaign, 110 days total). For a more detailed description of the study site and methodology, see Wood and Silver [33].

Statistical Analyses

The variance of each time series (e.g, temperature, moisture, respiration) was decomposed on a scale-by-scale basis using orthonormal wavelet transformation (Matlab version 7.0 [R2010a], Mathworks; Appendix S1). This spectral technique, analogous to Fourier analysis, breaks the process variance into pieces, each of which represents the contribution on a particular scale [42]. Given a time series X_t ($t = 0,2, ...,N-1$) that is regarded as a stochastic process with stationary increment, and a unit level Daubechies wavelet filter $h_{l,l}$ of width L, the wavelet variance at the j-scale $\tau_j = 2^{j-1}$ is defined as:

$$\hat{v}_X^2(\tau_j) = \frac{1}{M_j} \sum_{t=L_j-1}^{N-1} W_{j,t}^2 \qquad (1)$$

where $M_j = N - L_j + 1$ and $L_j = (2^j - 1)(L - 1) - 1$.

The coefficients W are computed as:

$$W_{j,t} = \sum_{l=0}^{L_j-1} h_{j,l} X_{t-l} \qquad (2)$$

We used the Mondal and Percival [43] method to compute an unbiased estimator of the wavelet variance for gappy series where the missing values are replaced by zeros (48% of data; Fig. 1):

$$\hat{u}_{XX}(\tau_j) = \frac{1}{M_j} \sum_{t=L_j-1}^{N-1} \sum_{l=0}^{L_j-1} \sum_{l'=0}^{L_j-1} h_{j,l} h_{j,l'} \hat{\beta}_{l,l'} X_{t-l} X_{t-l'} \qquad (3)$$

where $\hat{\beta}_{l,l'}^{-1} = \frac{1}{M_j} \sum_{t=L_j-1}^{N-1} \delta_{t-l} \delta_{t-l'}$ and δ_t assumes the value zero or

unity, with zero indicating that X_t is missing. In the bi-variate case the wavelet co-variance between two time series X and Y is defined as:

$$\hat{u}_{XY}(\tau_j) = \frac{1}{M_j} \sum_{t=L_j-1}^{N-1} \sum_{l=0}^{L_j-1} \sum_{l'=0}^{L_j-1} h_{j,l} h_{j,l'} \hat{\beta}_{l,l'} X_{t-l} Y_{t-l'} \qquad (4)$$

A normalized wavelet covariance (the analogy of the coefficient of correlation) can be obtained combining equation (3) and (4) to

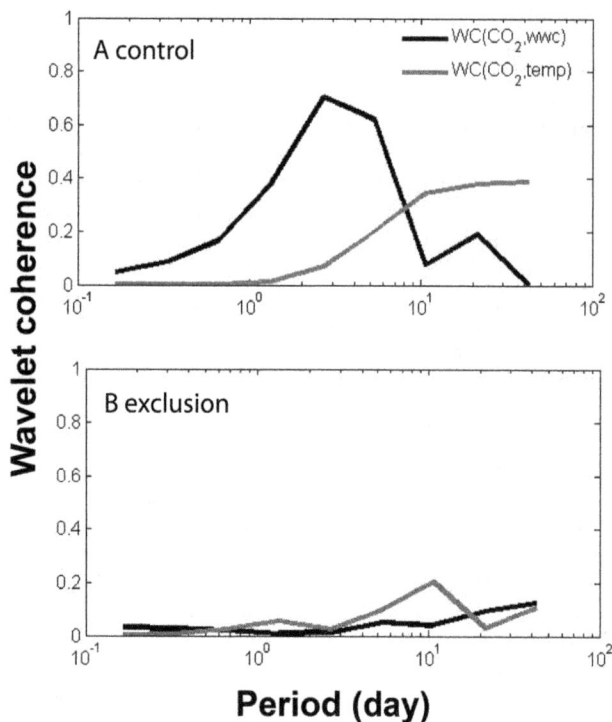

Figure 1. The wavelet coherence [44] between CO_2 efflux and soil moisture (black) and temperature (gray) in the (A) control and (B) exclusion plots.

form [44]:

$$WC_{XY}(\tau_j) = \frac{u_{XY}(\tau_j)}{\sqrt{u_{XX}(\tau_j)}\sqrt{u_{YY}(\tau_j)}} \quad (5)$$

The correlation among variables are explored in the following using the Haar-wavelet coherence defined in equation (5). We used regression analyses to determine relationships between CO_2 efflux (mean of three replicates per treatment) and mean soil characteristics (e.g., soil moisture and temperature). When significant diel variation was observed, regressions were performed using mean hourly values, all other regressions performed using mean daily values. All regressions were performed using SigmaPlot 10 (SigmaPlot for Windows, v. 7.101, 2001, SPSS Inc.).

Results

We found no significant diel periodicity in soil respiration. Soil CO_2 efflux did, however, display significant periodicity over daily to seasonal time-scales. Soil respiration in control plots showed high coherence with soil moisture for a broad range of time scales, with a peak correlation corresponding to a two-day periodicity (Fig. 1). This two-day periodicity is the timescale over which strong fluctuations in volumetric soil moisture occurred (Fig. 2A). Further analyses of the time series revealed a negative relationship between soil moisture and soil CO_2 efflux (Fig. 3). Soil respiration and temperature were correlated on a scale of weeks to months, with peak respiration occurring during the period of highest temperatures (Fig. 2).

Throughfall exclusion reduced volumetric soil moisture by an average of 29% relative to the controls. There was reduced

coherence between soil respiration and soil temperature, as well as with soil moisture in these plots. There was a significant parabolic relationship between mean daily volumetric soil moisture and mean daily soil CO_2 efflux when both treatments were included, with peak soil respiration occurring when volumetric soil moisture was approximately 0.375 m^3/m^3 (Fig. 4, $R^2 = 0.29$, $p < 0.0001$, $f = y_0 + a_* x + b_* x^2$). Variation in the residuals was significantly and positively related to soil temperature ($R^2 = 0.15$, $P < 0.0001$). The same relationship between soil moisture and soil CO_2 efflux was found when the control and exclusion plots were evaluated separately ($R^2 = 0.29$, $p < 0.0001$ [control]; $R^2 = 0.28$, $p < 0.0001$ [exclusion]). We found a significant, positive linear relationship between mean daily soil temperature and mean daily CO_2 efflux in the control plots, but not in the exclusion plots ($R^2 = 0.55$, $p < 0.0001$, Fig. 5).

Discussion

We found no significant diel variation in soil respiration in this forest. This is in contrast to findings in temperate and boreal forests, which have found that soil respiration varied with soil temperature and photosynthesis on diel time-scales [8,45,46, 47,48,49]. Significant diel variation has also been found in some tropical forest sites [8,50,51], but not all (this study, [9,21]). The lack of a consistent diel response of soil CO_2 efflux across tropical forest sites could be due to differences in the magnitude of the diel change in temperature across these forest sites, variability in the relative contribution of root versus microbial respiration to total soil respiration, or because other factors such as soil moisture status, exert a stronger control over total soil respiration in some systems than temperature. Furthermore, root and soil respiration have been found to demonstrate differential responses to environmental variables (e.g., soil moisture) [52,53,54]. Hence, while we observed no significant diel pattern of net soil respiration, partitioning soil respiration into its components (e.g., litter, root and soil) could reveal different results [53].

We found significant coherence between soil CO_2 efflux and soil moisture on a two-day time-scale. The periodicity of this relationship corresponds with large rainfall events that significantly increased volumetric soil moisture and lowered soil CO_2 efflux (e.g., Fig. 3). Rapid declines in soil CO_2 efflux in response to soil water saturation has been observed in seasonal forest in the Amazon [8] and in moist tropical forest in Panama [9]. The decline in soil CO_2 efflux in response to increased volumetric soil moisture could be the result of reduced diffusion of CO_2 from saturated soils [8,9,10]. Reduced soil CO_2 efflux could also be due to reduced soil microbial activity in low O_2 environments [25].

As expected, we found a significant parabolic relationship between soil moisture and soil CO_2 efflux with peak soil respiration occurring when volumetric soil moisture was at an intermediate value of approximately 0.375 m^3/m^3 (Fig. 5). This parabolic relationship between soil moisture and soil CO_2 efflux agrees with findings from other tropical forest sites [6,8,36]. Interestingly, the "tipping point" of the positive effect of soil moisture on CO_2 efflux is similar across tropical forests on clay soils, occurring at mean volumetric soil moisture values of approximately 0.35 m^3/m^3 (this study) [7,36] to 0.45 m^3/m^3 [6,8]. In many soils, when the soil moisture content is at about 40%, a small increase in soil moisture content leads to a large increase in soil resistance to the diffusion of gases, thereby reducing soil CO_2 emissions [9,12,55]. When tropical forests on sandy soils are considered, this tipping point is reduced (0.22 m^3/m^3) [7]. These findings would suggest that soil texture plays an

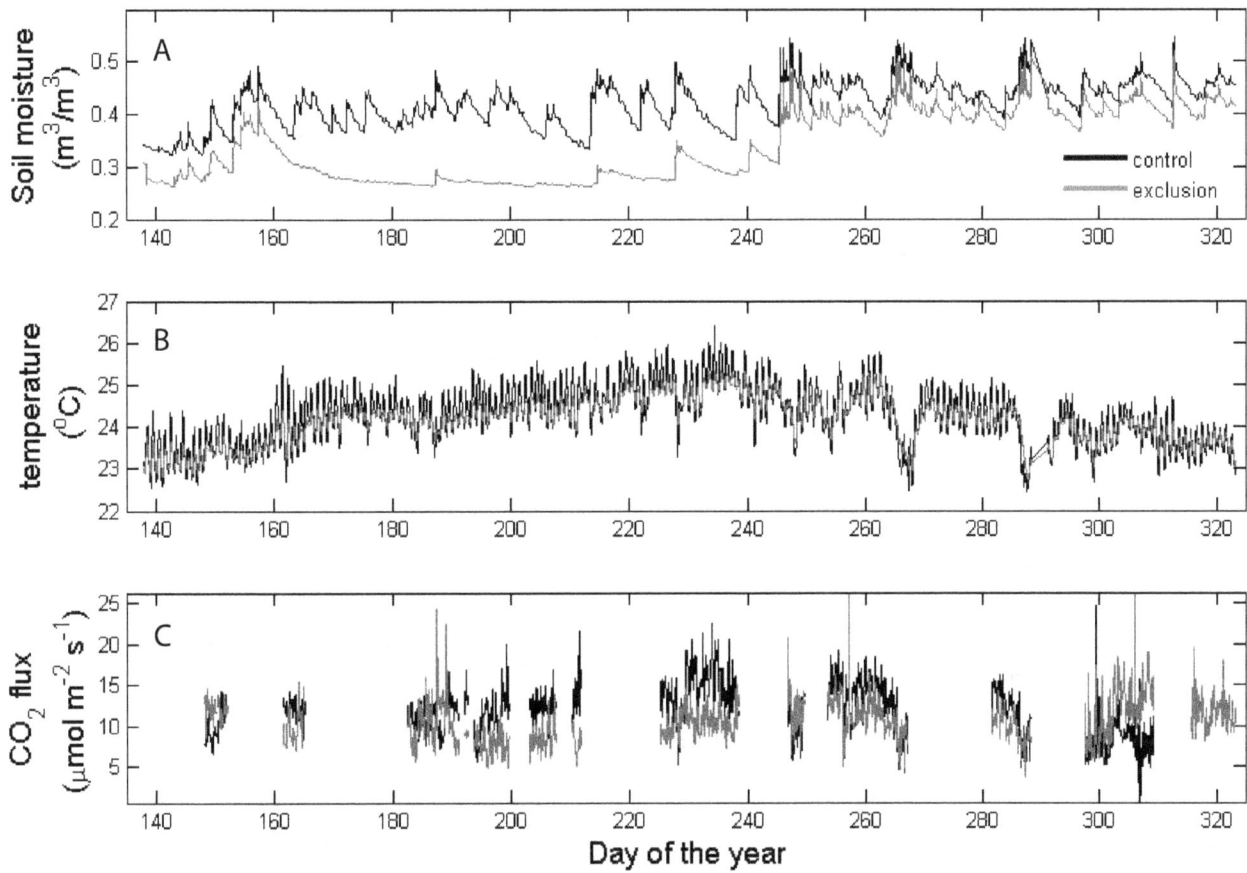

Figure 2. Time series for (A) soil moisture, (B) soil temperature, and (C) carbon dioxide (CO_2) flux over the 6-month study (June through December 2008) in the control (black) and exclusion (gray) plots. Throughfall exclusion shelters were in place from June through August (3-months).

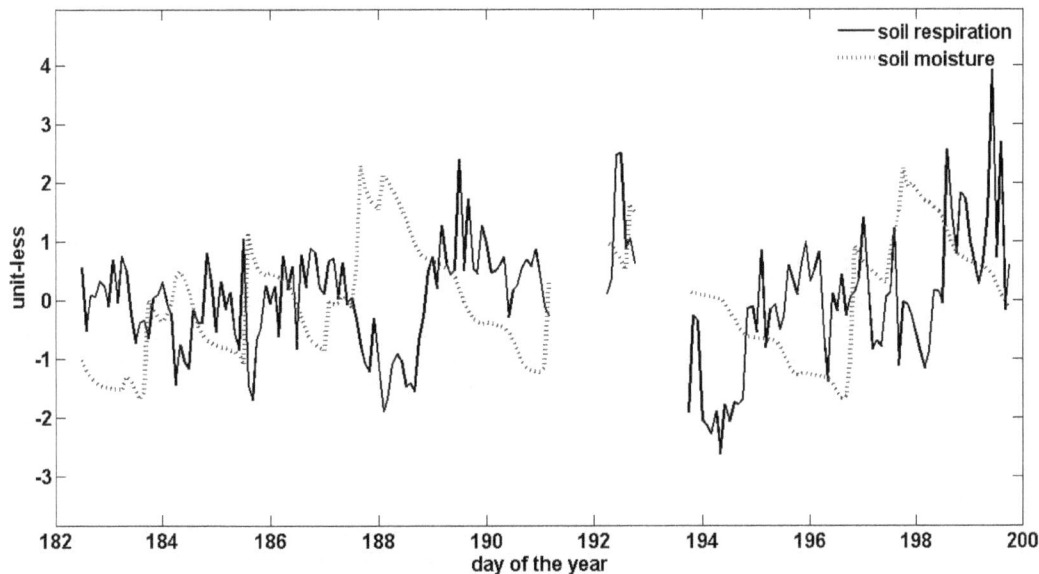

Figure 3. Plot of soil moisture and soil respiration for one of the eight field campaigns (18-days, June 30-July 18, 2008). Soil respiration declined following large rainfall events and the subsequent increase in soil moisture.

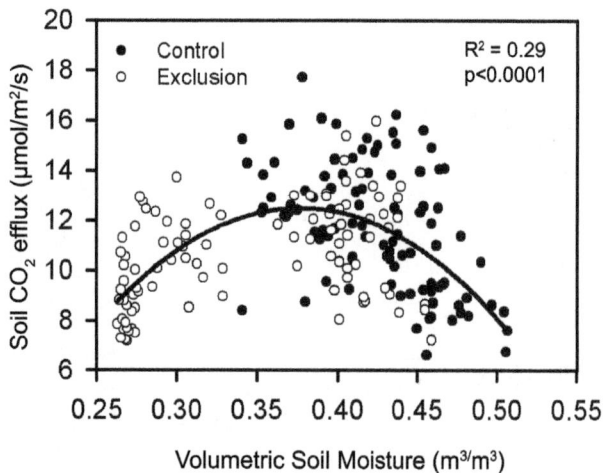

Figure 4. Regression between mean daily volumetric soil moisture and mean daily soil CO_2 efflux in both the control (solid circles) and exclusion (open circles) plots. The equation for the regression is $f = y_0 + a \cdot x + b \cdot x^2$.

important role in determining the tipping point of the positive effect of soil moisture on soil CO_2 efflux in tropical soils.

The observed trend of a positive coherence between soil respiration and temperature on seasonal timescales is intriguing given the low variability in temperature during the study period (Fig. 1; 2°C) [33]. Despite this low seasonal variability, there is evidence that soil respiration in tropical forest sites exhibits a positive response to relatively small increases in temperature over monthly to annual time-scales [6,7,8,10]. However, temperature and light tend to co-vary in tropical forests (Silver et al. *unpublished data*) [7], hence it is also possible that the observed relationship between temperature and soil respiration is driven by variation in light via the positive effect of light on photosynthesis and the resulting increase in carbohydrate allocation to roots rather than temperature. However, experimental manipulation of temperature in a field setting would be needed to distinguish the effects of temperature versus those of light availability on soil CO_2 efflux in tropical systems. Currently, no field-warming experiment has been conducted in a tropical forest [18,56,57].

Interestingly, the coherence between soil moisture and soil respiration on a two-day time-scale was reduced significantly in the exclusion plots (Fig. 1). This reduced coherence could be due to the filtering out of the effects of large rainfall events on soil moisture availability in the exclusion plots (Fig. 2; Days 160 to 240). This finding would suggest that considering the temporal variability of precipitation events in addition to the role of total precipitation inputs is important when evaluating moisture controls on soil CO_2 efflux. In addition to a reduced coherence between soil moisture and CO_2 efflux, there was also a reduction in the CO_2 response to temperature in the exclusion plots, which suggests that the positive effect of temperature on CO_2 efflux at weekly to monthly time scales is constrained by soil moisture. This result is supported by the significant positive relationship between soil temperature and soil CO_2 efflux in the control plots, but not the exclusions (Fig. 5). Interestingly, when we evaluated soil respiration on a weekly time-scale, we found no significant influence of soil moisture or temperature on soil respiration and no significant differences in soil respiration between the control and exclusion plots [33], which highlights the value of collecting soil CO_2 efflux measurements with high temporal resolution.

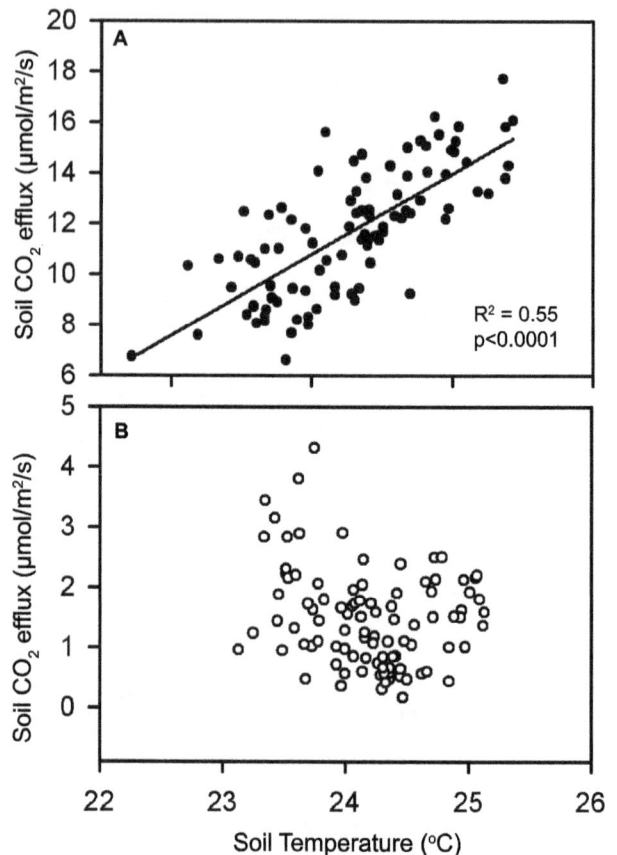

Figure 5. Regression between mean daily soil temperature and mean daily soil CO_2 efflux in the (A) control and (B) exclusion plots.

Overall, the reduced coherence of soil moisture and soil temperature with soil CO_2 efflux in the exclusion plots suggest that small reductions in soil moisture availability can result in moisture availability as a predominant limiting factor of soil CO_2 efflux in tropical soils, even in sites that receive relatively large rainfall inputs throughout the year.

Conclusions

Overall, higher soil moisture led to lower soil CO_2 emissions in this study. The reduction in CO_2 release could be the result of abiotic and biotic factors. The reduced water-filled pore space of the saturated soil may have decreased diffusion of CO_2 out of the soil, leading to lower CO_2 emissions [9,24]. Saturated soils also limit the diffusion of O_2 into the soil, which could have created anaerobic conditions that limit the production of CO_2 [23,25]. Continued dry down of soils has been shown to reduce CO_2 emissions from some tropical forests, but has had no effect in others [3,33,52,54,58]. Our results highlight the strong sensitivity of soil respiration to short-term dynamics in soil moisture and longer-term patterns in temperature or light availability in a humid tropical forest. Our results also show that the well-established relationship of soil respiration to temperature is changed when soil moisture is reduced. This finding would suggest that temperature exerts a positive control on soil respiration as long as soil moisture is not limiting. Determining which processes will dominate in tropical forests depends heavily on our ability to accurately predict how climate change will affect precipitation patterns and hydrologic cycles in these ecosystems.

Acknowledgments

We would like to thank Dr. Ariel E. Lugo for his invaluable support in Puerto Rico. We also thank C. Torrens, C. Estrada, H. Robles, and B. Quintero who helped in the laboratory and the field.

Author Contributions

Conceived and designed the experiments: TEW WLS. Performed the experiments: TEW WLS. Analyzed the data: MD. Contributed reagents/materials/analysis tools: TEW WLS MD. Wrote the paper: TEW WLS MD.

References

1. Jobbagy EG, Jackson RB (2000) The vertical distribution of soil organic carbon and its relation to climate and vegetation. Ecological Applications 10: 423–436.

2. Raich JW, Schlesinger WH (1992) The global carbon dioxide flux in soil respiration and its relationship to vegetation and climate. Tellus B 44: 81–99.

3. Davidson EA, Ishida FY, Nepstad DC (2004) Effects of an experimental drought on soil emissions of carbon dioxide, methane, nitrous oxide, and nitric oxide in a moist tropical forest. Global Change Biology 10: 718–730.

4. Conrad R (1996) Soil microorganisms as controllers of atmospheric trace gases (H2, CO, CH4, OCS, N2O, and NO). Microbiology Reviews 60: 609–640.

5. Davidson EA, Belk E, Boone RD (1998) Soil water content and temperature as independent or confounded factors controlling soil respiration in a temperate mixed hardwood forest. Global Change Biology 4: 217–227.

6. Schwendenmann L, Veldkamp E, Brenes T, O'Brien JJ, Mackensen J (2003) Spatial and temporal variation in soil CO2 efflux in an old-growth neotropical rain forest, La Selva, Costa Rica. Biogeochemistry 64: 111–128.

7. Sotta ED, Veldkamp E, Guimaraes BR, Paixao RK, Ruivo MLP, et al. (2006) Landscape and climatic controls on spatial and temporal variation in soil CO2 efflux in an Eastern Amazonian Rainforest, Caxiuana, Brazil. Forest Ecology and Management 237: 57–64.

8. Sotta ED, Meir P, Malhi Y, Nobre AD, Hodnett M, et al. (2004) Soil CO2 efflux in a tropical forest in the central Amazon. Global Change Biology 10: 601–617.

9. Kursar TA (1989) Evaluation of soil respiration and soil CO2 concentration in a lowland moist forest in Panama. Plant and Soil 113: 21–29.

10. Schwendenmann L, Veldkamp E (2006) Long-term CO2 production from deeply weathered soils of a tropical rain forest: evidence for a potential positive feedback to climate warming. Global Change Biology 12: 1878–1893.

11. Cleveland CC, Wieder WR, Reed SC, Townsend AR (2010) Experimental drought in a tropical rain forest increases soil carbon dioxide losses to the atmosphere. Ecology 91: 2313–2323.

12. Hall S, McDowell W, Silver W (2013) When Wet Gets Wetter: Decoupling of Moisture, Redox Biogeochemistry, and Greenhouse Gas Fluxes in a Humid Tropical Forest Soil. Ecosystems 16: 576–589.

13. Rowlings DW, Grace PR, Kiese R, Weier KL (2012) Environmental factors controlling temporal and spatial variability in the soil-atmosphere exchange of CO2, CH4 and N2O from an Australian subtropical rainforest. Global Change Biology 18: 726–738.

14. Vandecar KL, Lawrence D, Wood TE, Oberbauer SF, Das R, et al. (2009) Biotic and abiotic controls on diurnal fluctuations in labile soil phosphorus of a wet tropical forest. Ecology 90: 2547–2555.

15. Davidson EA, Janssens IA (2006) Temperature sensitivity of soil carbon decomposition and feedbacks to climate change. Nature 440: 165–173.

16. Knorr W, Prentice IC, House JI, Holland EA (2005) Long-term sensitivity of soil carbon turnover to warming. Nature 433: 298–301.

17. Holland EA, Neff JC, Townsend AR, McKeown B (2000) Uncertainties in the temperature sensitivity of decomposition in tropical and subtropical ecosystems: Implications for models. Global Biogeochemical Cycles 14: 1137–1151.

18. Wood TE, Cavaleri MA, Reed SC (2012) Tropical forest carbon balance in a warmer world: a critical review spanning microbial- to ecosystem-scale processes. Biological Reviews 87: 912–927.

19. Balser TC, Wixon DL (2009) Investigating biological control over soil carbon temperature sensitivity. Global Change Biology 15: 2935–2949.

20. Wofsy SC, Harriss RC, Kaplan WA (1988) Carbon Dioxide in the Atmosphere Over the Amazon Basin. Journal of Geophysical Research 93: 1377–1387.

21. Davidson EA, Verchot LV, Cattanio JH, Ackerman IL, Carvalho JEM (2000) Effects of soil water content on soil respiration in forests and cattle pastures of eastern Amazonia. Biogeochemistry 48: 53–69.

22. Singh JS, Gupta SR (1977) Plant decomposition and soil respiration in terrestrial ecosystems. The Botanical Review 43: 449–528.

23. Silver WL, Lugo AE, Keller M (1999) Soil oxygen availability and biogeochemistry along rainfall and topographic gradients in upland wet tropical forest soils. Biogeochemistry 44: 301–328.

24. Liptzin D, Silver W, Detto M (2011) Temporal Dynamics in Soil Oxygen and Greenhouse Gases in Two Humid Tropical Forests. Ecosystems 14: 171–182.

25. Orchard VA, Cook FJ (1983) Relationship between soil respiration and soil moisture. Soil Biology and Biochemistry 15: 447–453.

26. Dubinsky EA, Silver WL, Firestone MK (2010) Tropical forest soil microbial communities couple iron and carbon biogeochemistry. Ecology 91: 2604–2612.

27. DeAngelis KM, Silver WL, Thompson AW, Firestone MK (2010) Microbial communities acclimate to recurring changes in soil redox potential status. Environmental Microbiology 12: 3137–3149.

28. Vitousek PM (1984) Litterfall, nutrient cycling, and nutrient limitation in tropical forests. Ecology 65: 285–298.

29. Cleveland CC, Townsend AR, Taylor P, Alvarez-Clare S, Bustamante MMC, et al. (2011) Relationships among net primary productivity, nutrients and climate in tropical rain forest: a pan-tropical analysis. Ecology Letters 14: 939–947.

30. Chacon N, Silver W, Dubinsky E, Cusack D (2006) Iron Reduction and Soil Phosphorus Solubilization in Humid Tropical Forests Soils: The Roles of Labile Carbon Pools and an Electron Shuttle Compound. Biogeochemistry 78: 67–84.

31. Cleveland CC, Townsend AR (2006) Nutrient additions to a tropical rain forest drive substantial soil carbon dioxide losses to the atmosphere. Proceedings of the National Academy of Sciences of the United States of America 103: 10316–10321.

32. Liptzin D, Silver WL (2009) Effects of carbon additions on iron reduction and phosphorus availability in a humid tropical forest soil. Soil Biology and Biochemistry 41: 1696–1702.

33. Wood TE, Silver WL (2012) Strong spatial variability in trace gasdynamics following experimental drought in a humid tropical forest. Global Biogeochemical Cycles.

34. Stark JM, Firestone MK (1995) Mechanisms for soil moisture effects on activity of nitrifying bacteria. Applied and Environmental Microbiology 61: 218–221.

35. Allison SD (2005) Cheaters, diffusion and nutrients constrain decomposition by microbial enzymes in spatially structured environments. Ecology Letters 8: 626–635.

36. Chambers JQ, Tribuzy ES, Toledo LC, Crispim BF, Higuchi N, et al. (2004) Respiration from a Tropical Forest Ecosystem: Partitioning of Sources and Low Carbon Use Efficiency. Ecological Applications: 72–88.

37. Ino Y, Monsi M (1969) An experimental approach to the calculation of CO2 amount evolved from several soils. Japanese Journal of Botany 20: 153–188.

38. Edwards NT (1975) Effects of Temperature and Moisture on Carbon Dioxide Evolution in a Mixed Deciduous Forest Floor1. Soil Science Society of America Journal 39: 361–365.

39. Londo AJ, Messina MG, Schoenholtz SH (1999) Forest Harvesting Effects on Soil Temperature, Moisture, and Respiration in a Bottomland Hardwood Forest. Soil Science Society of America Journal 63: 637–644.

40. Holdridge LR (1967) Life Zone Ecology. San José, Costa Rica: Tropical Science Center.

41. Heartsill-Scalley T, Scatena FN, Estrada C, McDowell WH, Lugo AE (2007) Disturbance and long-term patterns of rainfall and throughfall nutrient fluxes in a subtropical wet forest in Puerto Rico. Journal of Hydrology 333: 472–485.

42. Cazelles B, Chavez M, Berteaux D, Ménard F, Vik J, et al. (2008) Wavelet analysis of ecological time series. Oecologia 156: 287–304.

43. Mondal D, Percival D (2010) Wavelet variance analysis for gappy time series. Annals of the Institute of Statistical Mathematics 62: 943–966.

44. Liu PC (1995) Wavelet spectrum analysis and ocean wind waves. In: Foufoula-Georgiou E, Kumar, P., editor. Wavelets in Geophysics. New York: Academic Press. pp. 151–166.

45. Högberg P, Nordgren A, Buchmann N, Taylor AFS, Ekblad A, et al. (2001) Large-scale forest girdling shows that current photosynthesis drives soil respiration. Nature 411: 789–792.

46. Kuzyakov Y, Gavrichkova O (2010) REVIEW: Time lag between photosynthesis and carbon dioxide efflux from soil: a review of mechanisms and controls. Global Change Biology 16: 3386–3406.

47. Tang JW, Baldocchi DD, Xu L (2005) Tree photosynthesis modulates soil respiration on a diurnal time scale. Global Change Biology 11: 1298–1304.

48. Detto M, Molini A, Katul G, Stoy P, Palmroth S, et al. (in press) Causality and Persistence in Ecological Systems: A Non-Parametric Spectral Granger Causality Approach. The American Naturalist.

49. Vargas R, Allen MF (2008) Environmental controls and the influence of vegetation type, fine roots and rhizomorphs on diel and seasonal variation in soil respiration. New Phytologist 179: 460–471.

50. Vargas R, Allen MF (2008) Diel patterns of soil respiration in a tropical forest after Hurricane Wilma. Journal of Geophysical Research 113: G03021.

51. Medina E, Zelwar M (1972) Soil respiration in tropical plant communities. In: Golley PM, Golley FB, editors. Tropical Ecology with an Emphasis on Organic Production. Athens, GA: University of Georgia. pp. 245–269.

52. Metcalfe DB, Meir P, Aragão LEOC, Malhi Y, da Costa ACL, et al. (2007) Factors controlling spatio-temporal variation in carbon dioxide efflux from surface litter, roots, and soil organic matter at four rain forest sites in the eastern Amazon. Journal of Geophysical Research 112: G04001.

53. Subke JA, Inglima I, Cotrufo MF (2006) Trends and methodological impacts in soil CO2 efflux partitioning: A metaanalytical review. Global Change Biology 12: 921–943.

54. Cattânio J, Davidson E, Nepstad D, Verchot L, Ackerman I (2002) Unexpected results of a pilot throughfall exclusion experiment on soil emissions of CO2, CH4, N2O, and NO in eastern Amazonia. Biology and Fertility of Soils 36: 102–108.

55. Grable AR (1966) Soil Aeration and Plant Growth. In: Norman AG, editor. Advances in Agronomy: Academic Press. pp. 57–106.

56. Corlett RT (2011) Impacts of warming on tropical lowland rainforests. Trends in Ecology and Evolution 26: 606–613.

57. Amthor JS, Hanson PJ, Norby RJ, Wullschleger SD (2010) A comment on "Appropriate experimental ecosystem warming methods by ecosystem, objective, and practicality" by Aronson and McNulty. Agricultural and Forest Meteorology 150: 497–498.

58. Sotta ED, Veldkamp E, Schwendenmann L, Guimaraes BR, Paixao RK, et al. (2007) Effects of an induced drought on soil carbon dioxide (CO2) efflux and soil CO2 production in an Eastern Amazonian rainforest, Brazil. Global Change Biology 13: 2218–2229.

8

Effects of Forest Age on Soil Autotrophic and Heterotrophic Respiration Differ between Evergreen and Deciduous Forests

Wei Wang[1]*, Wenjing Zeng[1], Weile Chen[1], Yuanhe Yang[3], Hui Zeng[2]

1 Department of Ecology, College of Urban and Environmental Sciences, and Key Laboratory for Earth Surface Processes of the Ministry of Education, Peking University, Beijing, China, **2** Shenzhen Graduate School, Key Laboratory for Urban Habitat Environmental Science and Technology, Peking University, Shenzhen, China, **3** Institute of Botany, The Chinese Academy of Sciences, Beijing, China

Abstract

We examined the effects of forest stand age on soil respiration (SR) including the heterotrophic respiration (HR) and autotrophic respiration (AR) of two forest types. We measured soil respiration and partitioned the HR and AR components across three age classes $\sim$15, $\sim$25, and $\sim$35-year-old *Pinus sylvestris* var. *mongolica* (Mongolia pine) and *Larix principis-rupprechtii* (larch) in a forest-steppe ecotone, northern China (June 2006 to October 2009). We analyzed the relationship between seasonal dynamics of SR, HR, AR and soil temperature (ST), soil water content (SWC) and normalized difference vegetation index (NDVI, a plant greenness and net primary productivity indicator). Our results showed that ST and SWC were driving factors for the seasonal dynamics of SR rather than plant greenness, irrespective of stand age and forest type. For $\sim$15-year-old stands, the seasonal dynamics of both AR and HR were dependent on ST. Higher Q_{10} of HR compared with AR occurred in larch. However, in Mongolia pine a similar Q_{10} occurred between HR and AR. With stand age, Q_{10} of both HR and AR increased in larch. For Mongolia pine, Q_{10} of HR increased with stand age, but AR showed no significant relationship with ST. As stand age increased, HR was correlated with SWC in Mongolia pine, but for larch AR correlated with SWC. The dependence of AR on NDVI occurred in $\sim$35-year-old Mongolia pine. Our study demonstrated the importance of separating autotrophic and heterotrophic respiration components of SR when stimulating the response of soil carbon efflux to environmental changes. When estimating the response of autotrophic and heterotrophic respiration to environmental changes, the effect of forest type on age-related trends is required.

Editor: Dafeng Hui, Tennessee State University, United States of America

Funding: This research was supported by the National Basic Research Program of China (No. 2010CB950600 and 2013CB956303), the Projects of National Natural Science Foundation of China (No. 31222011, 31270363 and 31070428) and supported by the Foundation for Innovative Research Groups of the National Natural Science Foundation of China (No. 31021001). The funders had no role in study design, data collection and analysis, decision to publish, or preparation of the manuscript.

Competing Interests: The authors have declared that no competing interests exist.

* E-mail: wangw@urban.pku.edu.cn

Introduction

Forest soil respiration (SR) is the primary pathway where plant-fixed CO_2 is released into the atmosphere [1–3]. This occurs from the root activity and their associated mycorrhizal fungi (below-ground autotrophic respiration, AR) and from heterotrophic respiration (HR) [4,5]. Quantifying forest SR, AR and HR components and their environmental controls requires an accurate evaluation of the response of the terrestrial carbon balance to future climate changes [8–10], based on the large annual exchange of carbon between forest ecosystems and the atmosphere [6,7].

Forest age was reported to play an important role in determining the distribution of ecosystem carbon pools, fluxes and carbon sequestration [11–13]. Forest stands of various ages evolve with different abiotic and biotic environments that can differentially respond to environmental changes [14]. Older stands accumulate aboveground litter and root inputs, possibly decreasing HR and increasing AR [15]. Soil carbon dynamics may become more complex with increasing stand age [16]. For instance, the Great Lakes forest chronosequence showed that changes in soil carbon stocks across different aged stands had different patterns, with old-growth stands accumulating carbon in the deep soil layers, but not surface soils [16]. Our limited knowledge in forest succession and carbon cycles has resulted in few large-scale ecosystem carbon models accounting for the change of forest metabolic rates with age [17]. To model the long-term forest carbon dynamics and its coupling with the climate system, we need to understand the response of forest ecosystems to the changing climate, including the role of stand age and the successional status on carbon dynamics [16].

Previous studies on the effect of stand age on forest carbon efflux focused on the measurements of total SR in a single forest type. Researchers have observed that total SR increased [18,19], decreased [15,20], was similar [21,22] or responded non-linearly [23,24] with forest age. However, the effect of forest age on SR may depend on the forest type [25]. For instance, deciduous forests are time-limited each year to photosynthesize and allocate carbon to storage and reproduction compared with evergreen forests [26]. These physiological and phenological differences between deciduous and evergreen forests [27] may modulate the SR response across different aged stands to environmental changes. Additionally, AR and HR may respond differently to environmental changes [28]. AR is strongly influenced by photosynthetic activity compared with HR [28,29,30,31]. Therefore, the ratio of AR and HR (of total SR) will modulate the response of SR across different aged stands to environmental changes [32,33]. For instance, Jassal et al. (2012) observed that SR in younger stands (~21 years) was more responsive to soil temperature (ST) and soil water content (SWC) compared with older stands (~60 years) [34]. We measured SR and partitioned the HR and AR components across three age classes (~15, ~25, and ~35 years) for the evergreen species, Pinus sylvestris var. mongolica (Mongolia pine) and the deciduous species, Larix principis-rupprechtii (larch) from June 2006 to October 2009 in a forest-steppe ecotone, northern China. We analyzed the relationships between SR, HR, AR and ST, SWC and NDVI (normalized difference vegetation index, indicative of plant greenness and net primary productivity). Our objective was to determine if forest type modulated the seasonal dynamics of SR and the AR and HR components and their responses to ST, SWC and NDVI across different aged stands. We tested the following hypotheses: (1) ST, SWC and NDVI will have significant effects on the seasonal dynamics of SR across varied aged stands, irrespective of forest type. This was based on the findings that SR depended on ST, SWC [35,36] and primary productivity [29,37]; (2) HR is more sensitive to ST and SWC than AR, irrespective of forest type. AR is controlled by NDVI, depending on stand age and forest type. The dependence of AR on NDVI was expected to decrease with stand age, because of the perennial life cycle, long carbon transport pathways and high storage capacity in older trees. However, AR dependence on NDVI may increase from deciduous to evergreen forest because of the greater dependence on seasonal accumulation and consumption of stored carbon in deciduous rather than evergreen forest [38].

Materials and Methods

Ethics Statement

The administration of the Saihanba Forestry Center gave permission for this research at each study site. We confirm that the field studies did not involve endangered or protected species.

Site description

The study was conducted at Saihanba Forestry Center, Hebei Province, northern China (117°12′–117°30′ E, 42°10′–42°50′ N, 1400 m a.s.l.) adjacent to the Beijing-Tianjin region. Our study site lay within a typical forest-steppe ecotone in a temperate region. The climate is semi-arid, semi-humid, with long cold winters (November to March) and a short spring and summer. The annual mean air temperature and precipitation from 1964 to 2004 were −1.4°C and 450.1 mm, respectively. The soils are predominantly sandy, accompanied by meadow and marsh-type. The soil has low nutrient content with the organic carbon levels at 0.71–1.88% and total nitrogen at 0.08–0.19%. The bulk density ranged from 0.74 to 1.06 g cm^{-3} with litter and fine root biomass increasing with stand age (Table 1). Primary forests were harvested using large-scale industrial logging techniques in the late 1900s. This area has been threatened from sandstorm since the 1950s. Consequently, the China Forestry Administration proposed the establishment of a large plantation in Saihanba to ensure the environmental safety of the Beijing-Tianjin region. Saihanba forestry was established in 1962 with the plantation now 94,700 hm^2, covering 76.8% of the area. This project has been the largest one in China with P. sylvestris var. mongolica (Mongolia pine) and L. principis-rupprechtii (larch) the most dominant species.

Table 1. Site characteristics of the six forest stands in this study.

Forest stands	Stand age (years)	Tree height (m)	DBH (cm)	Slope/aspect	Elevation (a.s.l.)	Soil texture	Tree density (stem ha^{-1})	SOC (%)	STN (%)	SBD (g cm^{-3})	LFB (g m^{-2})	FRB (g m^{-2})
Larix principis-rupprechtii	~15	3.34	3.1	0°/-	1582	Sandy soil	1900	0.94	0.09	1.06	1360	191
	~25	9.47	11.1	0°/-	1560	Sandy soil	2133	0.95	0.10	0.88	2207	198
	~35	15.67	21.4	0°/-	1500	Sandy soil	1825	1.88	0.18	0.74	3412	220
Pinus sylvestris var. mongolica	~15	2.70	4.3	1°/S	1509	Sandy soil	1800	0.71	0.08	0.98	1335	344
	~25	8.02	12.0	2°/S	1519	Sandy soil	2010	1.27	0.13	0.83	2798	483
	~35	14.70	20.7	1°/S	1502	Sandy soil	1760	1.60	0.19	0.85	2899	496

DBH = Diameter at Breast Height (1.3 m); NDVI = Normalized Difference Vegetation Index; SOC = Soil Organic Carbon; STN = Soil Total Nitrogen; SBD = Soil Bulk Density; LFB = Litter Floor Biomass; FRB = Fine Root Biomass.

Experimental design

To assess the effects of forest age on soil respiration (SR), autotrophic respiration (AR) and heterotrophic respiration (HR) across different forest types, we selected six stands with the same planting density (4995 stem/ha), including three age classes of Mongolia pine and larch. These were pure plantations established on former meadow grassland dominated by *Leymus chinensis*. In combination, these stands made up a chronosequence from 15 to 35 years old, with the oldest stand ready for intermediate cutting. Neither fertilization nor drainage works had been carried out since tree establishment in any of the stands. Moreover, the topography at all six stands is predominantly flat with a similar slope, position and elevation (Table 1). The selected Mongolia pine plantation has a stand area of 1.25 ha (15 years old), 5 ha (25 years old) and 10.14 ha (35 years old). The selected larch plantation had a stand area of 1.05 ha (15 years old), 3.67 ha (25 years old) and 5.11 ha (35 years old). The locations of the stands with different stand ages are shown in Figure 1. Detailed stand information is shown in Table 1. For each stand, three replicate plots were arranged in an area of 20 m×20 m. Five subsamples (i.e. SR collars) were randomly arranged in each plot. The distance between any two stands was ≤10 km, avoiding differences in climate and soil type. Stand age was obtained from forest management records and from core samples using an increment borer. Similar climate and soil properties among these stands create an ideal chronosequence to study age effects on soil carbon efflux.

Soil respiration (SR), soil temperature (ST) and water content (SWC)

SR was measured from June 2006 to October 2009 using a Li-8100 soil CO_2 flux system (LI-COR Inc., Lincoln, NE, USA). During the growing season, five polyvinyl chloride (PVC) collars (10 cm inside diameter, 6 cm height) were inserted 3 cm into the soil in each plot and left *in situ* throughout the study. The five PVC collars were placed in each plot, one in each of the four corners and the fifth in the middle. Live plants inside the collars were clipped at the soil surface 1 day before each measurement. SR was measured every 15–20 days. To minimize the daily variation in SR and represent the daily mean, measurements were made between 0800 and 1100 h [39,40]. For each measurement, the respiration rate was calculated as the mean of three plots per stand. During winter, longer soil collars (determined by snow depth, commonly >30 cm) were inserted into the soil surface and stabilized for 24 h before the SR measurement [41,42]. The duration of winter at our site was 5 months from November to late March with near consistent, continuous mean daily soil temperature <0.5°C at 5 cm [43]. The Li-8100 soil CO_2 flux system was kept in an isolated/heated container to maintain the temperature above freezing. Winter SR was measured monthly except in February.

During respiration measurements, ST was recorded in each collar at 5 cm soil depth with the Li-8100 temperature probe. Continuous measurements of ST were recorded at 30-min intervals with StowAway loggers (Onset Comp. Corp., Bourne, MA, USA) inserted in the soil at each site. SWC at 0–10 cm was measured inside the collars using time domain reflectometry (Soil moisture Equipment Corp., Santa Barbara, CA, USA). No data for SWC were obtained during the winter because the probe could not be fully inserted into the frozen soil.

Harmonic analysis of time-series AVHRR normalized difference vegetation index (NDVI) dataset

NDVI is derived from the red: near-infrared reflectance ratio:

$$NDVI = \frac{(NIR - VIS)}{(NIR + VIS)} \qquad (1)$$

where NIR and VIS represent the spectral reflectance measurements acquired in the near-infrared and visible (red) regions, respectively [44]. Due to their large areas, uniform distribution and sparse understory of both Mongolia pine and larch plantations, the NDVI of 16-Day L3Global 250 m product (MOD13 Q1) could well represent our plot measurements. We acquired the data from June 2006 to October 2009 using the website https://wist.echo.nasa.gov/api.

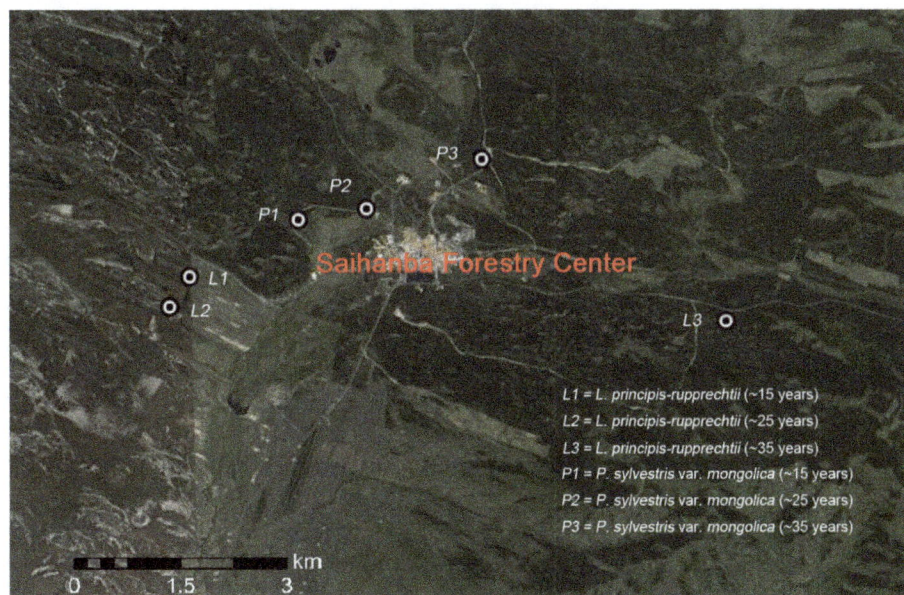

Figure 1. Location of sites showing different aged stands.

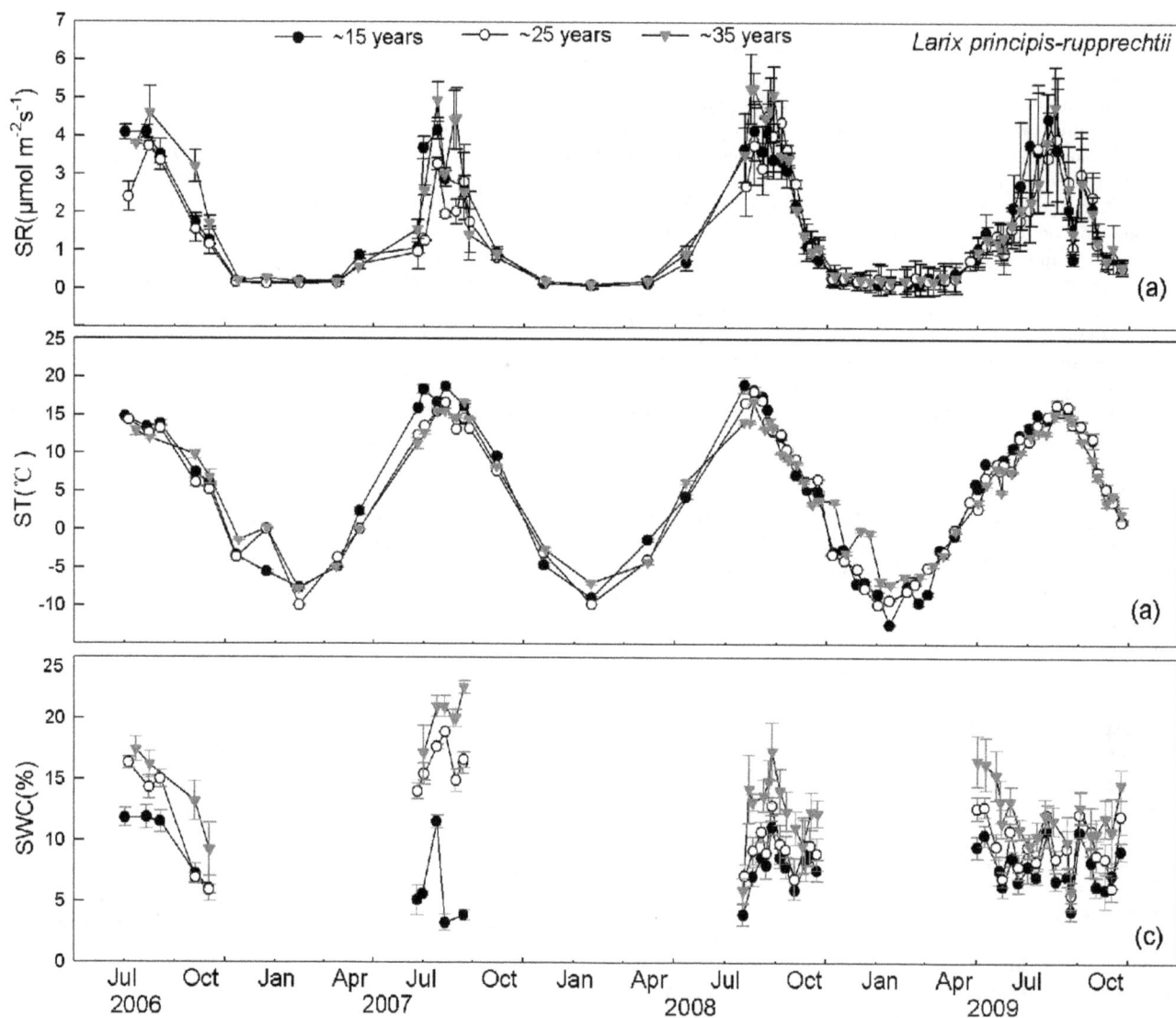

Figure 2. Seasonal changes of soil respiration (SR) and environmental factors across three stand ages of *Larix prinicipis*. SR (a), soil temperature at 5 cm depth (ST) (b) and soil water content at 10 cm depth (SWC) (c) are shown from June 2006 to October 2009.

Harmonic (Fourier) analysis was used to remove the excess noise remaining in the NDVI time-series from satellite-borne data sources and to obtain reasonably smooth continuous daily data [45]. The Fourier series analysis decomposes a signal into an infinite series of harmonic components. Each of these components is composed initially of a sine wave and a cosine wave of equal integer frequency. These two waves are then combined into a single cosine wave with a characteristic amplitude (size of the wave) and phase angle (offset of the wave) [46,47].

Heterotrophic respiration (HR) and autotrophic respiration (AR)

To detect the response of HR and AR to environmental changes, five additional soil collars per plot were placed with deep PVC collars (80 cm^2 and 70 cm deep) in October 2006. The 70-cm-long PVC collars isolated old plant roots, preventing new roots from growing inside the collars. This method has been successfully applied in various ecosystems to separate HR from the total SR [48–50].

To examine the transient response of dead root decomposition, we commenced the CO_2 efflux measurements above the PVC tubes immediately after installation. The interference of installation was eliminated after 1.5 years (from 2006) with the soil CO_2 efflux measured at the stands believed to represent HR. AR was calculated as the difference between SR and HR.

SR, its HR and AR components and their relationships with ST, SWC and NDVI

We examined the relationships between SR, HR, AR and ST by fitting exponential functions to the data from each stand using the following equation:

$$R = R_0 e^{\beta t} \qquad (2)$$

where R is observed SR (plot-wide averages measured periodically throughout the year), t is the concurrent ST (5 cm depth), with R_0 being the basal respiration at temperature of 0°C and β the fitted parameters obtained using least squares nonlinear regression with

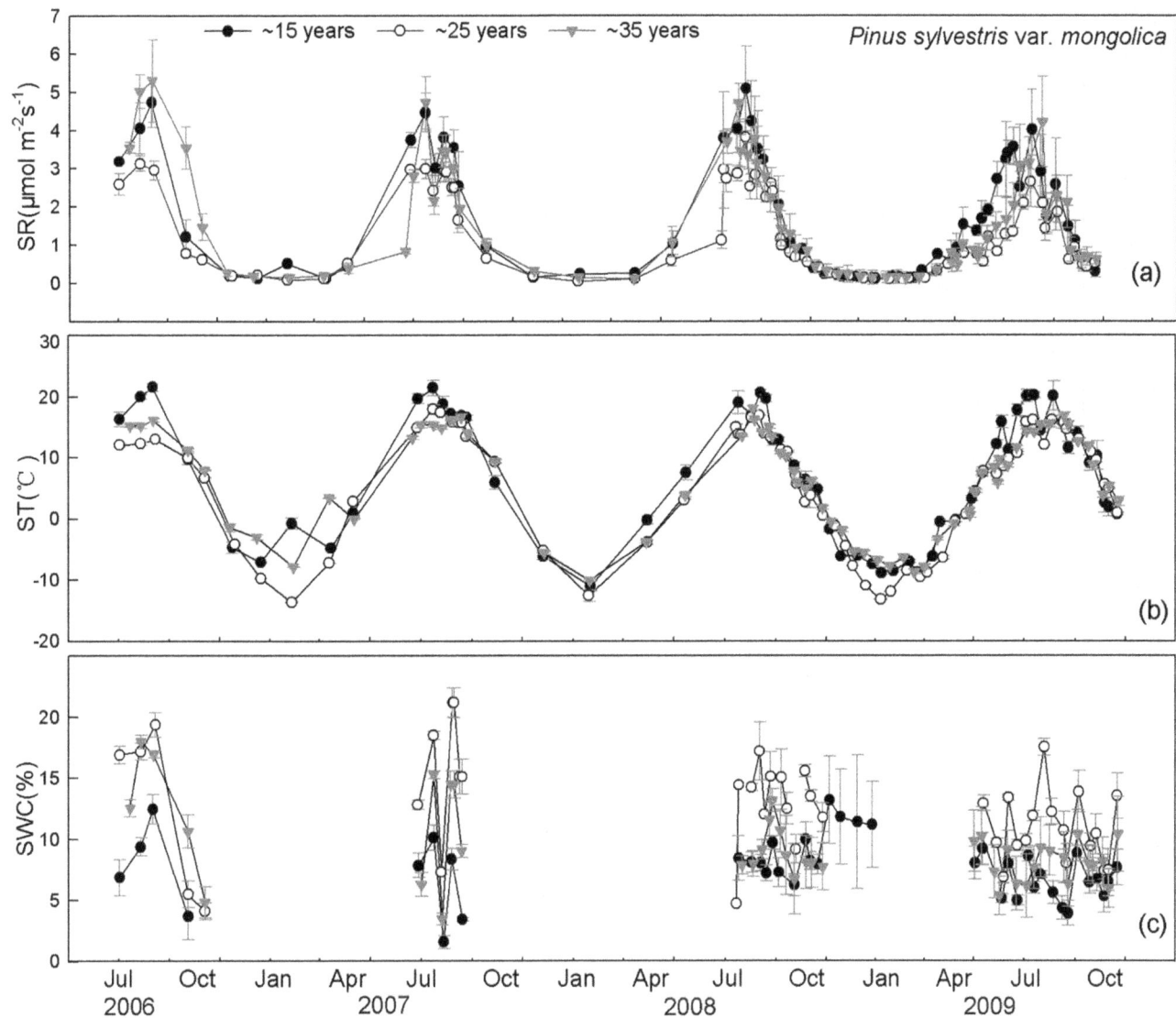

Figure 3. Seasonal changes of soil respiration (SR) and environmental factors across three stand ages of *Pinus sylvestris*. SR (a), soil temperature at 5 cm depth (ST) (b) and soil water content at 10 cm depth (SWC) (c) are shown from June 2006 to October 2009.

SigmaPlot V. 8.02. The β values were used to calculate apparent Q_{10} values, which describes the change in respiration rate over a 10°C increase in t using:

$$Q_{10} = e^{10\beta} \qquad (3)$$

In addition, we calculated R_{15}, the respiration values at a reference temperature of 15°C (without stimulation introduced by photosynthetic activity and without water limitations) [51] according to equation 4, using a Q_{10} coefficient calculated from equation (3). To remove the effect of ST, we examined the relationships between R_{15} and SWC and NDVI.

$$R_{15} = RQ_{10}^{(15-T)/10} \qquad (4)$$

where R_{15} is the respiration flux at a constant temperature of 15°C, R the measured respiration rate, Q_{10} from equation (3) and T the ST at a 5 cm depth.

Statistical analysis

To evaluate if SR, HR and AR significantly differed among different aged stands, forest type and measurement time, we used a three-way ANOVA. The relationships between SR, HR, AR and ST, SWC and NDVI were examined using regression analysis. All statistical analyses were performed with a significance level of 0.05 using SPSS (2009, ver. 18.0, SPSS Inc., Chicago, IL, USA).

Results

Seasonal dynamics of soil respiration (SR)

SR was significantly different among the different time measurements, irrespective of stand age and forest type (Table 2). SR was higher in summer and lower in winter, following the seasonal dynamics of ST, irrespective of stand age and forest type (Figs. 2, 3). Across different aged stands, the seasonal dynamics of SR were exponentially related to ST, possibly explaining the 88%–89% and 89%–95% variation in SR for larch and Mongolia pine, respectively (means for 4 years' measurements, Table 3). Apparent Q_{10} of SR across three stand ages ranged from 3.49–

Table 2. Results of the three-way ANOVA with soil respiration (SR), heterotrophic respiration (HR) and autotrophic respiration (AR) as the response variables respectively, and stand age (~15 years, ~25 years and ~35 years for SR, ~15 years and ~35 years for HR and AR), forest type (*Larix prinicipis* vs. *Pinus sylvestris*) and measurement time (from June 2006 to October 2009 for SR, from August 2008 to August 2009 for HR and AR) as factors.

Factor	SR					HR					AR				
	d.f.	Type III SS	Mean square	F	P	d.f.	Type III SS	Mean square	F	P	d.f.	Type III SS	Mean square	F	P
stand age	2	11.872	5.936	2.975	0.052	1	0.351	0.351	0.426	0.516	1	0.418	0.418	1.258	0.266
forest type	1	0.423	0.423	0.212	0.646	1	0.774	0.774	0.94	0.336	1	5.047	5.047	15.192	<0.01
time	3	34.367	11.456	5.741	<0.01	1	3.469	3.469	4.211	<0.05	1	1.289	1.289	3.879	0.053
stand age × forest type	2	2.071	1.035	0.519	0.596	1	0.012	0.012	0.015	0.904	1	2.344	2.344	7.056	<0.01
stand age × time	6	4.808	0.801	0.402	0.878	1	1.202	1.202	1.459	0.231	1	0.338	0.338	1.017	0.317
forest type × time	3	3.716	1.239	0.621	0.602	1	0.064	0.064	0.078	0.781	1	1.336	1.336	4.021	<0.05
stand age × forest type × time	6	2.39	0.398	0.2	0.977	1	0.771	0.771	0.935	0.337	1	0.11	0.11	0.332	0.567

The data in ~25-year-old stands were absent because of the artificially damaged collars for separating AR and HR during the experiments.

Table 3. Q_{10} of soil respiration (SR), basal respiration rate (R_0) and determination coefficient of exponential relationships between SR and soil temperature (ST), using $SR = R_0 \times e^{\beta T}$. L1, L2, L3 are ~15 years, ~25 years and ~35 years *Larix prinicipis* stand, respectively.

	2006				2007				2008				2009				Mean values for 4 years			
	R_0	β	Q_{10}	R^2	R_0	β	Q_{10}	R^2	R_0	β	Q_{10}	R^2	R_0	β	Q_{10}	R^2	R_0	β	Q_{10}	R^2
L1	0.40	0.17	5.37	0.99	0.39	0.11	3.06	0.90	0.43	0.14	4.06	0.91	0.48	0.12	3.25	0.90	0.45	0.13	3.53	0.89
L2	0.29	0.18	6.30	0.87	0.37	0.12	3.22	0.91	0.49	0.14	3.97	0.94	0.50	0.11	3.10	0.90	0.45	0.13	3.49	0.88
L3	0.31	0.22	8.94	0.98	0.40	0.14	3.86	0.91	0.41	0.18	5.87	0.92	0.52	0.13	3.56	0.93	0.46	0.15	4.31	0.89
P1	0.35	0.13	3.49	1.00	0.38	0.12	3.35	0.97	0.50	0.12	3.25	0.94	0.42	0.11	3.10	0.95	0.43	0.12	3.19	0.95
P2	0.45	0.12	3.39	0.85	0.37	0.12	3.29	0.97	0.46	0.12	3.46	0.91	0.35	0.10	2.72	0.91	0.40	0.11	3.10	0.89
P3	0.36	0.17	5.53	0.98	0.38	0.12	3.39	0.85	0.50	0.14	3.94	0.94	0.39	0.12	3.35	0.94	0.42	0.13	3.67	0.90

P1, P2, P3 are ~15 years, ~25 years and ~35 years *Pinus sylvestris* stand, respectively.

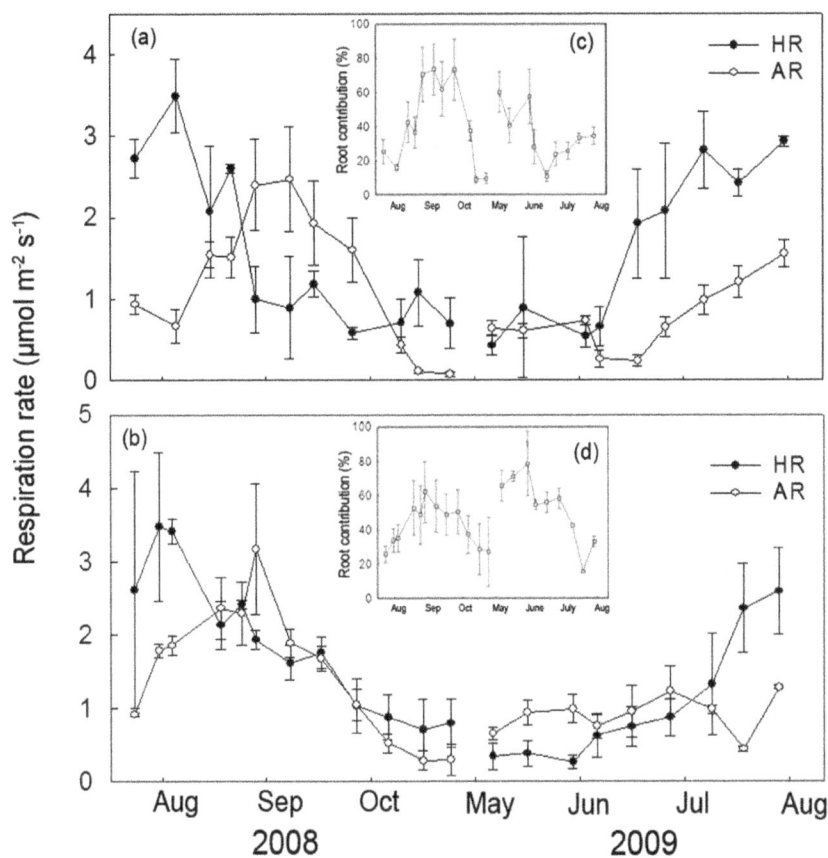

Figure 4. Seasonal changes in autotrophic and heterotrophic components of soil respiration (SR) for different aged stands of *Larix prinicipis.* Autotrophic respiration (AR) and heterotrophic respiration (HR) in ~15-year-old (a) and ~35-year-old stands (b) are shown from August 2008 to August 2009. Seasonal changes in the contribution of AR to SR are shown in inserted figures for ~15-year-old (c) and ~35-year-old (d) stands. The data in the ~25-year-old stand were absent because of the artificially damaged collars for separating AR and HR during the experiments.

4.31 and 3.10–3.67 for larch and Mongolia pine, respectively (means of 4 years' measurements, Table 3). Apparent Q_{10} of ~35-year-old stand was significantly higher compared with ~15 and ~25-year-old stands, irrespective of forest type (Table 3). Soil basal respiration (R_0) was similar across three stand ages of both forest types (Table 3).

SR_{15} showed no significant relationship with NDVI, irrespective of forest age and type ($P>0.05$ for all stands, Fig. S1). Conversely, a significant correlation was found between SR_{15} and SWC, which could explain the variation of 15%–26% and 10%–33% variation for larch and Mongolia pine, respectively (Fig. S2).

Seasonal dynamics of heterotrophic respiration (HR) and autotrophic respiration (AR)

HR was significantly different among different time measurements, irrespective of stand age and forest type (Table 2). The seasonal dynamics of HR were similar across different aged stands, showing higher values in summer and lower in spring and autumn (Figs. 4, 5). However, AR showed a more variable pattern. For larch, maximal AR occurred in early September with minimal values in late October, irrespective of stand age (Fig. 4a, 4b). The contribution of AR to total SR ranged from 9%–74% and 16%–79% (~15 and ~35-year-old-stands, respectively; Fig. 4c and d). HR, AR and the contribution of AR to total SR showed no significant differences between ~15 and ~35-year-old larch ($P>0.05$ for all the comparisons, Table 2). For Mongolia pine,

AR peaked in early August and late August (~15 and ~35-year-old-stands, respectively), with minimum AR in late October for both stands (Fig. 5a, 5b). The contribution of AR to total SR ranged from 16%–63% (Fig. 5c) for ~15 and 8%–79% for ~35-year-old stands (Fig. 5d). AR and the contribution of AR to total SR was significantly lower in the ~35 compared with ~15-year-old stand (39% *vs.* 22%; $P<0.05$). HR showed no significant differences between the two stands ($P>0.05$, Table 2). AR was significantly different between the two forest types, with a significant interaction effect among forest type, stand age and measurement time (Table 2).

The seasonal dynamics of AR and HR were dependent on ST in the ~15-year-old stand, which could explain the 22% and 69% variation in AR and HR, respectively for larch (Fig. 6a–6d). Additionally, this could explain the 68% and 61% variation in AR and HR (respectively) for Mongolia pine (Fig. 7a–7d). We found a higher Q_{10} for HR compared with AR in larch (3.85 *vs.* 2.98). In contrast, Mongolia pine showed a similar Q_{10} for HR compared with AR (2.61 *vs.* 2.76). As the stand age increased, the Q_{10} of both HR (3.85, ~15-year-old *vs.* 5.07, ~35-year-old) and AR (2.98, ~15-year-old *vs.* 3.23, ~35 years) increased in larch (Fig. 5a, 5b). For Mongolia pine, the Q_{10} of HR increased (2.61, ~15-year-old *vs.* 2.71, ~35-year-old), but AR showed no significant relationship with ST in ~35-year-old stand (Fig. 7e). In ~15-year-old stand, neither HR nor AR was controlled by SWC, irrespective of forest type. As stand age increased, HR correlated with SWC in Mongolia pine (Fig. 7f). In contrast, larch AR was significantly

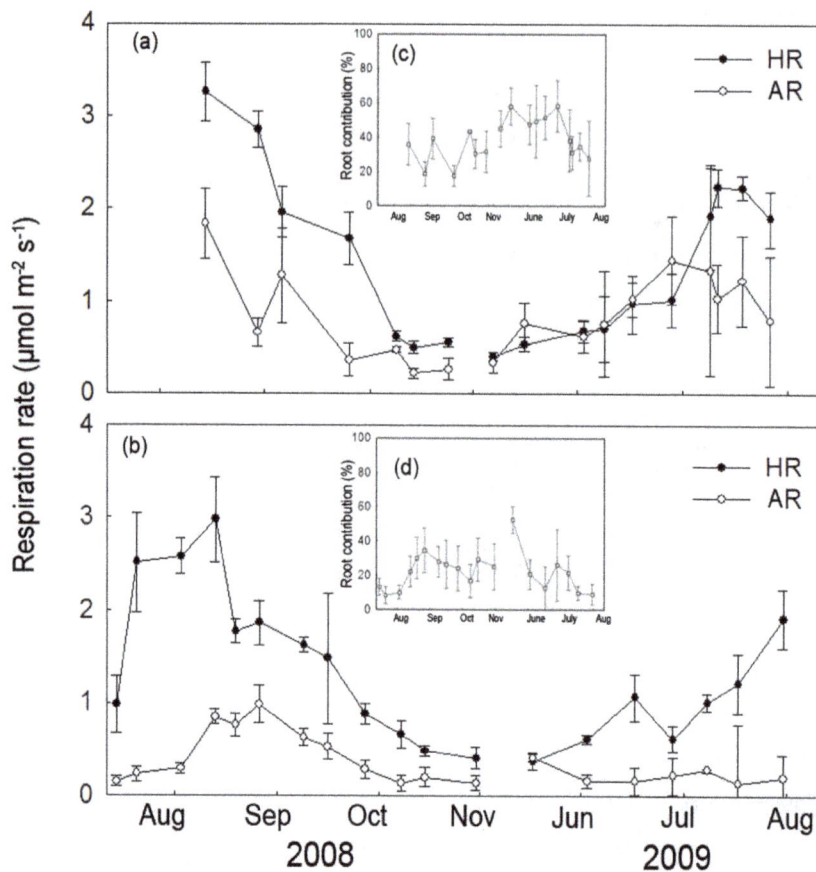

Figure 5. Seasonal changes in autotrophic and heterotrophic components of soil respiration (SR) for different aged stands of *Pinus sylvestris.* Autotrophic respiration (AR) and heterotrophic respiration (HR) in ~15-year-old (a) and ~35-year-old stands (b) are shown from August 2008 to August 2009. Seasonal changes in the contribution of AR to SR are shown in inserted figures for ~15-year-old (c) and ~35-year-old (d) stands. The data in the ~25-year-old stand were absent because of the artificially damaged collars for separating AR and HR during the experiments.

related with SWC ($P<0.01$, Fig. 6g). The dependence of AR on NDVI only occurred in ~35-year-old Mongolia pine (Fig. 6d, 6h, Fig. 7d, 7h).

Discussion

Drivers of seasonal dynamics for soil respiration (SR)

We hypothesized (1) that ST, SWC and NDVI will impose significant effects on the seasonal dynamics of SR across different aged stands, irrespective of forest type. However, we found that the seasonal dynamics of SR were principally driven by ST and SWC, not by NDVI, across different aged stands, irrespective of forest type. Forest SR is closely related to gross primary productivity (GPP) and leaf area index (LAI) [52,53], suggesting a coupling between the CO_2 assimilation in the forest canopy and CO_2 released from the soil. NDVI has shown a close correlation with GPP [54,55] and LAI [56]. As a convenient proxy for productivity, NDVI was expected to be positively associated with AR because of the dependence on photosynthesis [52]. We expected a positive relationship between SR and NDVI based on the higher contribution of AR to SR (Figs. 4, 5). We found strong positive relationships between SR and NDVI (Fig. S3); however, these relationships disappeared when we used R_{15} for both forest types across three different aged stands (Fig. S1). These results imply that the correlation between SR and NDVI may occur because of the significant relationship between ST and NDVI (Fig.

S4). Our results emphasized the significance of ST and SWC on simulating total SR for both forest types across different aged stands. Our results were inconsistent with recent evidence showing that SR was closely related to canopy photosynthesis at various timescales [57,58]. The discrepancies may occur because NDVI measures integrated past photosynthetic activity rather than present photosynthetic activity [59]. In addition, differences may have occurred because the seasonal change in foliar photosynthetic capacity and NDVI are not necessarily coincidence [53]. Moreover, the temporal dynamics of SR were reported to lag behind those of NDVI [33,60], so our monthly SR measurements may not reflect the real relationship between these two variables. Therefore, high-frequency measurements should be conducted to explore the control of temporal dynamics in SR.

Drivers of seasonal dynamics of heterotrophic respiration (HR) and autotrophic respiration (AR)

In our second hypothesis, we expected greater sensitivity in HR to variation in ST compared with AR. However, we observed a mixed result. In ~15-year-old stand, we found a higher Q_{10} for HR compared with AR in larch and a similar Q_{10} for AR with HR in Mongolia pine (Fig. 6a, 7a). The lower Q_{10} of AR compared with HR in larch may be attributed to differing plant physiology and phenology in Mongolia pine. Deciduous trees are time-limited each year to photosynthesize, allocating more carbon for aboveground growth during summer when ST is highest [26].

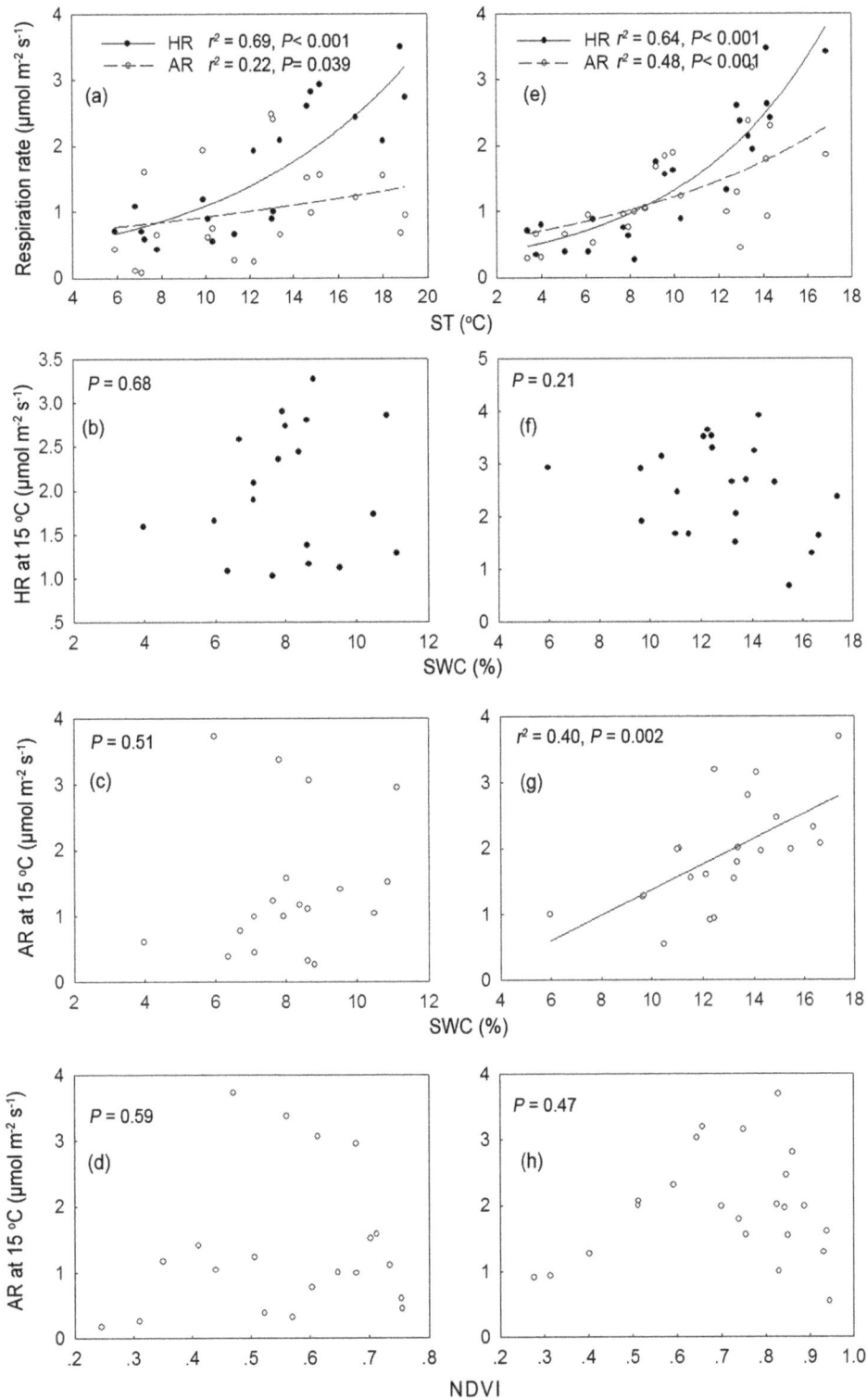

Figure 6. Relationships between autotrophic respiration (AR) and heterotrophic respiration (HR) and environmental factors across different aged stands of _Larix prinicipis._ The relationships are shown between autotrophic respiration (AR) and heterotrophic respiration (HR) and soil temperature (ST) for ~15-year-old (a), ~35-year-old stands (e), between soil water content (SWC) and HR_{15} for ~15-year-old (b), ~35-year-old stands (f), between SWC and AR_{15} for ~15-year-old (c) and ~35-year-old stands (g), and between NDVI and AR_{15} for ~15-year-old (d), and ~35-year-old (h) stands of _Larix prinicipis_. The data in the ~25-year-old stand were absent because of the artificially damaged collars for separating AR and HR during the experiments.

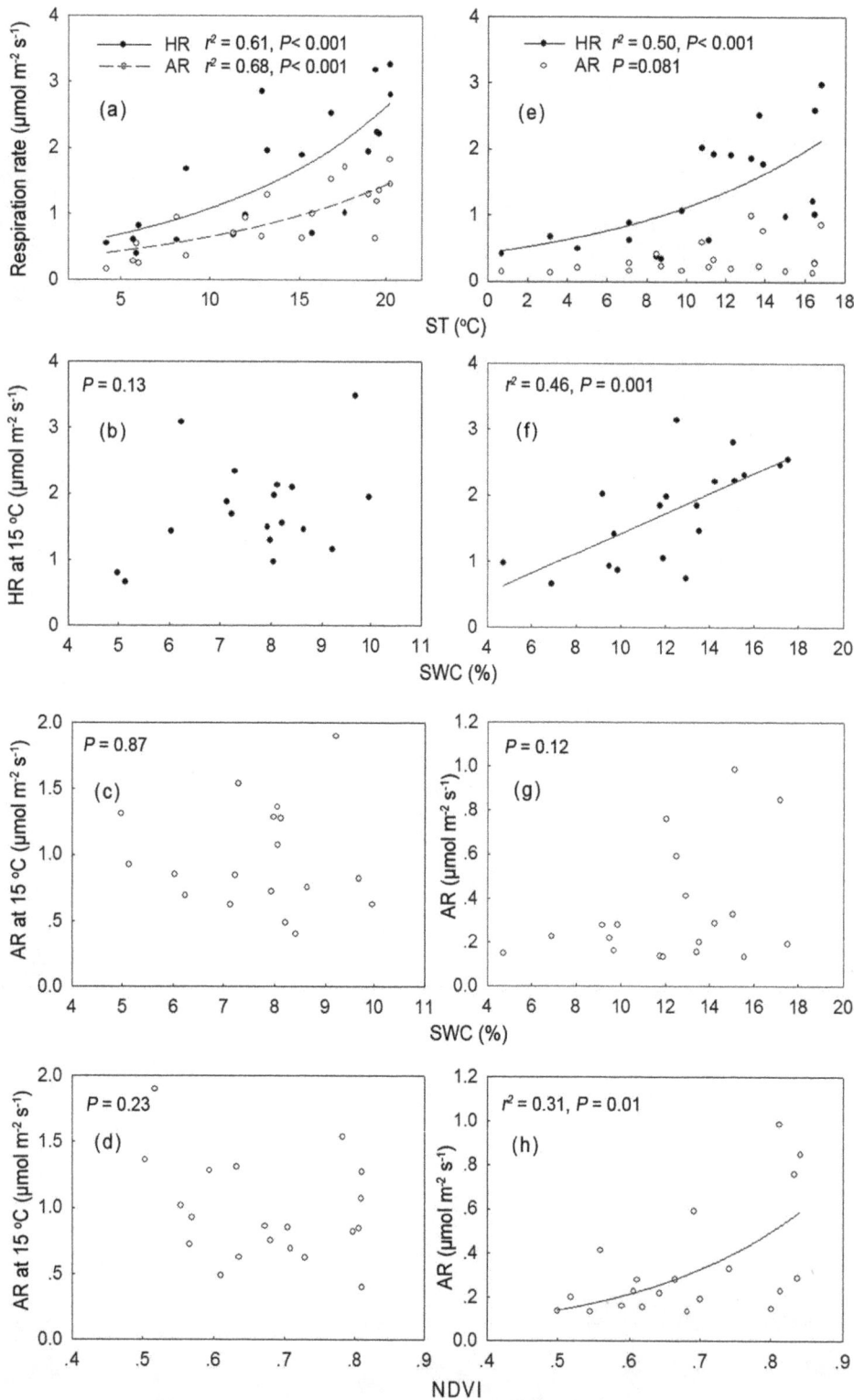

Figure 7. Relationships between autotrophic respiration (AR) and heterotrophic respiration (HR) and environmental factors across different aged stands of *Pinus sylvestris.* The relationships are shown between autotrophic respiration (AR) and heterotrophic respiration (HR) and soil temperature (ST) for ~15-year-old (a), ~35-year-old stands (e), between soil water content (SWC) and HR_{15} for ~15 years (b) and ~35 years (f); between SWC and AR_{15} for ~15-year-old (c) and ~35-year-old stands (g); between NDVI and AR_{15} for ~15-year-old stand (d), and between NDVI and AR for ~35-year-old stands (h) of *Pinus sylvestris.* The data in the ~25-year-old stand were absent because of the artificially damaged collars for separating AR and HR during the experiments.

Furthermore, peak root growth commonly occurs in spring and autumn for coniferous forests [61,62]. We observed the contribution of AR to total SR was higher in spring and autumn for larch. Therefore, the seasonal dynamics of AR in larch may be connected with those of root growth, resulting in lower Q_{10} of AR compared with HR. In contrast, Mongolia pine can photosynthesize year-round with little investment required in highly efficient photosynthetic activity. Consequently, the dynamics of AR in Mongolia pine stands may be associated with higher metabolic demand during leaf production [63]. We also found maximal AR in early summer in Mongolia pine (Fig. 5a), which was consistent with Lee et al.(2010), who observed maximal contribution of AR to total SR in summer for evergreen forest and in autumn for deciduous forest [64].

According to our second hypothesis, we expected that the seasonal dynamics of AR were more dependent on NDVI rather than HR. However, we found that AR was significantly affected by NDVI only in ~35-year-old Mongolia pine (Fig. 7h). We did not observe a significant relationship between AR and NDVI in larch, possibly because of greater carbon allocation in deciduous forest to storage and reproductive functions compared with evergreen forests [26]. The relationship between AR and NDVI in deciduous trees was more dependent on seasonal accumulation and consumption of stored C compared with evergreen trees [65]. The absence of a significant relationship in ~15-year-old Mongolia pine stand may be attributed to the soil water deficit, inducing a selection pressure favoring trees presenting high storage capability for reserve compounds to combat this stress [66,67]. Moreover, the carbon balance between growth and storage remained constant between age classes in evergreen forest [26]. Our results suggested that the correlation between AR and NDVI was dependent upon stand age, and this stand age-related effect was forest type-dependent.

We expected a higher dependence of HR on SWC compared with AR because AR was reported as less responsive to water variability compared with HR [68,69]. However, we found in ~15-year-old stand that neither HR nor AR was controlled by SWC, irrespective of forest type. This may occur because of the narrower variation in SWC in young forest characterized by high soil water evaporation [70]. With increased stand age, HR correlated with SWC in Mongolia pine (Fig. 7f). In contrast, larch AR was significantly related to SWC (Fig. 6g). Similar observations were reported by Lee et al. (2010), who found a significant correlation between AR and SWC in Quercus-dominated deciduous stand, not evergreen stands of Abies holophylla. One possible reason for this difference between Mongolia pine and larch in the dependence of AR and HR on SWC may be associated with reduced root longevity of deciduous tree species (<1 year) compared with evergreen species (<1–12 years) [71,72]. Higher turnover rates in deciduous tree roots may induce a rapid change in response to environmental fluctuations compared with evergreen trees. Therefore, our results suggest the response of AR and HR to SWC depended on stand age, with the effects of stand age dependent on forest type.

Conclusions

In summary, we tested the effects of forest type on SR, AR and HR across different aged stands in the same location to prevent confounding from climatic and edaphic conditions. We observed that ST and SWC were significant factors controlling the seasonal dynamics of total SR, irrespective of forest age and type. However, the response of AR and HR to ST, SWC and NDVI differed across various aged stands, with the effects of stand age dependent on the forest type. These results suggest that in stimulating the response of forest carbon efflux to environmental changes, we should consider the effects of stand age and the influence of forest type on this age-related trend. Furthermore, our study emphasized considering the response of HR and AR to environmental changes separately when predicting the response of soil carbon efflux in different aged forests to global climate changes. Future research should attempt an in-depth understanding of the effects of more functional types on carbon efflux across different aged stands.

Acknowledgments

We thank two anonymous reviewers for their constructive suggestions on an early version of this manuscript.

Supporting Information

Figure S1 Relationships between soil respiration at 15°C (SR_{15}) and NDVI across different aged stands. The relationships are shown for ~15 (a), ~25 (b), ~35-year-old stands (c) of Larix prinicipis, ~15 (d), ~25 (e), and ~35 year-old stands (f) of Pinus sylvestris.

Figure S2 Relationships between soil respiration at 15°C (SR_{15}) and soil water content (SWC) across different aged stands. The relationships are shown for ~15 (a), ~25 (b) and ~35-year-old stands (c) of Larix prinicipis, ~15 (d), ~25 (e) and ~35- year-old stands (f) of Pinus sylvestris.

Figure S3 Relationships between soil respiration (SR) and NDVI across different aged stands. The relationships are shown for ~15 (a), ~25 (b) and ~35- year-old stands (c) of Larix prinicipis, ~15 (d), ~25 (e) and ~35-year-old stands (f) of Pinus sylvestris.

Figure S4 Relationships between ST (soil temperature at 5 cm depth) and NDVI for non-growing season (dotted line) and growing season (solid line). The relationships are shown for ~15 (a) and ~25-year-old stands (b) of Larix prinicipis, ~15 (c), ~25 (d), and ~35-year-old stands (e) of Pinus sylvestris. The data of ~35-year-old stands L. prinicipis were absent because of the damage of StowAway loggers inserted in the soil.

Author Contributions

Conceived and designed the experiments: WW WC. Performed the experiments: WW WC. Analyzed the data: WW WZ WC. Contributed reagents/materials/analysis tools: WW YY HZ. Wrote the paper: WW WZ WC.

References

1. Bond-Lamberty B, Thomson A (2010) Temperature-associated increases in the global soil respiration record. Nature 464: 579–582.

2. Wang W, Chen W, Wang S (2010) Forest soil respiration and its heterotrophic and autotrophic components: Global patterns and responses to temperature and precipitation. Soil Biology & Biochemistry 42: 1236–1244.

3. Zhou YM, Li MH, Cheng XB, Wang CG, Fan AN, et al. (2010) Soil respiration in relation to photosynthesis of quercus mongolica trees at elevated CO_2. Plos one 5:e15134. doi:10.1371/journal.pone.0015134.

4. Hanson PJ, Edwards NT, Garten CT, Andrews JA (2000) Separating root and soil microbial contributions to soil respiration: A review of methods and observations. Biogeochemistry 48: 115–146.

5. Subke J-A, Inglima I, Cotrufo MF (2006) Trends and methodological impacts in soil CO_2 efflux partitioning: A metaanalytical review. Global Change Biology 12: 921–943.

6. Bonan GB (2008) Forests and climate change: Forcings, feedbacks, and the climate benefits of forests. Science 320: 1444–1449.

7. Luyssaert S, Schulze ED, Boerner A, Knohl A, Hessenmoeller D, et al. (2008) Old-growth forests as global carbon sinks. Nature 455: 213–215.

8. Cox PM, Betts RA, Jones CD, Spall SA, Totterdell IJ (2000) Acceleration of global warming due to carbon-cycle feedbacks in a coupled climate model. Nature 408: 184–187.

9. Davidson EA, Janssens IA (2006) Temperature sensitivity of soil carbon decomposition and feedbacks to climate change. Nature 440: 165–173.

10. Mahecha MD, Reichstein M, Carvalhais N, Lasslop G, Lange H, et al. (2010) Global convergence in the temperature sensitivity of respiration at ecosystem level. Science 329: 838–840.

11. Pregitzer KS, Euskirchen ES (2004) Carbon cycling and storage in world forests: biome patterns related to forest age. Global Change Biology 10: 2052–2077.

12. Zhou G, Liu S, Li Z, Zhang D, Tang X, et al. (2006) Old-growth forests can accumulate carbon in soils. Science 314: 1417–1417.

13. Yang Y, Luo Y, Finzi AC (2011) Carbon and nitrogen dynamics during forest stand development: a global synthesis. New Phytologist 190: 977–989.

14. Hasselquist NJ, Allen MF, Santiago LS (2010) Water relations of evergreen and drought-deciduous trees along a seasonally dry tropical forest chronosequence. Oecologia 164: 881–890.

15. Saiz G, Byrne KA, Butterbach-Bahl K, Kiese R, Blujdeas V, et al. (2006) Stand age-related effects on soil respiration in a first rotation Sitka spruce chronosequence in central Ireland. Global Change Biology 12: 1007–1020.

16. Tang J, Bolstad PV, Martin JG (2009) Soil carbon fluxes and stocks in a Great Lakes forest chronosequence. Global Change Biology 15: 145–155.

17. Coomes DA, Holdaway RJ, Kobe RK, Lines ER, Allen RB (2012) A general integrative framework for modelling woody biomass production and carbon sequestration rates in forests. Journal of Ecology 100: 42–64.

18. Smith DR, Kaduk JD, Balzter H, Wooster MJ, Mottram GN, et al. (2010) Soil surface CO_2 flux increases with successional time in a fire scar chronosequence of Canadian boreal jack pine forest. Biogeosciences 7: 1375–1381.

19. Odum EP (1969) Strategy of ecosystem development. Science 164: 262–270.

20. Martin JG, Bolstad PV (2005) Annual soil respiration in broadleaf forests of northern Wisconsin: influence of moisture and site biological, chemical, and physical characteristics. Biogeochemistry 73: 149–182.

21. Pypker TG, Fredeen AL (2003) Below ground CO_2 efflux from cut blocks of varying ages in sub-boreal British Columbia. Forest Ecology and Management 172: 249–259.

22. Yermakov Z, Rothstein DE (2006) Changes in soil carbon and nitrogen cycling along a 72-year wildfire chronosequence in Michigan jack pine forests. Oecologia 149: 690–700.

23. Wang CK, Bond-Lamberty B, Gower ST (2002) Soil surface CO2 flux in a boreal black spruce fire chronosequence. Journal of Geophysical Research-Atmospheres 108: DOI: 10.1029/2001JD000861.

24. Law BE, Sun OJ, Campbell J, Van Tuyl S, Thornton PE (2003) Changes in carbon storage and fluxes in a chronosequence of ponderosa pine. Global Change Biology 9: 510–524.

25. Campbell JL, Law BE (2005) Forest soil respiration across three climatically distinct chronosequences in Oregon. Biogeochemistry 73: 109–125.

26. Genet H, Breda N, Dufrene E (2010) Age-related variation in carbon allocation at tree and stand scales in beech (Fagus sylvatica L.) and sessile oak (Quercus petraea (Matt.) Liebl.) using a chronosequence approach. Tree Physiology 30: 177–192.

27. Falge E, Baldocchi D, Tenhunen J, Aubinet M, Bakwin P, et al. (2002) Seasonality of ecosystem respiration and gross primary production as derived from FLUXNET measurements. Agricultural and Forest Meteorology 113: 53–74.

28. Boone RD, Nadelhoffer KJ, Canary JD, Kaye JP (1998) Roots exert a strong influence on the temperature sensitivity of soil respiration. Nature 396: 570–572.

29. Hogberg P, Nordgren A, Buchmann N, Taylor AFS, Ekblad A, et al. (2001) Large-scale forest girdling shows that current photosynthesis drives soil respiration. Nature 411: 789–792.

30. Fang CM, Smith P, Moncrieff JB, Smith JU (2005) Similar response of labile and resistant soil organic matter pools to changes in temperature. Nature 433: 57–59.

31. Knorr W, Prentice IC, House JI, Holland EA (2005) Long-term sensitivity of soil carbon turnover to warming. Nature 433: 298–301.

32. Gong JR, Ge ZW, An R, Duan QW, You X, et al. (2012) Soil respiration in poplar plantations in northern China at different forest ages. Plant and Soil 360: 109–122.

33. Oishi AC, Palmroth S, Butnor JR, Johnsen KH, Oren R (2013) Spatial and temporal variability of soil CO_2 efflux in three proximate temperate forest ecosystems. Agricultural and Forest Meteorology 171: 256–269.

34. Jassal RS, Black TA, Nesic Z (2012) Biophysical controls of soil CO_2 efflux in two coastal Douglas-fir stands at different temporal scales. Agricultural and Forest Meteorology 153: 134–143.

35. Bond-Lamberty B, Wang CK, Gower ST (2004) A global relationship between the heterotrophic and autotrophic components of soil respiration? Global Change Biology 10: 1756–1766.

36. Gaumont-Guay D, Black TA, Griffis TJ, Barr AG, Morgenstern K, et al. (2006) Influence of temperature and drought on seasonal and interannual variations of soil, bole and ecosystem respiration in a boreal aspen stand. Agricultural and Forest Meteorology 140: 203–219.

37. Moyano FE, Kutsch WL, Rebmann C (2008) Soil respiration fluxes in relation to photosynthetic activity in broad-leaf and needle-leaf forest stands. Agricultural and Forest Meteorology 148: 135–143.

38. Kuzyakov Y, Gavrichkova O (2010) REVIEW: Time lag between photosynthesis and carbon dioxide efflux from soil: a review of mechanisms and controls. Global Change Biology 16: 3386–3406.

39. Pang X, Bao W, Zhu B, Cheng W (2013) Responses of soil respiration and its temperature sensitivity to thinning in a pine plantation. Agricultural and Forest Meteorology 171: 57–64.

40. Shi WY, Zhang JG, Yan MJ, Yamanaka N, Du S (2012) Seasonal and diurnal dynamics of soil respiration fluxes in two typical forests on the semiarid Loess Plateau of China: Temperature sensitivities of autotrophs and heterotrophs and analyses of integrated driving factors. Soil Biology and Biochemistry 52: 99–107.

41. Kurganova I, De Gerenyu VL, Rozanova L, Sapronov D, Myakshina T, et al. (2003) Annual and seasonal CO_2 fluxes from Russian southern taiga soils. Tellus Series B-Chemical and Physical Meteorology 55: 338–344.

42. Elberling B (2007) Annual soil CO_2 effluxes in the High Arctic: The role of snow thickness and vegetation type. Soil Biology & Biochemistry 39: 646–654.

43. Grogan P, Jonasson S (2006) Ecosystem CO_2 production during winter in a Swedish subarctic region: the relative importance of climate and vegetation type. Global Change Biology 12: 1479–1495.

44. Gamon JA, Field CB, Goulden ML, Griffin KL, Hartley AE, et al. (1995) Relationships between NDVI, canopy structure, and photosynthesis in 3 Californian vegetation types. Ecological Applications 5: 28–41.

45. Jakubauskas ME, Legates DR, Kastens JH (2001) Harmonic analysis of time-series AVHRR NDVI data. Photogrammetric Engineering and Remote Sensing 67: 461–470.

46. Verhoef W, Menenti M, Azzali S (1996) A colour composite of NOAA-AVHRR-NDVI based on time series analysis (1981–1992). International Journal of Remote Sensing 17: 231–235.

47. Azzali S, Menenti M (2000) Mapping vegetation-soil-climate complexes in southern Africa using temporal Fourier analysis of NOAA-AVHRR NDVI data. International Journal of Remote Sensing 21: 973–996.

48. Vogel JG, Valentine DW (2005) Small root exclusion collars provide reasonable estimates of root respiration when measured during the growing season of installation. Canadian Journal of Forest Research 35: 2112–2117.

49. Wan SQ, Hui DF, Wallace L, Luo YQ (2005) Direct and indirect effects of experimental warming on ecosystem carbon processes in a tallgrass prairie. Global Biogeochemical Cycles 19: DOI:10.1029/2004GB002315.

50. Zhou X, Wan S, Luo Y (2007) Source components and interannual variability of soil CO_2 efflux under experimental warming and clipping in a grassland ecosystem. Global Change Biology 13: 761–775.

51. Migliavacca M, Reichstein M, Richardson AD, Colombo R, Sutton MA, et al. (2011) Semiempirical modeling of abiotic and biotic factors controlling ecosystem respiration across eddy covariance sites. Global Change Biology 17: 390–409.

52. Janssens IA, Lankreijer H, Matteucci G, Kowalski AS, Buchmann N, et al. (2001) Productivity overshadows temperature in determining soil and ecosystem respiration across European forests. Global Change Biology 7: 269–278.

53. Ide R, Oguma H (2010) Use of digital cameras for phenological observations. Ecological Informatics 5: 339–347.

54. Ahrends HE, Etzold S, Kutsch WL, Stoeckli R, Bruegger R, et al. (2009) Tree phenology and carbon dioxide fluxes: use of digital photography at for process-based interpretation the ecosystem scale. Climate Research 39: 261–274.

55. Richardson AD, Braswell BH, Hollinger DY, Jenkins JP, Ollinger SV (2009) Near-surface remote sensing of spatial and temporal variation in canopy phenology. Ecological Applications 19: 1417–1428.

56. Tagesson T, Eklundh L, Lindroth A (2009) Applicability of leaf area index products for boreal regions of Sweden. International Journal of Remote Sensing 30: 5619–5632.

57. Ekblad A, Hogberg P (2001) Natural abundance of ^{13}C in CO_2 respired from forest soils reveals speed of link between tree photosynthesis and root respiration. Oecologia 127: 305–308.

58. Liu Q, Edwards NT, Post WM, Gu L, Ledford J, et al. (2006) Temperature-independent diel variation in soil respiration observed from a temperate deciduous forest. Global Change Biology 12: 2136–2145.

59. Gamon JA, Field CB, Goulden ML, Griffin KL, Hartley AE, et al. (1995) Relationships between NDVI, canopy structure, and photosynthesis in 3 Californian vegetation types. Ecological Applications 5: 28–41.

60. Bond-Lamberty B, Bunn AG, Thomson AM (2012) Multi-year lags between forest browning and soil respiration at high Northern latitudes. Plos one 7: e50441.

61. Cisneros-Dozal LM, Trumbore S, Hanson PJ (2006) Partitioning sources of soil-respired CO_2 and their seasonal variation using a unique radiocarbon tracer. Global Change Biology 12: 194–204.

62. Scott-Denton LE, Rosenstiel TN, Monson RK (2006) Differential controls by climate and substrate over the heterotrophic and rhizospheric components of soil respiration. Global Change Biology 12: 205–216.

63. Liang N, Hirano T, Zheng ZM, Tang J, Fujinuma Y (2010) Soil CO_2 efflux of a larch forest in northern Japan. Biogeosciences 7: 3447–3457.

64. Lee N-y, Koo J-W, Noh NJ, Kim J, Son Y (2010) Seasonal variation in soil CO_2 efflux in evergreen coniferous and broad-leaved deciduous forests in a cool-temperate forest, central Korea. Ecological Research 25: 609–617.

65. Kuptz D, Matyssek R, Grams TEE (2011) Seasonal dynamics in the stable carbon isotope composition delta^{13}C from non-leafy branch, trunk and coarse root CO_2 efflux of adult deciduous (*Fagus sylvatica*) and evergreen (*Picea abies*) trees. Plant Cell and Environment 34: 363–373.

66. Yordanov I, Velikova V, Tsonev T (2000) Plant responses to drought, acclimation, and stress tolerance. Photosynthetica 38: 171–186.

67. Chen G, Yang Y, Guo J, Xie J, Yang Z (2011) Relationships between carbon allocation and partitioning of soil respiration across world mature forests. Plant Ecology 212: 195–206.

68. Carbone MS, Winston GC, Trumbore SE (2008) Soil respiration in perennial grass and shrub ecosystems: Linking environmental controls with plant and microbial sources on seasonal and diel timescales. Journal of Geophysical Research-Biogeosciences 113: DOI: 10.1029/2007JG000611.

69. Muhr J, Borken W (2009) Delayed recovery of soil respiration after wetting of dry soil further reduces C losses from a Norway spruce forest soil. Journal of Geophysical Research-Biogeosciences 114: DOI: 10.1029/2009JG000998.

70. Drake P, Mendham D, White D, Ogden G, Dell B (2012) Water use and water-use efficiency of coppice and seedling *Eucalyptus globulus Labill.*: a comparison of stand-scale water balance components. Plant and Soil 350: 221–235.

71. Vogt KA, Grier CC, Vogt DJ (1986) Production, turnover, and nutrient dynamics of aboveground and belowground detritus of world forests. Advances in Ecological Research 15: 303–377.

72. Lyr H, Hoffman G (1967) Growth rates and growth periodicity of tree roots. International Review of Forestry Research 2: 181–236.

9

Precipitation Regime Shift Enhanced the Rain Pulse Effect on Soil Respiration in a Semi-Arid Steppe

Liming Yan[1,2], **Shiping Chen**[1]*, **Jianyang Xia**[3], **Yiqi Luo**[3]

1 State Key Laboratory of Vegetation and Environmental Change, Institute of Botany, Chinese Academy of Sciences, Beijing, China, **2** School of Life Sciences, Fudan University, Shanghai, China, **3** Department of Microbiology and Botany, University of Oklahoma, Norman, Oklahoma, United States of America

Abstract

The effect of resource pulses, such as rainfall events, on soil respiration plays an important role in controlling grassland carbon balance, but how shifts in long-term precipitation regime regulate rain pulse effect on soil respiration is still unclear. We first quantified the influence of rainfall event on soil respiration based on a two-year (2006 and 2009) continuously measured soil respiration data set in a temperate steppe in northern China. In 2006 and 2009, soil carbon release induced by rainfall events contributed about 44.5% (83.3 g C m^{-2}) and 39.6% (61.7 g C m^{-2}) to the growing-season total soil respiration, respectively. The pulse effect of rainfall event on soil respiration can be accurately predicted by a water status index (WSI), which is the product of rainfall event size and the ratio between antecedent soil temperature to moisture at the depth of 10 cm ($r^2 = 0.92$, $P<0.001$) through the growing season. It indicates the pulse effect can be enhanced by not only larger individual rainfall event, but also higher soil temperature/moisture ratio which is usually associated with longer dry spells. We then analyzed a long-term (1953–2009) precipitation record in the experimental area. We found both the extreme heavy rainfall events (>40 mm per event) and the long dry-spells (>5 days) during the growing seasons increased from 1953–2009. It suggests the shift in precipitation regime has increased the contribution of rain pulse effect to growing-season total soil respiration in this region. These findings highlight the importance of incorporating precipitation regime shift and its impacts on the rain pulse effect into the future predictions of grassland carbon cycle under climate change.

Editor: Bazartseren Boldgiv, National University of Mongolia, Mongolia

Funding: This study was supported partly by the "Strategic Priority Research Program" of the Chinese Academy of Sciences, Climate Change: Carbon Budget and Relevant Issues (XDA05050402), the National Natural Science Foundation of China (31170453), the National Basic Research Program of China (973 program, 2010CB833501), and a Selected Young Scientist Program of the State Key Laboratory of Vegetation and Environment Change to SC. The funders had no role in study design, data collection and analysis, decision to publish, or preparation of the manuscript.

Competing Interests: The authors have declared that no competing interests exist.

* Email: spchen@ibcas.ac.cn

Introduction

Global precipitation has been predicted to change with increasing intra-annual variability and more frequent extreme rainfall events [1,2]. Such shifts in precipitation regime could have profound impacts on belowground carbon (C) release, especially in the arid and semiarid ecosystems [3–6]. It has been widely reported that soil respiration in grassland ecosystems can increase significantly and immediately after a rainfall event, followed by a decrease with declining soil moisture [4,7]. This pulse effect of rainfall event on soil respiration has been suggested as an important contributor to ecosystem C release in grassland ecosystems [3,8,9]. Therefore, a better understanding of how the changing precipitation regime will affect rain pulse effect is important for predicting future grassland C feedbacks to climate change.

Previous studies have summarized that variations in soil respiration are determined by several factors, including soil temperature, soil water availability and carbon substrate supply [10–12]. Yet it is not clear which factors control the pulse effect of a single rainfall event on soil respiration. There have been both laboratory [13–15] and field [7,16,17] attempts trying to identify

the regulatory mechanisms of rain pulses on soil respiration. The magnitude of soil respiration response (e.g., soil carbon release) is positively correlated with rainfall event size [4,18,19]. Other studies (e.g. [20]) have observed that the antecedent soil water condition is also an important influencing factor. The pulse effect is intensified by the dry condition of the antecedent soil [20], and the effect is less obvious if the soil is wet before the rainfall event [21–23]. Soil temperature has been widely reported to regulate soil microbial activities and thus soil heterotrophic respiration in various ecosystems [22,24]. Other conditions, such as plant activity [25,26] and soil organic matter content [20,27], are also believed to influence the effects of rainfall event on soil respiration. Therefore, the predictability of rain pulse effect on soil respiration is still low and no effective approach or indicator has been developed so far.

In semi-arid ecosystems, water availability is the dominant factor regulating soil respiration [28–30]. Water availability also moderates the effects of the other factors on soil respiration such as temperature and substrate supply [30]. Water availability and its intra- and inter-annual variations are directly linked to both the intensity and the frequency of precipitation. Although IPCC [2] has reported a trend of significant change in both total amount and temporal patterns of precipitation, only a few studies have analyzed the general changes of precipitation regimes especially in semi-arid temperate grasslands [31]. Shifts in precipitation regime will alter not only the size of individual rainfall event, but also the length of the dry-spell duration and thus the antecedent soil water condition. Previous laboratory experiments have shown that both of these two factors can significantly influence the soil CO_2 release [32,33]. A growing body of works using field experiments have demonstrated that shifts in precipitation regime alone, even when the total rainfall amount does not change, can have large impacts on grassland soil CO_2 release [29,34,35] and its pulse responses to rainfall events [4,17]. However, to our knowledge, few studies have observed and quantified the dynamics of soil respiration after rainfall events [36,37]. Therefore, it is necessary to incorporate the changes in precipitation regime when evaluating the effect of rainfall event on C cycling in natural grassland ecosystems.

The area of grasslands contributed to about 40% of the total territory of China [38], and the temperate steppe in China is the third largest grassland area in the world [39]. In this study, we analyzed both a long-term (1953–2009) rainfall data set and two years (2006 and 2009) of continuously measured soil respiration data in a temperate steppe in northern China. We attempt to address the following questions: (1) What are the controlling factors of the rainfall event effect on soil respiration through the growing season in this ecosystem? (2) Has precipitation regime changed over the last ~60 yr in this semi-arid steppe? (3) How shifting precipitation regime would influence soil respiration responses to pulses in water availability?

Materials and Methods

2.1. Site description

Our field site (42°27′N, 116° 41′E), located Duolun County in the northeastern Inner Mongolia, China, belongs to Duolun Restoration Ecology Experimentation and Demonstration Station (DREEDS), Institute of Botany. No specific permissions were required for scientific researches. The site has been fenced in to exclude grazing since 2001, and no other management like mowing or fertilizing was applied. The mean altitude is about 1430 m above sea level. The vegetation was dominated by C3 grasses (e.g. *Stipa krylovii*, *Agropyron cristatum*, *Leymus chinensis*) and a semi-shrub species (*Artemisia frigida*). The soil in our study site is classified as Calcic-orthic Aridisol. The mean annual air temperature is 3.3°C and the mean annual precipitation is 377 mm, 95% (i.e. 358 mm) of which occurs during the growing season from April to October. Average air temperature was 3.13°C in 2006 and 3.09°C in 2009. Annual precipitation was 425 mm in 2006 and 185 mm in 2009, respectively.

2.2 Variable measurement

The growing season in this ecosystem usually starts from late April and ends in early October. Five PVC collars (20.3 cm in diameter and 8 cm in height) were inserted into the soil to a depth of 3 cm. The PVC collars were randomly distributed in the study site with 5 m–30 m distance between any two of them in late April, 2006, and covered an area of about 300 m^2. The site is flat so topography had little impact on the difference in measured soil respiration among replicates. Soil respiration rates were continuously measured during the growing season (from May to October) of 2006 and 2009 on a half-hour basis using an infra-red gas analyzer (LI-840, Li-Cor Inc., Lincoln, NE, USA) that was connected to five automatic measurement chambers. All living plants inside the chambers were clipped by hand weekly to exclude aboveground plant respiration. The clipped aboveground plants

were left in the chambers to include CO_2 efflux from the litter decomposition. The CO_2 concentrations in the chambers were recorded in CR1000 data logger (Campbell Scientific Inc., CSI, Utah, USA), and were processed by LoggerNet 3.1 (CSI, USA). Each measurement took 120 s. The soil respiration rates were determined from the time series of soil CO_2 efflux concentrations. Since the measuring system worked well during 2006 and 2009, only small data gaps (less than 2 h) existed in two years data sets and were filled by linear interpolation method.

Soil temperatures (°C) at a depth of 10 cm and volumetric soil moisture (%) at a depth of 0–10 cm were measured using 107 soil temperature probes (CSI, USA) and CS616 soil water probes (CSI, USA), respectively. Three combinations of soil temperature and water probes were placed close (about 1 m) to the soil respiration chambers. The mean values of half-hour soil temperature and moisture were recorded in a CR1000 data logger (CSI, USA) simultaneously.

The precipitation data of the past 57 years were provided by local meteorological stations in Duolun County. The precipitation data of the growing season of 2006 and 2009 were measured by a tipping bucket rain gauge (TE525MM, CSI) at the study site.

2.3. Statistical analyses

Considering the significant impact of soil temperature on the diurnal variation of soil respiration, we analyzed precipitation and soil respiration data on a daily time step. There was a rainfall event when precipitation was recorded, and a day without any precipitation was defined as a dry day. The rainfall event size was defined as the total amount of rainfall during a rainfall event. Note that one day may have more than one rainfall event and, at the other extreme, one rainfall event may last for a few days. An uninterrupted sequence of dry days preceded and succeeded by at least one rainfall event is referred to as a dry-spell event [40,41]. We added up the total precipitation amount and calculated the frequencies of different dry-spell durations (days) in growing seasons of the past 57 years (1953–2009). The rainfall and dry-spell events were then classified into different categories based on their size (rainfall event size: 0–2, 2–5, 5–10, 10–15, 15–20, 20–30, 30–40, 40–50, or >50 mm) and duration (dry-spell duration: 0–5, 5–10, 10–20, 20–30, or >30 days).

Since the size of individual rainfall events varies greatly, the rain pulse effect on soil respiration can last from several hours to a few weeks. Thus, if the rain pulse response lasted for at least three days, we would use daily soil respiration data with the following function to estimate the effect of a rainfall event on soil respiration [7]:

$$y_t = y_0 + ate^{-bt} \qquad (1)$$

where y_t is daily soil respiration rate after the rainfall event, y_0 is initial daily soil respiration before the rainfall event, t is the time (continuous but not a discrete daily time-step) after the rainfall event, and a and b are coefficients. However, if the rain pulse response lasted less than three days, we would compare the difference between post-rainfall soil respiration and the control at half-hourly time scale. The sum of all the differences over one day greater than 0±0.01 (g C m^{-2} day^{-1}) was defined as rainfall event. The calculation was applied for each chamber. Based on the measurements, there were totally 47 and 45 rainfall events during the growing season of 2006 and 2009, respectively. According to their durations, the half-hourly analyses were applied to 10 and 13 rainfall events in 2006 and 2009, respectively. We further defined the time for the soil respiration response to peak after a rainfall event as the peak time (T_{peak}), and calculated the duration of the pulse ($T_{duration}$) before soil respiration decreased to 99% of the antecedent soil respiration. The rain pulse effects were estimated by the accumulative differences during this period. These analyses were conducted using Matlab (Mathworks, Natick, MA).

Linear regression analyses were used to evaluate the relationships between rain pulse response patterns (T_{peak}, $T_{duration}$, pulse effect) of soil respiration, rain size and water status index (WSI) in the growing seasons of 2006 and 2009. The water status index (WSI) was defined as,

$$WSI = \text{Rainfall event size} \times \frac{ST_{pre}}{SM_{pre}} \quad (2)$$

where ST_{pre} and SM_{pre} are the soil temperature and moisture at 10 cm depth before a rainfall event. The index thus takes both rainfall event size and antecedent condition into consideration. The development of the WSI was based on the findings from previous studies, which have shown that larger rainfall event size [4] as well as drier [23] and warmer [22] pre-event soil conditions positively affect rain pulse effect. Growing-season total soil respiration was defined as the sum of daily soil respiration from May to October in each year.

Results

3.1. Growing-season total soil respiration and relative environmental factors

Soil respiration rates showed clear seasonal variations and reached maximum values in the middle of the growing seasons in both 2006 and 2009 (Fig. 1), with the growing-season total soil respiration as 187.2 ± 16.47 g C m^{-2} in 2006 and 155.9 ± 4.58 g C m^{-2} in 2009. Precipitation during the growing season (May to October) was slightly above the long-term average in 2006 (403 mm) but was far below the average in 2009 (168 mm) (Fig. 1). In both years, the heaviest rainfall occurred in the middle of the growing season (late July to early August; Fig 1). Accordingly, soil moisture fluctuated through the growing season in both years, with the highest soil moisture occurred in late July in both years (Fig. 1). Soil temperature had a pronounced seasonal dynamic in both years, with the highest temperature occurring in late July of 2006 and early August of 2009 (Fig. 1). Lower seasonal mean soil temperature (16.86°C) and higher soil moisture (11.26%) were found in 2006 than those of 2009 (17.14°C and 7.81%).

3.2. Pulse effects of soil respiration and its controlling factors

Equation (1) well fitted the temporal patterns of soil respiration following heavy rainfall events, with r^2 ranging from 0.91 to 0.99 ($P<0.05$; Fig. 2a–f). In 2009, for example, daily soil respiration rates increased by 72–160% within 2 days following rainfall events (Fig. 2d–f). For small rainfall events, daily average of soil respiration could not capture the rapid responses of soil respiration to rainfall events. For example, similar to the predicted curves in Figure 3, the rainfall event effect on soil respiration peaked within several hours of the end of small rainfall events (Figure S1 in File S1). After reaching the peak, the soil respiration rate gradually returned to its antecedent level in 2.93 to 27.9 days after rainfall events of 0.4 mm to 31 mm, respectively (estimated from the fitted curve in Fig. 2). Both the magnitude and duration of the pulse effect on soil respiration (T_{peak} and $T_{duration}$) were positively

correlated with rainfall event size (Fig. 3a–c). A higher sensitivity of rainfall event effect to rainfall event size was found in 2009 than in 2006 (Fig. 3c). It could be ascribed to the higher sensitivities of T_{peak} and $T_{duration}$ to rainfall event size in 2009 than in 2006 (Fig. 3a, b). It suggested that rainfall event size alone was not a good indicator for the effect of rainfall event on soil respiration in this ecosystem. When compared against WSI instead of rainfall event size, however, there were no significant differences in regressive curves between 2006 and 2009 ($P>0.1$) (Fig. 3d–f). The values of T_{peak} (about 3 days; $r^2 = 0.86$, $P<0.001$) and $T_{duration}$ (about 30 days; $r^2 = 0.94$, $P<0.001$) of the pulse responses increased logarithmically with the WSI (Fig. 3d and e). Accumulative soil respiration during the pulse processes increased linearly with the WSI (Pulse effect $= 0.1169 \times WSI$, $r^2 = 0.92$, $P<0.001$, Fig. 3f). Based on this relationship, we calculated the pulse effect of each rainfall event on soil respiration, and found the pulse effect contributed 44.5% and 39.6% of the measured growing-season total soil respiration in 2006 and 2009, respectively.

3.3. Long-term trend of the precipitation regime during 1953-2009

The mean precipitation during the growing season from 1953-2009 was 358 mm (Fig. 4b). The frequency of rainfall event occurrence decreased with increasing amount of rainfall (Fig. 4a). Long-term mean rainfall frequency with rainfall amount of 0–2 mm, 2–5 mm, 5–10 mm, 10–15 mm, 15–20 mm, 20–30 mm, 30–40 mm, 40–50 mm, and >50 mm per event was 48.4%, 21.8%, 14.0%, 7.3%, 3.8%, 2.8%, 1.0%, 0.4%, and 0.4%, respectively (Fig. 4a). We classified the duration of dry-spell into five groups (<5 days, 5–10 days, 10–20 days, 20–30 days, and > 30 days). The long-term mean occurrence frequency of the dry-spells was 71.7%, 21.4%, 5.9%, 0.8%, and 0.3% for the five groups, respectively (Fig. 4c).

During the 57 years, no clear temporal trend of the total growing-season precipitation was found (Figure S2 in File S1). The 40–50 mm ($r^2 = 0.16$, $P = 0.061$) and >50 mm ($r^2 = 0.23$, $P< 0.05$) rainfall events showed significant increasing tendencies over the years (Fig. S2k and l in File S1), while those of the smaller-sized rainfall events did not show any clear trend (Figure S2b–g in File S1). The mean dry-spell duration of the growing season increased from 1953 to 2009 ($P<0.001$; Figure S3a). During these 57 years, the frequency of dry spell events <5 days in duration significantly decreased ($r^2 = 0.28$, $P<0.001$; Figure S3b in File S1), while the dry-spell events of 5–10 days ($r^2 = 0.13$, $P = 0.003$) and 10–20 days ($r^2 = 0.11$, $P = 0.01$) occurred more frequently (Figure S3c–d in File S1). The frequency of dry-spell events longer than 20 days did not change during this time period (Figure S3e–f in File S1).

Discussion

4.1. Factors controlling rain pulse on soil respiration

In this study, significant pulse responses of soil respiration were found after rainfall events (Fig. 2), which was consistent with the results reported in other arid ecosystems [18,25]. The rain pulse induced soil respiration was 44.5% (83.3 g C m^{-2}) and 39.6% (61.7 g C m^{-2}) of the growing-season total soil respiration in 2006 and 2009, respectively. Since only few studies have quantified the contribution of rain pulse effect to growing-season total soil respiration, it is difficult to compare our results to those from other grasslands. However, some evidence show that the contribution of rain pulse to total soil respiration is lower in forest ecosystems. For example, Lee et al. [42] reported that the rainfall-induced soil CO_2 release accounted for 16–21% of the annual soil respiration in a deciduous forest in Japan, whereas Yuste et al. [43] estimated

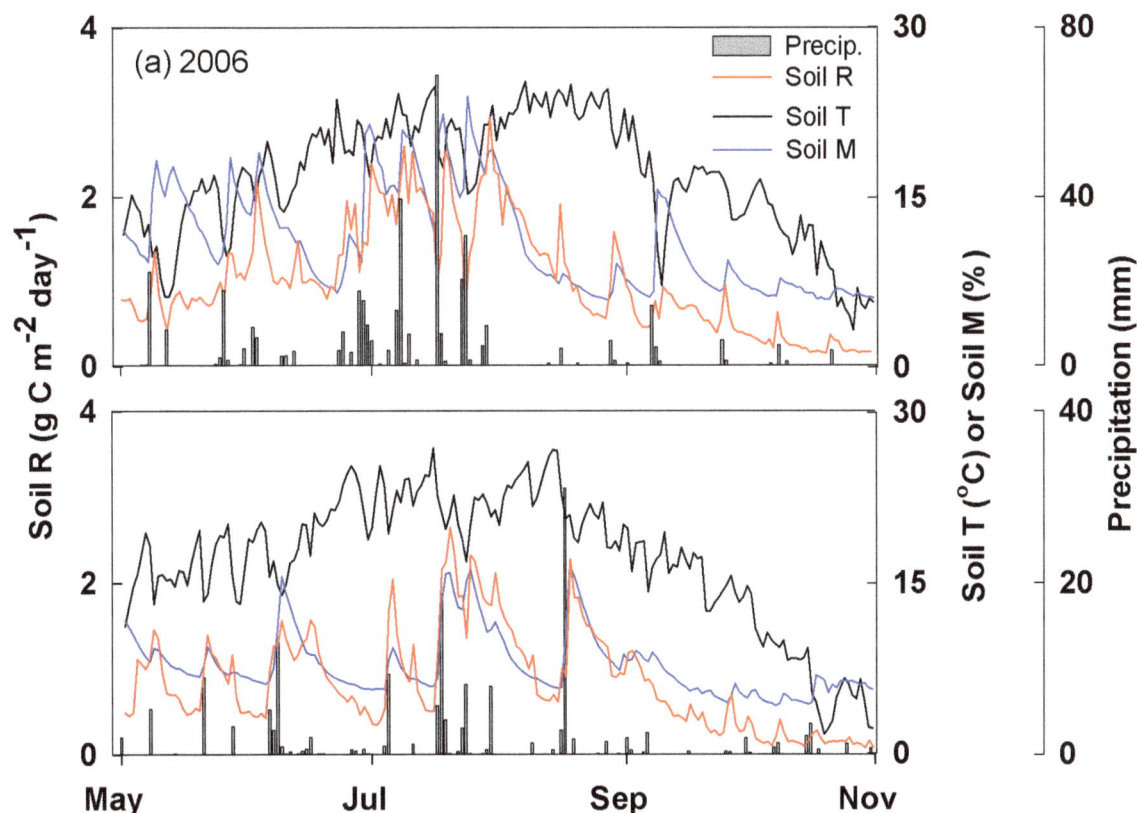

Figure 1. Seasonal patterns of precipitation (gray rectangle), soil moisture (black dashed line), soil temperature (black line) and soil respiration (gray line) during the growing seasons of 2006 (a) and 2009 (b).

that approximately 9–14% of the annual soil respiration in Belgian Campine region in Belgium was rainfall induced. The stronger rain pulse responses in grasslands could be the result of more severe water deficiency before rainfall events in grasslands and higher soil organic carbon content at the soil surface layer of grasslands than of shrublands or wood stands [36]. Therefore, the rain pulse effects on belowground C release is critical to ecosystem C balance in the semi-arid steppe ecosystem.

A rainfall event consists of at least three processes. Firstly, the percolating rainwater replaces the CO_2 in soil pore spaces. Secondly, the rainwater activates microbial activity and induces microbe-respired CO_2 in the shallow soil [44]. Thirdly, the rainwater also promotes the assimilation process by roots, which increases root respiration [3,45]. These processes have been confirmed by the positive relationship found between the rainfall event effect on soil respiration and precipitation in previous studies through field manipulative experiments [4,7,18]. Similar trends have also been found in our study. During the two growing seasons of 2006 and 2009, we observed that a larger rainfall event induced a greater pulse effect on soil respiration (Fig. 2a–c). Rather than using the limited water and resources in the soil to maintain activity, most microbes may simply become dormant in the dry soil [46]. The sudden increase in soil water availability after rains can induce microbial cell lysis or the rapid mineralization of cytoplasmic solutes and release the mineralized product into the surrounding environment to dispose of its osmolytes, which have accumulated during the dry period [27,47]. Consequently, rainfall event induced soil respiration largely comes from the decomposition of microbial cellular material. In addition, more rainfall means more water percolates to the rhizosphere and triggers root

activity and respiration [3]. Therefore, the dependence of pulse effect and its duration of soil respiration on precipitation amount in this study can be explained by the greater increase in water availability under heavier rainfall event (Fig. 5).

Although the effects of rainfall events on the changes in water availability before and after rain were similar between 2006 and 2009 (Fig. 5), a significant inter-annual difference in response patterns of soil respiration to rainfall event was found (Fig. 2a–c). It indicates that some other factors rather than rainfall event size regulate the effect of rain pulse on soil respiration. For example, many studies [20,23,48] have found that the same amount of water addition had greater effects on soil respiration under drier antecedent soil condition. Soil temperature also has significant effect on soil respiration pulse [5,24,36,]. In addition, vegetation, which influences both water and C cycles, are important in regulating the effect of rain pulse on soil respiration [49,50]. In this study, we introduced the WSI to predict the effect of rain pulse on soil respiration in two growing seasons with contrasting precipitation regimes (Fig. 3d–f). It should be noted that this empirical index does not explain the underlying mechanisms of the response of soil respiration to rainfall events, which have been studied by some hourly-scale stochastic models [37]. However, it highlights the importance of precipitation regime on the pulse response of soil respiration in semi-arid areas. Since the WSI was developed from only two growing-season measurements and based on five chambers, this index needs to be tested in other sites and years before it is used as an indicator for modeling and predicting rain pulse effect on soil respiration in semi-arid grasslands.

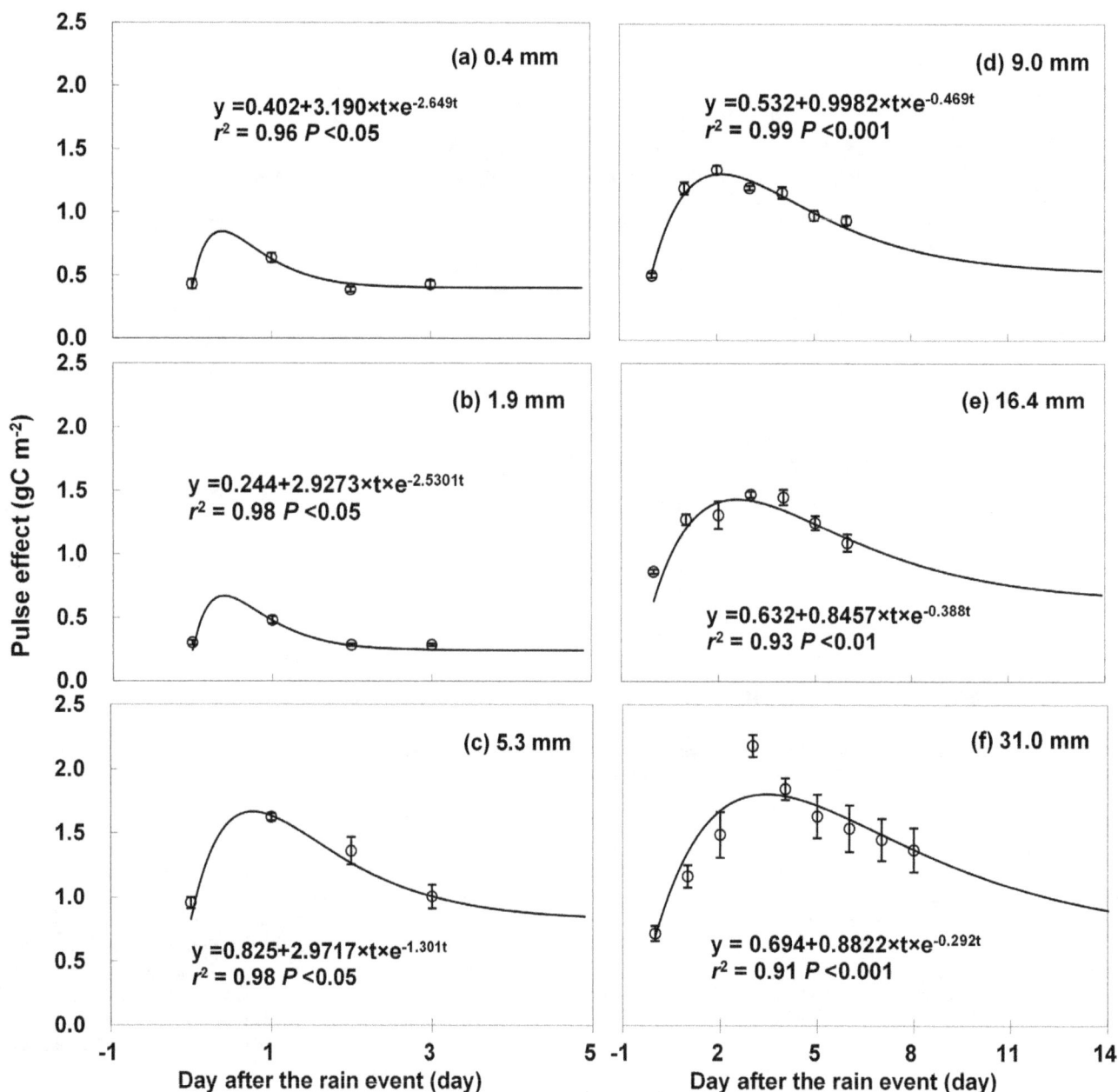

Figure 2. The response of soil respiration to rain pulses under different rain sizes. Open circles are the measured data and shown as mean ± standard deviation. Exponential equations in the form of $y = y_0 + ate^{-bt}$ are used to fit the data (represented by solid lines). Rain size is shown in each panel. The 0 in the x-axis represents the day before the rainfall event.

4.2. Precipitation regime shift and its influence on soil respiration

The long-term precipitation records showed no clear trend of total precipitation amount during a growing season but a higher frequency of heavy rainfall events in our study region (Figure S2 in File S1). It was consistent with the modeling predictions of greater precipitation intensity globally in the future [1,51]. That is, more precipitation will fall in a given daily rainfall event, leading to more extreme rainfall events in the future. With similar total precipitation amount, higher frequency of heavy rainfall events means longer dry-spells during the growing season (Figure S3 in File S1). Such a shift in precipitation regime has been widely

observed [40,52,53], although changes in precipitation amount would be greatly different among regions across the globe [2].

We found that a large proportion of the rain pulse effect on soil respiration was determined by the WSI, which considered the influences of rain size and antecedent soil temperature and moisture. The longer dry-spells are likely to lower the antecedent soil moisture. As a consequence, belowground C release is expected to be greater after a rainfall event compared to soils in rainfall regimes with a lower frequency of long dry-spells. The robustness of the WSI in explaining rain pulse effect on soil respiration needs to be examined in other types of grasslands. If the WSI is effective in most grassland ecosystems, it suggests not only total amount of annual precipitation but also the temporal

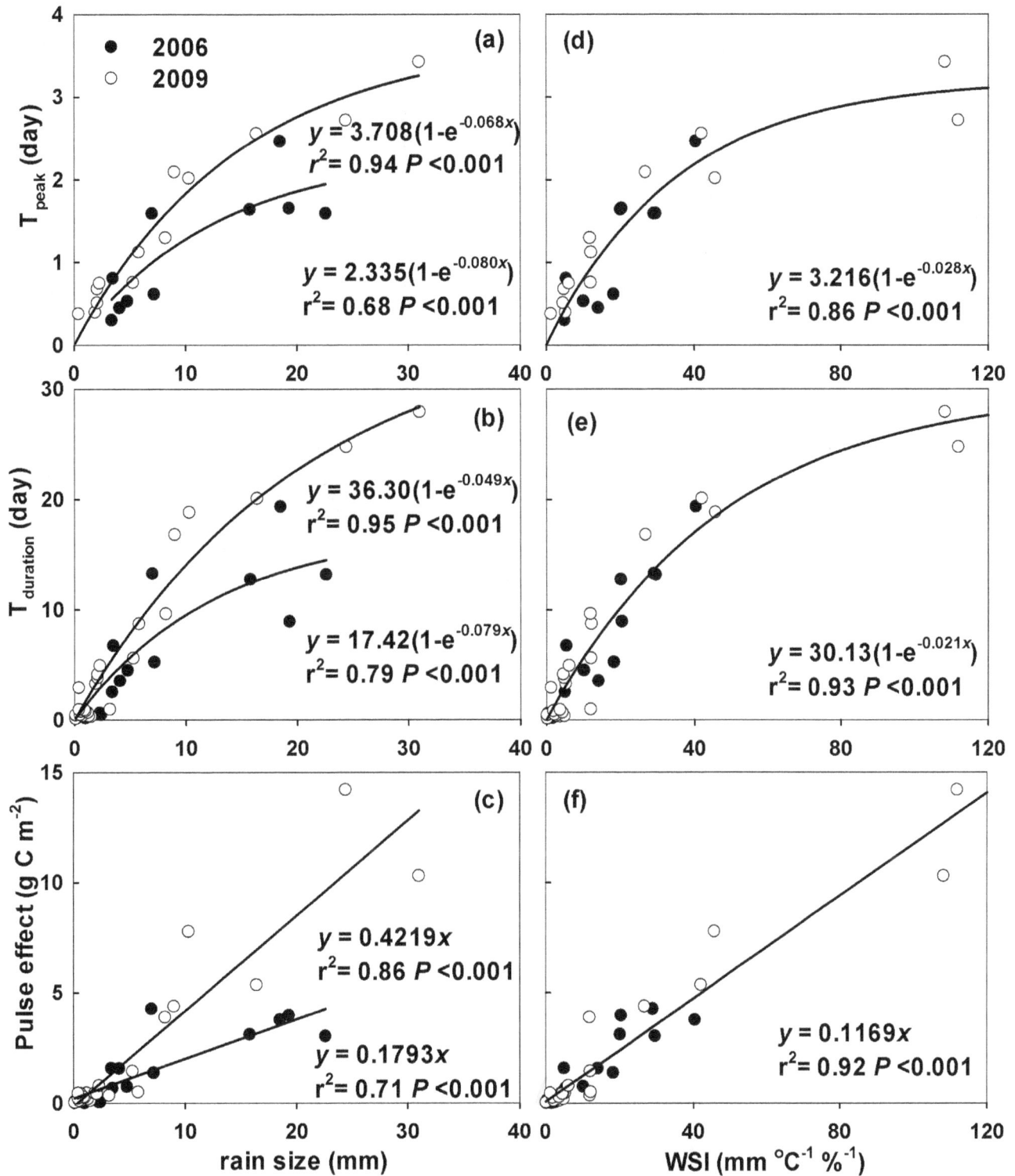

Figure 3. The relationship between dependent variables including the peak time (T$_{peak}$) and the duration (T$_{duration}$) of the pulses, and the pulse effects on soil respiration, and independent variables including rain size (a, b, c) and WSI (rain size × the ratio of antecedent soil temperature to soil moisture; d, e, f) in 2006 (solid circle) and 2009 (open circle), respectively.

distribution of rainfall events must be taken into consideration when predicting the belowground C release and evaluating C budget of an ecosystem undergoing climate change.

4.3. Implications for modeling soil carbon release

Currently, process-based ecosystem models are common tools for simulating and predicting the future states of terrestrial C cycle [2]. Most of these models are established at hourly or daily time

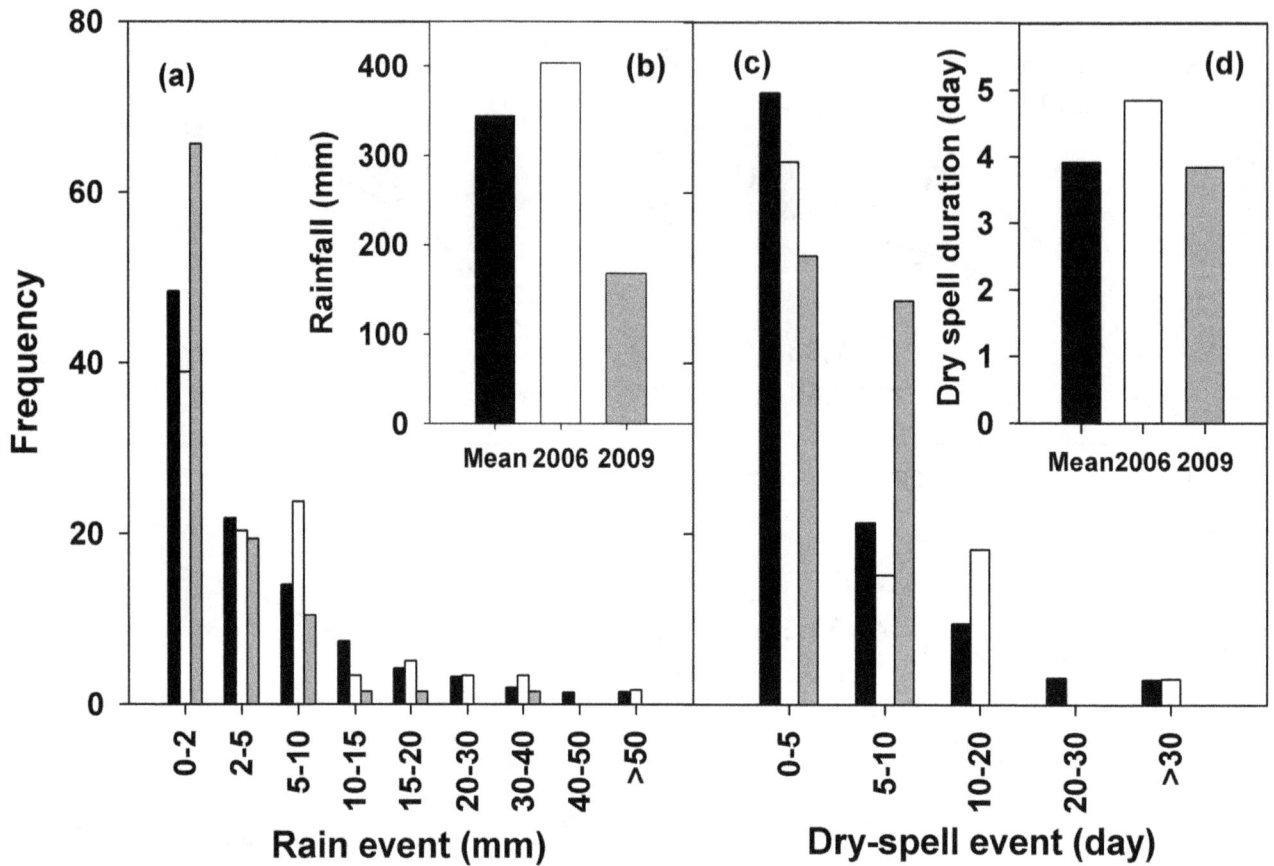

Figure 4. The frequency distribution of different class-categories of rainfall events (a) and dry-spell events (c) during the growing seasons of 2006 and 2009. The long-term mean annual precipitations from 1953 to 2009 are also represented here by black columns. The insert panels show the precipitation during the growing season (b) and the mean dry-spell duration during the growing season (d) in 2006 and 2009.

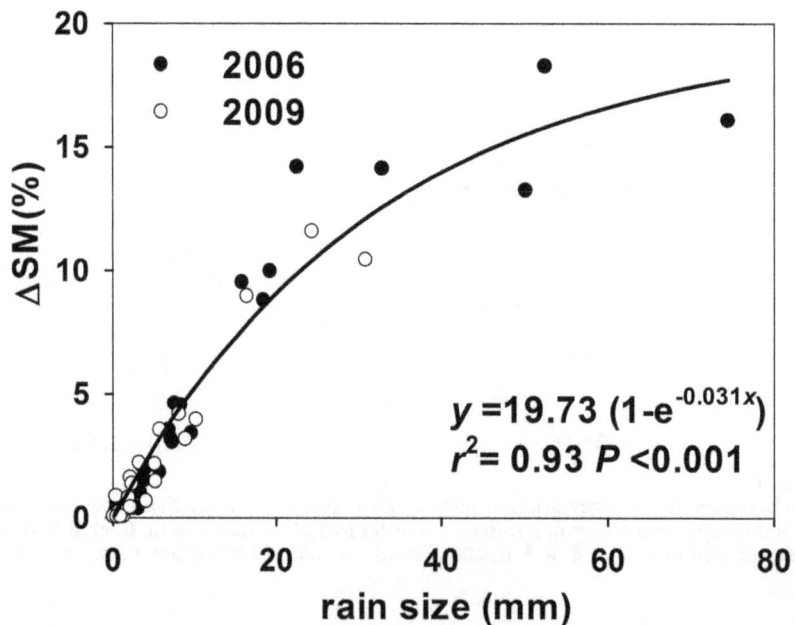

$$y = 19.73\,(1-e^{-0.031x})$$
$$r^2 = 0.93 \ P < 0.001$$

Figure 5. The relationships between the rain size and the change in water availability after a rainfall event (ΔSM; the changes in the largest volumetric soil moisture after a rainfall event from the antecedent volumetric soil moisture) in 2006 (open circle) and 2009 (solid circle) with the coefficient (r^2) and the significance level (P-value).

scales without incorporating the immediate pulse responses of soil respiration to rainfall events. Therefore, rain pulse effect on soil respiration is one of the key sources of the uncertainty in the modeled ecosystem C cycle, especially in grasslands [8,36]. In this study, we found that the rain pulse effect on soil respiration and its duration increased with the product of rain size and antecedent soil temperature-to-moisture ratio. However, the relationship between the WSI and rain pulse effect on soil respiration $(y = 0.1169 \times \text{WSI})$ as well as its duration $(y = 30.13(1 \text{-} e^{-0.021x}))$ was obtained from the data of only one grassland site during two growing seasons. The parameters in the regression model might vary with soil texture and soil organic matter content, which supplies C substrate for respiration [5,20,27]. Therefore, the applicability of the WSI we proposed in this study still needs to be tested in broader spatial scales and at more locations. Parameter information that constrains the relationship between the WSI and rain pulse effect on soil respiration could be useful for improving our ability to predict soil C cycle under future precipitation regimes.

There is evidence that CO_2 from carbonates may [54] or may not [55] contribute significantly to total CO_2 release when soils are moistened. Since the soil in our site is calcic, it is still unclear whether rain pulse effect will increase CO_2 release from inorganic C sources. Future study may pay more attention to the pulse effect of rainfall events on not only the decomposition of soil organic matters but also CO_2 release from inorganic C sources.

Conclusions

In this study, we presented a water status index (WSI) as a good indicator for predicting the effect of rain pulse on soil respiration in a temperate steppe in northern China. Long-term precipitation data showed a significant increase in frequency of large rainfall events and dry-spell duration with time. Although there are no measurements of soil temperature and moisture from the historical record to compare against measurements made in 2006 and 2009, the longer dry-spell may leads to larger antecedent ratio between

temperature:moisture in the soils, and thus greater rain pulse effect on soil respiration. In this study, rain pulse effect contributed to about 39–44.5% of the growing-season total soil respiration, which is much larger than previous studies in this [4] and other [23] ecosystems. It indicates that the effect of rain pulses on soil respiration cannot be neglected in future climate-carbon cycle modeling. Future research could test the robustness of WSI in other regions before it is used as a simple approach to estimate the rain pulse effect on soil respiration in grassland ecosystems.

Supporting Information

File S1 Supporting figures. Figure S1, Half-hourly soil respiration rate (SR, μmol m^{-2} s^{-1}) under different small rainfall events. Rainfall size is shown in each panel. The 0 in the x-axis is the beginning time of a rainfall. **Figure S2**, The precipitation amount during the growing season (a) and the occurrence frequency of the nine class-categories of precipitation during the growing season (b–l) from 1953 to 2009. **Figure S3**, The mean dry-spell duration during the growing season (a) and the occurrence frequency of the five class-categories of the dry-spell duration during the growing season (b–f) from 1953 to 2009.

Acknowledgments

We thank Dr. Bazartseren Boldgiv and two anonymous reviewers for their valuable comments and suggestions. Thanks, Ms. Jin Chen, for polishing the English language of an early version of this manuscript.

Author Contributions

Conceived and designed the experiments: SC LY. Analyzed the data: LY SC. Contributed to the writing of the manuscript: LY SC. Conducted the filed measurement: LY. Provided constructive suggestions on data analysis: JX YL. Provided substantial comments and revised the early version of manuscript: JX YL.

References

1. Dore MHI (2005) Climate change and changes in global precipitation patterns: What do we know? Environment International 31: 1167–1181.
2. IPCC (2007) Climate change 2007: the physical science basis. In: Solomon S, Qin D, Manning M, Chen Z, Marquis M, Averyt KB, Tignor M, Miller HL, editors, Contribution of Working Group I to the Fourth Assessment Report of the Intergovernmental Panel on Climate Change. Cambridge: Cambridge University Press. pp. 30–33.
3. Huxman TE, Snyder KA, Tissue D, Leffler AJ, Ogle K, et al. (2004) Precipitation pulses and carbon fluxes in semiarid and arid ecosystems. Oecologia 141: 254–268.
4. Chen SP, Lin GH, Huang JH, Jenerette GD (2009) Dependence of carbon sequestration on the differential responses of ecosystem photosynthesis and respiration to rain pulses in a semiarid steppe. Global Change Biology 15: 2450–2461.
5. Kim DG, Vargas R, Bond-Lamberty B, Turetsky MR (2012) Effects of soil rewetting and thawing on soil gas fluxes: a review of current literature and suggestions for future research. Biogeosciences 9: 2459–2483.
6. Liu R, Cieraad E, Li Y (2013) Summer rain pulses may stimulate a CO2 release rather than absorption in desert halophyte communities. Plant and Soil 373: 799–811.
7. Liu XZ, Wan SQ, Su B, Hui DF, Luo YQ (2002) Response of soil CO_2 efflux to water manipulation in a tallgrass prairie ecosystem. Plant and Soil 240: 213–223.
8. Lee X, Wu HJ, Sigler J, Oishi C, Siccama T (2004) Rapid and transient response of soil respiration to rain. Global Change Biology 10: 1017–1026.
9. Ma S, Baldocchi DD, Hatala JA, Detto M, Yuste JC (2012) Are rain-induced ecosystem respiration pulses enhanced by legacies of antecedent photodegradation in semi-arid environments? Agricultural and Forest Meteorology 154–155: 203–213.
10. Högberg P, Nordgren A, Buchmann N, Taylor AFS, Ekblad A, et al. (2001) Large-scale forest girdling shows that current photosynthesis drives soil respiration. Nature 411: 789–792.
11. Davidson EA, Janssens IA (2006) Temperature sensitivity of soil carbon decomposition and feedbacks to climate change. Nature 440, 165–173.
12. Wan S, Norby RJ, Ledford J, Weltzin JF (2007) Responses of soil respiration to elevated CO_2, air warming, and changing soil water availability in a model old-field grassland. Global Change Biology 13: 2411–2424.
13. Griffiths E, Birch HF (1961) Microbiological Changes in Freshly Moistened Soil. Nature 189: 424.
14. Kieft TL, Soroker E, Firestone MK (1987) Microbial biomass response to a rapid increase in water potential when dry soil is wetted. Soil Biology and Biochemistry 19: 119–126.
15. Borken W, Xu YJ, Davidson EA, Beese A (2002) Site and temporal variation of soil respiration in European beech, Norway spruce, and Scots pine forests. Global Change Biology 8: 1205–1216.
16. Schimel JP, Gulledge JM, Clein-Curley JS, Lindstrom JE, Braddock JF (1999) Moisture effects on microbial activity and community structure in decomposing birch litter in the Alaskan taiga. Soil Biology and Biochemistry 31: 831–838.
17. Chen SP, Lin GH, Huang JH, He M (2008) Responses of soil respiration to simulated precipitation pulses in semiarid steppe under different grazing regimes. Journal of Plant Ecology-UK 1: 237–246.
18. Sponseller RA (2007) Precipitation pulses and soil CO_2 flux in a Sonoran Desert ecosystem. Global Change Biology 13: 426–436.
19. Vargas R, Collins SL, Thomey ML, Johnson JE, Brown RF, et al. (2012) Precipitation variability and fire influence the temporal dynamics of soil CO_2 efflux in an arid grassland. Global Change Biology 18: 1401–1411.
20. Cable JM, Ogle K, Williams DG, Weltzin JF, Huxman TE (2008) Soil texture drives responses of soil respiration to precipitation pulses in the Sonoran Desert: Implications for climate change. Ecosystems 11: 961–979.
21. Bowling DR, Grote EE, Belnap J (2011) Rain pulse response of soil CO_2 exchange by biological soil crusts and grasslands of the semiarid Colorado Plateau, United States. Journal of Geophysical Research -Biogeosciences 116: G03028, doi:10.1029/2011JG001643.

22. Carbone MS, Still CJ, Ambrose AR, Dawson TE, Williams AP, et al. (2011) Seasonal and episodic moisture controls on plant and microbial contributions to soil respiration. Oecologia 167: 265–278.

23. Wu HJ, Lee X (2011) Short-term effects of rain on soil respiration in two New England forests. Plant and Soil 338: 329–342.

24. Jager G, Bruins EH (1975) Effect of Repeated Drying at Different Temperatures on Soil Organic-Matter Decomposition and Characteristics, and on Soil Microflora. Soil Biology and Biochemistry 7: 153–159.

25. Smart DR, Penuelas J (2005) Short-term CO_2 emissions from planted soil subject to elevated CO_2 and simulated precipitation. Applied Soil Ecology 28: 247–257.

26. Vargas R, Allen MF (2008) Environmental controls and the influence of vegetation type, fine roots and rhizomorphs on diel and seasonal variation in soil respiration. New Phytologist 179: 460–471.

27. Fierer N, Schimel JP (2003) A proposed mechanism for the pulse in carbon dioxide production commonly observed following the rapid rewetting of a dry soil. Soil Science Society of America Journal 67: 798–805.

28. Riveros-Iregui DA, Emanuel RE, Muth DJ, McGlynn BL, Epstein HE, et al. (2007) Diurnal hysteresis between soil CO_2 and soil temperature is controlled by soil water content. Geophysical Research Letters 34: L17404, doi:10.1029/2007GL030938.

29. Liu WX, Zhang Z, Wan SQ (2009) Predominant role of water in regulating soil and microbial respiration and their responses to climate change in a semiarid grassland. Global Change Biology 15: 184–195.

30. Yan LM, Chen SP, Huang JH, Lin GH (2011) Water regulated effects of photosynthetic substrate supply on soil respiration in a semiarid steppe. Global Change Biology 17: 1990–2001.

31. Liu BH, Xu M, Henderson M, Qi Y (2005) Observed trends of precipitation amount, frequency, and intensity in China, 1960–2000. Journal of Geophysical Research-Atmospheres 110: D08103, doi:10.1029/2004JD004864.

32. Fierer N, Schimel JP (2002) Effects of drying-rewetting frequency on soil carbon and nitrogen transformations. Soil Biology and Biochemistry 34: 777–787.

33. Miller AE, Schimel JP, Meixner T, Sickman JO, Melack JM (2005) Episodic rewetting enhances carbon and nitrogen release from chaparral soils. Soil Biology and Biochemistry 37: 2195–2204.

34. Knapp AK, Fay PA, Blair JM, Collins SL, Smith MD, et al. (2002) Rainfall variability, carbon cycling, and plant species diversity in a mesic grassland. Science 298: 2202–2205.

35. Harper CW, Blair JM, Fay PA, Knapp AK, Carlisle JD (2005) Increased rainfall variability and reduced rainfall amount decreases soil CO_2 flux in a grassland ecosystem. Global Change Biology 11: 322–334.

36. Xu LK, Baldocchi DD, Tang JW (2004) How soil moisture, rain pulses, and growth alter the response of ecosystem respiration to temperature. Global Biogeochemical Cycles 18: GB4002.

37. Daly E, Oishi AC, Porporato A, Katul GG (2008) A stochastic model for daily subsurface CO_2 concentration and related soil respiration. Advances in Water Resources 31: 987–994.

38. Ni J (2002) Carbon storage in grasslands of China. Journal of Arid Environments 50: 205–218.

39. Lee R, Yu FF, Price KP, Ellis J, Shi PJ (2002) Evaluating vegetation phenological patterns in Inner Mongolia using NDVI time-series analysis. International Journal of Remote Sensing 23: 2505–2512.

40. Gong DY, Shi PJ, Wang JA (2004) Daily precipitation changes in the semi-arid region over northern China. Journal of Arid Environments 59: 771–784.

41. Paulo AA, Pereira LS (2006) Drought concepts and characterization: Comparing drought indices applied at local and regional scales. Water International 31: 37–49.

42. Lee MS, Nakane K, Nakatsubo T, Mo WH, Koizumi H (2002) Effects of rainfall events on soil CO_2 flux in a cool temperate deciduous broad-leaved forest. Ecology Research 17: 401–409.

43. Yuste JC, Janssens IA, Ceulemans R (2005) Calibration and validation of an empirical approach to model soil CO_2 efflux in a deciduous forest. Biogeochemistry 73: 209–230.

44. Sharkhuu A, Plante AF, Enkhmandal O, Casper BB, Helliker BR, et al. (2013) Effects of open-top passive warming chambers on soil respiration in the semi-arid steppe to taiga forest transition zone in Northern Mongolia. Biogeochemistry 115: 333–348.

45. Reynolds JF, Kemp PR, Ogle K, Fernandez RJ (2004) Modifying the 'pulse-reserve' paradigm for deserts of North America: precipitation pulses, soil water, and plant responses. Oecologia 141: 194–210.

46. Boot CM, Schaeffer SM, Schimel JP (2013) Static osmolyte concentrations in microbial biomass during seasonal drought in a California grassland. Soil Biology and Biochemistry 57: 356–361.

47. Schimel J, Balser TC, Wallenstein M (2007) Microbial stress-response physiology and its implications for ecosystem function. Ecology 88: 1386–1394.

48. Yan LM, Chen SP, Huang JH, Lin GH (2010) Differential responses of auto- and heterotrophic soil respiration to water and nitrogen addition in a semiarid temperate steppe. Global Change Biology 16: 2345–2357.

49. Kim DG, Mu S, Kang S, Lee D (2010) Factors controlling soil CO_2 effluxes and the effects of rewetting on effluxes in adjacent deciduous, coniferous, and mixed forests in Korea. Soil Biology and Biochemistry 42: 576–585.

50. Vargas R, Detto M, Baldocchi DD, Allen MF (2011) Multiscale analysis of temporal variability of soil CO_2 production as influenced by weather and vegetation. Global Change Biology 16: 1589–1605.

51. Meehl GA, Arblaster JM, Tebaldi C (2005) Understanding future patterns of increased precipitation intensity in climate model simulations. Geophysical Research Letters 32: L18719.

52. Schmidli J, Frei C (2005) Trends of heavy precipitation and wet and dry spells in Switzerland during the 20th century. International Journal of Climatology 25: 753–771.

53. New M, Hewitson B, Stephenson DB, Tsiga A, Kruger A, et al. (2006) Evidence of trends in daily climate extremes over southern and west Africa. Journal of Geophysical Research-Atmospheres 111: D14102, doi:10.1029/2005JD006289.

54. Tang JW, Baldocchi DD (2005) Spatial-temporal variation in soil respiration in an oak-grass savanna ecosystem in California and its partitioning into autotrophic and heterotrophic components. Biogeochemistry 73: 183–207.

55. Inglima I, Alberti G, Bertolini T, Vaccari FP, Gioli B, et al. (2009) Precipitation pulses enhance respiration of Mediterranean ecosystems: the balance between organic and inorganic components of increased soil CO_2 efflux. Global Change Biology 15: 1289–1301.

10

Linking Annual N$_2$O Emission in Organic Soils to Mineral Nitrogen Input as Estimated by Heterotrophic Respiration and Soil C/N Ratio

Zhijian Mu[1]*, Aiying Huang[2], Jiupai Ni[3], Deti Xie[3]

1 Chongqing Key Laboratory of Soil Multi-scale Interfacial Processes, College of Resources & Environment, Southwest University, Chongqing, China, **2** College of Agronomy & Biotechnology, Southwest University, Chongqing, China, **3** Chongqing Engineering Research Center for Agricultural Non-point Source Pollution Control in Three -Gorges Region, College of Resources & Environment, Southwest University, Chongqing, China

Abstract

Organic soils are an important source of N$_2$O, but global estimates of these fluxes remain uncertain because measurements are sparse. We tested the hypothesis that N$_2$O fluxes can be predicted from estimates of mineral nitrogen input, calculated from readily-available measurements of CO$_2$ flux and soil C/N ratio. From studies of organic soils throughout the world, we compiled a data set of annual CO$_2$ and N$_2$O fluxes which were measured concurrently. The input of soil mineral nitrogen in these studies was estimated from applied fertilizer nitrogen and organic nitrogen mineralization. The latter was calculated by dividing the rate of soil heterotrophic respiration by soil C/N ratio. This index of mineral nitrogen input explained up to 69% of the overall variability of N$_2$O fluxes, whereas CO$_2$ flux or soil C/N ratio alone explained only 49% and 36% of the variability, respectively. Including water table level in the model, along with mineral nitrogen input, further improved the model with the explanatory proportion of variability in N$_2$O flux increasing to 75%. Unlike grassland or cropland soils, forest soils were evidently nitrogen-limited, so water table level had no significant effect on N$_2$O flux. Our proposed approach, which uses the product of soil-derived CO$_2$ flux and the inverse of soil C/N ratio as a proxy for nitrogen mineralization, shows promise for estimating regional or global N$_2$O fluxes from organic soils, although some further enhancements may be warranted.

Editor: Shuijin Hu, North Carolina State University, United States of America

Funding: This research was supported by the National Natural Science Foundation of China (grant number 41371211) and the National Major Science and Technology Projects for Water Pollution Control and Management (grant number 2012ZX07104-003). The funders had no role in study design, data collection and analysis, decision to publish, or preparation of the manuscript.

Competing Interests: The authors have declared that no competing interests exist.

* E-mail: muzj01@gmail.com

Introduction

Although organic soils occupy only 3% of the Earth's land area, they contain approximately 40% (610 Pg) of the terrestrial soil organic carbon (SOC) [1]. Climate warming and human disturbance such as drainage and cultivation are expected to accelerate carbon decomposition in organic soils, and the decomposition of SOC can facilitate the release of mineral nitrogen which can then be utilized by denitrifying and nitrifying bacteria to produce the potent greenhouse gas N$_2$O [2,3]. N$_2$O emissions from organic soils under agricultural use in Nordic countries were on average four times higher than those from mineral soils, indicating that N$_2$O derived from SOC decomposition dominates overall fluxes [4]. However, no consistent and quantitative relationship has been reported for N$_2$O emission and organic carbon decomposition in organic soils.

Organic carbon and nitrogen in soils, plant and microbial biomass are usually covalently bonded at relatively constant ratios. It is thus logical to expect that N$_2$O and CO$_2$ originated from SOC decomposition should be closely linked. Some studies have indeed found a significant relationship between soil N$_2$O and CO$_2$ emissions at the site level [5,6]. This relationship, however, was weaker when data were pooled across sites or ecosystems[7,8]. The

variability of soil C/N ratio may be one of the important factors undermining the correlation for organic soils. The C/N ratio in organic soils ranges from 50~100 in weakly decomposed peat to 12~35 in highly decomposed peat [9]. The supply of mineral nitrogen from SOC decomposition is the outcome of two concurrent and oppositely directed microbial processes – nitrogen mineralization and immobilization [10]. Soils with a high C/N ratio may be characterized by rapid immobilization of nitrogen and soils with a low C/N ratio by higher net nitrogen mineralization and a surplus of available NH$_4^+$ and NO$_3^-$ [11]. A negative relationship has accordingly been shown for C/N ratio of soils and N$_2$O fluxes [9]. Similar to the relationship between N$_2$O and CO$_2$ emissions, the correlation of N$_2$O emission with soil C/N ratio tended to be weak when the data from different sites at larger scales were included [4,12], which makes it difficult to scale up N$_2$O fluxes by CO$_2$ emissions or C/N ratio alone from individual sites to regional scales. In view of the coupling of soil carbon and nitrogen processes and the bridging function of C/N ratio, we hypothesized that a combination of soil CO$_2$ emission and C/N ratio would likely provide better measurements of N$_2$O emission at larger scales. In fact, Mu et al. [13] have linked N$_2$O flux to soil mineral nitrogen as estimated by CO$_2$ emission and C/ N ratio for agricultural mineral soils. To our knowledge, no such kind of attempt has ever been made for organic soils. The aim of this study was therefore to determine: 1) if N_2O flux from organic soils is related to soil mineral nitrogen input estimated from heterotrophic respiration divided by soil C/N ratio (a derived measure of soil nitrogen mineralization) plus fertilizer nitrogen; and 2) whether or not the relationship is sufficiently robust to serve as an approach for estimating N_2O flux from organic soils.

Materials and Methods

Data source

To test the hypothesis, we collected journal-published data of N_2O and CO_2 emissions measured simultaneously in the fields on peatlands or histosols for which the carbon and nitrogen content or ratio of the organic matter in the upper layers of the soil has been reported. Occasional and short-period flux measurements were not used and only data on annual emissions were considered. For long-term measurements, we used annual estimates rather than multi-year averages to reflect temporal variability. Annual emissions were directly reported by authors or estimated from points in the figures of publications. The final dataset comprised of 122 field measurements from 28 geographical sites (Table S1). Of all data, only 12 measurements at 9 sites were from the tropical regions and the rest were from the temperate regions. Most of the flux measurements were made using closed chamber technique with sampling frequency varying from 1–3 times per week to once per month. Other factors such as soil pH and water table level, if reported, were also recorded in the database. Readers should refer to the original papers for a more complete presentation of the data.

Estimation of soil mineral nitrogen input

The CO_2 emission measured in bare soils can be taken as the proxy of SOC decomposition or heterotrophic respiration [14]. There are limited studies in which CO_2 emission was measured in bare soils (Table S1). For the CO_2 emissions measured in soils with plants, the contribution of heterotrophic respiration or SOC decomposition was estimated using the following equation adapted from Bond-Lamberty and Thomson [15]:

$$R_h = 10e^{[0.22 + 0.87\ln(R_t/10)]} \qquad (1)$$

where R_h is heterotrophic respiration and R_t is total soil respiration (kg C ha^{-1} yr^{-1}).

The nitrogen mineralization rate from soil organic matter was then calculated using the following equation [13]:

$$N_m = R_h/S_{CN} \qquad (2)$$

where N_m is the gross nitrogen mineralization (kg N ha^{-1} yr^{-1}) and S_{CN} is soil C/N ratio.

The mineralized nitrogen from soil organic matter decomposition and the inorganic nitrogen from chemical fertilizers constitute the total input of soil mineral nitrogen (N_{mf}). Atmospheric nitrogen deposition, as another important external source of soil mineral nitrogen, was not considered for our study since there were few papers reporting it.

Statistical analysis

The dataset in the current study is of unbalanced nature with observations collected from peer-reviewed papers rather than from systematically designed experiments. Accordingly, the effects of soil mineral nitrogen input and other variables on N_2O flux were analyzed using the mixed model-REML estimation method of SAS/MIXED procedure (version 9.3), which is suitable for handling unbalanced data. The values of N_2O flux were first natural-log transformed to normalize their distribution and then analyzed by the following model:

$$\ln(f_{N2O}) = \text{constant} + \ln(N_{mf}) + pH + WT + NS_i + E\cos ys_j$$
$$+ NS_i \times \ln(N_{mf}) + E\cos ys_j \times \ln(N_{mf}) + E\cos ys_j \times WT$$

where f_{N2O} is the N_2O flux; N_{mf}, pH, WT, NS_i and $Ecosys_j$ are the fixed effects of mineral nitrogen input, soil pH, water table level, nitrogen source (i is mineralized nitrogen only or a combination of mineralized nitrogen and inorganic nitrogen from chemical fertilizers), and ecosystem type (j is forest or non-forest type), respectively. A preliminary check of the data showed that the general trend of N_2O flux in forest system differed from grass and cropland, so the ecosystems were simply classified into two subclasses as forest and non-forest. Some two-factor interactions were also included in the model. A significant level of $p = 0.05$ was used to determine if a given variable or interactive effect was kept in the model to further seek solutions for fixed effects. Four negative values of N_2O flux reported by Inubushi et al. [16] and Mojeremane et al. [17] can not be subjected to log-transformation and were not included in the analysis. In addition to determination coefficient (i.e., R^2 value), concordance between observed N_2O fluxes and model fits was also analyzed using Lin's concordance correlation coefficient (CCC, Stata SE 12.0) to assess the goodness-of-fit of the finalized models. The resulting CCC was interpreted using the benchmarks described by Klevens et al. [18] as follows: <0.20 is considered virtually no agreement; 0.21–0.40 is considered slight; 0.41–0.60 is considered fair; 0.61–0.80 is considered moderate; and 0.81–0.99 is substantial.

Results

As shown in Table 1, soil pH, soil mineral nitrogen source (NS) and ecosystem type did not affect the annual N_2O flux ($p>0.05$), while the input of soil mineral nitrogen (N_{mf}) and water table level (WT) had significant effects on N_2O flux ($p<0.01$). The F value of N_{mf} was the biggest, indicating the input of soil mineral nitrogen was the main factor controlling N_2O emission in organic soils. The two-factor interactive effects between NS, N_{mf}, WT and ecosystem type on N_2O flux were not statistically significant ($p>0.05$).

Only the significant variables were then kept in the model to solve the estimates for their effects. Two models with different combinations of independent variables are shown in Table 2. The first model was the simplest one with N_{mf} as the single independent variable. The second model was expanded by adding the effect of water table level. The 95% confidence intervals of the estimated effect of N_{mf} were overlapped for different models. The models indicated that N_2O flux was positively correlated with N_{mf} and negatively with water table level. Using the estimated effects and the variables in the dataset allowed a comparison between predicted and observed annual N_2O fluxes from organic soils. The variable N_{mf} explained up to 69% of the variability in the overall data of observed N_2O fluxes (Fig. 1), while the addition of water table level increased the explanatory ability to 75% (Fig. 2). When the overall data were further divided by ecosystem types, the performance of models was somewhat different (Fig. 1 & 2). For forest, the determination coefficient (R^2) was nearly stable at the value of 0.63 for both models. In contrast, the introduction of water table level into models slightly improved the fitted results for

Table 1. Results of type III tests of fixed effects.

Effect	Numerator DF	Denominator DF	F Value	Pr>F
N_{mf}	1	96	13.16	0.0005
pH	1	96	1.43	0.2344
WT	1	96	5.15	0.0255
NS	1	96	0.10	0.7472
Ecosystem	1	96	0.70	0.4040
NS×N_{mf}	1	96	0.11	0.7426
WT×N_{mf}	1	96	3.20	0.0767
Ecosystem×N_{mf}	1	96	0.21	0.6506
Ecosystem×WT	1	96	2.17	0.1437

N_{mf}, the mineral nitrogen input to soil; WT, water table level; NS, the source of soil mineral nitrogen.

non-forest systems with R^2 values increasing from 0.59 to 0.69. This indicated that the input of mineral nitrogen was the most important predictor of N_2O flux, while water table level was a weak predictor of N_2O flux and appeared to be dependent on ecosystem type.

The slope of regression lines in Fig. 1 & 2 ranged from 0.50 to 0.75, indicating that the relationship strays from the ideal 1:1 line. Therefore the concordance correlation coefficient (CCC) between observed and predicted N_2O fluxes was calculated to measure robustness of the models. For the overall data with log-transformation, the concordance was substantial with the CCC ranging from 0.82 to 0.86 for the two models. When the log-transformed data were converted to actual N_2O fluxes, however, the cluster of fluxes greater than 15.0 kg N ha^{-1} yr^{-1} was found to be distinctly underestimated. The CCC for this cluster of data

ranged from −0.002 to 0.16 and showed virtually no agreement, suggesting that some important factors responsible for these high fluxes were not accounted for by the models. For the rest of the data (103 fluxes out of 118), the CCC (ranging from 0.63 to 0.68) still showed a moderate concordance.

The variable N_{mf} in the models can be decomposed into soil heterotrophic respiration (R_h), C/N ratio and inorganic nitrogen rate from chemical fertilizer (N_f). The mixed procedure analysis indicated that each of these components of N_{mf} had a significant influence on N_2O flux ($p<0.001$), with R_h and N_f being positively related to N_2O flux and C/N ratio negatively related to N_2O flux. Soil carbon and nitrogen contents, which could replace the variable of C/N ratio, were also significantly negatively or positively correlated with N_2O flux ($p<0.001$). The fitting efficiency between observed and predicted N_2O fluxes by models

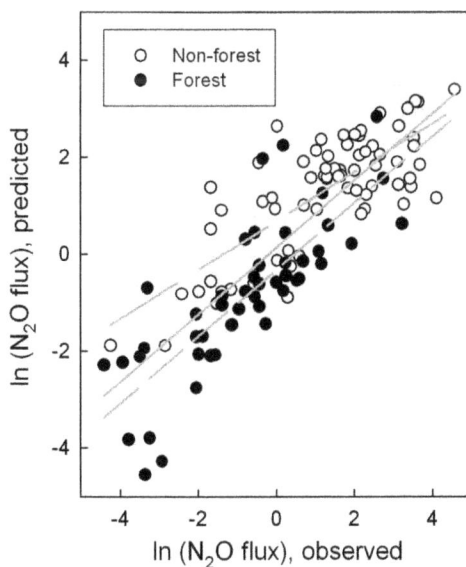

Figure 1. Correlation of observed fluxes of N_2O from organic soils and predicted values (kg N_2O-N ha^{-1}) by model 1 as presented in Table 2: ln(N_2O flux) = 1.8685 ln(N_{mf})−9.0314. Solid line shows linear regression fit for the overall data: y = 0.69x+0.13, R^2 = 0.69. Long-dashed line shows linear regression fit for the non-forest system: y = 0.50x+0.68, R^2 = 0.59. Short-dashed line shows linear regression fit for the forest system: y = 0.69x−0.33, R^2 = 0.63.

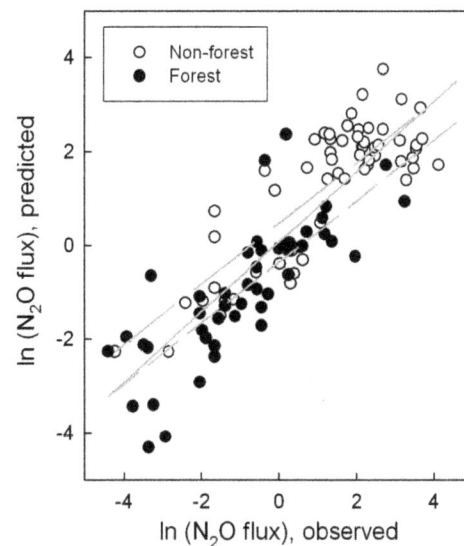

Figure 2. Correlation of observed fluxes of N_2O from organic soils and predicted values (kg N_2O-N ha^{-1}) by model 2 as presented in Table 2: ln(N_2O flux) = 1.5374 ln(N_{mf})−0.0221 WT−8.2334. Solid line shows linear regression fit for the overall data: y = 0.75x+0.08, R^2 = 0.75. Long-dashed line shows linear regression fit for the non-forest system: y = 0.65x+0.48, R^2 = 0.69. Short-dashed line shows linear regression fit for the forest system: y = 0.65x−0.35, R^2 = 0.63.

Table 2. Solutions for fixed effects of the models with log-transformed N_2O flux as dependent variable.

Model	Effect	Estimate	SE	DF	t Value	Pr>|t|	95% confidence Lower	Upper
1	Intercept	−9.0314	0.5952	116	−15.17	<.0001	−10.2102	−7.8526
	ln (N$_{mf}$)	1.8685	0.1157	116	16.15	<.0001	1.6393	2.0976
2	Intercept	−8.2334	0.6222	103	−13.23	<.0001	−9.4675	−6.9994
	ln (N$_{mf}$)	1.5374	0.1479	103	10.39	<.0001	1.2441	1.8308
	WT	−0.0221	0.0053	103	−4.14	<.0001	−0.0326	−0.0115

N$_{mf}$, the mineral nitrogen input to soil (kg N ha^{-1} yr^{-1}); WT, water table level (cm).

using the above-mentioned components of N_{mf} as inputs were nearly the same as those of models using N_{mf} itself (data not shown).

Discussion

Previous studies have linked N_2O flux directly to either CO_2 flux or soil C/N ratio [5,8,9]. In this study, soil CO_2 emission and C/N ratio were combined to estimate mineral nitrogen input, and the latter accounted for up to 69% of the variability of N_2O fluxes from organic soils with various properties, land management practices and climates. Soil CO_2 flux or C/N ratio alone explained only 49% and 36% of the overall variability of N_2O fluxes, respectively (Fig. 3). This suggests the necessity of combining soil CO_2 flux and C/N ratio for predicting N_2O flux on a large scale. Of course, soil CO_2 flux and C/N ratio can be independently incorporated into the same models, but the interpretation of such models would be relatively complicated and evasive since there are various mechanisms which may explain the control of CO_2 flux and C/N ratio over N_2O flux [8,9,19]. In contrast, the quotient of soil CO_2 flux and C/N ratio can well represent in theory the gross nitrogen mineralization [20], and the implication of models using such a quotient as input is straightforward and self-evident in the importance of mineral nitrogen input for regulating soil N_2O flux. There is no significant difference in the influence of different sources of mineral nitrogen on N_2O flux (Table 1), suggesting that the simplified models might also be suitable for evaluating the effect of mineral nitrogen from other sources such as atmospheric deposition, though this idea needs further verification.

A negative relationship between N_2O flux and groundwater level has been observed for individual sites [21,22], and still holds at a large scale as shown in this study. This is logical simply because high moisture with increasing water table level can limit N_2O emission from soils due to the low availability of nitrate and/ or efficient reduction of N_2O to N_2 through denitrification [16,23], while the lowering of water table increases oxygen penetration into the peat and enhances the decomposition of organic matter, as indicated by the negative relationship between heterotrophic respiration and water table level ($R^2 = 0.31$, $p<0.0001$). It has been reported that the control of soil water content or water table level over N_2O flux is important only when soil is not nitrogen limiting [24,25]. In this study, the percentage of observations with N_{mf} greater than 150 kg N ha^{-1} was only 19% for forest, but up to 87% for non-forest systems (Table S1). This suggests that forest soil is nitrogen limiting when compared with non-forest systems, which may be responsible for the insensitivity of N_2O flux to water table level for forest systems (Fig.1& 2). Besides the input of mineral nitrogen, forest differs from non-forest systems in many other factors, such as vegetation, below-/above-ground biomass, litter fall, soil compaction, and land management practices, all of which can influence N_2O flux but are not considered here due to limited and unsystematic information in literature sources of the current dataset. To fill the gap, ecosystem type was used as a proxy variable that we tried to incorporate into models; however, statistical analysis showed that its effect was not significant (Table 1).

It should be acknowledged that the models described here were dependent on simplifying assumptions that can introduce error. That is, the gross nitrogen mineralization was estimated from carbon mineralization and soil C/N ratio by assuming that the rate of carbon mineralization is the same as the rate of respiration and the C/N ratio of mineralized organic matter is the same as that of the bulk soil organic matter. In fact, carbon and nitrogen mineralization from soils originates from decomposable fractions

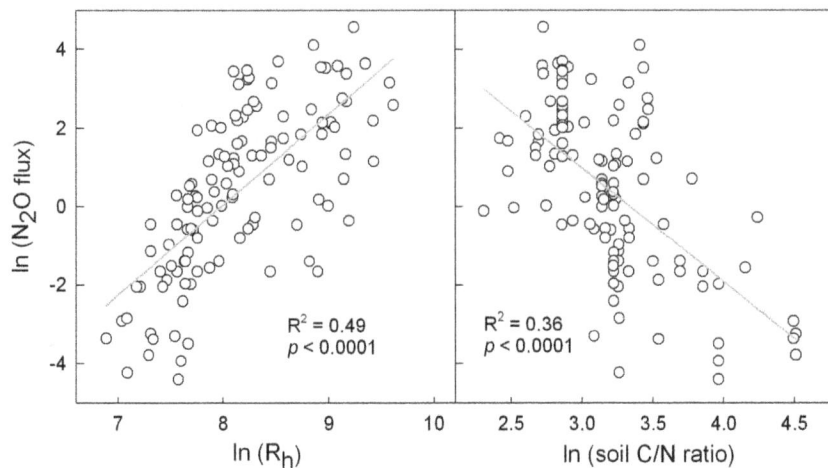

Figure 3. Correlation of observed fluxes of N$_2$O with estimated soil heterotrophic respiration (R$_h$, left panel) and C/N ratio (right panel).

of organic matter with different C/N ratios [26]. Most likely, the ratio of carbon evolved/nitrogen mineralized is much wider than the bulk soil carbon to nitrogen ratio [27,28]. This indicates that gross nitrogen mineralization might be over- or under-estimated if bulk soil C/N ratio was used in equation 2. The respiration process is also not exactly identical to carbon mineralization. The amount of carbon that is ultimately lost through respiration depends on how effectively the decomposer community converts mineralized carbon to biomass [29]. Similarly, the amount of nitrogen that is ultimately available to denitrifier or nitrifier for producing N$_2$O depends on how effectively the decomposer community converts mineralized nitrogen to biomass and plants compete with microbes for mineral nitrogen [10,20]. Empirical relationships have been established between nitrogen and carbon mineralization in studies performed usually under laboratory conditions [30]. Different organic matter fractions or their C/N ratios, and varying microbial use efficiency of carbon and nitrogen have also been proposed to predict nitrogen release [20,29]. However, these relationships are strongly dependent on the experimental conditions in which they have been established. Moreover, the current dataset is based on the in situ measurements in the field environment and contains only the basic information of respiratory carbon and bulk soil C/N ratio, thus necessitating the above-mentioned assumptions to estimate mineralized nitrogen. Such simplifications and assumptions may bring uncertainties, but it is necessary in some cases to understand the general trends and probabilistic nature of the environment [31].

N$_2$O emission from soils is of small magnitude and highly variable in space and time, and is thus very difficult to estimate. The measurement of soil N$_2$O flux also requires intricate techniques along with a lot of time and labor. In contrast, soil CO$_2$ emission is controlled primarily by soil temperature and moisture, and is relatively easy to measure or predict [32,33]. In addition, the estimates of soil respiration are currently more widely available than those of soil N$_2$O emission. The models developed in this study showed a promising approach to estimating N$_2$O emission from organic soils by using soil C/N ratio and CO$_2$ emission data derived from measurements or biogeochemical modeling. It should be mentioned, however, that several aspects of the information in the current dataset might impose uncertainties on these models. First, soil heterotrophic respiration was simply estimated from total soil respiration using a universal relationship

between them [15], but the relative contribution of organic matter decomposition or heterotrophic respiration would vary over time and depend on root respiration of the growing plants [8]. Second, the majority of the global organic soils are distributed in the boreal and sub-arctic regions and about 10%–15% in the tropical countries [1,3], but most of the current data came from northern Europe, indicating that the models developed in the present study might be biased to the temperate regions.

Conclusion

A fairly large number of data were collected to explore the relationship between annual N$_2$O emission and multiple variables for organic soil by a mixed-model analysis, and the input of soil mineral nitrogen was found to be the most useful predictor for N$_2$O flux. Soil mineral nitrogen was supposed to be composed of organic nitrogen mineralization as estimated by CO$_2$ emission and soil C/N ratio, thus providing a possibility for upscaling N$_2$O emission from organic soils by use of regional soil databases including information on C/N ratio and carbon storage change or CO$_2$ emission data. The approach proposed here may have validity as a whole, but needs further evaluation and advancement before practical application due to uncertainties associated with simplifying assumptions and a regionally unbalanced data source. A better understanding of the processes of carbon and nitrogen mineralization and their stoichiometric relationship as well as additional experimental data from organic soils outside of temperate Europe regions will help to improve the relationship established in this study.

Author Contributions

Conceived and designed the experiments: ZJM. Performed the experiments: ZJM AYH. Analyzed the data: ZJM. Contributed reagents/materials/analysis tools: JPN DTX. Wrote the paper: ZJM.

References

1. Page SE, Rieley J, Banks CJ (2011) Global and regional importance of the tropical peatland carbon pool. Global Change Biol 17: 798–818.

2. Goldberg SD, Knorr KH, Blodau C, Lischeid G, Gebauer G (2010) Impact of altering the water table height of an acidic fen on N_2O and NO fluxes and soil concentrations. Global Change Biol 16: 220–233.

3. Frolking S, Talbot J, Jones MC, Treat CC, Kauffman JB, et al. (2011) Peatlands in the Earth's 21st century climate system. Environ Rev 19: 371–396.

4. Maljanen M, Sigurdsson BD, Guðmundsson J, Óskarsson H, Huttunen JT, et al. (2010) Greenhouse gas balances of managed peatlands in the Nordic countries – present knowledge and gaps. Biogeosciences 7: 2711–2738.

5. Garcia-Montiel DC, Melillo JM, Steudler PA, Neill C, Feigl BJ, et al. (2002) Relationship between N_2O and CO_2 emissions from the Amazon Basin. Geophys Res Lett 29: Art.No. 1090.

6. Chatskikh D, Olesen JE (2007) Soil tillage enhanced CO_2 and N_2O emissions from loamy sand soil under spring barley. Soil Till Res 97: 5–18.

7. Keller M, Varner R, Dias JD, Silva H, Crill P, et al. (2005) Soil–atmosphere exchange of nitrous oxide, nitric oxide, methane, and carbon dioxide in logged and undisturbed forest in the Tapajos national forest, Brazil. Earth Interact 9: Art. No. 23.

8. Xu XF, Tian HQ, Hui DF (2008) Convergence in the relationship of CO_2 and N_2O exchanges between soil and atmosphere within terrestrial ecosystems. Global Change Biol 14: 1651–1660.

9. Klemedtsson L, Von Arnold K, Weslien P, Gundeersen P (2005) Soil CN ratio as a scalar parameter to predict nitrous oxide emissions. Global Change Biol 11: 1142–1147.

10. Luxhøi J, Bruun S, Stenberg B, Breland TA, Jensen LS (2006) Prediction of gross and net nitrogen mineralization-immobilization-turnover from respiration. Soil Sci Soc Am J 70: 1121–1128.

11. Bengtsson G, Bengtson P, Månsson KF (2003) Gross nitrogen mineralization-, immobilization-, and nitrification rates as a function of soil C/N ratio and microbial activity. Soil Biol Biochem 35: 143–154.

12. Ojanen H, Minkkinen K, Alm J, Penttila T (2010) Soil – atmosphere CO_2, CH_4 and N_2O fluxes in boreal forestry-drained peatlands. Forest Ecol Manage 260: 411–421.

13. Mu ZJ, Huang AY, Kimura SD, Jin T, Wei SQ, et al. (2009) Linking N_2O emission to soil mineral N as estimated by CO_2 emission and soil C/N ratio. Soil Biol Biochem 41: 2593–2597.

14. Hanson PJ, Edwards NT, Garten CT, Andrews JA (2000) Separating root and soil microbial contributions to soil respiration: A review of methods and observations. Biogeochemistry 48: 115–146.

15. Bond-Lamberty B, Thomson A (2010) A global database of soil respiration data. Biogeosciences 7: 1915–1926.

16. Inubushi K, Furukawa Y, Hadi A, Purnomo E, Tsuruta H (2003) Seasonal changes of CO_2, CH_4 and N_2O fluxes in relation to land-use change in tropical peatlands located in coastal area of South Kalimantan. Chemosphere 52: 603–608.

17. Mojeremane W, Rees RM, Mencuccini M (2012) The effects of site preparation practices on carbon dioxide, methane and nitrous oxide fluxes from a peaty gley soil. Forestry 85: 1–15.

18. Klevens J, Trick WE, Kee R, Angulo F, Garcia D, et al. (2011) Concordance in the measurement of quality of life and health indicators between two methods of computer-assisted interviews: self-administered and by telephone. Qual Life Res 20: 1179–1186.

19. Rochette P, Tremblay N, Fallon E, Angers DA, Chantigny MH, et al. (2010) N_2O emissions from an irrigated and non-irrigated organic soil in eastern Canada as influenced by N fertilizer addition. Eur J Soil Sci 61: 186–196.

20. Murphy DV, Recous S, Stockdale EA, Fillery IRP, Jensen LS, et al. (2003) Gross nitrogen fluxes in soil: Theory, measurement and application of ^{15}N pool dilution techniques. Adv Agron 79: 69–118.

21. Regina K, Silvola J, Martikainen PJ (1999) Short-term effects of changing water table on N_2O fluxes from peat monoliths from natural and drained boreal peatlands. Global Change Biol 5: 183–189.

22. Danevcic T, Mandic-Mulec I, Stres B, Stopar D, Hacin J (2010) Emissions of CO_2, CH_4 and N_2O from southern European peatlands. Soil Biol Biochem 42: 1437–1446.

23. Maljanen M, Shurpali N, Hytönen J, Mäkiranta P, Aro L, et al. (2012) Afforestation does not necessarily reduce nitrous oxide emissions from managed boreal peat soils. Biogeochemistry 108: 199–218.

24. Smith KA, Thomson PE, Clayton PE, McTaggart IP, Conen F (1998) Effects of temperature, water content and nitrogen fertilization on emissions of nitrous oxide by soil. Atmos Environ 32: 3301–3309.

25. Weslien P, Klemedtsson AK, Borjesson G, Klemedtsson L (2009) Strong pH influence on N_2O and CH_4 fluxes from forested organic soils. Eur J Soil Sci 60: 311–320.

26. Springob G, Kirchmann H (2003) Bulk soil C to N ratio as a simple measure of net N mineralization from stabilized soil organic matter in sandy arable soils. Soil Biol Biochem 35: 629–632.

27. Sollins P, Spycher G, Glassman CA (1984) Net nitrogen mineralization from light- and heavy-fraction forest soil organic matter. Soil Biol Biochem 16: 31–37.

28. Kader MA, Sleutel S, Begum SA, D'Haene K, Jegajeevagan K, et al. (2010) Soil organic matter fractionation as a tool for predicting nitrogen mineralization in silty arable soils. Soil Use Manage 26: 494–507.

29. Manzoni S, Taylor P, Richter A, Porporato A, Ågren GI (2012) Environmental and stoichiometric controls on microbial carbon-use efficiency in soils. New Phytol 196: 79–91.

30. Nicolardot B, Recous S, Mary B (2001) Simulation of C and N mineralization during crop residue decomposition: a simple dynamic model based on the C:N ratio of the residues. Plant Soil 228: 83–103.

31. Yan XY, Yagi K, Akiyama H, Akimoto H (2005) Statisical analysis of the major variables controlling methane emission from rice fields. Global Change Biol 11: 1131–1141.

32. Lloyd J, Taylor JA (1994) On the temperature-dependence of soil respiration. Funct Ecol 8: 315–323.

33. Raich JW, Potter CS, Bhagawati D (2002) Interannual variability in global soil respiration, 1984-94. Global Change Biol 8: 800–812.

11

Organic Matter and Water Addition Enhance Soil Respiration in an Arid Region

Liming Lai[1], Jianjian Wang[1,2], Yuan Tian[3], Xuechun Zhao[1], Lianhe Jiang[1], Xi Chen[3], Yong Gao[4], Shaoming Wang[5], Yuanrun Zheng[1]*

1 Key Laboratory of Resource Plants, Beijing Botanical garden, West China Subalpine Botanical Garden, Institute of Botany, Chinese Academy of Sciences, Beijing, China, 2 University of Chinese Academy of Sciences, Beijing, China, 3 Xinjiang Institute of Ecology and Geography, Chinese Academy of Sciences, Urumqi, Xinjiang, China, 4 Inner Mongolia Agricultural University, Hohhot, Inner Mongolia, China, 5 Key Laboratory of Oasis Ecological Agriculture of Xinjiang Bingtuan, Shihezi, Xinjiang, China

Abstract

Climate change is generally predicted to increase net primary production, which could lead to additional C input to soil. In arid central Asia, precipitation has increased and is predicted to increase further. To assess the combined effects of these changes on soil CO_2 efflux in arid land, a two factorial manipulation experiment in the shrubland of an arid region in northwest China was conducted. The experiment used a nested design with fresh organic matter and water as the two controlled parameters. It was found that both fresh organic matter and water enhanced soil respiration, and there was a synergistic effect of these two treatments on soil respiration increase. Water addition not only enhanced soil C emission, but also regulated soil C sequestration by fresh organic matter addition. The results indicated that the soil CO_2 flux of the shrubland is likely to increase with climate change, and precipitation played a dominant role in regulating soil C balance in the shrubland of an arid region.

Editor: Wei-Chun Chin, University of California, Merced, United States of America

Funding: This work was supported by China National Key Basic Research Program (2009CB825103, 2012CB956204). The funders had no role in study design, data collection and analysis, decision to publish, or preparation of the manuscript.

Competing Interests: The authors have declared that no competing interests exist.

* E-mail: zhengyr@ibcas.ac.cn

Introduction

Soil respiration (R_S) is considered the second largest terrestrial carbon (C) flux [1]. This large flux is estimated at $\sim 98 \pm 12$ Pg/yr, which is an order of magnitude larger than fossil fuel combustion [2], and atmospheric concentration of CO_2 has increased at 1.9 ppm/yr during the last 10 years [3]. Therefore, any alterations in soil CO_2 efflux could potentially exacerbate greenhouse gas induced climate warming [4]. The response of R_S to climate change is a critical component in predicting the possible changes in the global C cycle and the climate feedbacks [5]. Generally, elevated CO_2 is assumed to stimulate primary production and C input into soil [6], and additional soil C sequestration could mitigate climate change [7]. As the decomposition of soil organic matter is a temperature dependent process, global warming can also increase soil CO_2 flux to the atmosphere [5]. Thus, there could be positive or negative feebacks between the elevated CO_2 and soil C sequestration, and how R_S changes under elevated CO_2 will determine whether the additional C input will be sequestered in soil. Although many simulation experiments [8,9] and modelling analyses [10] have suggested that R_S should change with climate, the responses of R_S are not well understood [2,11].

Soil temperature, moisture and substrate concentration have long been identified as the main abiotic controlling factors of R_S [4,12] and are used as the fundamental parameters in R_S simulating models [4,13]. Temperature can affect almost all biochemical and physiological aspects of the respiration process [14]. The R_S-temperature relationship has been suggested to be a critical component in predicting global C cycle feedbacks to climate warming [4,5]. The temperature sensitivity of R_S is often described by Q_{10}, the factor by which R_S changes with a 10°C rise in temperature. The global mean Q_{10} value was reported at ~ 1.5 [2,15]. Despite the many studies across different temporal and spatial patterns on this subject [11,16], some aspects remain unclear due to the complexity of the R_S process and the effects of environmental factors [12,17]. One important question is the effects of soil temperature with depth on Q_{10} value estimation, because of the phase shift of temperature fluctuations with soil depth increasing [17,18]. Therefore, the temperature measurements at different soil depths should be considered in field experiments.

Soil moisture plays a crucial role in microbial growth and activity [19]. The activities of the microbes need a certain range of soil water content, and will be depressed as the soils continue to dry [8]. At low soil water potentials, the contact with available substrate and physiological performance of microbes are limited [20]. All these could inhibit the biochemical processes underground and so lower the R_S [19]. According to the Fourth Assessment Report of IPCC, significantly increased precipitation has been observed in central Asia [3]. The changed precipitation is anticipated to have great effects on plant growth, plant physiology and ecosystem productivity, and will influence R_S dynamics [7,21].

Under elevated atmospheric CO_2 concentrations, the stimulated plant productivity can supply more substrate for soil microbes. It has been shown that addition of a small amount of substrate (e.g.

cellulose) to soil can strongly increase the total soil respiration and remain constant after the additional substrate exhaustion, which would lead to a soil C loss [22]. However, some studies suggest that increased C inputs could enhance soil C sequestration in terrestrial ecosystems [23,24]. Therefore, it is necessary to clarify the balance between the organic matter input to the soil and the R_S flux for a better understanding of C cycling under climate change.

Arid regions occupy ~20% of the global terrestrial surface [25], and the amount of soil C stored in arid ecosystems is huge [26]. Although much research has been conducted to investigate the climate change soil C cycle, arid regions have received relatively less attention [27]. Arid ecosystems have been predicted to be one of the most responsive ecosystem types to global climate change [28]. Further, the significant potential C sink capacity in soils of arid ecosystems through restorative management makes arid regions more important for the sequestration of soil C in view of global climate change [26].

The Xinjiang Uygar Autonomous Region (XUAR) in northwest China covers over one sixth of China's land area and includes the majority of the country's arid areas [29]. Widely distributed saline/alkaline soils and low precipitation are two principal characteristics of this region [30]. During the past 50 years, the precipitation in XUAR has significantly increased [31], and this increasing trend is predicted to continue in the future [3]. According to an analysis on climate change of XUAR by Xu and Wei [32], the precipitation in the region had increased by 12.8%–28.8% since 1987. To the soil C pool, the climate changes might bring two contradictory processes: soils may receive more plant C input that will result in a net C sequestration; meanwhile, increasing soil organic C decomposition activities will also stimulate C output. As the CO_2 flux from R_S represents the major pathway for C exchange between the soil C pool and the atmosphere, identifying the response of R_S to the impacts of more plant C input and precipitation would help gain a better understanding of C cycling in arid regions under global climate change. Many studies have reported the direct effects of C input [24,33] and precipitation [27] on C cycling, but little is known about how R_S will be affected by the predicted changes in increased C input, precipitation and their interactions, especially in arid regions. An experiment consisting of fresh organic matter and water addition treatments in a shrubland in XUAR was conducted to answer following questions: (1) will the soil have a net

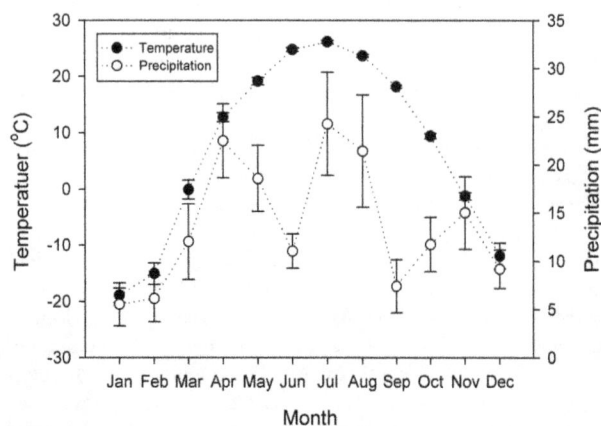

Figure 2. Daily courses of soil temperature at different measuring depths at the adjacent natural conditions (data of Jun-21 and 22). T0, T5, T10, T20, T30 and T40 represent temperature measured at 0, 5, 10, 20, 30 and 40 cm soil depth from the soil surface, respectively. Lowercase letters represent time-delay of daily highest temperature from T0, and values are 60, 240, 420, 740 and 960 sec minutes for a, b, c, d and e, respectively.

C gain or loss from C addition and water addition? (2) how will the relationships between R_S and environmental factors change under C addition and water addition? (3) which depth is most suitable for simulating the temperature sensitivity of R_S in arid regions?

Materials and Methods

Ethics Statement

All necessary permits were obtained for the field studies described. The study sites are managed by the Fukang Station of Desert Ecology, Xinjiang Institute of Ecology and Geography, Chinese Academy of Sciences.

Study site description

The study was conducted at the Chinese Academy of Sciences' Fukang Station of Desert Ecology (87° 56′ E, 44° 17′ N, elevation 461 m), an arid area located in the hinterland of northwest China. There is no grazing or other human disturbance existed in the study site. Mean annual temperature is 6.6°C with mean annual rainfall of 160 mm (Figure 1). The soil is clay-loam with high pH (9.38±0.03) and electrical conductivity (1.54±0.36 dS/m). The field site is *Reaumuria soongorica* dominated shrubland, which is a popular native vegetation species in the XUAR.

Experimental design

The experiment used a nested design with fresh added organic matter (FOM) and water (P) as the two controlled parameters. In early June 2011, 27 randomly distributed study plots (3×3 m) were established in a *R. soongorica* community with uniformly distributed vegetation. Adjacent plots were at least 2 m apart for mitigating buffering effects.

It was difficult to add FOM to the soil surface without being disturbed by the wind in spring. Soil layers deeper than 20 cm can only receive extremely small amounts of plant C and the SOC is relatively stable (Fontaine et al., 2007). Approximately 76% of feeder roots of the *R. soongorica* are located above 20 cm [30]. Therefore, the 20–30 cm layer was chosen as the FOM input layers.

FOM was added to the 20–30 cm deep soil to increase the SOC by 0% (F0, no FOM input), 5% (F1, 33.64 g/m²), and 10% (F2, 67.28 g/m²). Based on the long-term precipitation data from the nearby weather station of Fukang research station, three water

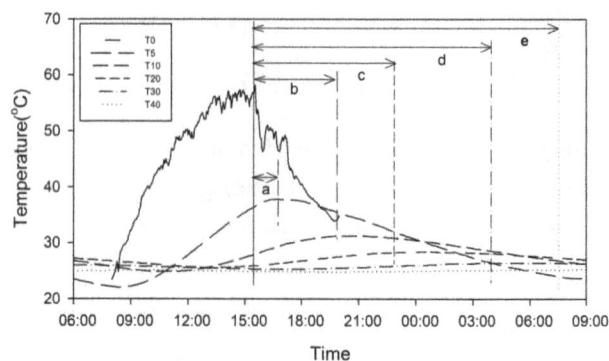

Figure 1. Mean values of temperature and precipitation of the study area (2004~2011). Vertical bars indicate standard errors of means (*n* = 8).

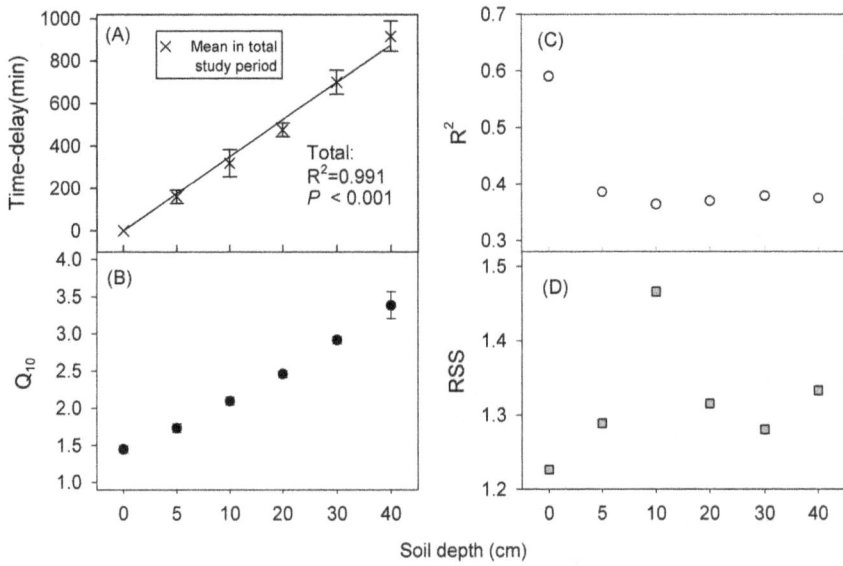

Figure 3. Dependence of the time-delay of daily highest temperature from T0 (mean ± SE), values of Q_{10}, R^2 and RSS on the depth of soil temperature measuring point. Sub-Fig (A), (B), (C) and (D) represent time-delay from T0, values of Q_{10}, R^2 and RSS, respectively. Q_{10}: the temperature sensitivity of R_S; RSS: residual sum of squares of the exponential function.

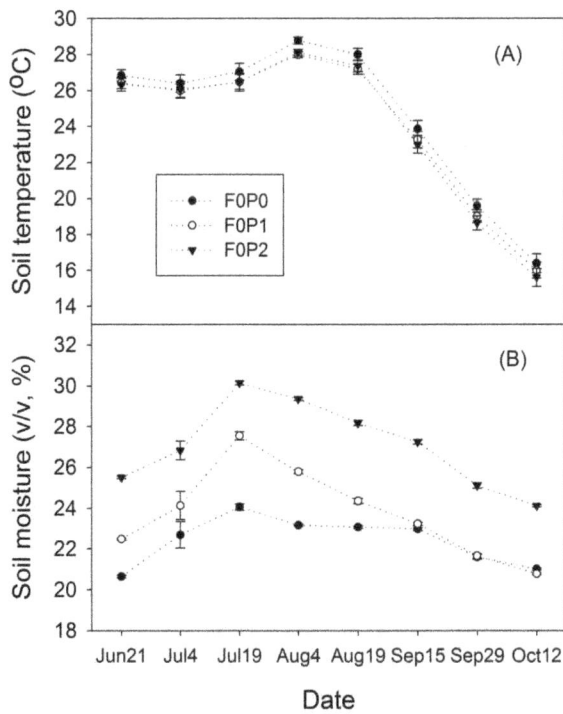

Figure 4. Mean values of temperature and soil moisture under different water addition regimes without FOM treatments. Sub-Fig (A) and (B) represent temperature and soil moisture, respectively. Vertical bars indicate standard errors of means (n = 3). FOM: fresh organic matter; F0: no FOM input; P0, P1, P2: no added water, 50% and 100% increase in water addition, respectively.

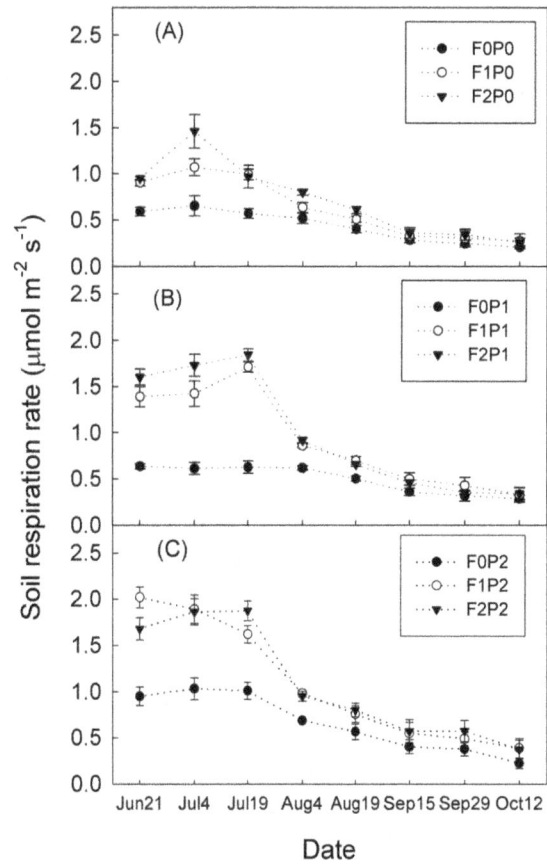

Figure 5. Daily means of soil respiration (mean ± SE) under different treatments. Sub-Fig (A), (B) and (C) represent soil respiration rate under P0, P1 and P2, respectively. Vertical bars indicate standard errors of means (n = 3). F1 and F2: 5% and 10% increase in soil organic carbon. Other abbreviations are same as Figure 4.

addition treatments were applied weekly at rates equivalent to 0% (P0, no additional irrigation), 50% (P1, approximately 2.43 mm), and 100% (P2, approximately 4.87 mm) increased precipitation. Two water addition regimes (50% and 100% of normal weekly precipitation) used in this study was set as about two and four times of average precipitation increase from 1987 to 2001 [32] in order to get much more significant effect of water addition for well understanding the potential results of water increase. At each water addition plot, water was added with the sprinkling can manually. A randomized block design was used for the plots, with the three FOM treatments nested within the three water addition treatments with three replications.

At the end of the growing season in 2010, plant litter was collected from an adjacent *R. soongorica* community site as the FOM for the experiment. After collection, impurities such as lumps of soil were removed and the plant material was dried at 65°C in an oven for 48 h, and then stored at 4°C until use. Before use in the field experiments, the plant materials were ground and sieved to 2 mm. The C, N content and the C: N of the FOM were 44.42±3.31% (mean ± SE, n = 5), 3.18±0.08% (n = 5) and 13.94±0.97 (n = 5), respectively.

To prepare the plots, dead plant tissues and litter from the soil surface was collected (a very small amount), and replaced after the initial plot treatments had been completed. Next, the top 0–20 cm soil was carefully removed in layers, taking care to keep the soils in their original shape and structure, and ensuring that the plants and roots were not badly affected. The FOM was then added. Finally, the top soil was replaced in its original position, and the gaps between the soil blocks were filled with soil from the 0–10 cm layers collected from an adjacent area. The F0 plots were treated the same as the added FOM treatment plots.

Soil CO_2 flux measurements

Soil respiration was measured with an LI-8100 automated soil CO_2 flux system equipped with an LI-COR 8100-103 chamber (LI-COR chamber volume of 4843 cm^3, Lincoln, NE, USA). At each plot, one PVC collar (20.3 cm inner diameter and 15 cm high) was inserted into the soil with 3 cm exposed above the soil

surface. Before setting up a PVC collar, litter on the soil surface was cleared. Measurements were conducted every two weeks from June to October. Besides the 27 plots with manual treatments, 5 more PVC collars were set at adjacent natural conditions (NC) as controls, similar to treatment sites and without human disturbances in recent 30 years. Measurements were made from 8:00 to 20:00 in two-hour rounds. To reduce the effects of temperature rise, the plots were separated into two parts and measurements conducted on two consecutive days with sunny conditions.

Soil temperature and moisture measurements

The soil temperature and soil water content (v/v, %) at 0, 5, 10, 20, 30 and 40 cm depth of each single irrigation controlled sites (F0P0, F0P1 and F0P2) were measured automatically with 5TE soil moisture sensors equipped with the EM50 data-logger (Decagon devices, USA).

Root biomass measurements

Due to the high spatial variability of the shrub roots, root samples were collected from one large square core (25×25 cm, and 60 cm deep) from each plot (27 plots) in October 2011. The roots were carefully separated from these samples and cleaned in water, retaining only apparently living material based on the color, texture, and shape of the roots. Fine roots were classified by their diameter (<2 mm). Then roots were dried at 70°C to a constant mass and weighed. Roots deeper than 60 cm were excluded because very few roots reach that depth.

Analysis of soil physical and chemical properties

Soil samples were collected from 0–30 cm and sieved to 2 mm immediately and stored at 4°C for microbial biomass C (MBC) analysis. MBC was determined using the fumigation-extraction method [34]. SOC was measured using the method described by Nelson and Sommers [35]. The pH (1:5 solid–water ratio) and EC (1:5 solid–water ratio) were determined with a Eutech PC700 pH/EC meter (Thermo Fisher Scientific Inc., Waltham, Massachusetts, USA).

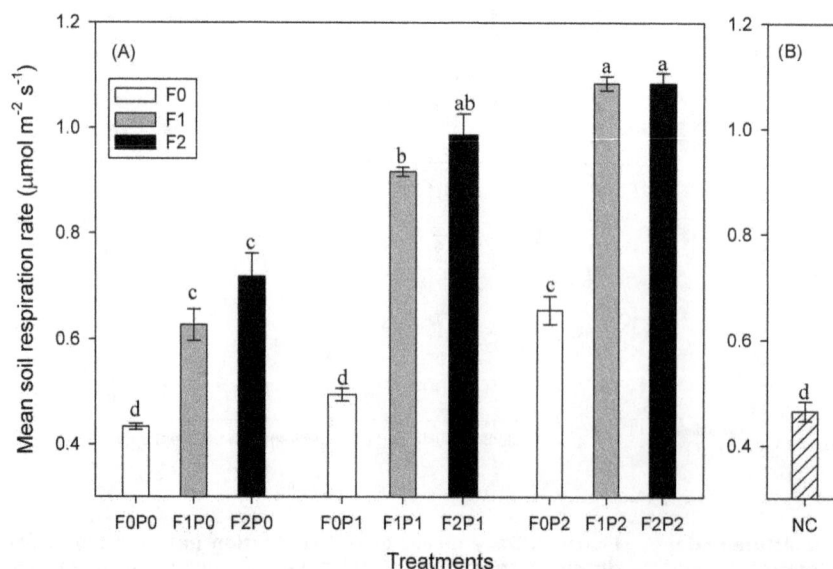

Figure 6. Mean values (mean ± SE) of soil respiration under different treatments and at the natural condition site in total study period. Sub-Fig (A) and (B) represent soil respiration under different treatments and natural site, respectively. Vertical bars indicate standard errors of means (n = 3). Bars with different lowercase letters are significantly different from each other at p<0.05. Abbreviations are same as Figure 5.

Table 1. Results (*F*-Values) of ANOVA with fresh organic matter (FOM) and water addition, and their interactions on soil respiration.

Source	df	F
FOM (F)	2	188.703***
Water addition (P)	2	99.136***
F×P	4	12.571***

***, *P*<0.001.

Data analysis

The temperature dependence of R_S was fitted with an exponential function [36]:

$$Rs = ae^{bT} \tag{1}$$

where a and b are fitted constants, and T is the soil temperature (°C).

From the equation 1, the Q_{10} was calculated as:

$$Q10 = \frac{ae^{b(T+10)}}{ae^{bT}} = e^{10b} \tag{2}$$

Where b is the fitted constant in equation 1.

A linear function was used to describe the relationship between R_S and soil moisture [37]:

$$Rs = c + dW \tag{3}$$

where c and d are fitted constants, and W is the soil moisture.

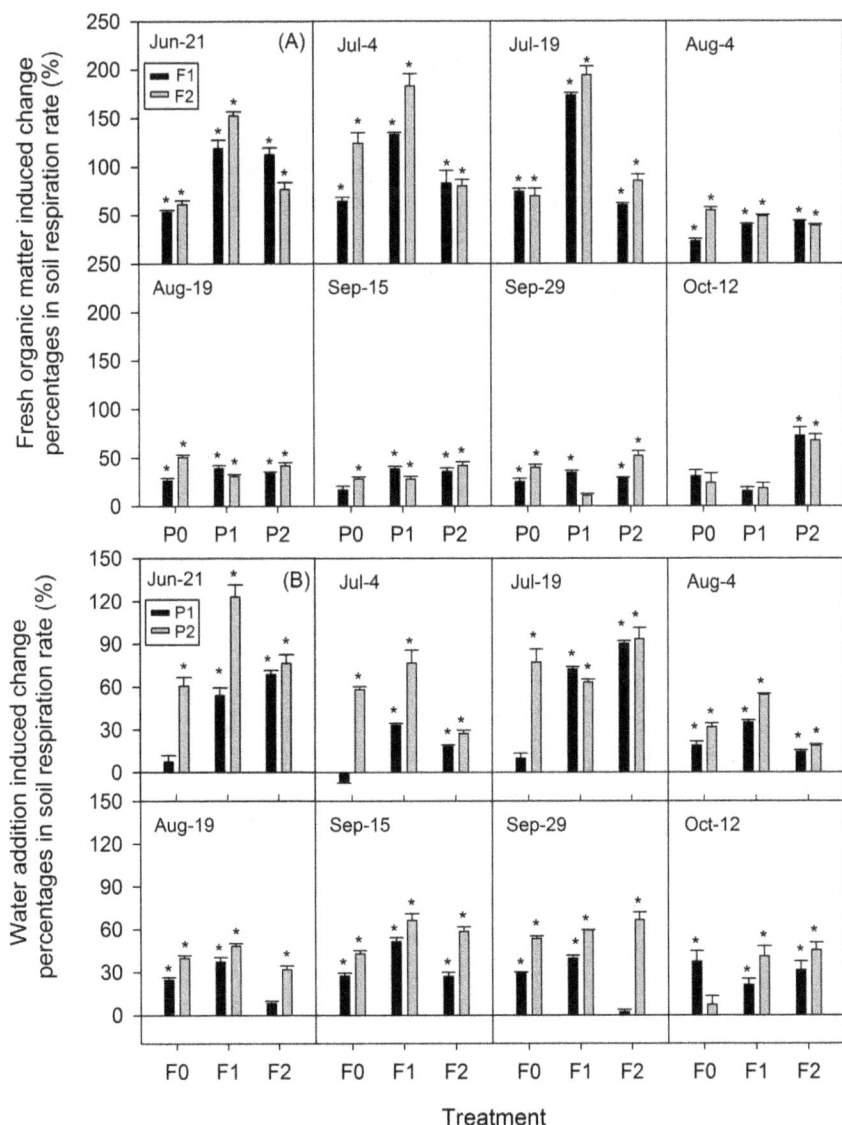

Figure 7. Fresh organic matter and water addition induced change percentages in soil respiration rate (mean ± SE). Sub-Fig (A) and (B) represent fresh organic matter and water addition induced change percentages, respectively. Vertical bars indicate standard errors of means (*n* = 3). * represent significant different compared with controls at *P*<0.01. Other abbreviations are same as Figure 4.

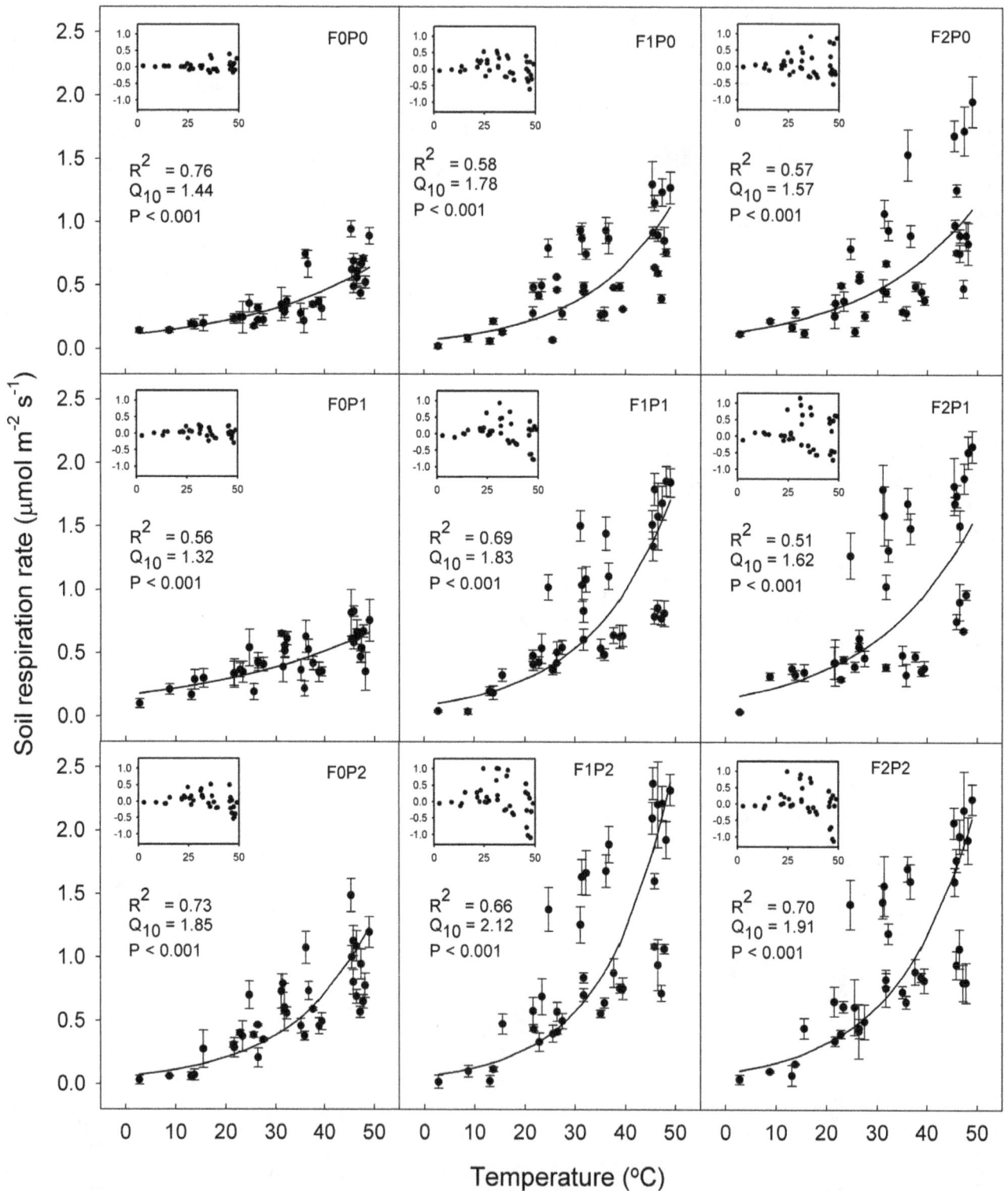

Figure 8. Relationship between soil respiration (mean ± SE) and soil temperature under different FOM and water addition treatments. An exponential was used to describe the relationship. Vertical bars indicate standard errors of means (*n* = 3). Inserts represent the residuals from the equation. Other abbreviations are same as Figure 4.

Table 2. Fitted relationships of soil respiration (μmol m^{-2} s^{-1}) with soil moisture (at 10 cm soil depth) (W, %) in all treatments.

	Functions	R^2	P
F0P0	$R_S = -0.55+4.18w$	0.43	***
F1P0	$R_S = -1.27+8.04w$	0.58	***
F2P0	$R_S = -1.81+10.72w$	0.64	***
F0P1	$R_S = -2.41+11.62w$	0.50	***
F1P1	$R_S = -7.31+32.95w$	0.45	***
F2P1	$R_S = -8.63+38.51w$	0.44	***
F0P2	$R_S = -5.46+23.99w$	0.38	***
F1P2	$R_S = -9.08+39.86w$	0.29	***
F2P2	$R_S = -9.25+40.52w$	0.33	***

F0, F1, F2: no FOM input, 5% and 10% increase in soil organic carbon; P0, P1, P2: no added water, 50% and 100% increase in water addition, respectively. Other abbreviations are same as Table 1.

An exponential-exponential function was used to describe effects of soil temperature and soil moisture on Rs [38]:

$$Rs = je^{mW}e^{nT} \qquad (4)$$

where j, m and n are fitted constants, W is the soil moisture, and T is the soil temperature (°C).

The model residuals or residual sum of squares (RSS) were used to evaluate the model performance.

To investigate the R_S changes induced by a single treatment factor, changes in R_S were calculated between F0 and F1, F0 and F2 under same water treatment; and R_S changes between P0 and P1, P0 and P2 under same FOM treatment. The changes were calculated as:

FOM (or Water) induced change percentages in

$$Rs = \frac{Rs' - Rsc}{Rsc} \times 100\% \qquad (5)$$

for FOM induced change percentages in R_S, the R_S' is soil respiration with F1 or F2 treatment, R_{SC} is soil respiration with F0 treatment, under the same water condition; for water induced changes percentages in R_S, R_S' is soil respiration in plots with P1 or P2 treatment, R_{SC} is soil respiration in plots with P0 treatment, under same FOM treatment condition.

All statistical linear and nonlinear regression analyses, multiple comparisons including one-way ANOVA and homogeneity of variance tests were performed with SPSS 13.0 software [39]. Multiple comparisons of means for different treatments were analyzed using Tukey's test.

Results

Effects of measurement depth on temperature sensitivity of soil respiration

R_S ranged from 0.02 to 0.85 μmol CO$_2$ m^{-2} s^{-1} with an average of 0.47$\pm$0.02 μmol CO$_2$ m^{-2} s^{-1} for the NC plots. The temperature varied from 2.77–60.16, 9.42–38.00, 13.04–32.64, 15.29–29.35, 16.64–28.87 and 17.55–28.11°C for the 0, 5, 10, 20, 30 and 40 cm soil layers, respectively. An increasing time delay of the daily highest value of soil temperature and lower

amplitudes of daily temperature with deeper soil depths was observed (e.g. temperature of Jun-21 and 22, Figure 2). The time delay of the temperature between 0 cm and other soil depths increased significantly with a linear relationship in the whole measuring period ($P<0.001$, Figure 3A).

Q_{10} calculated from seasonal values of R_S and temperature increased significantly with depth of the soil temperature measuring point, from 1.45$\pm$0.05 (at 0 cm) to 3.39$\pm$0.18 (at 40 cm) (Figure 3B). The R^2 and RSS (residual sum of squares) of the exponential relationship between R_S and temperature showed a better simulating effect at the 0 cm temperature, because the highest R^2 and lowest RSS were observed (Figure 3C, D). For reducing the uncertainty in simulating the relationship of R_S and temperature, due to time delay of the temperature at deeper soil layers, only 0 cm temperature was used for Q_{10} calculations. Further, the equations derived from 0 cm temperature had higher R^2 and lower RSS compared with that derived from temperature at other soil layers.

Temperature and soil moisture

During the study period (June–October), soil temperature at 0–30 cm had no significant differences under the three precipitation regimes, although slightly lower temperatures were observed in the plots with added water (Figure 4A). Water addition significantly increased soil moisture at 0–30 cm ($P<0.001$).

In contrast to the P0 plots, the relative increase in mean soil moisture values at P2 were 3.1–6.2% (v/v) through the whole study period, and the relative increase in mean soil moisture at P1 plot was 1.3–3.5% (v/v) from Jun 21 to Aug 19, but there were no significant differences between the soil moisture of P0 and P1 treatments during the last two months (Figure 4B).

Dynamics of soil respiration in the study period

R_S in all treatments showed a seasonal pattern with a single peak curve that occurred in July and began to decrease in August, partly due to the high temperature and lower soil moisture (Figure 4, Figure 5). Under single water addition, the R_S variations in F0P0, F0P1 and F0P2 plots were lower than that under FOM addition.

One-way ANOVA showed that there were significant differences among mean R_S values of the nine treatments ($P<0.05$; Figure 6A). They were highest at F2P2 and lowest at F0P0 plots, with values of 1.08$\pm$0.02 and 0.47$\pm$0.02 μmol m^{-2} s^{-1}, respectively.

No significant difference was found ($P>0.05$, Figure 6) for the mean R_S between the F0P0 (with manual disturbances) treatment and the adjacent natural condition site.

Effects of FOM addition on soil respiration

The FOM addition to soil significantly increased mean R_S of the total study period (Table 1, Figure 6A). Compared with R_S in plots without FOM addition for each precipitation treatment (F0P0, F0P1 and F0P2), the greatest increases induced by FOM addition were observed in the first two months of the experiment (Jun-21 to Jul-19), and varied from 52 to 194% ($P<0.01$, Figure 7A). The FOM addition-induced increase in R_S was larger than that induced by water addition at this period, and then the stimulating effects on R_S declined from Aug-4 to Oct-12 (Figure 7).

Effects of water addition on soil respiration

The water addition had a significant effect on variations of R_S (Table 1). During the whole study period, P1 and P2 treatments

Table 3. Fitted relationships of soil respiration ($\mu mol\ m^{-2}\ s^{-1}$) with soil moisture (at 10 soil depth) (W, %) and soil temperature (T, at soil surface, °C).

	Functions	R^2	P
F0P0	$R_S = 0.047e^{0.3007T}e^{4.407W}$	0.82	***
F1P0	$R_S = 0.014e^{0.0457T}e^{8.352W}$	0.65	***
F2P0	$R_S = 0.016e^{0.0297T}e^{10.897W}$	0.76	***
F0P1	$R_S = 8.05E^{-03}e^{0.020T}e^{13.332W}$	0.61	***
F1P1	$R_S = 2.32E^{-03}e^{0.051T}e^{15.926W}$	0.72	***
F2P1	$R_S = 1.59E^{-05}e^{0.025T}e^{39.975W}$	0.66	***
F0P2	$R_S = 5.27E^{-03}e^{0.056T}e^{10.273W}$	0.74	***
F1P2	$R_S = 0.013e^{0.072T}e^{6.275W}$	0.66	***
F2P2	$R_S = 8.12E^{-03}e^{0.067T}e^{10.035W}$	0.71	***

Other abbreviations are same as Table 1.

stimulated mean R_S by 14 and 51%, 46 and 73% and 37 and 51% under the F0, F1 and F2 treatments, respectively (Figure 6A). R_S changes induced by water addition were also higher during the first two months (except some values on Jul-4). Water addition induced higher increases in R_S on plots with FOM treatment (F1 and F2) than those without FOM treatment (F0) on most of measuring days (Figure 7B).

Variation in soil respiration related to temperature, soil moisture, microbial biomass and fine root biomass

Due to the depression of the mesophilic microbial community by the high temperature, relationship between R_S and the temperature was calculated used data below 50°C. The variations in R_S showed a significant exponential relationship with temperature for all treatments ($P<0.001$), which explained 51–76% of the variation in R_S (Figure 8). The residual distributions in all treatments indicated a high simulation of R_S at a relatively lower temperature range ($\sim<30°C$). Q_{10} for the nine treatments ranged from 1.32 to 2.12. Comparing with the Q_{10} for controls plots (F0P0), FOM and/or precipitation treatments increased the Q_{10} (except Q_{10} for F0P1).

When soil moisture was taken as a single controlling factor of R_S, the linear equations explained 29–64% of the R_S variation ($P<0.001$, Table 2) and the fitted equations were better in the treatments without water addition ($R^2 = 0.43$–0.64).

Soil moisture and temperature together could improve the correlation coefficients of the regression equation for R_S ($P<0.001$, $R^2 = 0.61$–0.82) in all treatments (Table 3). Similarly, the fitted relationships were better under plots without water addition.

When soil temperature, soil moisture, microbial biomass (MB) and fine root biomass (FRB) were used, the regression equation, $R_S = 0.501T + 0.023\ W + 0.033FRB - 0.002\ MB - 16.36$ ($P=0.018$, $R^2 = 0.92$), could be built and R^2 was improved. When stepwise regression was used to build an equation between R_S and the four factors, only root biomass significantly influenced Rs ($P<0.001$, $R^2 = 0.86$).

Discussion

Measurement depth and soil respiration

Poor understanding of the temperature sensitivity of R_S results in uncertainty in climate models [11,16]. In many studies, Q_{10} was calculated without accounting for the depth of the soil temperature measuring point [40]. For example, studies have measurements at 2 cm depth [17], 5 cm depth [38] and 10 cm depth [41]. Because heat conductance from the soil surface to below-ground needs a period of time, a time-lag exists between the temperature at the soil surface and other layers. Previous studies [18,37] stated that Q_{10} values generally increased with increasing soil depth, because of the lower temperature fluctuations in deeper soil layers. Similar results were also found in this study. Besides that, the highest R^2 and lowest RSS of the exponential relationship between R_S and temperature at the soil surface were found (Figure 3). Therefore, 0 cm (soil surface) was considered as the appropriate measuring point for field experiments in arid areas, which was also reported by Pavelka et al. [42].

FOM addition and soil respiration

Elevated atmospheric CO_2 concentration can enhance plant growth [6,43], and the enhanced plant productivity can increase soil C input. The importance of C substrate to R_S has been well documented [12,44]. In the shrubland, FOM addition increased both daily (Figure 5) and the total study period (Figure 6A) R_S averages. A similar stimulation of R_S by increased organic matter

Table 4. Soil C input–emission balanced with different FOM and water addition treatments during study period. Data are in contrast to the controls (F0P0).

Treatments	C added (g/m²)	Additional C lost as CO_2 (g/m²)	Soil C input–emission balance (g/m²)
F1P0	21.03	11.43±1.77	9.60±1.77
F2P0	42.05	16.83±2.56	25.22±2.56
F0P1	0	3.58±0.69	−3.58±0.69
F1P1	21.03	28.57±0.52	−7.54±0.52
F2P1	42.05	32.69±2.35	9.36±2.35
F0P2	0	13.08±1.56	−13.08±1.56
F1P2	21.03	38.48±0.79	−17.45±0.79
F2P2	42.05	38.47±1.17	3.58±1.17

Other abbreviations are same as Table 1.
C lost as CO_2 (g/m²) = mean soil respiration of the whole period ($\mu mol\ CO_2\ m^{-2}\ s^{-1}$) ×12 g × 1E^{-6} × time (s); Additional C lost as CO_2 (g/m²) = (C lost as CO_2 in other treatments) − (C lost as CO_2 in F0P0); Soil C input–emission balance (g/m²) = (addition C lost as CO_2) − (C added).

input to soil was also reported by Sulzman et al. [45]. One reason for this variability is that enhanced FOM input supplied more easily decomposed substrate for the soil microbes which would stimulate both the microbial activities and biomass [22]. Besides that, nutrients released from the FOM decomposition can enhance root growth [46], thus increasing CO_2 from root respiration.

During the study period, higher R_S-stimulating effects by FOM addition were observed in the first two months (Figure 7A). This can be attributed to the "priming effect" [47], which means that fresh organic C can supply energy and nutrients for soil microbes, and can therefore, accelerate soil organic C (SOC) mineralization [48]. Thereafter, although the FOM-induced change in R_S declined, it remained positive and constant till the end of the experiment (Fig 6A). This might be due to the fact that the FOM was exhausted during the first two months, and the "priming effect" was inhibited. Previous studies also reported similar results [22,44]. The results indicated that FOM addition can cause a sustained R_S increase, at least throughout this study period from June to October.

In the long-term incubation ranged from weeks to months, the priming effect can be considered as real priming effect (soil organic matter decomposition) if the amount of the primed C is higher than both microbial biomass and the added C [49]. In this study, the initial microbial biomass C was 20.58 g/m^2 in the 20–30 cm soil layer. Thus, the real priming effect was happened under the F1P1 and F1P2 treatments (Table 4), while under the other FOM addition treatments could not exclude the apparent priming effect.

The amount of FOM added in the F2 treatment was double that of the F1 treatment, however, the FOM-induced increase in R_S between these two treatments did not follow this relationship under the same water conditions. This demonstrated that the effect of FOM addition on the primed CO_2 amount can be varied with the amount of FOM [49].

Soil water addition and soil respiration

Soil moisture is generally considered to be a crucial limiting factor for growth and activities of roots and microbes [7,21] and the diffusion of soluble substrates and oxygen [19] thus indirectly affecting R_S. Consistent with previous reports [21,46], this study showed that precipitation treatments stimulated R_S of the shrubland. However, some field observations found that precipitation treatments (addition or reduction) did not affect R_S in a temperate forest [50] and a tropical rain forest [51]. Borken and Matzner [9] attributed these differences to the stock of plant available water in soil, and stated that R_S would not be affected by water treatments until the stock of plant available water was significantly changed. Therefore, there should be an optimum water content that can provide the greatest boost to R_S. Rey et al. [41] reported that there was a soil moisture threshold, and R_S flux increased with moisture content below this threshold, and remains steady above the threshold. Some authors stated that the optimum water range was from 30 to 50% [52]. In this study, the R_S was stimulated with an increase in soil moisture (Table 2) and soil moisture in all plots ranged from 20.6 to 30.1% (52.8 to 77.2% of the field moisture capacity) during the experimental period (Figure 4B). This indicated that the soil moisture threshold for R_S in the shrubland was >30%, and the predicted precipitation increase will stimulate R_S and release more CO_2 into the atmosphere, although we could not estimate the accurate optimum moisture for the arid soil communities. In arid land, the low soil water content is not ideal for plant growth and decomposers, and may suppress the response of temperature on respiration [53], R_S was more dependent on soil moisture. After the manipulation of water addition, the activities of plants and microbes will be

enhanced after the relief of water stress, which can increase R_S variation. Therefore, soil moisture could better explain the seasonal variations of R_S in drier treatments (Table 2).

Interactive effects among temperature, FOM input and water addition

Temperature has long been identified as a crucial controlling factor on R_S [4,14]. Variations of R_S are usually highly correlated with temperature in a positive exponential relationship [36]. Hence, it is often assumed that global warming will stimulate soil respiration and lead to a positive feedback loop between R_S and atmosphere CO_2 [2]. In this experiment, the variation of R_S also showed an exponential increase relationship with temperature for all treatments (Figure 8). The temperature sensitivity of R_S-Q_{10} has received research interest due to the concern about variations of R_S under global warming, and the global mean Q_{10} value was reported at ~1.5 [2,15]. In this study, the Q_{10} for the control was 1.44, which was closed to the global mean Q_{10} value (1.5). However, more water addition stimulated Q_{10} relative to the controls (Figure 8). Similar results that low soil water content can decrease the temperature sensitivity of R_S have been reported by others, and the reason was attributed to the limitation in diffusion of solute substrate [5,8,37]. Besides water addition, FOM input was also found to have a stimulating effect on Q_{10} in the study (Figure 8). Similar results that additional FOM amplified the stimulated effects of higher temperature, which was ideal for the increasing of root and microbial biomass, have been reported by previous studies [46]. Arid ecosystems have been predicted to be one of the most responsive ecosystem types to global climate change [28]. Similarly, the result indicated that R_S in arid land might change to be more sensitive to global warming if the soil received more organic C and precipitation.

As described above, R_S is affected interactively by many factors. In these studies, it was found that interactions of soil moisture and temperature better predict variations in R_S for all treatments than single factors (Table 3). Similar results have been found in earlier works [37,54].

Given that both FOM input and water addition showed positive effects on R_S, their interactive effects were assumed to be larger. In the results, the synergistic effects of these two treatments on R_S increases were found in the first two months of the experiment, thus the effect of FOM addition was larger in watered treatments (Figure 7A), and the effect of water addition was larger in FOM addition treatments (Figure 7B). It can be explained that soil in arid regions only receives a small amount of plant litter input because vegetation is sparse and water is limited [26]. The interactive effects of these two treatments could relieve the stress from lack of substrate availability [19] and therefore cause larger R_S changes than the single treatment. However, the synergistic effects decreased afterwards, possibly attributed to exhaustion of the FOM, because the FOM input was more easily decomposed than the SOC [48]. Fontaine et al. [22] stated that supply of organic matter increased the populations of microbes, which could survive on SOC after organic matter exhaustion. This could explain the stability of the stimulated effects of FOM at plots without water addition after FOM exhaustion during the last three months (Figure 7A).

Organic C input and precipitation are predicted to increase in arid regions. In contrast to the controls (F0P0), the soil C input–emission balanced was calculated (Table 4). In our prior study about the effects of treatments (FOM and water addition) on the soil organic C pool, it showed that the SOC in soil of F2 treatment plots were slightly higher than in soil of other treatment plots, but the differences were not significant [55]. In this study, the positive

C input–emission balance was found in F2 treatment plots under the same water addition regime (Table 4). Besides that, results of the soil C input–emission balance also indicated precipitation increase played a dominant role in determining whether the soil C pool will get a net sequestration from more organic C input. It was found that water addition not only caused a negative C balance at plots without FOM treatment, but also lowered the soil C sequestration that resulted from FOM input. Furthermore, a soil priming effect caused by organic C input was also reported [22,44], but it only happened under better soil water conditions.

Conclusions

As organic C input to soil and precipitation are predicted to increase in arid regions, a better understanding of how R_S responds to these changes is essential to predict how soil C dynamics may change with global climate change. In the study, it was concluded that the temperature at the soil surface but not other soil depths can better simulate the relationship between R_S and temperature. The FOM and water addition treatments stimulated R_S and its temperature sensitivity. In the shrubland of an arid region, precipitation, rather than FOM, played a more

dominant role on soil C balance. Water addition enhanced soil C emission, and although soil could get net sequestration by FOM addition without precipitation addition, soil sequestration became negative when precipitation increased under F0 and F1 treatments. These results indicated that the soil CO_2 flux of this shrubland is likely to increase with climate change, and the critical role of precipitation in soil C balance had implications to future studies conducted at the arid region.

Acknowledgments

We acknowledge the Fukang Station of Desert Ecology, Xinjiang Institute of Ecology and Geography, Chinese Academy of Sciences for providing the study site and technical advice. We appreciated Prof. Chin, Prof. Murphy and the other two anonymous experts for insightful comments and great efforts to improve this manuscript, and we thank Yan Zhao, Zhongdong Lan and Yuanli Li for their help with the fieldwork and laboratory analysis.

Author Contributions

Conceived and designed the experiments: XC YG SW YZ. Performed the experiments: LL JW YT XZ. Analyzed the data: LL LJ YZ. Wrote the paper: LL YZ.

References

1. Raich JW, Potter CS, Bhagawati D (2002) Interannual variability in global soil respiration, 1980–94. Glob Change Biol 8: 800–812.
2. Bond-Lamberty B, Thomson AM (2010) Temperature-associated increases in the global soil respiration record. Nature 464: 579–582.
3. Solomon S, Qin D, Manning M, Chen Z, Marquis M, et al. (2007) Technical summary. In: Climate Change 2007: The Physical Science Basis. Contribution of Working Group I to the Fourth Assessment Report of the Intergovernmental Panel on Climate Change (eds Solomon S, Qin D, Manning M, Chen Z, Marquis M, et al.), 20–91. Cambridge University Press, Cambridge, UK/New York, NY, USA.
4. Cox PM, Betts RA, Jones CD, Spall SA, Totterdell IJ (2000) Acceleration of global warming due to carbon-cycle feedbacks in a coupled climate model. Nature 408: 184–187.
5. Davidson EA, Janssens IA (2006) Temperature sensitivity of soil carbon decomposition and feedbacks to climate change. Nature 440: 165–173.
6. Gill RA, Polley HW, Johnson HB, Anderson LJ, Maherali H, et al. (2002) Nonlinear grassland responses to past and future atmospheric CO_2. Nature 417: 279–282.
7. Schindlbacher A, Wunderlich S, Borken W, Kitzler B, Zechmeister-Boltenstern S, et al. (2012) Soil respiration under climate change: prolonged summer drought offsets soil warming effects. Glob Change Biol 18: 2270–2279.
8. Reichstein M, Subke JA, Angeli AC, Tenhunen JD (2005) Does the temperature sensitivity of decomposition of soil organic matter depend upon water content, soil horizon, or incubation time? Glob Change Biol 11: 1754–1767.
9. Borken W, Matzner E (2009) Reappraisal of drying and wetting effects on C and N mineralization and fluxes in soils. Glob Change Biol 15: 808–824.
10. Heimann M, Reichstein M (2008) Terrestrial ecosystem carbon dynamics and climate feedbacks. Nature 451: 289–292.
11. Jones CD, Cox PM, Huntingford C (2003) Uncertainty in climate-carbon-cycle projections associated with the sensitivity of soil respiration to temperature. Tellus 55B: 642–648.
12. Wan SQ, Norby RJ, Ledford J, Weltzin J (2007) Responses of soil respiration to elevated CO_2, air warming, and changing soil water availability in a model old-field grassland. Glob Change Biol 13: 2411–2424.
13. Trumbore S (2006) Carbon respired by terrestrial ecosystems-recent progress and challenges. Glob Change Biol 12: 141–153.
14. Luo YQ, Zhou XH (2006) Soil respiration and the environment. San Diego: Academic/Elsevier press.
15. Mahecha MD, Reichstein M, Carvalhais N, Lasslop G, Lange H, et al. (2010) Global convergence in the temperature sensitivity of respiration at ecosystem level. Science 329: 838–840.
16. Fierer N, Colman BP, Schimel JP, Jackson RB (2006) Predicting the temperature dependence of microbial respiration in soil: a continental-scale analysis. Global Biogeochem Cycle 20: GB3026.
17. Gaumont-Guay D, Black TA, Griffis TJ, Barr AG, Jassal RS, et al. (2006) Interpreting the dependence of soil respiration on soil temperature and water content in a boreal aspen stand. Agr Forest Meteorol 140: 220–235.
18. Graf A, Weihermüller L, Huisman JA, Herbst M, Bauer J, et al. (2008) Measurement depth effects on the apparent temperature sensitivity of soil respiration in field studies. Biogeosciences Discuss 5: 1867–1898.
19. Schimel J, Balser TC, Wallenstein MD (2007) Microbial stress response physiology and its implications for ecosystem function. Ecology 88: 1386–1394.

20. Yuste JC, Baldocchi DD, Gershenson A, Goldstein A, Misson L, et al. (2007) Microbial soil respiration and its dependency on carbon inputs, soil temperature and moisture. Glob Change Biol 13: 2018–2035.
21. Wilcox CS, Ferguson JW, Fernandez GCJ, Nowak RS (2004) Fine root growth dynamics of four Mojave Desert shrubs as related to soil moisture and microsite. J Arid Environ 56: 129–148.
22. Fontaine S, Bardoux G, Abbadie L, Mariotti A (2004) Carbon input to soil may decrease soil carbon content. Ecol Lett 7: 314–320.
23. Gifford RM (1994) The global carbon cycle: a viewpoint on the missing sink. Aust J Plant Physiol 21: 1–15.
24. Jastrow JD, Miller RM, Matamala R, Norby RJ, Boutton TW, et al. (2005) Elevated atmospheric carbon dioxide increases soil carbon. Glob Change Biol 11: 2057–2064.
25. Smith SD, Huxman TE, Zitzer SF, Charlet TN, Housman DC, et al. (2000) Elevated CO_2 increases productivity and invasive species success in an arid ecosystem. Nature 408: 79–82.
26. Lal R (2009) Sequestering carbon in soils of arid ecosystems. Land Degrad Develop 20: 441–454.
27. Sponseller RA (2007) Precipitation pulses and soil CO_2 flux in a Sonoran Desert ecosystem. Glob Change Biol 13: 426–436.
28. Melillo JM, McGuire AD, Kicklighter DW, Moore B, Vorosmarty CJ, et al. (1993) Global climate change and terrestrial net primary production. Nature 363: 234–240.
29. Zheng YR, Xie ZX, Jiang LH, Shimizu H, Drake S (2006) Changes in Holdridge Life Zone diversity in the Xinjiang Uygur Autonomous Region (XUAR) of China over the past forty years. J Arid Environ 66: 113–126.
30. Xu H, Li Y (2006) Water-use strategy of three central Asian desert shrubs and their responses to rain pulse events. Plant Soil 285: 5–17.
31. Qian W, Zhu Y (2001) Climate change in China from 1880 to 1998 and its impact on the environmental condition. Climatic Change 50: 419–444.
32. Xu G, Wei W (2004) Climate Change of Xinjiang and its Impact on Eco-Enviroment. Arid Land Geography, 27: 14–18. (In Chinese with English abstracts)
33. Fontaine S, Barot S, Barré P, Bdioui N, Mary B, et al. (2007) Stability of organic carbon in deep soil layers controlled by fresh carbon supply. Nature 450: 277–280.
34. Vance, E.D., Brookes, P.C., Jenkinson, D.S., 1987. An extraction method for measuring microbial biomass. Soil Biol. Biochem. 19: 703–707.
35. Nelson DW, Sommers LE (1982) Total carbon, organic carbon, and organic matter. In: Page AL, Miller RH, Keeney DR, eds. Methods of soil analysis. Madison, WI, USA: American Society of Agronomy and Soil Science Society of America, pp. 101–129.
36. Lloyd J, Taylor JA (1994) On the temperature dependence of soil respiration. Funct Ecol 8: 315–323.
37. Davidson EA, Belk E, Boone RD (1998) Soil water content and temperature as independent or confounded factors controlling soil respiration in a temperate mixed hardwood forest. Glob Change Biol 4: 217–227.
38. Lavigne MB, Foster RJ, Goodine G (2004) Seasonal and annual changes in soil respiration in relation to soil temperature, water potential and trenching. Tree Physiol 24: 415–424.
39. SPSS. 2000. SPSS for windows, version 13.0. SPSS, Chicago, USA.

40. Hashimoto S (2005) Q_{10} values of soil respiration in Japanese forests. J For Res 10: 409–413.

41. Rey A, Pegoraro E, Tedeschi V, De Parri I, Jarvis PG, et al. (2002) Annual variation in soil respiration and its components in a coppice oak forest in Central Italy. Glob Change Biol 8(9): 851–866

42. Pavelka M, Acosta M, Marek MV, Kutsch W, Janous D (2007) Dependence of the Q_{10} values on the depth of the soil temperature measuring point. Plant Soil 292: 171–179.

43. Zheng Y, Xie Z, Rimmington GM, Yu Y, Gao Y, et al. (2010) Elevated CO_2 accelerates net assimilation rate and enhance growth of dominant shrub species in a sand dune in central Inner Mongolia. Environ Exp Bot 68: 31–36.

44. Sayer EJ, Powers JS, Tanner EVJ (2007) Increased Litterfall in Tropical Forests Boosts the Transfer of Soil CO_2 to the Atmosphere. PLoS ONE 2: e1299.

45. Sulzman EW, Brant JB, Bowden RD, Lajtha K (2005) Contribution of aboveground litter, belowground litter, and rhizosphere respiration to total soil CO_2 efflux in an old growth coniferous forest. Biogeochemistry 73: 231–256.

46. Xiao C, Janssens IA, Liu P, Zhou Z, Sun OJ (2007) Irrigation and enhanced soil carbon input effects on below-ground carbon cycling in semiarid temperate grasslands. New Phytol 174: 835–846.

47. Bingemann CW, Varner JE, Martin WP (1953) The effect of the addition of organic materials on the decomposition of an organic soil. Soil Sci Soc Amer J 17: 34–38.

48. Fontaine S, Mariotti A, Abbadie L (2003) The priming effect of organic matter: a question of microbial competition? Soil Biol Biochem 35: 837–843.

49. Blagodatskaya Kuzyakov (2008) Mechanisms of real and apparent priming effects and their dependence on soil microbial biomass and community structure: critical review. Biol Fertil Soils 45: 115–131.

50. Hanson PJ, O'Neill EG, Chambers MLS, Riggs JS, Joslin JD, et al. (2003) Soil respiration and litter decomposition. In: North American temperate deciduous forest responses to changing precipitation regimes (Hanson PJ, Wullschleger SD, eds.), pp. 163–189, Springer, New York.

51. Davidson EA, Ishida FY, Nepstad DC (2004) Effects of an experimental drought on soil emissions of carbon dioxide, methane, nitrous oxide, and nitric oxide in a moist tropical forest. Glob Change Biol 10: 718–730.

52. Ilstedt U, Nordgren A, Malmer A (2000) Optimum soil water for soil respiration before and after amendment with glucose in humid tropical acrisols and a boreal mor layer. Soil Biol Biochem 32: 1591–1599.

53. Reichstein M, Papale D, Valentini R, Aubinet M, Bernhofer C, et al. (2007) Determinants of terrestrial ecosystem carbon balance inferred from European eddy covariance flux sites. Geophys Res Lett 34: L01402.

54. Lai L, Zhao X, Jiang L, Wang Y, Luo L, et al. (2012) Soil respiration in different agricultural and natural ecosystems in an arid region. PLoS ONE 7: e48011.

55. Lai L, Li Y, Tian Y, Jiang L, Zhao X, et al. (2013) Effects of Added Organic Matter and Water on Soil Carbon Sequestration in an Arid Region. PLoS ONE 8: e70224.

12

Effects of Warming and Clipping on Ecosystem Carbon Fluxes across Two Hydrologically Contrasting Years in an Alpine Meadow of the Qinghai-Tibet Plateau

Fei Peng*, Quangang You, Manhou Xu, Jian Guo, Tao Wang, Xian Xue*

Key Laboratory of Desert and Desertification, Chinese Academy of Sciences, Cold and Arid Regions Environmental and Engineering Research Institute, Chinese Academy of Sciences, Lanzhou, China

Abstract

Responses of ecosystem carbon (C) fluxes to human disturbance and climatic warming will affect terrestrial ecosystem C storage and feedback to climate change. We conducted a manipulative experiment to investigate the effects of warming and clipping on soil respiration (Rs), ecosystem respiration (ER), net ecosystem exchange (NEE) and gross ecosystem production (GEP) in an alpine meadow in a permafrost region during two hydrologically contrasting years (2012, with 29.9% higher precipitation than the long-term mean, and 2013, with 18.9% lower precipitation than the long-tem mean). Our results showed that GEP was higher than ER, leading to a net C sink (measured by NEE) over the two growing seasons. Warming significantly stimulated ecosystem C fluxes in 2012 but did not significantly affect these fluxes in 2013. On average, the warming-induced increase in GEP (1.49 μ mol m^{-2}s^{-1}) was higher than in ER (0.80 μ mol m^{-2}s^{-1}), resulting in an increase in NEE (0.70 μ mol m^{-2}s^{-1}). Clipping and its interaction with warming had no significant effects on C fluxes, whereas clipping significantly reduced aboveground biomass (AGB) by 51.5 g m^{-2} in 2013. These results suggest the response of C fluxes to warming and clipping depends on hydrological variations. In the wet year, the warming treatment caused a reduction in water, but increases in soil temperature and AGB contributed to the positive response of ecosystem C fluxes to warming. In the dry year, the reduction in soil moisture, caused by warming, and the reduction in AGB, caused by clipping, were compensated by higher soil temperatures in warmed plots. Our findings highlight the importance of changes in soil moisture in mediating the responses of ecosystem C fluxes to climate warming in an alpine meadow ecosystem.

Editor: Ben Bond-Lamberty, DOE Pacific Northwest National Laboratory, United States of America

Funding: Financial support came from the Foundation for Excellent Youth Scholars of CAREERI, CAS (351191001), National Natural Science Foundation of China (41301210, 41201195 and 41301211), and Chinese Academy of Sciences (Hundred Talents Program). The funders had no role in study design, data collection and analysis, decision to publish, or preparation of the manuscript.

Competing Interests: The authors have declared that no competing interests exist.

* Email: pengguy02@yahoo.com (FP); xianxue@lzb.ac.cn (XX)

Introduction

Global mean temperature has increased by 0.76°C since the year 1850 and is predicted to rise an additional 1.8–4°C by the end of the 21st century [1]. Elevated global temperature can substantially impact the global carbon (C) budget, resulting in positive or negative feedbacks to global climate change [2,3]. The balance between C fixed by photosynthesis and C emitted to the atmosphere through plant and heterotrophic respiration determines the rate of terrestrial C storage [4].

Studies have shown that global warming could stimulate both ecosystem C uptake and emission across various terrestrial biomes [5]. However, the response of net C balance to warming is highly variable because of different temperature and soil moisture sensitivities in the processes that control C uptake and emission [6]. It is generally assumed that the terrestrial ecosystem might act as a net C source under a global warming scenario because the processes controlling ecosystem C emission are more sensitive to higher temperatures than the processes controlling C uptake [3,7,8]. However, some evidence indicates that warming could

increase net C uptake, and global C models project enhanced terrestrial CO_2 uptake in response to warming through the middle of this century [9,10]. Current and completed experimental studies that have investigated warming effects have focused mostly on net primary productivity (NPP), biomass and soil respiration [5,11], from which the change in the C balance change was estimated. However, responses of gross primary production (GPP) and ecosystem respiration (ER), the major components net ecosystem exchange (NEE), to warming in field experiments [12] have received less attention in the alpine area [13].

Mowing (clipping) or grazing in grasslands, which account for 20% of the land use of the global terrestrial ice-free surface, may have substantial effects on ecosystem C fluxes, especially on a short-term basis [14]. Clipping would result in rapid changes in nutrient cycling [15], vegetation cover, plant community composition [16], and soil microclimate [17]. Collectively, these processes appear to stimulate the rate of ecosystem C cycling, however, their impacts on the net C balance are inconsistent [18,19].

Carbon stored in permafrost at high latitudes and in mountain areas is one of the major components of the terrestrial C pool. It is

Table 1. Results (F-values) of a three-way ANOVA on the effects of warming (W), clipping (C), measuring month (M), and their interactions on soil respiration (Rs), ecosystem respiration (ER), net ecosystem exchange (NEE) and gross ecosystem production (GEP).

	M	W	C	M×W	M×C	W×C	M×C×W
Rs	88.3**	3.9^	0.1	0.1	0.3	4.2*	1.1
ER	21.8**	8.3**	0.2	0.4	0.1	0.1	0.4
NEE	43.0**	4.8**	0.8	1.2	0.4	0.1	0.1
GEP	44.3**	9.5**	0.0	0.7	0.1	0.1	0.1
Rs/ER	1.5	0.3	0.1	2.5**	1.8	4.0**	0.1
ER/GEP	13.3**	0.4	0.5	2.1^	0.9	2.1	0.8
AGB	117.0**	3.8^	22.2**	1.2	2.3^	0.3	2.1^
RB	1.0	1.3	2.0	0.1	0.1	3.0^	0.2
AGB/RB	26.6**	10.3**	19.6**	6.6**	8.1**	13.0**	8.5**

Significance: ^, P<0.1; *, P<0.05; **, P<0.01.

estimated that soils in the permafrost regions store as much as 1672 Pg C ($1\ Pg = 10^{15}$ g), which is equivalent to double the atmospheric C pool [20,21]. Ecosystems in permafrost regions are C sinks because microbial decomposition of soil organic matter is inhibited under low annual mean temperature, and there is limited availability of organic C in frozen soil [22–24]. Altered growing season length, and changes in plant growth, ecosystem energy exchange and land use, together with the thawing of permafrost under a changing climate are projected to enhance the capability of ecosystem C uptake [21]. However, these altered dynamics do not appear to be able to compensate for the C released from thawing permafrost, resulting in ecosystems in permafrost regions acting as positive feedback to global change [21].

The Qinghai-Tibet Plateau (QTP) is experiencing a "much greater than average" increase in surface temperature, based on data observed at meteorological stations [25] and predictions from coupled climate-carbon models [1]. Grassland in the QTP is the largest vegetation unit of the Eurasian continent and covers an area of approximately 2.5 million km^2 [26]. Grazing is the most prevalent land use practice in the grassland. Results from eddy covariance measurements showed that the alpine meadow in the QTP is a weak C sink with annual variations [23,24]. Several studies have examined responses of ER and aboveground biomass to warming and clipping [17,27,28], in which the alpine meadow is thought to be a net C sink based on C balance calculations [28]. However, no field experiment has been conducted in the permafrost region of the QTP to measure the response of NEE, which provides a direct measure of the C balance. We conducted a two-year warming and clipping experiment to investigate how NEE and its components (GPP and ER) respond to warming and clipping, and how the associated changes in soil moisture, soil temperature, above- and belowground biomass affect the responses of ecosystem C fluxes in the permafrost region of the QTP.

Materials and Methods

Experimental site

The study site is situated in the region of the Yangtze River source, inland of the QTP near the Beilu River research station (34°49′N, 92°56′E, no specific permissions were required for activities in this location) at an altitude of 4635 m. This area has a typical alpine climate: mean annual temperature is −3.8°C and monthly air temperature ranges from −27.9°C in January to 19.2°C in July. Mean annual precipitation is 290.9 mm, of which over 95% falls during the warm growing season (May to October). Mean annual potential evaporation is 1316.9 mm, mean annual relative humidity is 57%, and mean annual wind velocity is 4.1 m s^{-1} [29]. The study site is a winter-grazed range, dominated by alpine meadow vegetation: *Kobresia capillifolia*, *K. pygmaea*, and *Carex moorcroftii*, with a mean plant height of 5 cm. Plant roots occur mainly within the 0–20 cm soil layer, and average soil organic C is 1.5%. The soil development is weak, and the soil belongs to alpine meadow soil (Chinese soil taxonomy), or is classified as a Cryosol according to World Reference Base, with a Mattic Epipedon at a depth of approximately 0–10 cm, and an organic-rich layer at a depth of 20–30 cm [30]. The parent soil material is of fluvioglacial origin and is composed of 99% sand. The Mattic Epipedon lowers the saturated soil water content, but increases soil water storage, and plant roots are dense and compressed within this layer. Permafrost thickness observed near the experimental site is 60–200 m and the depth of the active layer is 2.0–3.2 m [29,31]. However, because of climatic warming, the thickness of the active layer has been increasing at a rate of 3.1 cm y^{-1} since 1995 [32]. The experimental field is on a mountain slope

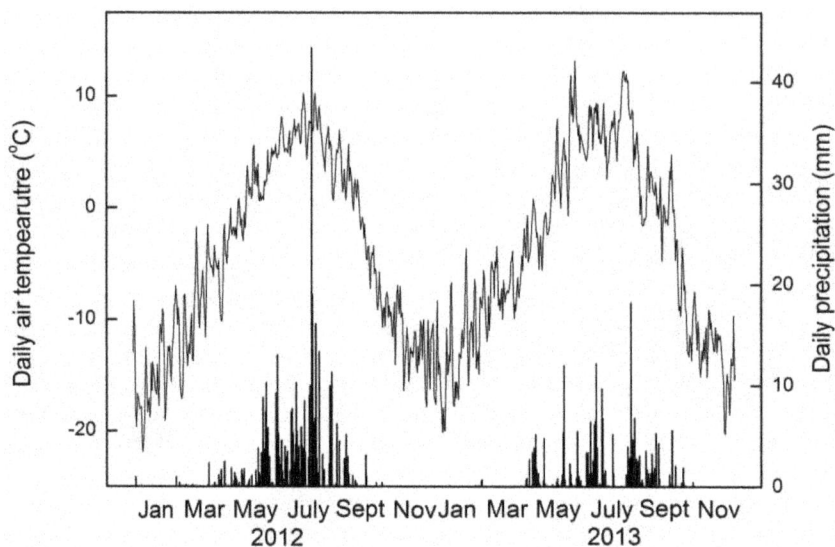

Figure 1. Daily precipitation (columns) and daily mean air temperature (line) in 2012 and 2013. Data are from the micro-meteorological station adjacent (approx. 100 m) to the experimental plots.

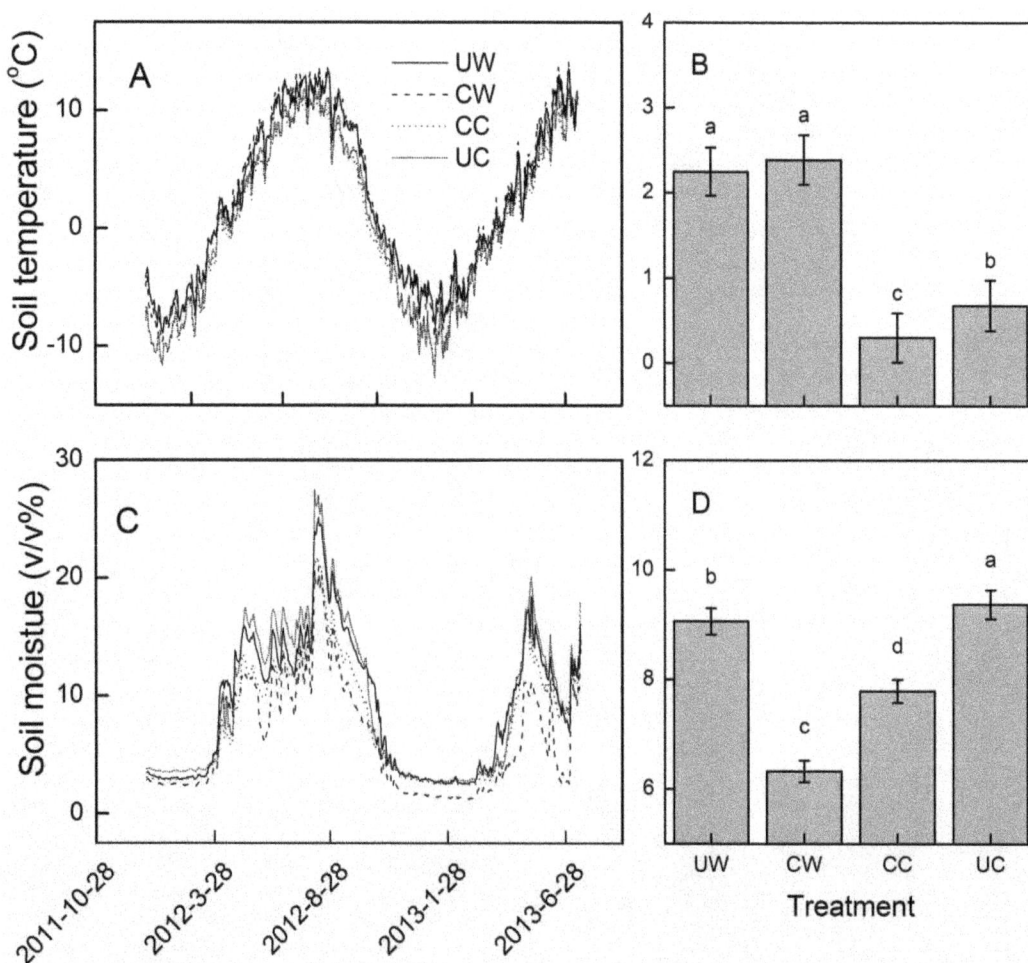

Figure 2. Daily soil temperature measured at a depth of 5 cm and volumetric soil moisture measured at a depth of 0–10 cm (A, C), and effects of warming and clipping on average (2012–2013) soil temperature (B) and soil moisture (d). UW, unclipped warming; CW, clipped warming; CC, clipped control; UC, unclipped control.

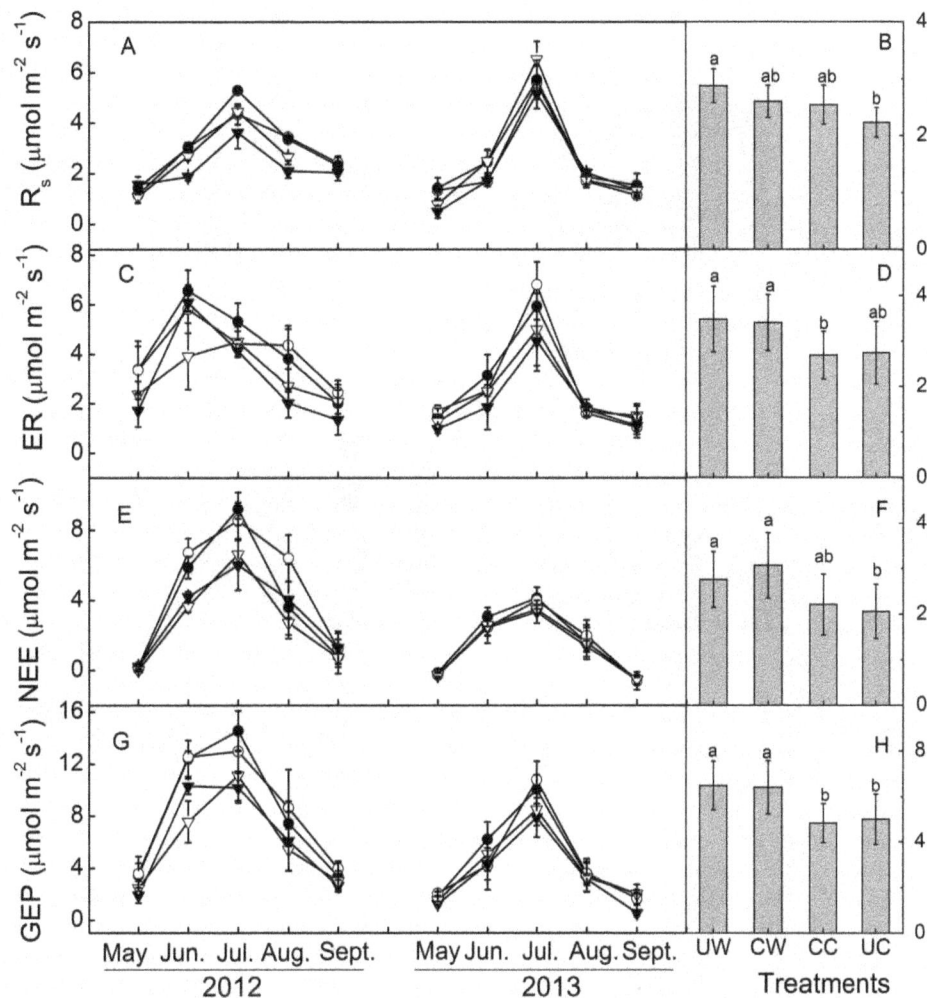

Figure 3. Mean monthly and overall means of soil respiration (Rs, A, B), ecosystem respiration (ER, C, D), net ecosystem exchange (NEE, E, F) and gross ecosystem production (GEP, G, H). Clipping was conducted in September 2011 and 2012. UW, unclipped warming; CW, clipped warming; CC, clipped control; UC, unclipped control.

with a mean incline of 5°. Detailed information about the soil properties is presented in Table 1.

Experimental design and measurement protocols

Experimental design. A two factorial experimental design (warming and clipping) was used with five replicates in each of the four treatments, i.e. unclipped control (UC), clipped control (CC), unclipped warming (UW) and clipped warming (CW). In total, 20 plots (2×2 m) were used in a complete randomized block distribution in the field. Plots were selected for homogeneity of topography, soil texture, aboveground biomass, and species composition. In each warmed plot, one $165 \text{ cm} \times 15$ cm infrared heater (MR-2420, Kalglo Electronics Inc., Bethlehem, PA, USA) was suspended in the middle of each plot at a height of 1.5 m above the ground with a radiation output of 150 watts m^{-2}. The heating has been operating continuously since July 1st 2010. To simulate the shading effects of the heaters, one "dummy" heater,

made of a metal sheet with the same shape and size as the heaters, was also installed in the control plot.

Plants in the clipped plots were clipped at the soil surface on an annual basis, usually in last September. The rotational grazing system is that two ranches for each family, one is for the summer grazing and another is for the winter grazing. The rotational use of ranches is implemented usually in September. We consulted to the owner and they ensured that grassland of our study site is a winter grazing ranch. In our study, the clipping treatment was conducted in last September. One sheep needs 1.46×10^6 g grass per year [33] and the carrying capacity for alpine meadow is 1.39 head of sheep per hectare. Based on those data, about 203 g m^{-2} grass per year would be grazed for the alpine meadow. The aboveground biomass in September 2013 was about 320 g m^{-2}. In the clipping treatment, biomass cut was <320 g m^{-2}. Therefore we believe that our clipping treatment provided a reasonable simulation of local grazing practices.

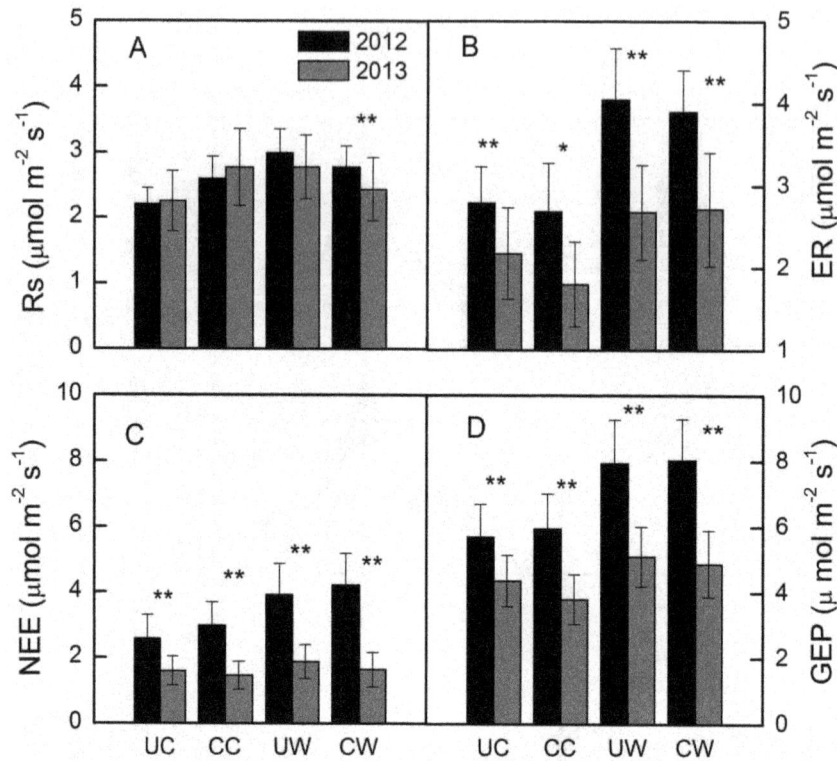

Figure 4. Mean annual soil respiration (Rs), ecosystem respiration (ER), net ecosystem exchange (NEE) and gross ecosystem production (GEP) under unclipped control (UC), clipped control (CC), unclipped warming (UW) and clipped warming (WC) treatments in 2012 and 2013. Symbols above the bars represent significant differences at p<0.05 (*) and p<0.01 (**).

Measurement protocol. Air temperature, water vapor pressure and relative humidity were monitored automatically at a height of 20 cm above the soil surface in the center of each plot using a Model HMP45C probe (Campbell Scientific Inc., Bethlehem, PA, USA). Nine thermistors were installed to monitor soil temperatures at depths of 5, 15, 30, 60, 100, 150, 200, 250 and 300 cm. All the probes were connected to a CR1000 datalogger (Campbell Scientific Inc.). Data recorded every 10 min were averaged and reported as daily values. Pavelka et al. (2007) stated that for grassland ecosystems, surface soil temperature is the most suitable depth for measuring soil temperature because of the optimized regression coefficient between surface soil temperature and soil respiration [34]. Therefore, we used soil temperature measured at a depth of 5 cm in the following analyses.

An EnviroSmart sensor (Sentek Pty Ltd., Stepney, Australia), which used frequency domain reflection, was used to monitor volumetric soil moisture at depths of 0–10, 10–20, 20–40, 40–60 and 60–100 cm. These soil moisture data were also recorded using a CR1000 datalogger. When analyzing the relationships between C fluxes and soil moisture, we used the daily average soil moisture data that were collected when ecosystem C flux measurements were conducted.

Soil respiration (Rs) was measured by using Licor-6400-09 (Lincoln, NE, USA) on PVC collars 5 cm in height and 10.5 cm in diameter, which were permanently inserted 2–3 cm into the soil in the center of each plot. Small living plants were cut at the soil surface at least one day before measurements to eliminate the effect of respiration from aboveground biomass [35]. ER and NEE were measured with a transparent chamber ($0.5 \times 0.5 \times 0.5$ m) attached to an infrared gas analyzer (IRGA, Licor-6400, Lincoln, NE, USA). The transparent chamber is a custom-designed chamber made of Polytetrafluoroethene (4 mm in thickness) with light transmittance about 99%. During measurements, a foam gasket was placed the chamber and the soil surface to minimize leaks. One small fan ran continuously to mix the air inside the chamber during measurements. Nine consecutive recordings of CO_2 concentration were taken in each plot at 10 s intervals during a 90 s period. Following the measurement of NEE, the chamber was vented for several minutes and covered with an opaque cloth for measuring ER, as the opaque cloth eliminated light (and hence photosynthesis). CO_2 flux rates were determined from the time-course of the CO_2 concentrations used to calculate NEE and ER. The method used was similar to that reported by Steduto et al. (2002) [36] and Niu et al. (2008) [37]. Gross ecosystem productivity (GEP) was the calculated as the sum of NEE and ER. Rs, NEE and ER were measured in each plot on a monthly basis from May to September in 2012 and 2013.

Aboveground biomass (AGB) was obtained from a step-wise linear regression with AGB as the dependent variable, and coverage and plant height as independent variables. 100 small plots (30 cm×30 cm) were included in the regression analysis (AGB = 22.76×plant height +308.26×coverage −121.80, $R^2 = 0.74$, $P < 0.01$). Coverage of each experimental plot was measured using a 10 cm×10 cm frame in four diagonally divided subplots replicated eight times. Plant height was measured 40 times by a ruler and averaged for each experimental plot. A biomass index was used as the ratio of the derived biomass on any given date to the maximum biomass during the entire study period [38]. Root biomass (RB) was obtained from soil samples that were air-dried for one week and passed through a 2- mm diameter sieve to remove large particles. Roots were separated from the soil by washing, and a 0.25-mm diameter sieve was used to retrieve fine

Table 2. Results (F-values) of a three-way ANOVA on the effects of warming (W), clipping (C), measuring month (M), and their interactions on soil respiration (Rs), ecosystem respiration (ER), net ecosystem exchange (NEE) and gross ecosystem production (GEP) in contrasting years.

	M	W	C	M×W	M×C	W×C	M×C×W
2012							
Rs	54.2**	11.6**	0.8	2.2^	0.3	5.1*	1.4
ER	13.9**	9.4**	0.1	0.6	0.4	0.01	1.2
NEE	46.9**	8.6**	2.6	2.0	1.3	0.0	0.04
GEP	40.3**	12.2**	0.2	0.9	0.6	0.1	0.2
ER/GEP	16.6**	0.05	0.2	0.6	0.1	0.8	0.8
AGB	33.9**	2.1	2.0	0.6	0.4	0.5	0.2
RB	7.2**	0.7	1.0	0.5	0.09	2.8	0.2
AGB/RB	2.4^	1.2	2.2	0.5	0.2	1.6	0.2
2013							
Rs	78.3**	0.1	0.02	0.8	0.6	3.3^	0.6
ER	31.0**	2.6	0.2	1.1	0.2	0.2	0.3
NEE	51.2**	1.1	0.3	0.3	0.1	0.01	0.1
GEP	44.2**	3.2^	0.9	0.7	0.4	0.2	0.3
ER/GEP	22.1**	9.0**	11.0**	13.9**	8.3**	8.8**	10.9**
AGB	89.9**	0.06	20.1**	1.3	0.8	0.5	1.1
RB	2.9*	0.7	1.2	0.2	0.3	0.8	0.2
AGB/RB	20.7**	1.5	6.2*	1.5	2.3^	2.6	2.7*

Significance: ^, $P<0.1$; *, $P<0.05$; **, $P<0.01$.

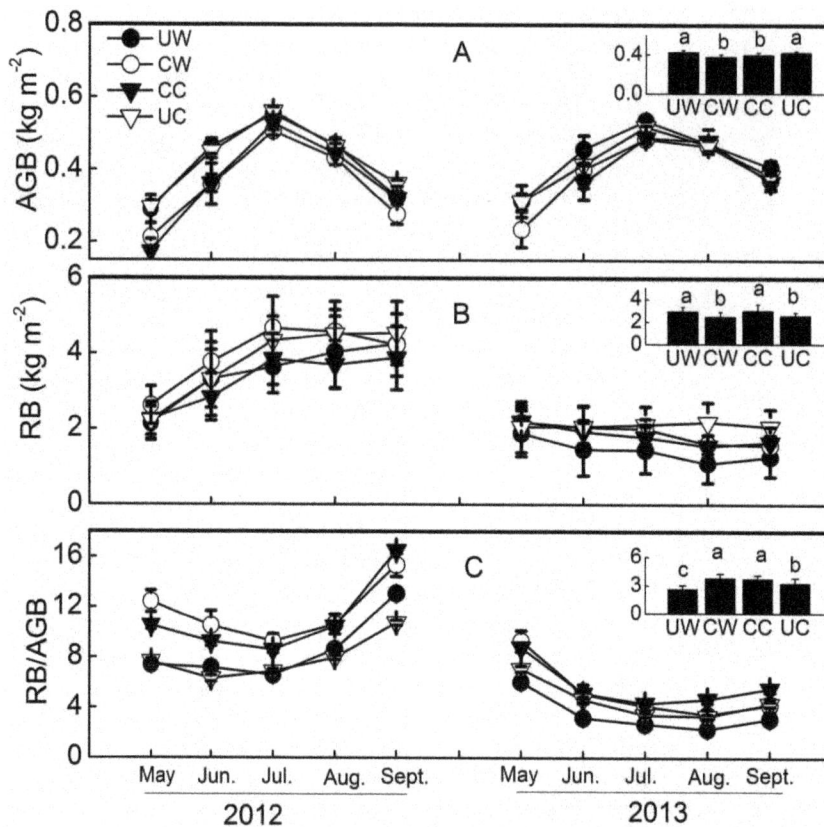

Figure 5. Temporal variations and overall means (inserted panels) of aboveground biomass (AGB, A), root biomass (RB, B) and the ratio of RB to AGB (RB/AGB, C). Clipping was conducted in September 2011 and 2012. See Figures 2 and 3 for notes and abbreviations.

roots. Living roots were separated from dead roots by their color and consistency [39]. Separated roots were dried at 75°C for 48 h.

Data analysis

Temperature and soil moisture data used in analyses were from January 1st 2012 to July 18th 2013 because a power failure prevented data from being collected from July 19th 2013 onwards. The effect of the warming and clipping treatments on soil temperature (5 cm), soil moisture (0–10 cm), Rs, ER, NEE, GEP, AGB and RB were determined with a three-way analysis of variance (ANOVA) using SPSS Version 18.0. (SPSS, Inc., Chicago, IL, USA).

Relationships between C fluxes and soil microclimate (soil temperature and soil moisture) were examined using daily soil microclimate data that were collected when ecosystem C fluxes

were measured. Linear regression analyses were used to examine the relationships of C fluxes with abiotic (soil moisture and soil temperature) and biotic factors (monthly AGB and RB).

Results

Microclimate

In comparison to the long-term average (1981–2008) mean annual air temperature (MAT, −5.1°C), higher MAT values were recorded in 2012 and 2013 (−3.5°C and −3.8°C, respectively). Annual precipitation in 2012 (420.1 mm, Fig. 1) was higher than the long-term mean annual precipitation (294.5 mm), but it was lower than the long-term mean in 2013 (238.8 mm) (Fig. 1).

Experimental warming significantly elevated annual mean soil temperature (Figs. 2A and 2B, $P<0.05$). In unwarmed plots, the average daily soil temperature at 5 cm depth was 0.65°C and

Table 3. Fitted quadratic models of the relationships between ecosystem respiration (ER), net ecosystem exchange (NEE), gross ecosystem production (GEP) and soil moisture (θ, v/v%, 10 cm). Max. F, θ represents the value of θ when ER, NEE and GEP are at their maximum.

	ER/μmol m^{-2}s^{-1}	NEE/μmol m^{-2}s^{-1}	GEP/μmol m^{-2}s^{-1}
Fitted model	ER = $-0.058\theta^2 + 1.71\theta - 7.72$	NEE = $-0.076\theta^2 + 2.37\theta - 12.53$	GEP = $-0.139\theta^2 + 4.21\theta - 21.02$
R^2	0.31	0.45	0.36
p	0.007	<0.001	<0.001
Max. F, θ/%	14.7%	15.6%	15.1%

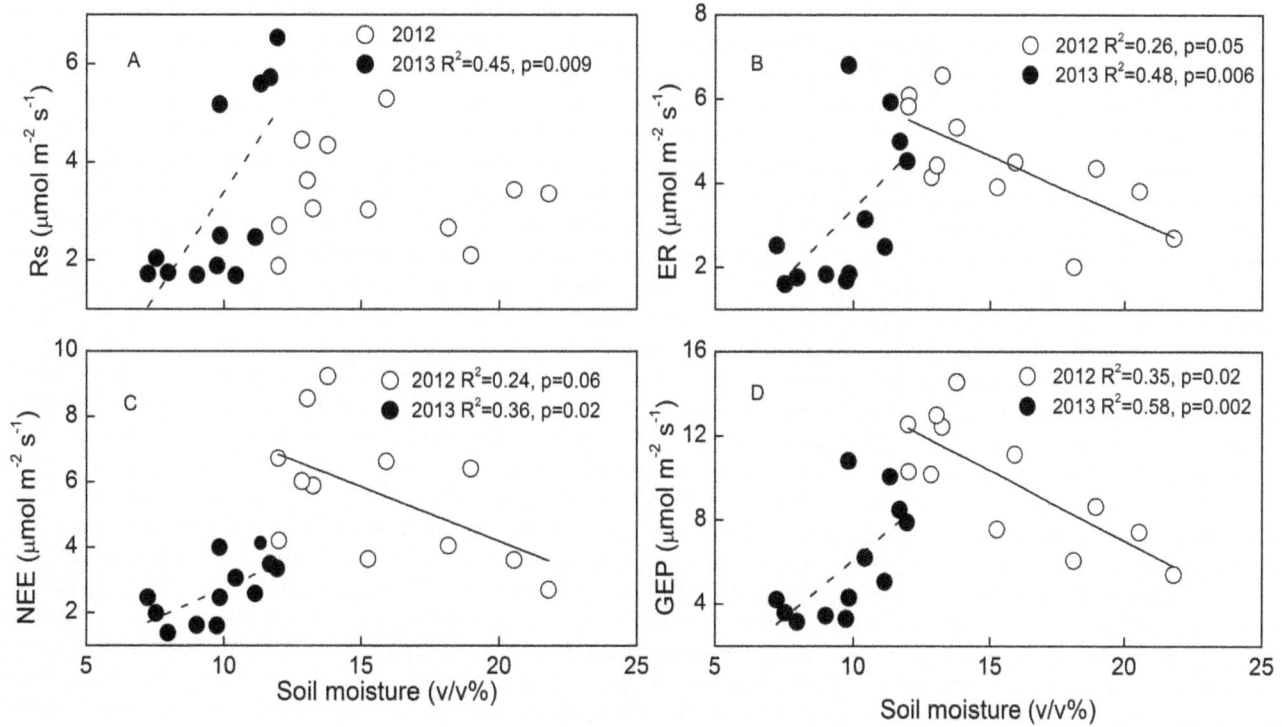

Figure 6. The relationships between soil moisture (0–10 cm) and ecosystem C fluxes: soil respiration (Rs), ecosystem respiration (ER), net ecosystem exchange (NEE) and gross ecosystem productivity (GEP) in 2012 (hollow circles) and 2013 (solid circles), respectively. Soil moisture and ecosystem C fluxes data were the average for all plots in each month. Data for both years were collected from June to August.

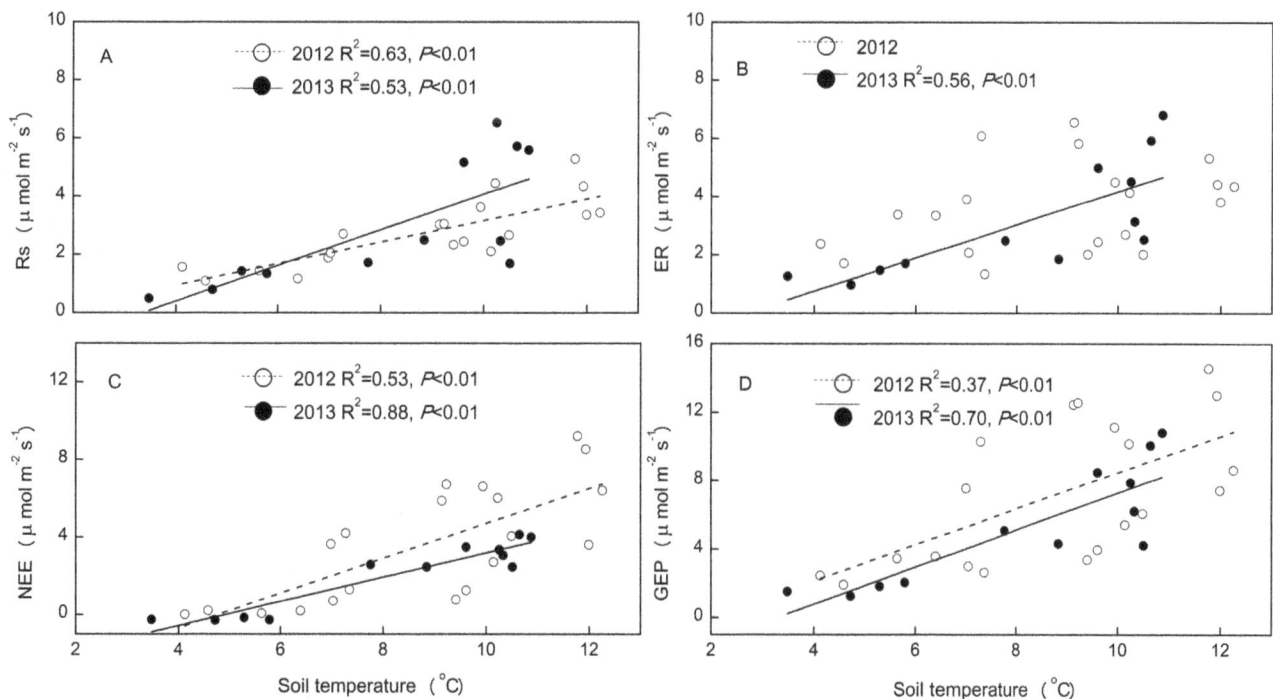

Figure 7. The relationships between soil temperature (5 cm) and ecosystem C fluxes: soil respiration (Rs), ecosystem respiration (ER), net ecosystem exchange (NEE) and gross ecosystem productivity (GEP). Soil temperature and ecosystem C fluxes data were the average of all plots in each month.

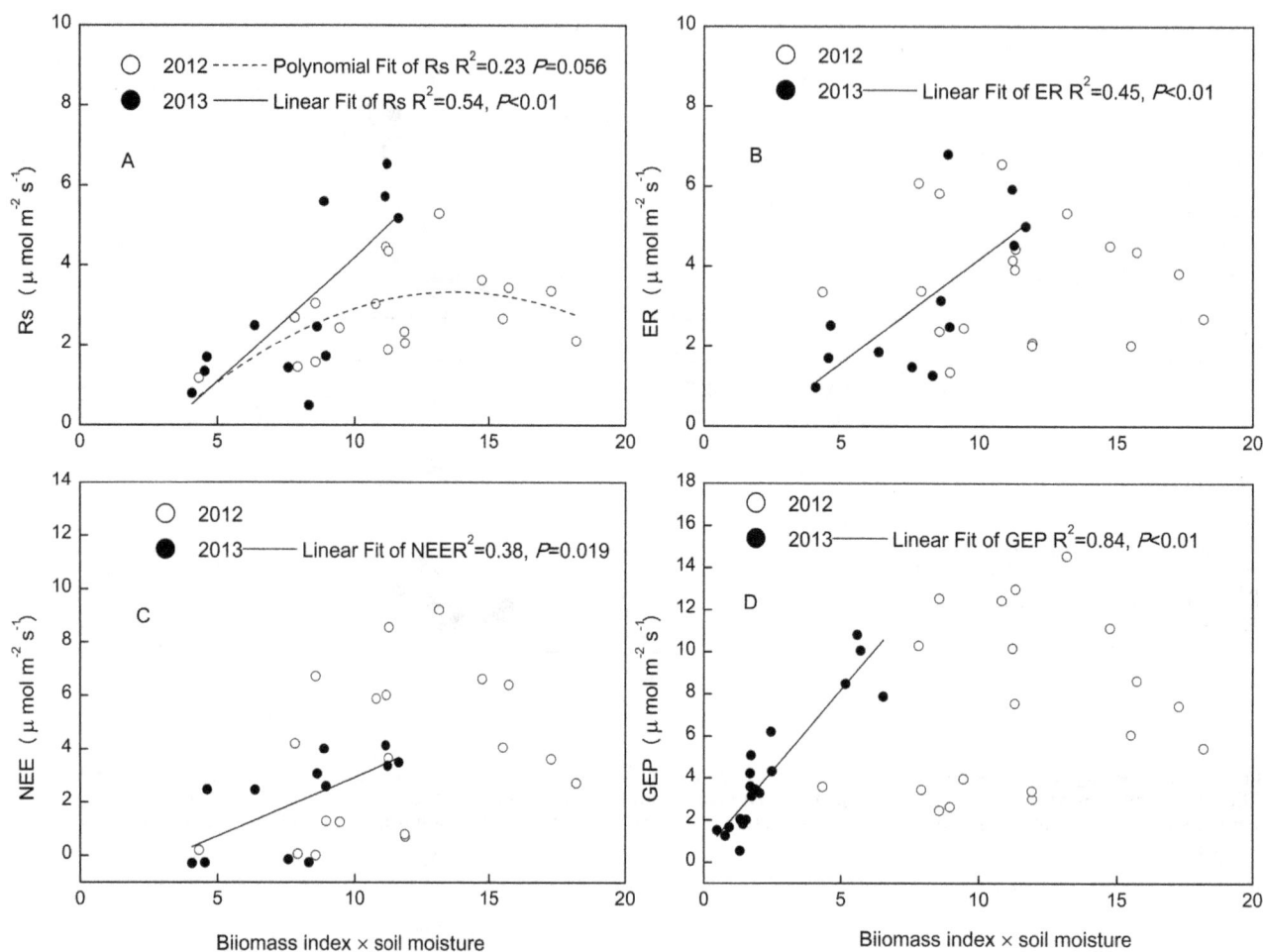

Figure 8. The relationships between changes in ecosystem C fluxes: soil respiration (Rs), ecosystem respiration (ER), net ecosystem exchange (NEE) and gross ecosystem productivity (GEP), and the changes in the product of abveground biomass index and soil moisture during 2012 and 2013.

0.14°C in 2012 and 2013, respectively. Warming significantly increased the soil temperature by 1.96°C (2012) and 1.59°C (2013) (Fig. 2B). The average daily soil temperature in unclipped plots was 1.69°C and 0.99°C in 2012 and 2013, respectively, but was unaffected by clipping. Volumetric soil moisture measured over 0–10 cm fluctuated greatly over the study period (Fig. 2C). The average daily soil moisture in unwarmed plots was 9.31 (v/v%) and 7.22 (v/v%), and warming significantly reduced soil moisture by 7.4% and 17.4% ($P<0.05$) in 2012 and 2013, respectively (Fig. 2D). The average daily soil moisture in unclipped plots was 10.08 (v/v%) and 7.61 (v/v%), and clipping significantly decreased it by 28.3% and 36.5% in 2012 and 2013, respectively (Fig. 2D).

Warming and clipping effects on C fluxes

The temporal dynamics of Rs, ER, NEE, and GEP followed the seasonal patterns of air and soil temperature in both years, which peaked in mid-growing season (Figs. 1 and 3). Substantial inter-annual variations in ecosystem C fluxes were observed in this study (Fig. 3). The annual average ER, NEE and GEP were all significantly higher in 2012 than in 2013 (Fig. 4) in all treatments, but higher Rs in 2012 was only observed in the CW treatment (Fig. 4A). On average, NEE, ER and GEP were 47%, 22% and 34% higher, respectively, in 2012 than in 2013.

Warming significantly increased NEE ($P = 0.03$), whereas no significant effects of clipping ($P = 0.37$) or its interaction with warming ($P = 0.83$) were detected (Table 1). When analyzed separately by year using a three-way ANOVA, warming only significantly increased NEE in 2012, by 28.5% (Table 2). Warming induced an enhanced growing season mean NEE in 2012, which was lower in clipped (17%) than in unclipped plots (30%). Measuring date had a significant effect on NEE, but the interaction of measuring date with warming or clipping had no effect on NEE in either year (Table 2).

Similar to NEE, average GEP was significantly increased by warming ($P = 0.003$) but not by clipping ($P = 0.97$) or by their interaction ($P = 0.87$, Table 1). When analyzed separately by year using a three-way ANOVA, warming significantly increased average GEP (Table 2) by 2.13 μ mol m^{-2}s^{-1} in 2012 and marginally enhanced it by 0.82 μ mol m^{-2}s^{-1} in 2013. The increased GEP caused by warming was lower in clipped (23%) than in unclipped plots (28.3%) in 2012, but GEP was higher in clipped (22.6%) than in unclipped plots (13.4%) in 2013.

Warming also significantly increased average Rs ($P = 0.052$) and ER ($P = 0.005$), but clipping had no significant effect on average Rs ($P = 0.73$) or ER ($P = 0.66$, Table 1). Similar to NEE, when analyzed separately by year using a three-way ANOVA, the effect of warming on ER and Rs was only significant in 2012 (Table 2),

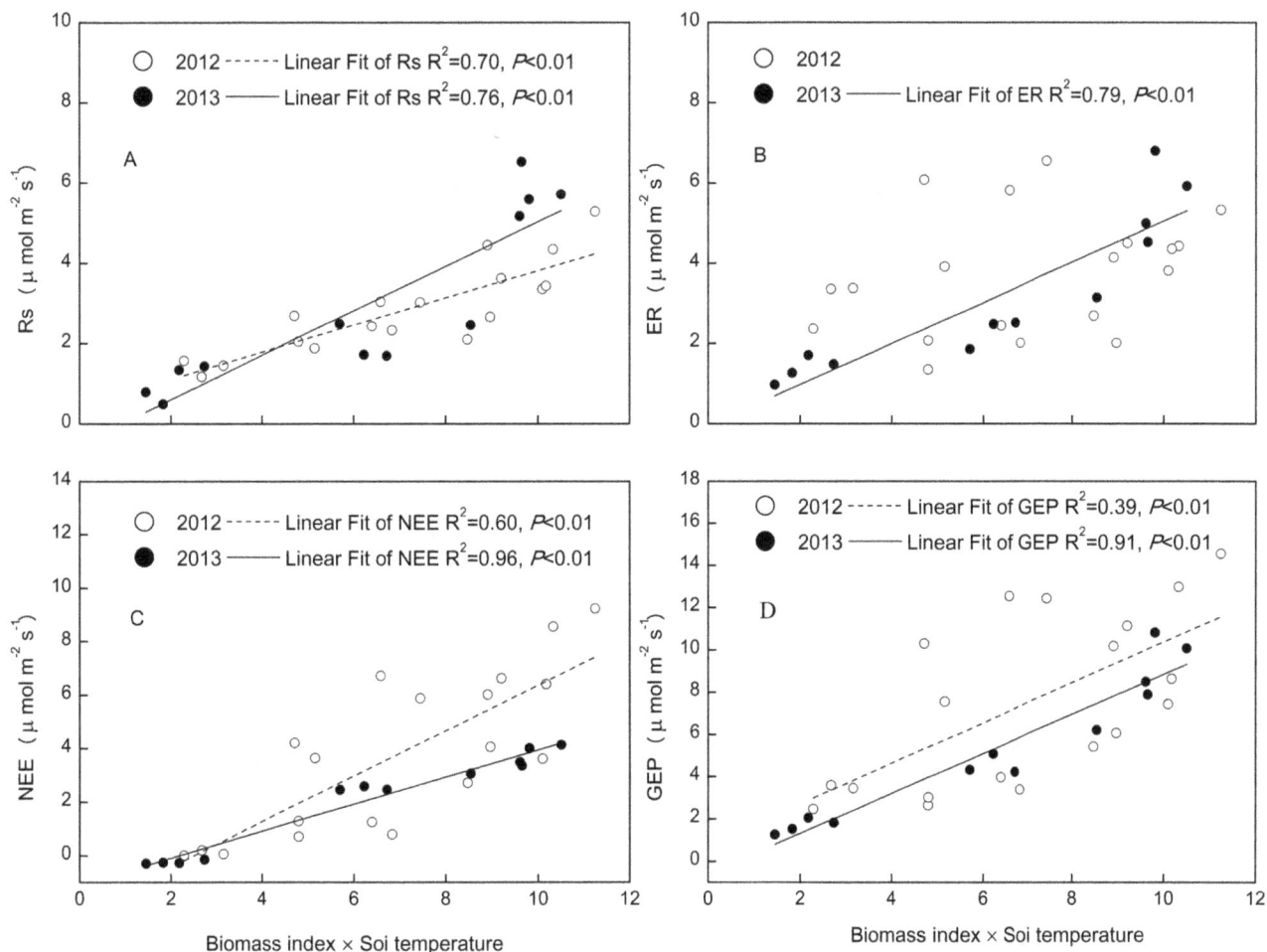

Figure 9. The relationships between changes in ecosystem C fluxes: soil respiration (Rs), ecosystem respiration (ER), net ecosystem exchange (NEE) and gross ecosystem productivity (GEP), and the changes in the product of aboveground biomass index and soil temperature during 2012 and 2013.

which increased by 25.8% and 17%, respectively. The interaction between warming and clipping had a significant effect on Rs but not on ER (Table 1). The average increase in Rs caused by warming was lower in clipped (5.9%) than in unclipped plots (27.4%).

There was no significant effect of warming or clipping on the ER to GEP ratio (ER/GEP, $P = 0.51$ and $P = 0.46$ for 2012 and 2013, respectively, Table 1). When analyzed separately by year using a three-way ANOVA, ER/GEP was significantly affected by warming, clipping, measurement date and their interactions in 2013 (Table 2).

Warming and clipping effects on biomass

Similar to the inter-annual variation of ecosystem C fluxes, RB was significantly lower in 2013 (by a value of 50.2%) than in 2012, whereas there was no difference in AGB over the two growing seasons (Fig. 5).

Warming marginally increased AGB ($p = 0.053$) and clipping significantly reduced AGB ($p < 0.001$), but there was no significant effect on RB ($p = 0.26$ and $p = 0.16$ for warming and clipping, respectively, Table 1). When analyzed separately by year using a three-way ANOVA, warming had no significant effect on AGB or RB in either year, whereas clipping significantly reduced AGB in

2013 (Table 2). The reduction in AGB by clipping was higher in unwarmed (14.4%) plots than in warmed plots (10.3%) in 2013.

Impacts of biotic and abiotic factors on ecosystem C fluxes

Over the two growing seasons, there was no clear relationship between ER, NEE, GEP and soil moisture, but there was a quadratic relationship between these variables and soil moisture when May and September data were excluded (Table 3). The optimal soil moisture for ER, NEE and GEP was about 15% (Table 3). When plotted separately, ER, NEE, and GEP decreased linearly with increasing soil moisture in 2012, and increased linearly with increasing soil moisture in 2013 (Fig. 6).

The temperature response curves for Rs, ER, NEE and GEP in 2012 were quite similar to those recorded in 2013. However, the slope of relationship between Rs and soil temperature was higher in 2012 than in 2013, but that between NEE and soil temperature was smaller in 2012 than in 2013 (Fig. 7).

The statistical interaction term of above-ground biomass index and soil moisture showed polynominal relationship only with Rs in 2012 (Fig. 8a), whereas it linearly correlated with all the ecosystem C fluxes in 2013 (Fig. 8). The interaction term of above-ground biomass and soil temperature explained more variation in ecosystem C fluxes than did the interaction term of above-ground

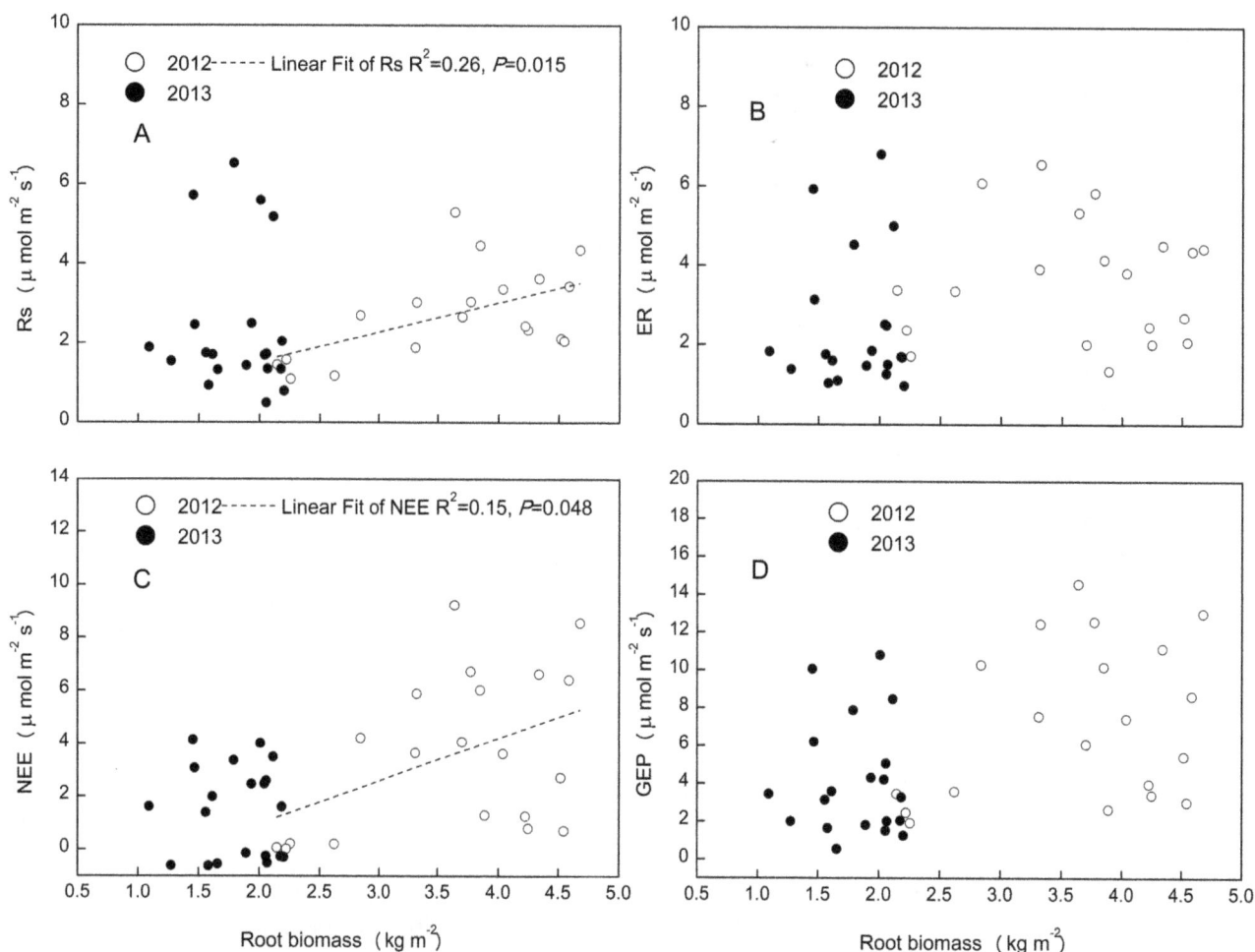

Figure 10. Relationship between root biomass and ecosystem C fluxes: soil respiration (Rs), ecosystem respiration (ER), net ecosystem exchange (NEE) and gross ecosystem productivity (GEP).

biomass and soil moisture (Figs. 8 and 9). The fitting slope between Rs and the interaction of above-ground biomass and soil temperature was higher in 2013 than in 2012 (Fig. 9a), but that between NEE and the interaction of above-ground biomass and soil temperature was smaller in 2013 than in 2012 (Fig. 9c). Root biomass only had effect on ecosystem C fluxes in wet year (Fig. 10).

Discussion

C fluxes and their inter-annual variation

In this alpine meadow ecosystem studied in the QTP, a higher uptake of C (GEP) than release (ER) resulted in a net C sink (Figs. 3–4). This result is similar to that reported in seasonally frozen areas in the QTP [23,24], and in some arctic ecosystems [7].

Higher precipitation and associated higher soil moisture in 2012 than in 2013, and the similar MAT between the two years suggest that drought reduced annual ER, NEE and GEP in 2013. Our results are in agreement with those from temperate grassland ecosystems [37,40,41]. The quadratic relationship between ecosystem C fluxes and soil moisture (Table 3) at the temporal scale support the above findings, as ecosystem C fluxes were positively related to soil moisture in 2013 (Fig. 6). There was greater fluctuation in GEP than ER: GEP was 34% lower in 2013

than in 2012, compared to ER, which was 22% lower in 2013 than in 2012. These results are consistent with those from boreal and temperate forests [42] and temperate grasslands [40], which indicate that GEP is more sensitive to inter-annual climatic variation than ER in alpine meadow ecosystems. Differences in the magnitudes of the inter-annual variation in GEP and ER could be explained by differences in the slopes between these variables when plotted against soil moisture (Fig. 6). The greater dependence of GEP than ER on soil moisture across both years suggests that GEP is more sensitive to changes in soil moisture than ER. Despite the significant inter-annual difference in ER, NEE and GEP, Rs did not differ significantly between the two years. This indicates that annual variations in Rs may be controlled by other factors, such as soil temperature, as Rs had the highest slope when plotted against this variable (Fig. 7).

Drought typically reduces aboveground biomass in grasslands [43,44]. However, in our study, no significant reduction in AGB was observed in 2013. One reason could be the various reactions of different species to drought [45,46], and this compensation may hold community AGB constant. The 50.2% decrease in RB in 2013 could be attributed to the reduction in soil moisture because RB in alpine ecosystems is positively correlated with annual precipitation [47]. The relative reduction rate of RB in our study (50.2% reduction in RB and 2.1 m^3 m^{-3} reduction in soil moisture) was higher than in a temperate grassland ecosystem

(23% reduction in RB and 5.2 m^3 m^{-3} reduction in soil moisture) [48]. The divergent responses of RB in the surface and deep soil layers [46,49] could cancel each other out and therefore lead to a lower relative decrease in RB across the whole soil profile.

However, only one wet and dry year was included in this study. The wet year was followed directly by a dry year, which might lowers the effect of drought because soil drought might lag the meteorological drought.

Main effects of warming

Experimental warming stimulated GEP more than ER (1.49 µ mol $m^{-2}s^{-1}$ vs. 0.79 µ mol $m^{-2}s^{-1}$), leading to an increase in NEE in the warming treatment of this alpine meadow ecosystem in a permafrost area of the QTP. Warming effects on ecosystem C exchange are likely modulated by soil water regimes [7,37]. For example, Oberbauer (2007) reported that higher soil moisture in wet tundra limited increases in ER relative to increases in GEP under warming conditions, indicating the dependence of the warming effect on hydrological conditions. In the current alpine meadow ecosystem, differences in the responses of NEE to warming between 2012 and 2013 (Table 2) differed from results from a temperate steppe, in which NEE demonstrated no change under a warming treatment over two hydrologically contrasting years [40]. Soil moisture showed positive impacts on C fluxes in 2013 and negative impacts on fluxes in 2012 (Fig. 6). ABG and RB were positively correlated with ecosystem C fluxes (Figs. 8, 9). As there were no significant effects of warming on AGB or RB in either year (Tables 1, 2), the significant increase in NEE in warmed plots in 2012 could be attributed to the higher stimulation of GEP (2.09 µ mol $m^{-2}s^{-1}$) than ER (1.07 µ mol $m^{-2}s^{-1}$).

The insensitivity of NEE to warming in 2013 could be attributed to the effect of the soil moisture deficit on GEP and ER (Table 2, Fig. 6). The positive responses of GEP and ER to warming (Table 1) are consistent with those in a tundra ecosystem [7], but differ from those in a subalpine meadow ecosystem, where soil moisture stress induced by warming reduced ER [27,40]. ER is composed of Rs and respiration of AGB. Therefore, the significant increase in ER in 2012 could be attributed mainly to the stimulation in Rs (Table 2), as AGB was insensitive to warming (Table 2). This indicates that the response of ER to warming was determined by Rs even though AGB respiration is the major component of ER in alpine meadow ecosystems [50]. Rs is composed of root respiration and microbial decomposition of soil organic matter [51]. There was no significant change in RB at a depth of 0–10 cm in the warming treatment (Tables 1, 2), which suggests that the response of ecosystem C emission to warming is determined by soil organic matter decomposition. The non-significant response of ER to warming in 2013 likely resulted from lower soil moisture (less than 15%, the optimal soil moisture for ecosystem C fluxes, Table 3), and the warming-induced reduction in soil moisture (Fig. 2). This is because the negative effects of drought and warming induced soil water stress on Rs and ER, which could override the positive effect of warming on these variables, which has been shown for a Montane meadow [52] and a subalpine meadow ecosystem in the QTP [27]. The marginal increase in GEP in 2013 (Table 2) likely resulted from a change in the species composition, which was observed in an open top chamber warming experiment nearby our study site, where coverage of grass and sedges decreased but that of forbs increased with warming [46]. The increased forbs biomass could ameliorate the negative impact of warming-induced soil moisture stress and the effect of lower AGB on GEP [40].

Although experimental warming tends to have a positive effect on plant productivity across ecosystems, experiments in grasslands indicate that clear increases in plant productivity in response to warming are relatively rare [53]. We did not detect a significant change in RB. We attributed this to the fact that we sampled the RB at a depth of 0–10 cm, which was constrained by a reduction in soil water, whereas the RB at a depth of 10–50 cm was significantly stimulated by warming. In contrast to the response of forbs, sedges and grass [28] may have cancelled each other out, leading to the non-significant change in AGB in this alpine meadow ecosystem. AGB and RB were positively correlated with ecosystem C fluxes (Figs. 8, 9). There was no significant change in AGB or RB with warming, whereas ecosystem C fluxes were significantly stimulated by warming (Tables 1, 2). It is possible that this resulted from the large seasonal variation in AGB and RB compared to the relatively smaller warming-induced changes in these biomass pools.

Main effects of clipping

There was no significant effect of clipping on C fluxes, which contrasted with results from other studies where increases in GEP, ER, and NEE have been reported for a temperature steppe [54] and tallgrass prairie [13], and decreases in GEP, ER, and NEE have been reported for a Swiss grassland [55]. The negative impact of clipping on ecosystem C fluxes is attributed to the grass being cut in the middle of the growing season, which may reduce the green leaf area and thus C fluxes [55]. Positive effects of clipping on C fluxes may result primarily from improved light conditions with the removal of standing litter [54] and compensatory growth from clipping [56]. In the current study, we clipped the plants in late September once they had started to senesce, and this could be one reason for the non-significant effect of clipping on GEP, NEE and ER. In addition, soil temperature has been found to influence CO_2 exchange in alpine meadow ecosystems [24], and we did not detect a significant effect of clipping on soil temperature (Fig. 2). Besides temperature, biomass also affects C fluxes in an alpine meadow ecosystem [24], as was observed in our study (Fig. 8). The significant decrease in AGB in 2013 under the clipping treatment with a non-significant change in C fluxes, suggests that soil temperature is the major factor controlling the response of ecosystem C fluxes to clipping.

Conclusion

Ecosystem C fluxes responded positively to elevated temperature, with a higher relative increase in GEP than in ER, leading to a net C gain in this alpine meadow ecosystem. Clipping and its interaction with warming had no significant effect on ecosystem C fluxes because clipping did not significantly affect soil temperature. In addition, this study was conducted during two hydrologically contrasting years (wet in 2012 and dry in 2013), which provided a unique opportunity to understand how drought affects ecosystem C fluxes and their response to warming and clipping in an alpine meadow ecosystem. In the dry year, positive effects of warming on ecosystem C fluxes were cancelled by lower soil moisture. However, we caution that our study encompassed only a single wet and dry year, and thus our inferences of drought need to be supported by future research. Our findings will improve our understanding of the response of ecosystem C fluxes to the combined effects of climate change factors and human activities in an alpine meadow ecosystem in the permafrost region of the QTP.

Acknowledgments

Authors thank Prof. Yongzhi, Liu, Hanbo Yun, Guilong Wu, and Yuanwu Yang for their help in setting up the field experiment.

Author Contributions

Conceived and designed the experiments: FP XX. Performed the experiments: FP QY MX JG. Analyzed the data: FP. Contributed to the writing of the manuscript: FP TW XX.

References

1. IPCC (2007) Climate change 2007: The physical Science Basis Contributin of Working group I to the Fourth Assessment Report of the Intergovernmental Panel on Climate Change. In: S S., D Qin, M Manning, Z Chen, M Marquis, K. B Averyt, M Tignor and H. L Miller, editors. Cambridge, United Kingdom/ New York, NY USA: Cambridge University Press. 749–766.
2. Luo YQ, Wan SQ, Hui DF, Wallance LL (2001) Acclimatization of soil respiration to warming in a tall grass prairie. Nature 413: 622–625.
3. Melillo JM, Steudler PA, Abler JD (2002) Soil warming and carbon-cycle feedbacks to the climate system. Science 298: 2173–2175.
4. Friedlingstein P, Cox P, Betts R, Bopp L, von Bloh W, et al. (2006) Climate-Carbon Cycle Feedback Analysis: Results from the C4MIP Model Intercomparison. J. Climate 19: 3337–3353.
5. Rustad LE, Campbell JL, Marion GM, Norby RJ, Mitchell MJ, et al. (2001) A meta-analysis of the response of soil respiration, net nitrogen mineralization, and aboveground plant growth to experimental ecosystem warming. Oecologia 126: 543–562.
6. Peñuelas J, Gordon C, Llorens L, Nielsen T, Tietema A, et al. (2004) Nonintrusive Field Experiments Show Different Plant Responses to Warming and Drought Among Sites, Seasons, and Species in a North-South European Gradient. Ecosystems 7: 598–612.
7. Oberbauer SF, Tweedie CE, Welker JM, Fahnestock JT, Henry GHR, et al. (2007) Tundra CO2 fluxes in response to experimental warming across latitudinal and moisture gradients. Ecol. Monogr. 77: 221–238.
8. Kirschbaum MF (1995) The temperature dependence of soil organic mater decomposition, and the effect of global warming on soil organic C storage. Soil Biolo. Biochem. 27: 753–760.
9. Cramer W, Bondeau A, Woodward FI, Prentice IC, Betts RA, et al. (2001) Global response of terrestrial ecosystem structure and function to CO2 and climate change: results from six dynamic global vegetation models. Global Change Biolo. 7: 357–373.
10. Canadell JG, Le Quéré C, Raupach MR, Field CB, Buitenhuis ET, et al. (2007) Contributions to accelerating atmospheric CO2 growth from economic activity, carbon intensity, and efficiency of natural sinks. PNAS 104: 18866–18870.
11. Wu ZT, Dijkstra P, Koch GW, PeÑUelas J, Hungate BA (2011) Responses of terrestrial ecosystems to temperature and precipitation change: a meta-analysis of experimental manipulation. Global Change Biolo. 17: 927–942.
12. Lu M, Zhou XH, Yang Q, Li H, Luo YQ, et al. (2013) Responses of ecosystem carbon cycle to experimental warming: a meta-analysis. Ecology.
13. Niu SL, Sherry RA, Zhou XH, Luo YQ (2013) Ecosystem carbon fluxes in responses to warming and clipping in a tallgrass prairie. Ecosystems.
14. Bahn M, Knapp M, Garajova Z, Pfahringer N, Cernusca A (2006) Root respiration in temperate mountain grasslands differing in land use. Global Change Biolo. 12: 995–1006.
15. Ross DJ, Tate KR, Scott NA, Feltham CW (1999) Land-use change: effects on soil carbon, nitrogen and phosphorus pools and fluxes in three adjacent ecosystems. Soil Biolo. Biochem. 31: 803–813.
16. Klein J, Harte J, Zhao X (2004) Experimental warming causes large and rapid species loss, dampened by simulated grazing, on the Tibetan Plateau. Ecol. Lett. 7: 1170–1179.
17. Luo CY, Xu GP, Chao ZG, Wang SP, Lin XW, et al. (2010) Effect of warming and grazing on litter mass loss and temperature sensitivity of litter and dung mass loss on the Tibetan plateau. Global Change Biol. 16: 1606–1617.
18. Derner JD, Boutton TW, Briske DD (2006) Grazing and ecosystem carbon storage in the North American Great Plains. Plant Soil 280: 77–90.
19. Niu SL, Sherry RA, Zhou XH, Wan SQ, Luo YQ (2010) Nitrogen regulation of the climate-carbon feedback: evidence from a long-term global change experiment. Ecology 91: 3261–3273.
20. Tarnocai C, Canadell JG, Schuur EAG, Kuhry P, Mazhitova G, et al. (2009) Soil organic carbon pools in the northern circumpolar permafrost region. Global Biogeochem. Cy. 23: GB2023.
21. Schuur EAG, Bockheim J, Canadell JG (2008) Vulnerability of permafrsot carbon to climate change:implication for the global carbon cycle. BioScience 58: 701–714.
22. Harden JW, Sundquist ET, Stallard RF, Mark RK (1992) Dynamics of soil carbon during deglaciation of the Laurentide ice sheet. Science 258: 1921–1924.
23. Kato T, Tang Y, Gu S, Cui X, Hirota M, et al. (2004) Carbon dioxide exchange between the atmosphere and an alpine meadow ecosystem on the Qinghai-Tibetan Plateau, China. Agr. Forest Meteor. 124: 121–134.
24. Kato T, Tang Y, Gu S, Hirota M, Du M, et al. (2006) Temperature and biomass influences on interannual changes in CO2 exchange in an alpine meadow on the Qinghai-Tibetan Plateau. Global Change Biol. 12: 1285–1298.
25. Liu XD, Chen BD (2000) Climatic warming in the Tibetan Plateau during recent decades. Inter. J. Climatol. 20: 1729–1742.
26. Zheng D, Zhang QS, Wu SH (2000) Mountain Geoecology and sustainable development of the Tibetan Plateau. Dordrecht, Netherlands: Kluwer Academic Publishers.
27. Lin XW, Zhang ZH, Wang SP, Hu YG, Xu GP, et al. (2011) Response of ecosystem respiration to warming and grazing during the growing seasons in the alpine meadow on the Tibetan plateau. Agr. Forest Meteor. 151: 792–802.
28. Li N, Wang GX, Yang Y, Gao YH, Liu GS (2011) Plant production, and carbon and nitrogen source pools, are strongly intensified by experimental warming in alpine ecosystems in the Qinghai-Tibet Plateau. Soil Biol. Biochem. 43: 942–953.
29. Lu Z, Wu Q, Yu S, Zhang L (2006) Heat and water difference of active layers beneath different surface conditions near Beiluhe in Qinghai-Xizang Plateau. J. Glaciol. Geogryol. 28: 642–647.
30. Wang G, Wang Y, Li Y, Cheng H (2007) Influences of alpine ecosystem responses to climatic change on soil properties on the Qinghai-Tibet Plateau, China. Catena 70: 506–514.
31. Pang Q, Cheng G, Li S, Zhang W (2009) Active layer thickness calculation over the Qinghai-Tibet Plateau. Cold Regions Science and Technology 57: 23–28.
32. Wu QB, Liu YZ (2004) Ground temperature monitoring and its recent change in Qinghai-Tibet Plateau. Cold Reg. Sci. Technol. 38: 85–92.
33. Yang ZL, Yang GH (2000) Potential productivity and livestock carrying capacity of high-frigid grassland in China. Resources Sci. 22: 72–77.
34. Pavelka M, Acosta M, Marek MV, Kutsch W, Janous D (2007) Dependence of the Q_{10} values on the depth of soil temperature measuring point. Plant Soil 292: 171–179.
35. Zhou XH, Wan SQ, Luo YQ (2007) Source components and interannual variability of soil CO2 efflux under experimental warming and clipping in a grassland ecosystem. Global Change Biol. 13: 761–775.
36. Steduto P, Çetinkökü Ö, Albrizio R, Kanber R (2002) Automated closed-system canopy-chamber for continuous field-crop monitoring of CO2 and H2O fluxes. Agr. Forest Meteor.111: 171–186.
37. Niu S, Wu M, Han Y, Xia J, Li L, et al. (2008) Water-mediated responses of ecosystem carbon fluxes to climatic change in a temperate steppe. New Phytol. 177: 209–219.
38. Lawrence BF, Bruce GJ (2005) Interacting effects of temperature, soil moisture and plant biomass production on ecosystem respiration in a northern temperate grassland. Agr. Forest Meteor. 130: 237–253.
39. Yang Y, Fang J, Ji C, Han W (2009) Above- and belowground biomass allocation in Tibetan grasslands. J. Veg. Sci. 20: 177–184.
40. Xia JY, Niu SL, Wan SQ (2009) Response of ecosystem carbon exchange to warming and nitrogen addition during two hydrologically contrasting growing seasons in a temperate steppe. Global Change Biol. 15: 1544–1556.
41. Williams M, Law BE, Anthoni PM, Unsworth MH (2001) Use of a simulation model and ecosystem flux data to examine carbon-water interactions in ponderosa pine. Tree Physiol. 21: 287–298.
42. Barr AG, Griffis TJ, Black TA, Lee X, Staebler RM, et al. (2002) Comparing the carbon budgets of boreal and temperate deciduous forest stands. Can. J. Forest Res. 32: 813–822.
43. Kahmen A, Perner J, Buchmann N (2005) Diversity-dependent productivity in semi-natural grasslands following climate perturbations. Funct. Ecol. 19: 594–601.
44. Gilgen AK, Buchman N (2009) Responses of tempeartue grasslands at different altitudes to simulated summer drought differed but scaled with annual precipitation. Biogeosciences 6: 2525–2539.
45. Sebastià M-T (2007) Plant guilds drive biomass response to global warming and water availability in subalpine grassland. J. Appl. Ecol. 44: 158–167.
46. Li N, Wang GX, Yang Y, Gao YH, Liu LA, et al. (2011) Short-term effects of temperature enhancement on community structure and biomass of alpine meadow in the Qinghai-Tibet Plateau. Acta Ecol. Sinica 31: 895–905.
47. Li XJ, Zhang XZ, Wu JS, Shen ZX, Zhang YJ, et al. (2011) Root biomass distribution in alpine ecosystem of the northern Tibet Plateau. Environ. Earth Sci. 64: 1911–1919.
48. De Boeck HJ, Lemmens CMHM, Gielen B, Bossuyt H, Malchair S, et al. (2007) Combined effects of climat warming and plant diversity loss on above- and below-ground productivity. Environ. Exp.l Bot. 60: 95–104.
49. Xu MH, Peng F, Xue X, You QG, Guo J (2014) All-year warming and autumnal clipping lead to the downward movement of the root biomass, carbon and total nitrogen in the soil of an alpine meadow. Environ. and Exp. Bot.: in press.
50. Zhang PC, Tang YH, Hirota M, Yamamoto A, Mariko S (2009) Use of regression method to partition sources of ecosystem respiration in an alpine meadow. Soil Biol. Biochem. 41: 663–670.
51. Hanson PJ, Edwards NT, Garten CT, Andrews JA (2000) Separating root and soil microbial contributions to soil respiratin: a review of methods and observations. Biogeochem. 48: 115–146.

52. Saleska S, Harte K, Torn M (1999) The effect of experimental ecosystem warming on CO_2 fluxes in a montane meadow. Global Change Biol. 5: 125–141.

53. Dukes JS, Chiariello NR, Cleland EE, Moore LA, Shaw MR, et al. (2005) Responses of Grassland Production to Single and Multiple Global Environmental Changes. PLoS Biol 3: e319.

54. Niu SL, Wu M, Han Y, Xia J, Zhang Z, et al. (2010) Nitrogen effects on net ecosystem carbon exchange in a temperate steppe. Global Change Biol. 16: 144–155.

55. Rogiers N, Eugster W, Furger M, Siegwolf R (2005) Effect of land management on ecosystem carbon fluxes at a subalpine grassland site in the Swiss Alps. Theor. Appl. Climatol. 80: 187–203.

56. Zhao W, Chen S-P, Lin G-H (2008) Compensatory growth responses to clipping defoliation in Leymus chinensis (Poaceae) under nutrient addition and water deficiency conditions. Plant Ecol. 196: 85–99.

13

Early Spring, Severe Frost Events, and Drought Induce Rapid Carbon Loss in High Elevation Meadows

Chelsea Arnold, Teamrat A. Ghezzehei, Asmeret Asefaw Berhe*

School of Natural Sciences, University of California Merced, Atwater, California, United States of America

Abstract

By the end of the 20th century, the onset of spring in the Sierra Nevada mountain range of California has been occurring on average three weeks earlier than historic records. Superimposed on this trend is an increase in the presence of highly anomalous "extreme" years, where spring arrives either significantly late or early. The timing of the onset of continuous snowpack coupled to the date at which the snowmelt season is initiated play an important role in the development and sustainability of mountain ecosystems. In this study, we assess the impact of extreme winter precipitation variation on aboveground net primary productivity and soil respiration over three years (2011 to 2013). We found that the duration of snow cover, particularly the timing of the onset of a continuous snowpack and presence of early spring frost events contributed to a dramatic change in ecosystem processes. We found an average 100% increase in soil respiration in 2012 and 2103, compared to 2011, and an average 39% decline in aboveground net primary productivity observed over the same time period. The overall growing season length increased by 57 days in 2012 and 61 days in 2013. These results demonstrate the dependency of these keystone ecosystems on a stable climate and indicate that even small changes in climate can potentially alter their resiliency.

Editor: Benjamin Poulter, Montana State University, United States of America

Funding: Funding for this study was provided by University of California Merced startup funds and Graduate Research Council grants to AAB and TAG. The funders had no role in study design, data collection and analysis, decision to publish, or preparation of the manuscript.

Competing Interests: The authors have declared that no competing interests exist.

* Email: AABerhe@UCMerced.edu

Introduction

Magnitude and timing of extreme weather events have recently gained attention for their potential to alter ecosystem processes [1–4]. The presence of more extreme weather events has increased concerns over the ability of natural ecosystems to respond to such rapid changes [5]. Extreme interannual change in weather (for example, from a very "wet" to a very "dry" year and vice versa) may trigger rapid carbon loss from an ecosystem [6,7]. For high elevation mountain ecosystems in particular, the seasonal timing of the accumulation and melting of the snowpack is crucial for supplying abundant water to low-lying communities and high-elevation forests [8]. It is also essential for promoting meadow productivity [9] and soil carbon storage [6,10]. It is expected that earlier snowmelt will result in drying of meadow soils over the course of the growing season. This drying may lead to increased carbon storage through an increase in the net primary productivity of the system, but it can also lead to a loss of carbon through increased rates of decomposition. Whether the ecosystem remains a sink for or shifts to a source of atmospheric carbon dioxide (CO_2) will have large implications on the ability of the meadow to filter, store, and release water to the river systems. Prolonged conditions that result in a significant loss of carbon can eventually trigger a tipping point to an ecological regime shift in the meadow.

The coupled hydrological and biogeochemical cycles in high elevation meadows are influenced by the depth and duration of the annual winter snowpack that acts as an insulating blanket during the winter [11]. Not only does the winter snowpack protect the meadow soils from large temperature fluctuations and winter desiccation, it also functions to recharge the meadow soils during the spring snowmelt [12]. There is a vital two-way relationship between hydrology and soil organic matter (SOM) dynamics in such high elevation systems. Hydrology exerts a strong control on storage, stability, and composition of SOM [13] in the meadow soils, while SOM dynamics controls the ability of the meadows to provide ecosystem services such as filtering, storing and releasing water to the river systems. Without these wetland systems to slow the passage of water from the snowpack to the streams, the watersheds become less resilient to flood pulses [14]. The essential nature of those ecosystem services warrants a "keystone" status of mountain meadows in terms of mountain hydrology. A keystone species is one that has a disproportionately large impact to an ecosystem in comparison to its abundance. Mountain meadows, though small in aerial extent in the Sierra Nevada, are an essential component of the mountain water cycle. Watersheds that have lost meadow functioning due to degradation have limited water storage capacity and ability to attenuate floods [15]. Degradation of meadows results in a flashy system, where the surface and shallow subsurface flows in the watershed respond rapidly to precipitation events. Furthermore, high-elevation meadows, which are hotspots of biodiversity [16] and function as breeding grounds for many organisms in the central Sierra Nevada Mountains and

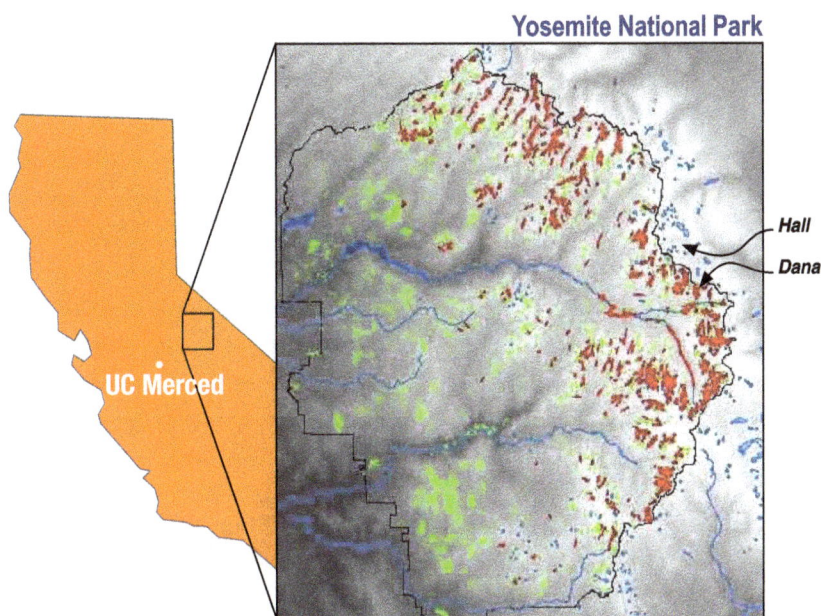

Figure 1. Map of study sites along the boundary of Yosemite National Park, California. Polygons represent the extent of meadow area in Yosemite National Park, with subalpine meadows (>3 hectares and between 2600–3200 m) highlighted in red.

other similar ecosystems, are likely to be a key indicator of the overall health of the watersheds.

While interannual variations in snowpack depth and duration are normal in the Sierra Nevada [17], consecutive years with extreme water conditions can significantly increase or decrease the overall length of the summer growing season and duration of snow free days, which will directly affect soil carbon storage. Previously, hydrologic modeling research in the Sierra Nevada has highlighted the sensitivity of that region's watersheds to earlier onset of spring and increased duration of low flows [18]. It was shown that some watersheds that are highly vulnerable to an increase in duration of low flows with climate warming, also occupy the largest mountain meadow area. An increase in the duration of low flows can cause meadows systems to dry down significantly, causing feedbacks to ecosystem processes such as primary productivity and soil respiration. If the trend in the onset of spring [19] continues, and meadows dry down earlier in the growing season, we can expect an increase in the decomposition of soil organic matter as the normally saturated soils become aerobic. This could also potentially impact river systems through a reduction in the ability of meadows to contribute to baseflow as they dry down. In addition, the timing of the onset of snowcover in the early winter can impact meadow soils and biota due to the widely fluctuating soil and air temperatures. The meadow soils in theses systems remain at $0°C$ as soon as snow accumulates in a continuous snowpack. This insulating layer protects overwintering biota and prevents drying of the meadow soil. Colder winter temperatures coupled to lack of continuous snowpack renders meadow soils and biota susceptible to severe desiccation which will impact ecosystem processes such as soil respiration and productivity in the following summer growing season.

An increase in interannual variation in the onset of spring has the potential to dramatically affect the balance between carbon storage and loss. This would occur mainly through changes to the input of carbon from above and belowground biomass [20], and loss through soil respiration and leaching [21,22] in meadow soils. A period of rapid carbon loss for the organic-rich high-elevation

meadow soils can trigger a positive feedback loop that contributes to declining soil moisture [23], further organic matter decomposition and reduced plant productivity through changes in soil structure [24].

In mountain meadow ecosystems, mean changes in the timing of spring snowmelt have already been shown to influence plant phenology [25,26], interactions between plants and pollinators [27], and longer term changes in meadow vegetation community structure [28]. In order to examine how the timing and duration of snow cover and presence of early season frost events can influence ecosystem processes, (net primary productivity and soil respiration), we monitored changes in surface carbon dioxide flux and above ground net primary productivity over three consecutive summers (2011 to 2013) in two high elevation meadows in the Central Sierra Nevada mountain range of California.

Methods

Methodology

The objective of this study was to track ecological responses of high elevation meadows to extreme seasonality. We combine field-based measurements of soil respiration and aboveground net primary productivity with remote sensing techniques to gauge how the amount and timing of precipitation, and seasonality impact meadow systems.

Site description

Our study was conducted in two subalpine meadows with different hydrologic regimes located at the crest of the Sierra Nevada mountain range along the boundary of Yosemite National Park (YNP) (Figure 1). Both meadows were formed as a direct result of past glaciation. Their resulting geomorphic position in the landscape remains conducive to high water tables throughout much of the growing season. One meadow is located in the Harvey Monroe Hall Research Natural Area (Hall RNA) at 3200-m elevation on a large medial moraine on the eastern side of the central Sierra Nevada. The mean daily temperatures range from

Table 1. Historic ranking of SWE in Dana Meadow.

The top 12 driest years on record (1927–2012)				The top 12 wettest years on record (1927–2012)			
Rank	Driest	SWE (cm)	% of normal	Rank	Wettest	SWE (cm)	% of normal
1	**2012**	**19.2**	**25**	1	1969	147.32	192.0
2	1977	19.81	25.8	2	1995	133.35	173.8
3	1976	30.23	39.4	3	2006	122.43	159.6
4	1931	33.27	43.4	4	1986	121.16	157.9
5	**2012**	**37.59**	**49.0**	5	1978	120.40	156.9
6	1990	38.61	50.3	6	1998	119.89	156.3
7	1964	39.88	52.0	**7**	**2011**	**119.38**	**155.6**
8	1929	40.64	53.0	8	1952	118.36	154.3
9	1994	44.70	58.3	9	1956	115.57	150.7
10	1992	45.21	58.9	10	1980	113.54	148.0
11	1939	45.47	59.3	11	2005	106.93	139.4
12	1930	46.74	60.9	12	1993	105.41	137.4

Recent years that were in the top 10 wettest or driest years are in bold.

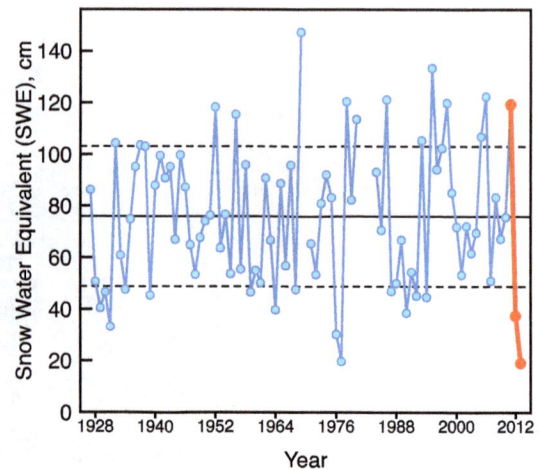

Figure 2. Historic SWE record for Dana meadows (Yosemite National Park) with 2011–2013 highlighted in red.

−4.9°C to 12.9°C [29]. The soils in the Hall RNA are characterized as Inceptisols with the suborders Andic Cryumbrepts and Lithic Cryumbrepts [30]. For a contrasting type of meadow, we chose Dana Meadows, which is located at a 3000-m elevation along YNP's Tioga Pass Road in a U-shaped glacial valley with hummocky ablation till. Dana Meadows exhibits mean temperatures similar to those of the Hall RNA and an average precipitation of 1000 mm/year. The soils in Dana Meadows are classified as Inceptisols with the suborders Xeric Dystrocryepts and Vitrandic Eutrocryepts [31]. The research permits for these two study sites were granted by the United States Department of Interior–National Parks, Yosemite National Park for study site Dana Meadows (37.893100N, −119.256900W), and the United States Department of Agriculture–Forest Service, Pacific Southwest Region for study site Hall RNA (37.958056N, −119.296111W). The field studies did not involve endangered or protected species.

Field methods

In July 2010, transects were established along a hydrologic gradient at two locations in the Hall RNA and two locations in Dana Meadows. The hydrologic gradient in the meadow was established using vegetation associations as a proxy for water table depth [32]. In all four transects, the same vegetation type was utilized to identify each meadow region: *Carex filifolia* in the xeric sites, *Ptilagrostis kingii* in the mesic sites and *Carex scopulorum/Carex subnigricans* in the hydric sites. Three replicate soil collars were inserted in the soil at depths of approximately 3–5 cm (the variation was due to differences in soil characteristics) at three hydrologically distinct regions of the transect (designated as: dry, intermediate, and wet). The collars are located approximately 2 meters apart. The soil carbon dioxide (CO_2) efflux was measured using a LI-COR 8100A portable infrared gas analyzer (LI-COR Biosciences, Lincoln, Nebraska USA), fitted with a portable 10-cm soil respiration chamber. After a 45-second pre-purge, one-minute measurements were recorded and were followed by a 30-second post-purge. Weekly measurements were recorded during the first half of the growing season, followed by biweekly measurements through September. All measurements were taken at mid-day from collars with vegetation left intact. In each of the 4 transects, there were 6 collars in 2011, and 18 collars in 2012/2013. Above ground productivity was estimated by harvesting the total biomass

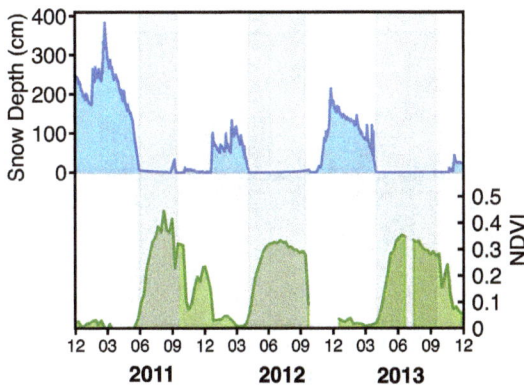

Figure 3. Snow depth (top panel) from Dana Meadows and NDVI (weighted mean average of all subalpine meadows in Yosemite National Park). Gray panels denote the growing season in the meadows as defined by the first day the meadow is snow free and the date where the NDVI crosses a threshold of 0.3.

in six 20 cm square quadrats in each region (dry, intermediate and wet) of the transect at peak production each year. Vegetation samples were oven dried at 50°C, and weighed to determine biomass. Historical and current meteorological data were obtained from the California Department of Water Resources station for Dana Meadows (ID: DAN). Meteorological data utilized for this study include, maximum air temperature, minimum air temperature, snow depth and snow water equivalent (SWE). SWE is the amount of water contained in a unit of snowpack. The April 1 SWE is an important metric for water resource managers in California. It represents the time where historically there has already been the maximum snowpack accumulation for the year, and thus represents the amount of available water to downstream users. Soil temperature and water content were measured at one site in the Hall RNA. Decagon 5TM sensors (Decagon Devices, Inc.) were inserted at 5, 15 and 25 cm below the soil surface. They were continuously monitored using a Decagon EM-50 datalogger.

Satellite-based remote sensing imagery

The Terra/MODIS surface reflectance (MOD09Q1.5) 8-day L3 global 250-m product was downloaded directly from the Land Processes Distributed Active Archive Center of the United States Geological Survey (https://lpdaac.usgs.gov). This level 3 surface reflectance product, which had been radiometrically corrected and georeferenced, provided a measure of the surface reflectance at the ground level in the absence of atmospheric scattering or absorption. The data were projected in a custom sinusoidal projection specific to the MODIS imagery. The eight-day composite images represented the maximum surface reflectance value for that time period and minimized the impacts of clouds and aerosols.

Processing MODIS imagery to NDVI

In the first stage of processing, the MODIS product MOD09Q1.5 was reprojected from a custom sinusoidal projection to the California Albers projection. The latter is a version of the Albers Equal Area projection optimized for statewide calculations. Bands 1 (620–670 nm) and 2 (841–876 nm) were utilized to calculate the NDVI over the entire MODIS image. The following equation was used: NDVI = (band 2 − band 1)/(band 2 + band 1). The resulting NDVI product was resampled down to 30 m and was used to produce an average NDVI for the entire meadow

Table 2. Mean maximum air temperature (°C) in Dana Meadow.

Month	2013	2012	2011	2010	Historic Mean (2000–2010)
JAN	3.12	4.78	4.93		2.21
FEB	3.21	2.82	0.89		2.44
MAR	6.81	3.64	2.47		5.35
APR	9.11	8.06	5.78		6.62
MAY	10.75	11.86	7.67		11.24
JUN	15.97	16.28	13.48		15.44
JUL	20.05	19.87	18.39		19.62
AUG	18.57	21.33	19.48		18.72
SEPT	15.11	13.24	17.17		14.59
OCT	10.43		11.94		7.06
NOV	7.00	6.17	3.68		5.04
DEC	4.68	−0.88	2.53	1.76	1.83
DJF	**1.82**	**3.38**	**2.53**		
MAM	**8.89**	**7.85**	**5.31**		
JJA	**18.20**	**19.16**	**17.12**		
SON	**10.85**	**9.70**	**10.93**		

Mean seasonal temperatures (DJF, MAM, JJA, SON) are highlighted in gray. Historic mean maximum air temperatures from 2000–2010 are referenced in far right column.

Table 3. Mean minimum air temperature (°C) in Dana Meadow.

	2013		2012		2011		2010	Historic Mean (2000–2010)
JAN	−11.20		−7.85		−9.48			−10.92
FEB	−12.64	**−12.12**	−11.07	**−9.69**	−12.57	**−10.45**		−11.40
MAR	−8.01		−9.57		−11.00			−10.11
APR	−6.33		−7.18		−9.54			−8.56
MAY	−2.03	**−5.46**	−3.39	**−6.71**	−7.35	**−9.30**		−3.75
JUN	1.53		0.44		−2.20			0.34
JUL	4.87		3.46		2.29			4.25
AUG	2.85	**3.08**	5.89	**3.26**	2.76	**0.95**		2.84
SEPT	0.17		−3.06		1.56			−0.93
OCT	−4.89		−6.36		−2.62			−4.59
NOV	−5.78	**−3.50**		**−4.71**	−8.43	**−3.16**		−7.51
DEC	−10.09		−12.51		−10.16			−10.37

Mean seasonal temperatures ((DJF, MAM, JJA, SON) are highlighted in gray. Historic mean minimum air temperatures from 2000–2010 are referenced in far right column.

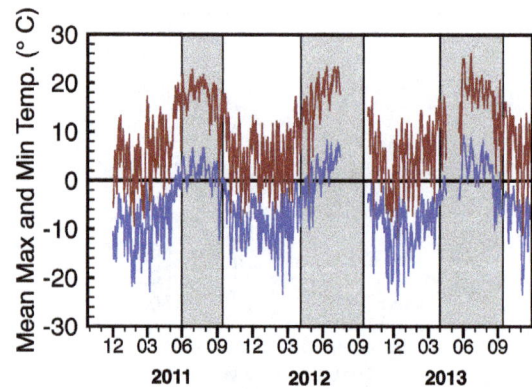

Figure 4. Maximum (red) and minimum (blue) daily air temperatures for Dana Meadow in 2011, 2012 and 2013. Gray panels denote the growing season in the meadow as defined by the first day the meadow is snow free and the date where the NDVI crosses a threshold of 0.3.

polygon region. The meadow polygons, resulting derived data layer and the associated metadata are currently being prepared as a spatial data product by the U.S. Geological Survey's Western Ecological Research Center at the Yosemite Field Station [33].

Statistical Analysis

Soil respiration rates for each collar were integrated over time to determine the cumulative CO_2 efflux for the growing season. Missing data was filled in via linear interpolation between the prior sampling date and the next date sampled. Repeated measures analysis of variance (RM-ANOVA) was used to determine significant differences between the effects of moisture class across the three years for both ANPP and cumulative CO_2 flux data. The different sites were utilized as replicates. If the RM-ANOVA model was significant, a Tukey's post hoc test ($p<0.05$) was used to assess differences between means. In addition, in order to determine the effect of year within a moisture region of the meadow, a subset of data was created for dry, intermediate and wet sites and a one-way RM-ANOVA model was utilized to determine significance within moisture classes across years. If the model was significant, a Tukey's post hoc ($p<0.05$) was used to determine differences between means. Data was tested for normality prior to analysis using the Shapiro-Wilks test. All statistical analyses were conducted using R statistical software (r-project.org).

Results and Discussion

Meteorological Data

The last several years in California have been marked by extreme seasonal weather on either ends of the spectrum. The 2011 water year (October 2010 through September 2011) was the seventh-wettest year on record (1929–2012) in YNP, with the April 1 SWE in Dana meadows reaching 156% of the 50 year mean (1951–2000) (Table 1). The 2012 water year was the fifth-driest year on record with only 49% of the mean SWE and the 2013 water year ranked the driest year on record with 25% of the mean SWE (Table 1). Looking at the entire historic record of Dana Meadows SWE, there is an increase after 1969 in the number of years with SWE values greater or less than one standard deviation from the mean (Figure 2). This apparent increase in the SWE variability corresponds to trends found in increase in the variability

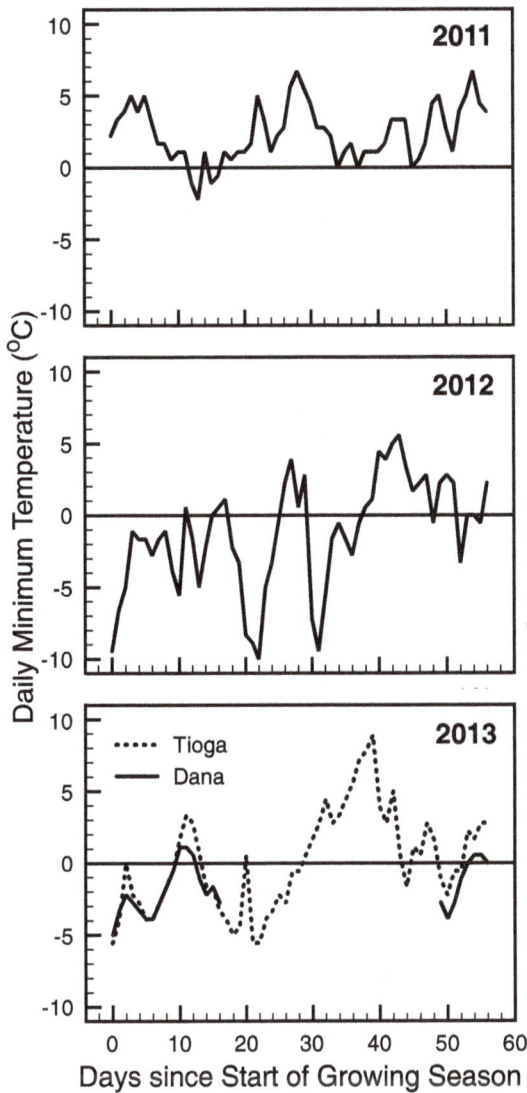

Figure 5. Daily minimum temperatures were used to compare frost events in the first sixty days since the start of the growing season (2011–2013) in Dana Meadow. Growing season was determined by the first day that the meadow was snow free each year. Dotted line represents data from a nearby meteorological station (Station id:TES) at north end of meadow was used for missing data in 2013.

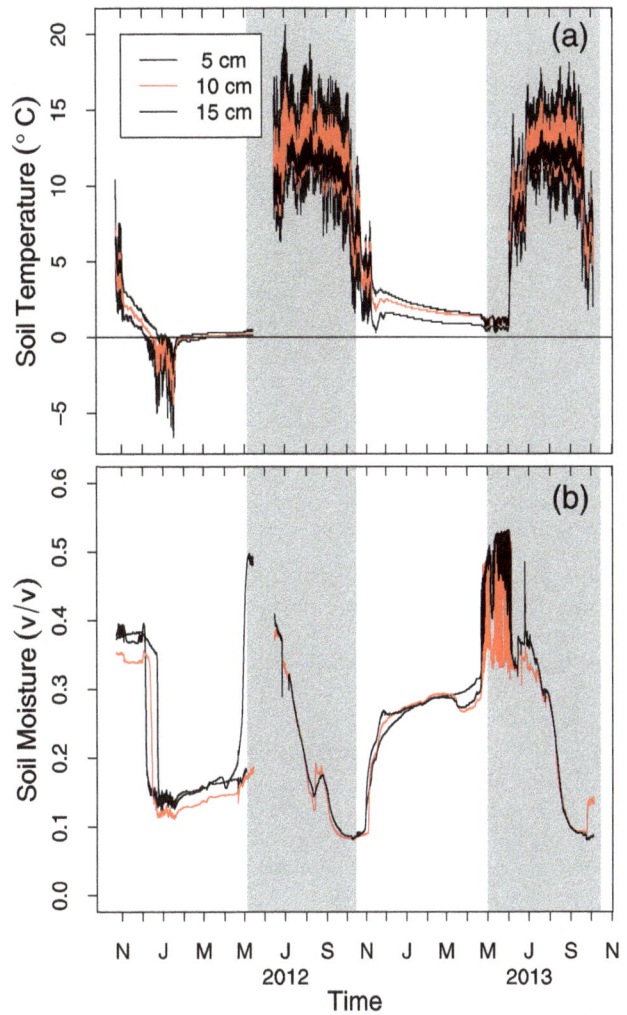

Figure 6. Time series of soil temperatures and volumetric water content for a dry meadow site from November 2011 to November 2013. The shaded panels indicate the growing season in 2012 and 2013.

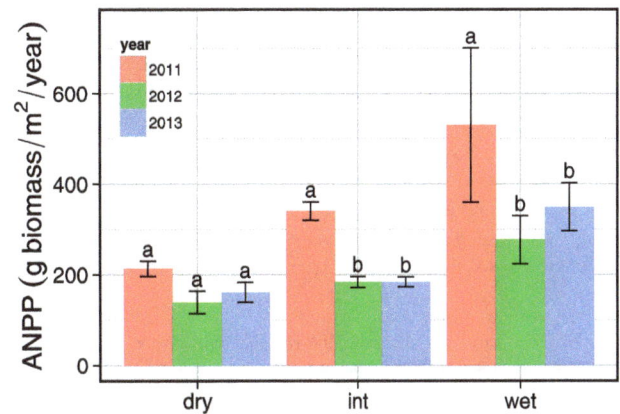

Figure 7. Mean annual aboveground net primary productivity averaged across all four sites for each moisture region in 2011, 2012, and 2013. Error bars represent standard error among sites. Letters denote significant differences in homogenous groups across years as determined by a Tukey post hoc test (p<0.05).

of streamflow in Central California around the same time period [34].

In ecosystems dependent upon the enduring winter snowpack to insulate them from freezing events, the timing of the first day of continual snow cover for the winter can be critical to biological communities [35]. Likewise, the duration of that snow cover and timing of subsequent spring melt plays an essential role in microbial turnover [36,37], plant phenology [38], and meadow hydrology [39,40]. In addition, recent research has shown that winter warming in arctic ecosystems is contributing to a decline in plant productivity during the subsequent summer growing season [41]. Not only was the depth of snowpack distinctly different in the three consecutive years, but also the duration of snow cover differed greatly in all three years (Figure 3). The water year 2012 was especially anomalous with no continual snow cover until mid January. A significant ice storm occurred over the bare soils on

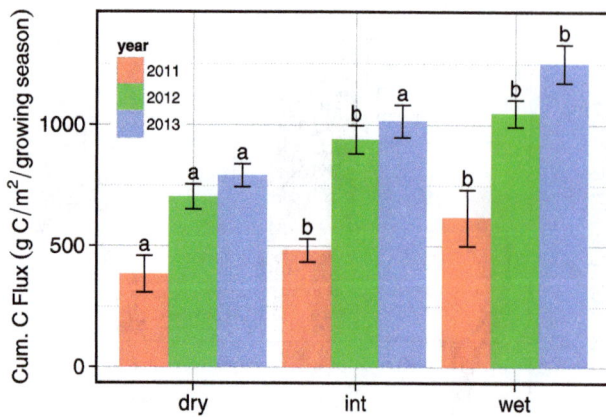

Figure 8. Mean cumulative carbon flux averaged across all four sites for each moisture region in 2011, 2012, and 2013. Error bars represent standard error among sites. Letters denote significant differences in homogenous groups across years as determined by a Tukey post hoc test (p<0.05).

December 4, 2012, with widespread needle damage to conifers noted at the study site after snowmelt.

With the exception of January, the average maximum air temperature in Dana Meadows was warmer in 2013 and 2012 through July of each year, as compared to 2011, with the spring (Mar–May) mean temperature increasing by 2–3°C (Table 2). This early warming is a contributing factor to the onset of an early spring in those years. The mean monthly minimum temperatures show warmer spring and summer temperatures in 2012 and 2013 as compared to 2011 (Table 3).

Ecosystem Response

The extreme seasonal changes from 2011 and 2012/2013 caused dramatic shifts in the onset of spring and in the number of snow-free days in the meadows of YNP. In Dana Meadows, the first snow-free day in 2012 and 2013 occurred 57–61 days earlier than in 2011, and the growing season increased by 35–37%, from approximately 106 days in 2011 to 163/167 days in 2012/2013. The documented shift to an earlier onset of spring in the Sierra Nevada [19] appears to have altered the response of the meadow ecosystems; rather than increasing their productivity, the earlier spring has rendered the meadows more sensitive to late winter/early spring frost events [11,28,42,43].

The maximum and minimum temperatures in the meadow show a clear seasonal trend, with very few frost events occurring within a normal growing season (Figure 4). The seasonal trends are similar between years, but the time point when the growing season is initiated is critical for assessing potential frost impacts on newly sprouting vegetation. As the snow melts, it causes saturation of the meadow soil and plants respond rapidly to this moisture and

available nutrients by sending up green shoots. This leaves them susceptible to freezing temperatures. Tranquillini (1964) has shown that high elevation plants are very frost resistant, however notes that plants dependent on an insulating snowpack are susceptible to frost damage even in minor frost events [44]. If spring arrives earlier, as in 2012 and 2013, there is an increased likelihood of a severe frost event to damage newly sprouting vegetation. This pattern was evident during two frost events that occurred in 2012 after the snow had cleared from the meadow (Figures 4 and 5). The first event occurred over a four-day period that peaked on May 27, when the temperature dropped to −10°C. This event occurred approximately 20 days into the growing season. The second event occurred over three days beginning on June 5 that included a low temperature of −9.4°C. Because meadows undergo a rapid greening within days of a snowmelt, frost can damage sensitive meadow species and reduce overall productivity [43,45]. In 2013, there were several frost events that occurred on May 19 and 20 and on May 22 and 23 with a low at −5.5°C.

Frost damage to vegetation was apparent on a larger scale using satellite-derived Normalized Difference Vegetation Index (NDVI) mapping. The NDVI time series for the three years is shown in Figure 3. A rapid greening was apparent at the beginning of the growing season for both years; however, instead of reaching a peak in 2012, the NDVI plateaued before the vegetation senesced in mid-summer, which could indicate that the meadow vegetation was stressed and never reached maximum greenness in 2012. Normal meadow NDVI ranges from 0 during snowcover to a maximum of 0.45–0.65 at peak production and then falls to around 0.3 during senescence. In 2011, the peak NDVI occurred around 0.45, but in 2012 and 2013, the peak NDVI was just above the senescence value. Since the meadow soils begin the season saturated due to snowmelt and dry down over the growing season, it is unlikely that the plateau was caused by the meadows drying earlier in the growing season. If this were the case, we would expect to see a peak NDVI soon after the rapid increase at the beginning of the season. The pattern in 2013 is similar, though there is a small peak in early season NDVI, but the overall pattern is much lower than in 2011, indicating that although aboveground productivity did recover slightly from 2012, the meadow was still in a stressed state. There is an anomalous peak of NDVI that occurs between November 2011 and January 2012. This value was reflective of the senesced vegetation and bare ground that lacked a continuous snowpack until mid January 2012.

Another potential explanation for the decreased peak NDVI in 2012 is the winter desiccation of the meadow soils that may have damaged overwintering roots. Figure 6 shows a time series of soil temperature and volumetric water content at three depths in the soil from a drier region of the meadow. The soil temperatures dropped well below freezing for an extended period of time in early 2012 before the snowpack began to accumulate in mid-January. This drop in soil temperature triggered a desiccating event in the soil, as seen in the lower panel. Volumetric soil

Table 4. One-Way RM-ANOVA results from homogeneous moisture regions of the meadow.

	Dry	Intermediate	Wet
2011–2012	P<0.003	P<0.0007	P<0.006
2011–2013	P<0.001	P<0.0004	P<0.002
2012–2013	P = 0.798	P = 0.888	P = 0.667

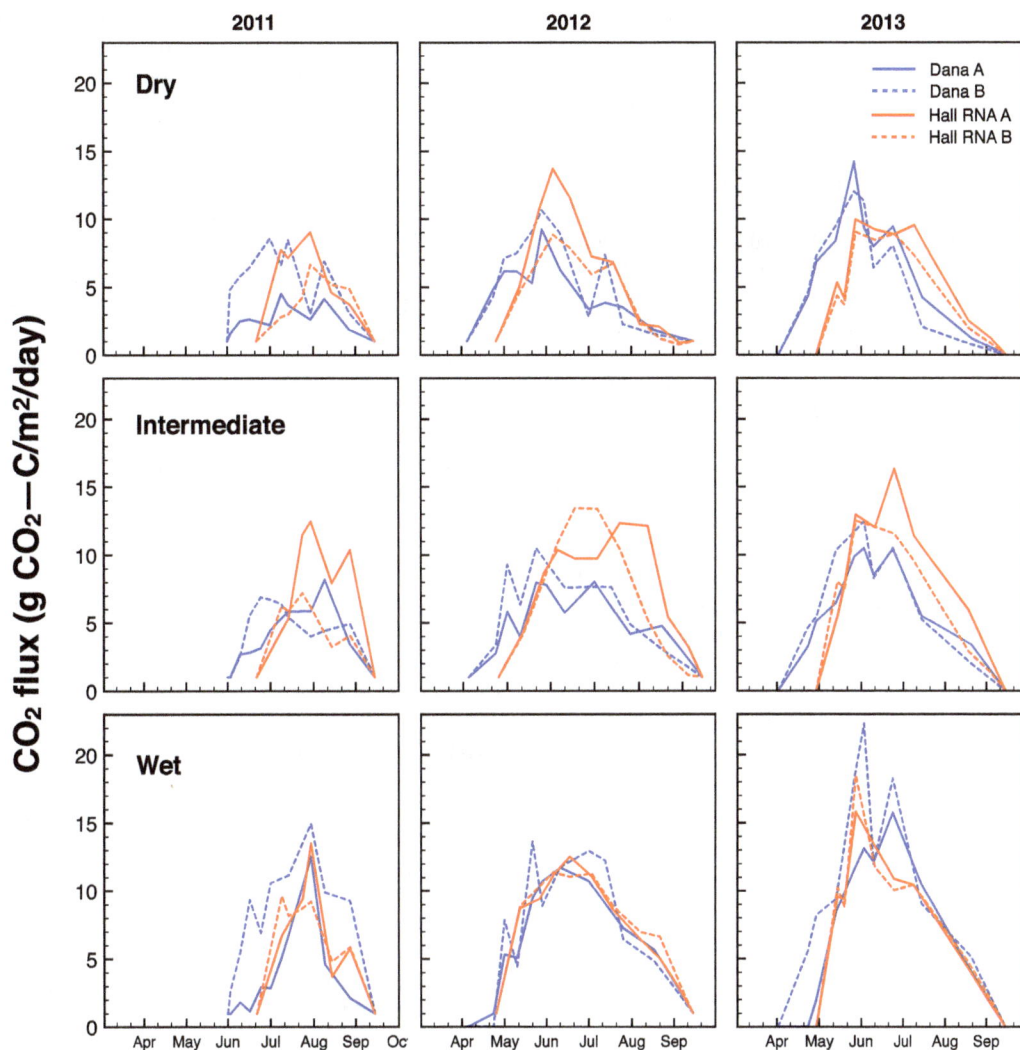

Figure 9. LI-COR surface CO_2 flux data for 2011, 2012, and 2013 in Dana Meadows and the Hall RNA.

moisture content was subsequently extremely low before snowpack accumulated. In a similar situation, Bokhorst et al have found that winter warming in the arctic, where snow melts off of the surface renders plants less productive the following spring [46].

The prolonged growing season in 2012 and 2013 should have led to increased plant productivity if water and temperature were not limiting, but the productivity actually declined by an average of 39% in all the regions of the meadow (Figure 7). Averaged across all sites, the drop in productivity from 2011 to 2012 was the most significant of all three years (p<0.000001). The aboveground productivity in 2013 was still significantly lower than 2011 (p< 0.0001) and didn't significantly increase from 2012, indicating that the system is still at a stressed state and not recovering rapidly. However, since belowground biomass was not quantified in this study, it is unknown if the meadow systems are adapting to the change in seasonality by putting more energy into belowground biomass. However, due the fact that meadows already allocate a greater proportion of their carbon inputs (60–80%) to roots, it is unlikely that a shift to more belowground production could offset the carbon losses via soil respiration in 2012 and 2013. Studies have also shown that 90% of the fixed carbon is re-respired in the same season in peats and fens [47]. There is a delicate balance

between productivity and carbon loss via respiration and unless the plants drastically increased the belowground biomass in response to the early spring, the losses via soil respiration will override the carbon inputs from the additional belowground biomass.

Different hydrologic regions of the meadows (dry – wet) responded differently to the change in duration and amount of snow cover (Figure 7). There was no significant difference in the productivity of the dry regions of the meadow across the three years. However, both the intermediate and wet regions showed significant reduction in biomass from 2011–2012 (intermediate, p<.0001; wet, p<0.0001). Both regions of the meadow still had significantly lower productivity in 2013 than 2011 (intermediate, p<0.0001; wet, p<0.0001), but no significant change from 2012 to 2013.

The mean cumulative carbon flux shows an increasing trend over time (Figure 8). Averaged across all moisture regions of the meadows, the mean cumulative carbon flux from the meadows was significant with respect to year (RM ANOVA model; year,p< 0.00001, moisture, p<0.00001, moisture:year, p = 0.447). However, this was driven mainly by the significance between 2011–2012 (p<0.00001) and 2011–2013 (p<0.00001). While 2013 continued to have larger cumulative fluxes in all regions of the meadow, they were not significantly different than 2012 (p = 0.481). There was no interaction between moisture region and year. Comparing the individual moisture regions of the meadow, there was significant change in the cumulative carbon flux of the dry, intermediate and wet regions of the meadow from 2011 to 2012 and 2011 to 2013, but no significant change from 2012 to 2013 (Table 4).

This result suggests that easily decomposable soil organic matter rapidly decomposed with the shift in environmental conditions in 2012 and 2013 [48–51]. The short growing season in 2011 and wet conditions throughout the growing season effectively reduced the overall soil carbon efflux in all regions of the meadow. Although the 2011 water year was extremely wet, and the meadows experienced little drying, this had no apparent effect on the available moisture in 2012. One explanation for this finding may be the dryness caused by a lack of snowpack in December 2011 through mid-January 2012; when the snow finally accumulated in January, the meadows were extremely dry underneath the snowpack (Figure 6). These dry soil conditions at the start of the 2012 growing season and below average snowpack coupled to above average surface temperatures throughout the 2012 growing season led to the extreme drying of the meadow soils and subsequently large carbon fluxes. The 2013 water year was the driest on record for Dana Meadows and there was a continuing trend of high mean cumulative carbon flux from the meadows. Although the cumulative flux was similar in both 2012 and 2013, there was a difference in the timing of peak soil carbon efflux. In 2013, the peak soil carbon efflux occurred early in the growing season and rapidly declined in all regions of the meadow, whereas in 2012, the peak occurred in the middle of the growing season (Figure 9). This shift indicates how responsive these ecosystems are to seasonal variation. The sustained magnitude of the cumulative carbon flux from the system over two summers has resulted in a loss of over 6% of the total carbon stock in the meadows we studied.

With climate extremes occurring at an increasing frequency around the world, our data demonstrate that sensitive ecosystems respond rapidly to the changes in seasonality and may reach a tipping point sooner rather than later. Multiple years of ecosystem stresses such as frost or drought can potentially cause a regime shift in vegetation with ramifications to the cycling of carbon in these systems. The magnitude of loss was significant given the small areal extent of these meadows, which is not proportional to their importance to overall ecosystem functioning and keystone position on the landscape. If the frequency of extreme events continues in this region, coupled to a decline in meadow aboveground productivity, we can expect carbon stocks in the meadows to rapidly decline, leading to meadow degradation and a reduction in ecosystem services in these watersheds.

Acknowledgments

The authors wish to thank J.R. Matchett, Eric L. Berlow, Samuel J. Traina, and John Harte for their comments on earlier versions of this manuscript, and Jesseca Burkhart for help in the field.

Author Contributions

Conceived and designed the experiments: CA TAG AAB. Performed the experiments: CA TAG AAB. Analyzed the data: CA TAG AAB. Contributed reagents/materials/analysis tools: CA TAG AAB. Contributed to the writing of the manuscript: CA TAG AAB.

References

1. Walther G-R, Post E, Convey P, Menzel A, Parmesan C, et al. (2002) Ecological responses to recent climate change. Nature 416(6879): 389–395.
2. Holmgren M, Stapp P, Dickman CR, Gracia C, Graham S, et al. (2006) Extreme climatic events shape arid and semiarid ecosystems. Frontiers in Ecology and the Environment 4: 87–95.
3. Jentsch A, Beierkuhnlein C (2008) Research frontiers in climate change: effects of extreme meteorological events on ecosystems. Comptes Rendus Geoscience 340: 621–628.
4. Jentsch A, Kreyling J, Elmer M, Gellesch E, Glaser B, et al. (2011) Climate extremes initiate ecosystem-regulating functions while maintaining productivity. Journal of Ecology 99: 689–702.
5. Craine JM, Nippert JB, Elmore AJ, Skibbe AM, Hutchinson SL, et al. (2012) Timing of climate variability and grassland productivity. Proceedings of the National Academy of Sciences 109: 3401–3405.
6. Galvagno M, Wohlfahrt G, Cremonese E, Rossini M, Colombo R, et al. (2013) Phenology and carbon dioxide source/sink strength of a subalpine grassland in response to an exceptionally short snow season. Environmental Research Letters 8: 025008.
7. Reichstein M, Bahn M, Ciais P, Frank D, Mahecha MD, et al. (2013) Climate extremes and the carbon cycle. Nature 500: 287–295.
8. Trujillo E, Molotch NP, Goulden ML, Kelly AE, Bales RC (2012) Elevation-dependent influence of snow accumulation on forest greening. Nature Geoscience 5: 705–709.
9. Baptist F, Flahaut C, Streb P, Choler P (2010) No increase in alpine snowbed productivity in response to experimental lengthening of the growing season. Plant Biology 12: 755–764.
10. Aurela M, Laurila T, Tuovinen J-P (2004) The timing of snow melt controls the annual CO2 balance in a subarctic fen. Geophysical Research Letters 31: L16119.
11. Inouye DW (2008) Effects of climate change on phenology, frost damage, and floral abundance of montane wildflowers. Ecology 89: 353–362.
12. Loheide II SP, Lundquist JD (2009) Snowmelt-induced diel fluxes through the hyporheic zone. Water Resour Res 45: W07404.
13. Heimann M, Reichstein M (2008) Terrestrial ecosystem carbon dynamics and climate feedbacks. Nature 451: 289–292.
14. Hammersmark CT, Rains MC, Mount JF (2008) Quantifying the hydrological effects of stream restoration in a montane meadow, northern California, USA. River Research and applications 24: 735–753.
15. Brown KA (2013) Groundwater storage in a mountain meadow northern Sierra Nevada California [Ph.D.]. Stanislaus: California State University.
16. Myers N, Mittermeier RA, Mittermeier CG, Da Fonseca GA, Kent J (2000) Biodiversity hotspots for conservation priorities. Nature 403: 853–858.
17. Kapnick S, Hall A (2009) Observed changes in the Sierra Nevada snowpack: potential causes and concerns. Prepared for the CEC and Cal/EPA ECE-500-2009-016-D: 09-016.
18. Null SE, Viers JH, Mount JF (2010) Hydrologic response and watershed sensitivity to climate warming in California's Sierra Nevada. PLoS One 5: e9932.
19. Cayan DR, Kammerdiener SA, Dettinger MD, Caprio JM, Peterson DH (2001) Changes in the onset of spring in the western United States. Bulletin-American Meteorological Society 82: 399–416.
20. Chivers M, Turetsky M, Waddington J, Harden J, McGuire A (2009) Effects of experimental water table and temperature manipulations on ecosystem CO2 fluxes in an Alaskan rich fen. Ecosystems 12: 1329–1342.
21. Xiang S-R, Doyle A, Holden PA, Schimel JP (2008) Drying and rewetting effects on C and N mineralization and microbial activity in surface and subsurface California grassland soils. Soil Biology and Biochemistry 40: 2281–2289.
22. Alm J, Schulman L, Walden J, Nykänen H, Martikainen PJ, et al. (1999) Carbon balance of a boreal bog during a year with an exceptionally dry summer. Ecology 80: 161–174.
23. Orchard VA, Cook F (1983) Relationship between soil respiration and soil moisture. Soil Biology and Biochemistry 15: 447–453.
24. Stephens JC, Allen Jr LH, Chen E (1984) Organic soil subsidence. Man-induced land subsidence Reviews in Engineering Geology VI Geological Society of America: 107–122.
25. Wipf S, Stoeckli V, Bebi P (2009) Winter climate change in alpine tundra: plant responses to changes in snow depth and snowmelt timing. Climatic Change 94: 105–121.
26. Price MV, Waser NM (1998) Effects of experimental warming on plant reproductive phenology in a subalpine meadow. Ecology 79: 1261–1271.
27. Thomson JD (2010) Flowering phenology, fruiting success and progressive deterioration of pollination in an early-flowering geophyte. Philosophical Transactions of the Royal Society B: Biological Sciences 365: 3187–3199.
28. Forrest J, Inouye DW, Thomson JD (2010) Flowering phenology in subalpine meadows: Does climate variation influence community co-flowering patterns? Ecology 91: 431–440.

29. Taylor DW (1984) Vegetation of the Harvey Monroe Hall Research Natural Area, Inyo National Forest, California: Unpub. report for the US Forest Service, Pacific Southwest Forest and Range Experiment Station.

30. United States Department of Agriculture (n.d.) Staff SS United States General Soil Map (STATSGO2). Natural Resource Conservation Service, United States Department of Agriculture.

31. United States Department of Agriculture (n.d.) Staff SS Soil Survey Geographic (SSURGO) Database. Natura Resource Convservation Service, United States Department of Agriculture.

32. Allen-Diaz BH (1991) Water table and plant species relationships in Sierra Nevada meadows. American Midland Naturalist: 30–43.

33. Berlow EL, Knapp RA, Ostoja SM, Williams RJ, McKenny H, et al. (2013) A network extension of species occupancy models in a patchy environment applied to the Yosemite toad (Anaxyrus canorus). PLoS One 8: e72200.

34. Pagano T, Garen D (2005) A recent increase in western US streamflow variability and persistence. Journal of Hydrometeorology 6: 173–179.

35. Schimel JP, Bilbrough C, Welker JM (2004) Increased snow depth affects microbial activity and nitrogen mineralization in two Arctic tundra communities. Soil Biology and Biochemistry 36: 217–227.

36. Lipson D, Schadt C, Schmidt S (2002) Changes in soil microbial community structure and function in an alpine dry meadow following spring snow melt. Microbial ecology 43: 307–314.

37. Nemergut DR, Costello EK, Meyer AF, Pescador MY, Weintraub MN, et al. (2005) Structure and function of alpine and arctic soil microbial communities. Research in Microbiology 156: 775–784.

38. Walker MD, Ingersoll RC, Webber PJ (1995) Effects of interannual climate variation on phenology and growth of two alpine forbs. Ecology: 1067–1083.

39. Bales RC, Molotch NP, Painter TH, Dettinger MD, Rice R, et al. (2006) Mountain hydrology of the western United States. Water Resources Research 42: 8432.

40. Jordan RP (1978) The snowmelt hydrology of a small alpine watershed. [Ph.D.]. University of British Columbia: University of British Columbia.

41. Bokhorst S, Bjerke J, Street L, Callaghan T, Phoenix G (2011) Impacts of multiple extreme winter warming events on sub-Arctic heathland: phenology, reproduction, growth, and CO2 flux responses. Global Change Biology 17: 2817–2830.

42. Inouye D (2001) The ecological and evolutionary significance of frost in the context of climate change. Ecology Letters 3: 457–463.

43. Inouye DW, Morales MA, Dodge GJ (2002) Variation in timing and abundance of flowering by Delphinium barbeyi Huth (Ranunculaceae): the roles of snowpack, frost, and La Niña, in the context of climate change. Oecologia 130: 543–550.

44. Tranquillini W (1964) The physiology of plants at high altitudes. Annual Review of Plant Physiology 15: 345–362.

45. Neuner G, Hacker J (2012) Ice formation and propagation in alpine plants. Plants in Alpine Regions: 163–174.

46. Bokhorst SF, Bjerke JW, Tømmervik H, Callaghan TV, Phoenix GK (2009) Winter warming events damage sub-Arctic vegetation: consistent evidence from an experimental manipulation and a natural event. Journal of Ecology 97: 1408–1415.

47. Clymo RS (1983) Peat. In: Gore, A.J.P. (Ed.), Ecosystems of the World, 4A. Mires: swamp, bog, fen and moor. Amsterdam: Generall Studies, Elsevier.

48. Laiho R (2006) Decomposition in peatlands: reconciling seemingly contrasting results on the impacts of lowered water levels. Soil Biology and Biochemistry 38: 2011–2024.

49. Davidson EA, Janssens IA (2006) Temperature sensitivity of soil carbon decomposition and feedbacks to climate change. Nature 440: 165–173.

50. Berhe AA, Suttle KB, Burton SD, Banfield JF (2012) Contingency in the Direction and Mechanics of Soil Organic Matter Responses to Increased Rainfall. Plant and Soil 358: 371–383.

51. Berhe AA, Kleber M (2013) Erosion, deposition, and the persistence of soil organic matter: mechanistic considerations and problems with terminology. Earth Surface Processes and Landforms 38: 908–912.

14

Modeling Spatial Patterns of Soil Respiration in Maize Fields from Vegetation and Soil Property Factors with the Use of Remote Sensing and Geographical Information System

Ni Huang[1], Li Wang[1]*, Yiqiang Guo[2], Pengyu Hao[1], Zheng Niu[1]

1 The State Key Laboratory of Remote Sensing Science, Institute of Remote Sensing and Digital Earth, Chinese Academy of Sciences, Beijing, China, **2** Land Consolidation and Rehabilitation Center, Ministry of Land and Resources, Beijing, China

Abstract

To examine the method for estimating the spatial patterns of soil respiration (R_s) in agricultural ecosystems using remote sensing and geographical information system (GIS), R_s rates were measured at 53 sites during the peak growing season of maize in three counties in North China. Through Pearson's correlation analysis, leaf area index (LAI), canopy chlorophyll content, aboveground biomass, soil organic carbon (SOC) content, and soil total nitrogen content were selected as the factors that affected spatial variability in R_s during the peak growing season of maize. The use of a structural equation modeling approach revealed that only LAI and SOC content directly affected R_s. Meanwhile, other factors indirectly affected R_s through LAI and SOC content. When three greenness vegetation indices were extracted from an optical image of an environmental and disaster mitigation satellite in China, enhanced vegetation index (EVI) showed the best correlation with LAI and was thus used as a proxy for LAI to estimate R_s at the regional scale. The spatial distribution of SOC content was obtained by extrapolating the SOC content at the plot scale based on the kriging interpolation method in GIS. When data were pooled for 38 plots, a first-order exponential analysis indicated that approximately 73% of the spatial variability in R_s during the peak growing season of maize can be explained by EVI and SOC content. Further test analysis based on independent data from 15 plots showed that the simple exponential model had acceptable accuracy in estimating the spatial patterns of R_s in maize fields on the basis of remotely sensed EVI and GIS-interpolated SOC content, with R^2 of 0.69 and root-mean-square error of 0.51 μmol CO_2 m^{-2} s^{-1}. The conclusions from this study provide valuable information for estimates of R_s during the peak growing season of maize in three counties in North China.

Editor: Ben Bond-Lamberty, DOE Pacific Northwest National Laboratory, United States of America

Funding: This work was supported by the National Natural Science Foundation of China (41301498), the Public Service Sectors (Ministry of Land and Resources) Special Fund Research (201311127), the Special Foundation for Young Scientists of the State Laboratory of Remote Sensing Science (13RC-07), and the Major State Basic Research Development Program of China (2013CB733405). The funders had no role in study design, data collection and analysis, decision to publish, or preparation of the manuscript.

Competing Interests: The authors have declared that no competing interests exist.

* Email: wangli@radi.ac.cn

Introduction

Soil CO_2 efflux from terrestrial ecosystems to the atmosphere has been considered the second largest global carbon flux and is a vital component of ecosystem respiration [1]. In recent decades, significant progress has been made in identifying the biophysical factors that influence soil respiration (R_s) to predict soil CO_2 emission accurately in time and space [2–4].

The majority of R_s arises from root and microbial tissue. Therefore, understanding the spatial and temporal changes of these sources will facilitate the modeling of R_s. However, the large spatial and temporal heterogeneity of root and microbial activity within the landscape and the covariation of potentially important factors (i.e., temperature and water content) pose great challenges to the development of mechanistically based models that account for spatial and temporal variability in R_s [2]. Thus, many different

statistical models of R_s have been developed on the basis of data collected from different ecosystems [5]. Numerous studies have established R_s models based on soil temperature, soil moisture, or both [6,7]. Aside from soil temperature and moisture, plant productivity proxies [e.g., leaf area index (LAI), canopy chlorophyll content (Chl_{canopy}), and plant biomass] [8–10] and soil properties [e.g., soil organic carbon (SOC) content, soil total nitrogen (STN) content, and soil C and N ratio (soil C/N)] [11,12] also potentially influence R_s and are often included in models of R_s. However, most of the factors that affect variations in R_s tend to be derived through field measurements [13]. Furthermore, direct observation of these variables across long time spans or large spatial scales is expensive because of the required manpower and material resources. A simple method to derive data related to variations in R_s is necessary to facilitate the determination of the spatial and temporal distribution of R_s.

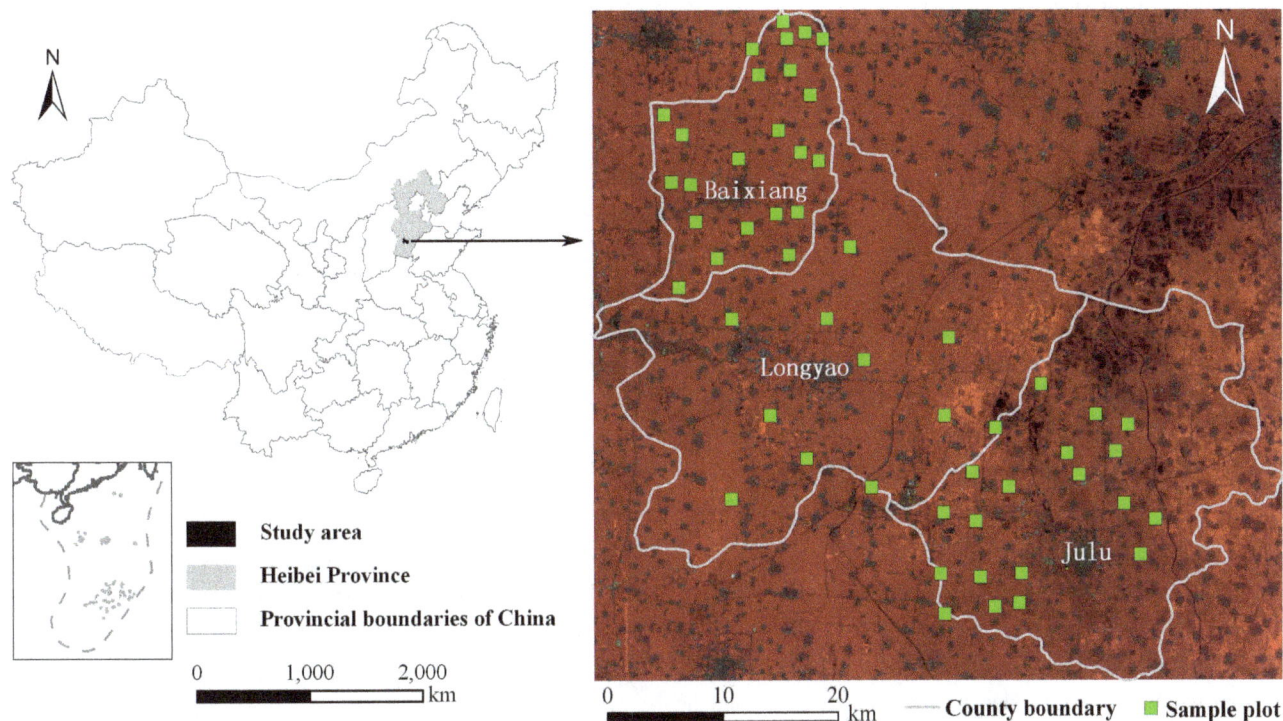

Figure 1. Spatial location of the sample plots for field experiments in three counties in North China. The box in the bottom left corner of Figure 1 shows the South China Sea islands.

Remote sensing and geographical information system (GIS) provide powerful tools for data acquisition, spatial analysis, and graphical display [14–16]. In the field of global change research, significant advances have been made in the development and application of remote sensing and GIS. These advances include land cover and land-use changes [17,18], environmental vulnerability and risk assessment [19,20], ecological restoration and management [21–23], and terrestrial ecosystem carbon cycle [24–26]. However, applying the data derived from remote sensing and GIS into R_s modeling remains controversial, especially for remote sensing data, because remotely sensed data in principle are independent measurements of site properties, not functionally important variables (e.g., soil temperature, soil moisture, and plant growth variables) that control R_s [3,27,28]. On the basis of statistical analysis of field experiments, previous studies found that remotely sensed vegetation indices (VIs) correlate with R_s in crop sites that lack drought stress [10] and can be used to model the spatial patterns of R_s during the peak growing season of alpine grasslands in the Tibetan Plateau [26]. However, few studies explore the potential of remote sensing and GIS data for estimating the spatial patterns of R_s in agricultural land, which may be affected by more complex factors than natural grasslands because of the influence of human activity. Although modern agriculture has successfully increased food production, the processes involved have profoundly affected the global carbon cycle through tillage, drainage and conversion of natural to agricultural ecosystems [29,30]. Therefore, a simple method should be identified to study the spatial characteristics of R_s in agricultural ecosystems.

This study aims to examine a potential new approach for estimating the spatial patterns of R_s during the peak growing season of maize by using remote sensing and GIS technology in Baixiang, Longyao and Julu Counties, which are typical agricul-

tural areas in the north plain of China. Studying the spatial characteristics of soil CO_2 efflux in maize fields will contribute to eco-agricultural development.

Materials and Methods

Ethics Statement

No specific permissions were required for the 53 sample plots in this study. We confirmed that the field studies did not involve endangered or protected species, and the specific location of the sample plots was provided in the manuscript (Fig. 1).

Study Site

The study site is situated within three counties (Baixiang, Longyao and Julu) in Southern Hebei Province of North China (Fig. 1). The total area of the study site is 1.64×10^3 km^2. This area is located in the North China Plain with a flat open terrain, single landform type, and a mean elevation of 30 m above sea level. Calcareous alluvial soil with high capacity to retain water and fertilizer is the main soil type in the study area. The study site is suitable for farming, and maize is the main crop. The climate is continental monsoon with four distinct seasons and adequate light and heat resources. Long records of meteorological data near the study site (http://cdc.cma.gov.cn) indicate that the mean annual temperature is 13.5°C with the coldest temperatures in January and the hottest in July. The mean annual precipitation is 502.8 mm, but precipitation is distributed unevenly in the four seasons with the greatest precipitation occurring in summer (362.5 mm). Therefore, drought influences agricultural development, and agriculture mainly involves irrigation in this study site.

Fifty-three sample plots located in the maize fields were identified within the study site (Fig. 1). The distance between any two sample plots was larger than 2 km. Each sample plot (greater than 100 m×100 m) has a large maize area, flat terrain, and maize under uniform growing conditions. All measurements were performed from August 11, 2013 to August 20, 2013, which corresponded to the tassel stage and peak growing period of maize. During the 10 days of field measurements, continuous measurements were performed, except on August 12 because of a minor precipitation event. Therefore, all field measurements required 9 days.

Field measurements

Soil respiration measurements. In each sample plot, R_s was measured by using a soil respiration chamber (LI-6400-09; LiCor, Lincoln, Nebraska, USA) connected to a portable photosynthesis system (LI-6400; LiCor, Lincoln, Nebraska, USA). The soil respiration chamber was mounted on a PVC soil collar that was sharpened at the bottom. Each PVC collar (5 cm long, 11 cm inside diameter) was inserted 2 cm to 3 cm into the ground and was installed at least 24 h prior to performing any measurements. To reduce the difference in root biomass, soil collars were placed in three locations on the basis of their distance to the maize plant: near a maize plant, inter-plant, and inter-row. Two collars were placed in each of the three positions for each R_s measurement. At least three to four consecutive measurements on each collar were performed to prevent any systematic error in the R_s estimates. An average R_s value was used for each collar, and the average value from six collars was used to represent the R_s value at plot level. Each R_s measurement was conducted between 09:00 h and 15:00 h (local time) because fluxes measured during this time interval are usually representative of the daily mean flux.

Soil temperature and soil moisture measurements. After the soil respiration measurement on a PVC soil collar in each plot, soil temperature and soil moisture were measured in this collar to minimize sample difference. Soil temperature was measured at a 10 cm depth (T_{s10}) by using a ground thermometer. Volumetric soil moisture at a depth of 0 cm to 20 cm (SM_{20}) was determined by using a portable time domain reflectometry probe (HydroSense, Campbell, USA). Thus, six soil temperature and moisture measurements were performed in each plot. The average value was used to represent soil temperature or soil moisture at the plot level.

Maize biophysical parameter measurements. LAI was measured by using an LAI-2000 (LI-COR Inc., Lincoln, Nebraska). In each plot, six representative positions were selected for LAI measurement, and in every position, two repeated measurements were performed. Leaf chlorophyll content (Chl_{leaf}) was determined by using a portable chlorophyll meter (SPAD-502, New Jersey, USA). Fully expanded leaves, which depended on the height of the maize plant, were randomly selected from three locations that corresponded to the upper, middle, and lower parts

of the maize plant. For each leaf location, 10 SPAD values were randomly collected. The vertical leaf area distribution in maize canopy was analyzed by measuring the area of each green leaf from the bottom to the top of eight randomly distributed maize plants with the use of an area meter (LI-3100, LI-COR, Lincoln, Nebraska). The area-weighted mean SPAD reading was used to derive Chl_{leaf}. However, the SPAD reading was in arbitrary units rather than in actual amounts of chlorophyll per unit area of the leaf tissue. A transform relationship exists between the SPAD readings and the actual chlorophyll content in maize [31]. To convert the SPAD readings to chlorophyll content per unit leaf area ($\mu g\ cm^{-2}$), this study used the transform relationship ($Chl_{leaf} = 0.95 \times SPAD\ reading - 3.25$) derived by Wu et al. [32] in maize plots, and the same SPAD meter was employed in this study. Chl_{canopy} was then determined by using the following equation:

$$Chl_{canopy} = Chl_{leaf} \times GLAI \qquad (1)$$

where Chl_{canopy} is the canopy chlorophyll content ($g\ m^{-2}$), Chl_{leaf} is the leaf chlorophyll content of maize ($g\ m^{-2}$), and GLAI represents the green leaf area per unit ground area.

In each sample plot, three representative maize plants were harvested for aboveground biomass (AGB) measurement. These fresh maize plants were sealed in a plastic bag and immediately transported to a nearby laboratory for subsequent analysis. Thereafter, fresh samples were oven dried at 65°C until the mass of the sample reached a constant weight. The AGB in each plot can be derived by multiplying the average dry weight per plant (g plant^{-1}) and the average plant density of maize (plants m^{-2}).

Soil property measurements. Soil within the six PVC collars in each plot was destructively sampled after measuring R_s, soil temperature and soil moisture. Soil was sampled to a depth of approximately 20 cm by a cylindrical soil driller (4 cm diameter, 20 cm height), in which fine root biomass and microbial activity are the highest [33,34]. These collected soil samples were sealed in plastic bags and stored at room temperature while being transported to the laboratory. Six collected soil samples in each plot were uniformly mixed to form a composite sample for laboratory analysis. The composite sample was air-dried in the laboratory to a constant weight for soil chemical analyses. The air-dried soil samples were ground to pass through a 0.2 mm sieve after any visible plant tissues and debris were manually removed. The SOC content was estimated by using the standard Mebius method [35]. The STN content was analyzed by using the Kjeldahl digestion procedure [36]. In this study, soil C/N was calculated by the ratio of SOC and STN contents.

Table 1. Calculation for vegetation indices [a].

Vegetation index	Formula	Reference
Normalized difference vegetation index	$NDVI = \dfrac{R_{Nir} - R_{Red}}{R_{Nir} + R_{Red}}$	Rouse et al. [47], Gamon et al. [48]
Modified soil adjusted vegetation index	$MSAVI = \dfrac{2R_{Nir} + 1 - \sqrt{(2R_{Nir} + 1)^2 - 8(R_{Nir} - R_{Red})}}{2}$	Qi et al. [49]
Enhanced vegetation index	$EVI = 2.5 \times \dfrac{R_{Nir} - R_{Red}}{1 + R_{Nir} + 6 \times R_{Red} - 7.5 \times R_{Blue}}$	Huete et al. [50]

[a]R_{Blue}, R_{Red}, and R_{Nir} are reflectance of blue, red, and NIR band in the HJ-1A CCD optical image, respectively.

Table 2. Pearson's correlation among soil respiration and factors affecting soil respiration in maize fields during the peak growing season in three counties in North China.

	R_s	T_{s10}	SWC_{20}	Chl_{canopy}	LAI	AGB	SOC content	STN content	Soil C/N
R_s	1.00	-0.27	-0.18	0.54***	0.75***	0.59***	0.76***	0.59***	-0.23
T_{s10}		1.00	0.18	-0.15	-0.28	-0.27	-0.49**	-0.66***	0.51**
SWC_{20}			1.00	-0.17	-0.05	-0.07	0.16	-0.00	0.06
Chl_{canopy}				1.00	0.83***	0.81***	0.26	0.20	-0.18
LAI					1.00	0.76***	0.44**	0.38*	-0.28
AGB						1.00	0.45**	0.34*	-0.15
SOC content							1.00	0.78***	-0.29
STN content								1.00	-0.79***
Soil C/N									1.00

R_s is the daily mean soil respiration rate ($\mu mol \ CO_2 \ m^{-2} \ s^{-1}$), T_{s10} is the soil temperature at 10 cm depth (°C), SWC_{20} is the soil water content at 0 cm to 20 cm depth ($m^3 \ m^{-3}$), Chl_{canopy} is the canopy chlorophyll content ($g \ m^{-2}$), LAI is the leaf area index, AGB is the aboveground biomass ($kg \ m^{-2}$), SOC content is the soil organic carbon content ($kg \ kg^{-1}$), STN content is the soil total nitrogen content ($g \ kg^{-1}$), and soil C/N is the soil C: N ratio. Significance levels:

*$p<0.05$,
**$p<0.01$,
***$p<0.001$.

Spatial data acquisition

Maize classification data. This study aimed to derive the spatial distribution of R_s in maize fields based on the field measurements at the plot scale. Maize classification data is necessary to spatially extrapolate R_s at the plot scale to the whole study area. Multi-temporal normalized difference vegetation index (NDVI) data collected over the growing season were used to classify maize at the study site [37–39]. Clouds are common occurrences in the study area during the growing season. Thus, obtaining a time sequence of cloud-free scenes is difficult. Two types of satellite data were used to establish the time-series NDVI data. One was the Operational Land Imager (OLI) image of Landsat 8, and the other was the small constellation for environmental and disaster mitigation (HJ-1A and B) charge coupled device (CCD) image [40–42]. Five scenes of OLI images acquired on May 3, 2013, May 19, 2013, July 6, 2013, October 10, 2013, and October 26, 2013 were downloaded from the U.S. Geological Survey (http://earthexplorer.usgs.gov/). Three HJ-1A and B CCD optical images acquired on June 6, 2013, August 17, 2013, and September 15, 2013 were downloaded from the China Center for Resource Satellite Data and Applications (http://www. cresda.com). The two types of remote sensing images exhibit same spatial resolution (30 m). The 30 m spatial resolution is appropriate for classifying maize patterns in the study area given the relatively large field in the region, which could spatially corresponded to five or more 30 m pixels. The strong relationship of the NDVI with biophysical vegetation characteristics, such as LAI and green biomass [43,44], enables the discrimination of land cover types on the basis of their unique phenological responses. Before land-use classification, pre-processing (i.e., radiometric calibration, atmospheric correction and geometric correction) of OLI images and HJ-1A and B CCD optical images was accomplished by using the Environment for Visualizing Images (ENVI) software (Version 4.7, Research Systems Inc., Boulder, Colorado, USA) [45,46]. This process ensured the consistency between the two types of remote sensing data and the seasonality of the NDVI time series. The maximum likelihood classification method, integrated in the ENVI software, was applied to the eight-date NDVI time series that spanned one maize growing season of the study site.

Spectral vegetation index for vegetation biophysical parameter estimation. Three greenness indices, namely, NDVI, enhanced vegetation index (EVI), and modified soil adjusted vegetation index (MSAVI), were derived from the HJ-1A CCD optical image acquired on August 17, 2013 (Table 1) for vegetation biophysical parameter estimation. Previous studies reported that greenness VIs offer important and convenient measures for vegetation biophysical parameters, such as LAI and Chl_{canopy} [51–54]. Meanwhile, LAI and Chl_{canopy} are also found to be good indicators of plant canopy photosynthesis [55–57] and are used in the modeling of R_s [58]. To obtain the spatial patterns of vegetation biophysical parameters in maize fields, the spatial distribution of vegetation biophysical parameters over the whole study area was overlapped with the maize classification data.

Quantifying the spatial pattern of SOC content. Statistics and geostatistics have been widely applied to quantify the spatial distribution patterns of SOC at a regional scale [59–61]. Based on the theory of regionalized variables, geostatistics provides advanced tools to quantify the spatial features of soil parameters and to conduct spatial interpolation [62,63]. In this study, geostatistical analyses were performed by using the geostatistical analyst module of ArcGIS software (Version 9.3, 2008) to quantify the spatial pattern of SOC content. To obtain the spatial pattern of the SOC content in the maize fields, the spatial distribution of

Table 3. Spatial characteristics of soil respiration (R_s, µmol CO_2 m^{-2} s^{-1}), soil temperature at 10 cm depth (T_{s10}, °C), soil water content at 0 cm to 20 cm depth (SWC_{20}, m^3 m^{-3}), canopy chlorophyll content (Chl_{canopy}, g m^{-2}), leaf area index (LAI), aboveground biomass (AGB, kg m^{-2}), soil organic carbon content (SOC content, g kg^{-1}), soil total nitrogen content (STN content, g kg^{-1}) and soil C: N ratio (soil C/N) in maize fields during the peak growing season in three counties in North China.

Variables	Mean	Maximum	Minimum	CV (%)
R_s	5.43	7.33	2.64	15.45
T_{s10}	28.32	30.93	25.78	4.73
SWC_{20}	27.54	33.27	19.54	12.48
Chl_{canopy}	0.18	0.21	0.16	6.54
LAI	3.75	4.53	2.81	8.64
AGB	0.94	1.89	0.44	31.93
SOC content	11.86	17.26	6.40	16.71
STN content	1.25	1.78	0.53	24.47
Soil C/N	9.82	14.38	7.07	18.53

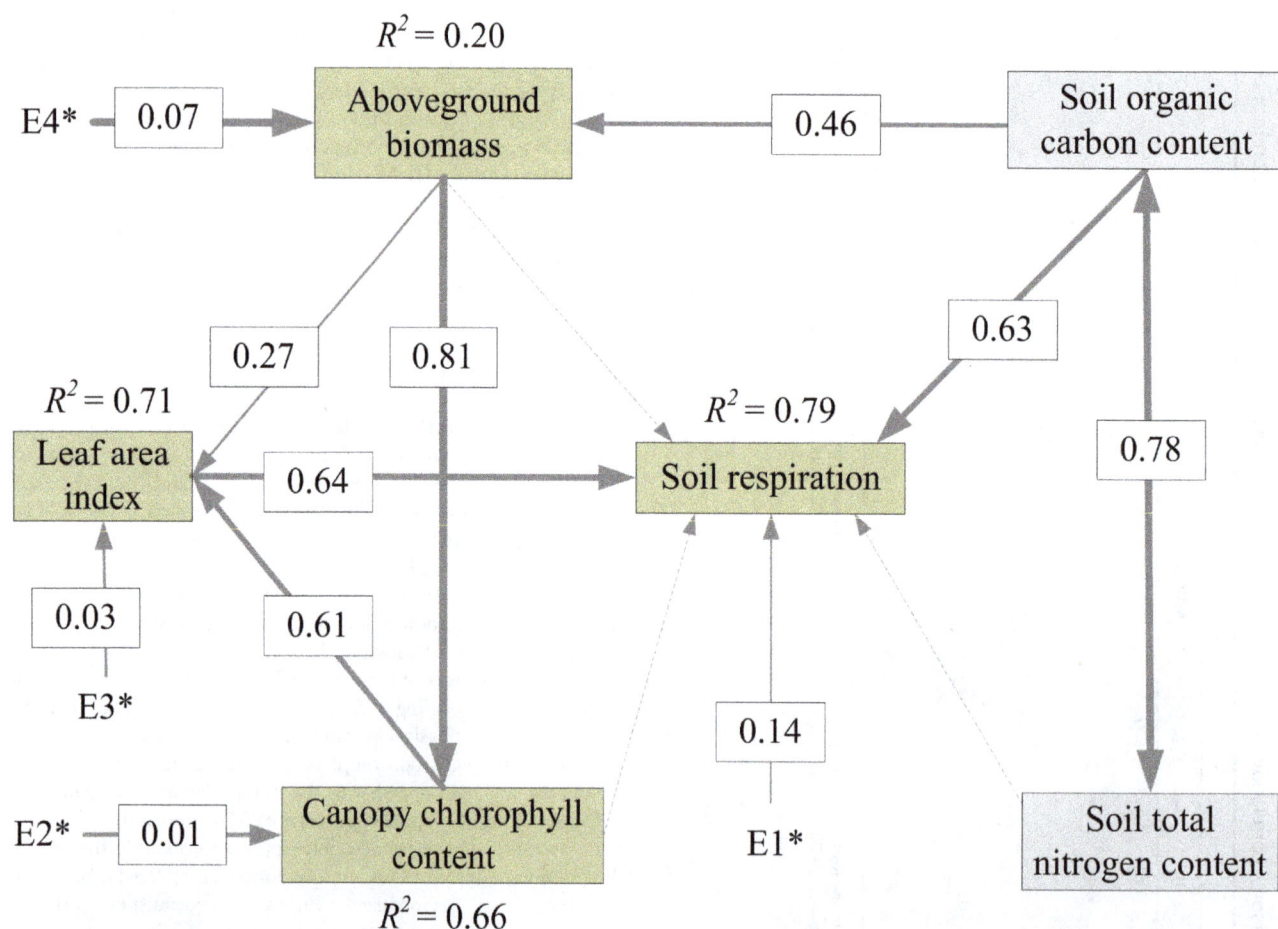

Figure 2. Final structural equation modeling (SEM) for soil respiration. Non-significant paths are shown in dashed line. The thickness of the solid arrows reflects the magnitude of the standardized SEM coefficients. Standardized coefficients are listed on each significant path. * represents error terms for observed variables, among them, E1, E2, E3, and E4 represent measurement errors for soil respiration, canopy chlorophyll content, leaf area index, and aboveground biomass, respectively.

Table 4. Total, direct, and indirect effects in the structural equation modeling.

Variable	Direct effect	Indirect effect	Total
Soil respiration			
Aboveground biomass	−0.10ns	0.46	0.36
Soil organic carbon content	0.63	0.16	0.79
Soil total nitrogen content	−0.09ns	0.30	0. 21
Leaf area index	0.64	-	0.64
Canopy chlorophyll content	−0.04ns	0.39	0.35
Aboveground biomass			
Soil organic carbon content	0.46	-	0.46
Soil total nitrogen content	−0.01ns	-	−0.01ns
Leaf area index			
Aboveground biomass	0.27	0.50	0.77
Soil organic carbon content	-	0.35	0.35
Soil total nitrogen content	-	−0.01ns	−0.01ns
Canopy chlorophyll content	0.61	-	0.61
Canopy chlorophyll content			
Aboveground biomass	0.81	-	0.81
Soil organic carbon content	-	0.37	0.37
Soil total nitrogen content	-	−0.01ns	−0.01ns

These effects were calculated using standardized path coefficients. Non-significant effects are indicated by "ns".

the SOC content over the whole study area was overlapped with the maize classification data.

Modeling spatial patterns of soil respiration

Identifying factors affecting spatial variability of soil respiration. The variables that explain the spatial variability of R_s are as follows: (1) soil properties, measured by SOC content, STN content and soil C/N; (2) environmental factors, encompassing T_{s10} and SM_{20}, and (3) plant photosynthesis proxy factors, including AGB, LAI and Chl_{canopy}. Pearson's correlation requires variables to be normally distributed and mutually independent. Each variable was tested for normal distribution by using the Shapiro–Wilk normality test and for randomness by the runs test of the Statistical Package for the Social Sciences (SPSS, Chicago, Illinois, USA). The results of the statistical analysis showed that each of these measured variables followed a normal distribution (Shapiro-Wilk, p>0.05) and showed randomness (runs test, p> 0.05). Thus, Pearson's correlation analysis, as implemented in the SPSS software, was used to screen important variables that influence R_s. Five variables with statistically significant correlation (p<0.05) with R_s, namely, SOC content, STN content, LAI, AGB, and Chl_{canopy}, were screened out (Table 2). However, these variables were cross-correlated [64–66] and included both direct and indirect effects. To solve this problem, structural equation modeling (SEM) was used to evaluate explicitly the causal relationships among these interacting variables [67–69] and to divide the total effects of variables on R_s into direct and indirect effects. On the basis of the theoretical knowledge on the major factors that influence spatial patterns of R_s at regional scales [8,13,26], we developed an SEM model to relate R_s to SOC content, STN content, LAI, AGB, and Chl_{canopy}. This SEM model was used to identify the direct effect factors for R_s estimation. The SEM model was fitted by using AMOS 18.0 for Windows [70]. After using the SEM, the fit indices, namely, comparative fit

index = 0.984 and goodness-of-fit index = 0.946. Thus, the theoretical model showed a good fit with the sample data.

Quantifying the spatial patterns of soil respiration in maize fields. In this study, the direct effect factors of R_s identified by SEM were used to estimate R_s. The spatial distribution data of these direct effect factors were first obtained on the basis of remote sensing or GIS to quantify the spatial patterns of R_s in maize fields. A simple exponential model that used the proxy data was then employed to estimate the spatial pattern of R_s during the peak growing season of maize. The accuracy of this method was examined by separating the observed data into two datasets through a random generator. One dataset consisted of 38 sample plots for analysis, whereas the other consisted of 15 for testing the accuracy of the R_s estimation.

Result

Spatial characteristics of soil respiration

Based on field-measured data at 38 plots, the daily mean R_s of maize during the peak growing season was 5.43 μmol CO_2 m^{-2} s^{-1} with a range of 2.64 μmol CO_2 m^{-2} s^{-1} to 7.33 μmol CO_2 m^{-2} s^{-1} and a coefficient of variation (CV) of 15.45% (Table 3). The spatial variability of soil temperature at 10 cm depth (T_{s10}) was relatively small at the study site with a CV of 4.73% and was far less than the spatial variation in soil water content at 0 cm to 20 cm depth (SWC_{20}). The AGB of maize showed greater spatial variability (CV = 31.93%) than LAI (CV = 8.64%) and Chl_{canopy} (CV = 6.54%).

Mean SOC content, STN content, and soil C/N at 0 cm to 20 cm depth in maize fields of the study site were 11.86 g kg^{-1} (ranged from 6.40 g kg^{-1} to 17.26 g kg^{-1}), 1.25 g kg^{-1} (ranged from 0.53 g kg^{-1} to 1.78 g kg^{-1}), and 9.82 (ranged from 7.07 to 14.38), respectively. Their CVs were not similar with the STN content which showed greater spatial variability than the SOC content and soil C/N.

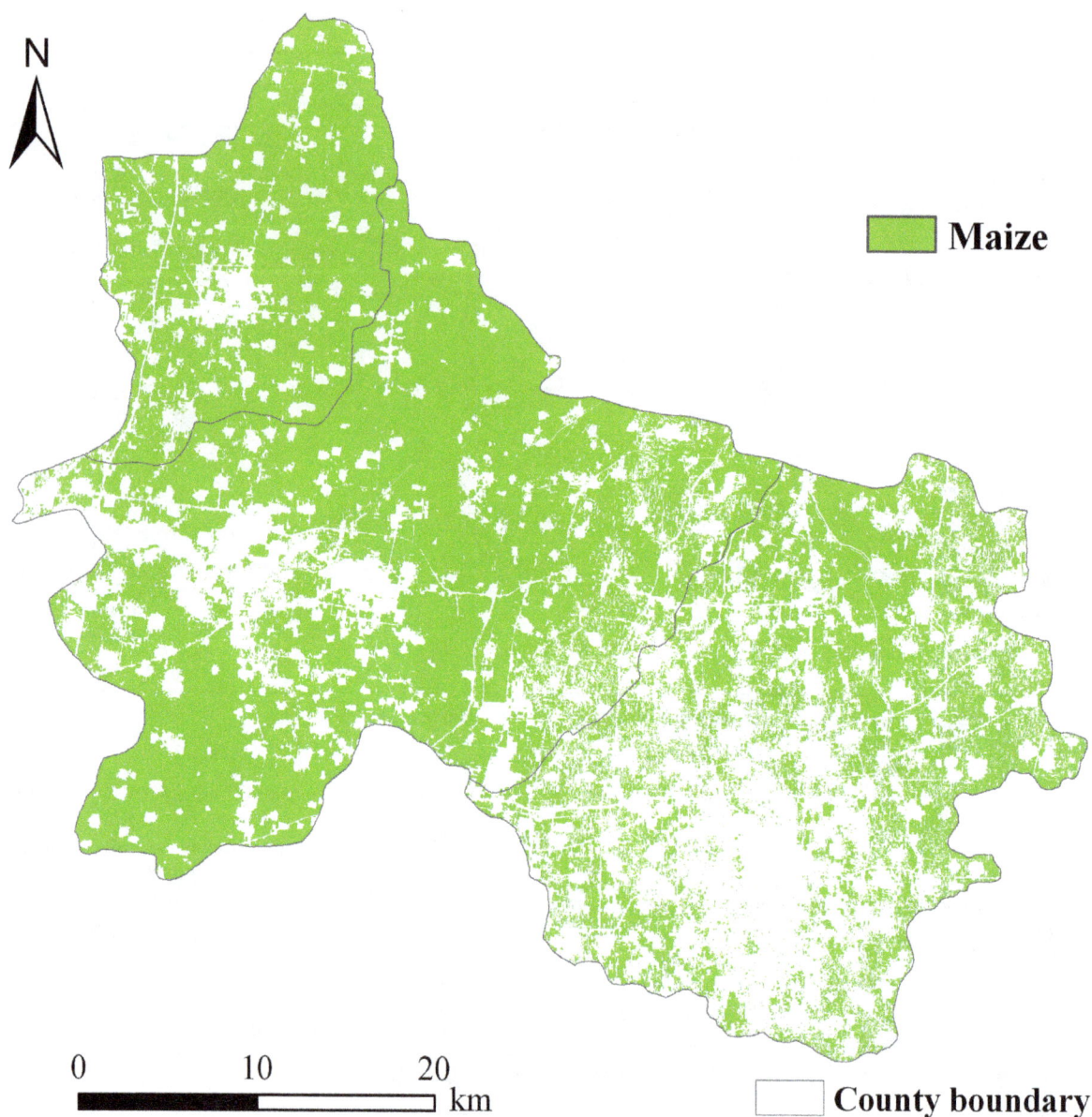

Figure 3. Maize classification map in three counties in North China.

Factors driving spatial variability of soil respiration

Based on Pearson's correlation analysis, five variables with significant correlation with R_s, namely, Chl_{canopy}, LAI, AGB, SOC content, and STN content, were selected (Table 2). However, the five selected variables were intercorrelated (Table 2), and their relationships with R_s combined both direct and indirect correlations. Thus, an SEM model was further used to evaluate the causal relationships among these interacting variables. The final SEM explained 79% of the variation in R_s (Fig. 2). The direct, indirect, and total effects of the variables are shown in Table 4. Among the five selected variables, LAI and SOC content directly affected R_s and can be used to predict R_s with relatively high accuracy ($R^2 = 0.79$). The other three variables (i.e., Chl_{canopy}, ABG, and STN content), despite having a significant correlation with R_s, only affected R_s indirectly through their direct relationship with SOC content and LAI. Thus, the two direct effect factors were used to estimate R_s, and the spatially distributed data proxies of these two factors were used to quantify the spatial patterns of R_s in maize fields during the peak growing season.

Spatial data used for soil respiration estimation

Maize classification. The maize classification map of the study area is shown in Figure 3. The classification accuracy for maize at the study site could not be quantitatively assessed because of the limitation of the sample data. However, 53 sample plots were all located in the maize classification map, and the county-level maize patterns classified in the map were consistent with the general maize patterns across the three counties. In addition, the classified maize area was close to the maize area reported by the China County Statistical Yearbook [71]. Thus, the classification accuracy of maize was believed to be reasonable, and the maize classification map was then used to predict the spatial pattern of R_s during the peak growing season of maize.

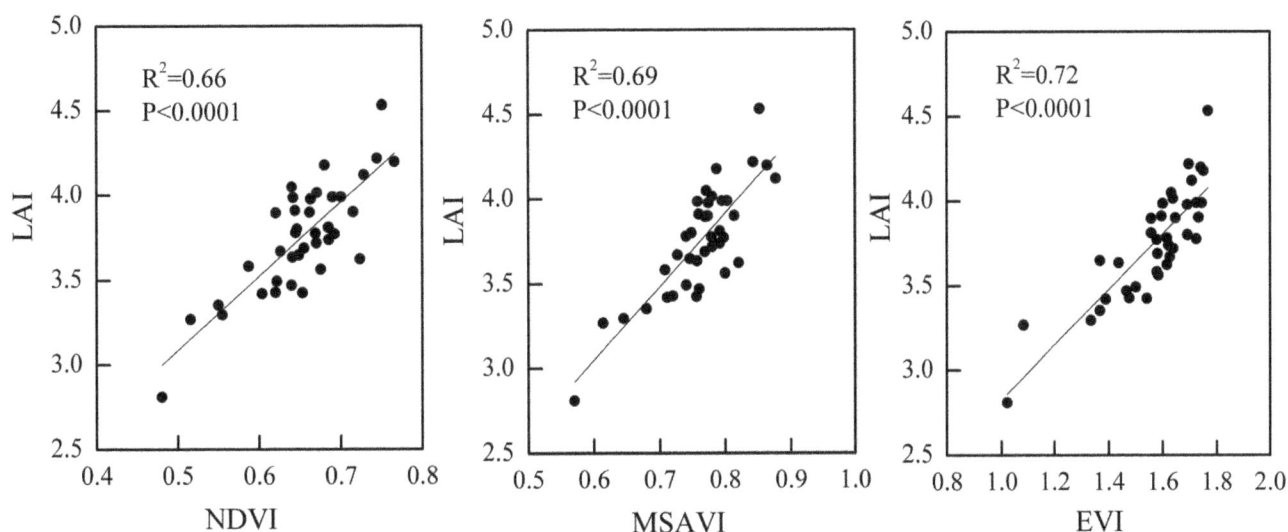

Figure 4. Linear relationships between three vegetation indices (VIs) and leaf area index (LAI) during the peak growing season of maize in three counties in North China (n = 38). The VIs are: normalized difference vegetation index (NDVI), enhanced vegetation index (EVI), and modified soil adjusted vegetation index (MSAVI).

LAI estimation from spectral vegetation index. Among the three greenness indices calculated from the optical image of HJ-1A satellite, EVI showed the best linear relationship with LAI, with a determination coefficient (R^2) of 0.72, followed by MSAVI and NDVI (Fig. 4). The explanation of LAI variance increased from 66% to 72% when EVI was used instead of NDVI for LAI estimation, and this increase was statistically significant (p<0.05). However, EVI and MSAVI did not exhibit a significant difference in explaining the variation in LAI, despite EVI having a slightly better relationship with LAI than MSAVI. Thus, EVI was used as a proxy for LAI to estimate R_s during the peak growing season of maize for simplicity. The spatially distributed EVI during the peak growing season of maize exhibited relatively small variability (Fig. 5). Overall, the EVI in the north and southwest parts of the study site (i.e., Baixiang and Longyao Counties) showed a high value. Relatively low EVI values mainly occurred in the southeast parts of the study site (i.e., Julu County), especially the northwest Julu County (Fig. 5).

Spatial distribution of SOC content. Kriging interpolation was performed by using ArcGIS 9.3 software to produce the spatial distribution map of the SOC content in maize fields of the study area. A cell size of 30 m×30 m was selected for the spatial interpolation to match the spatial resolution of images from OLI and HJ-1A/B. The final result of this spatial interpolation process is shown in Figure 6. Based on the spatial distribution map of the SOC content in maize fields, SOC content values were higher in the northwest and southwest parts of the study area than in the southeastern part.

Spatial distribution of soil respiration

The EVI and SOC content were used to estimate the spatial pattern of R_s during the peak growing season of maize on the basis of a simple exponential model. The geo-location information (latitude and longitude) of the 38 sample plots was used in the extraction of pixels. Pixels that contained these plots from the spatial distribution maps of EVI and SOC content data (Figs. 5 and 6) were extracted. These data were used to determine the model parameters by least-squares fitting. The resulting model was as follows:

$$R_s = 1.57 \times \exp(0.44 \times EVI + 0.05 \times SOC\ content) \quad (2)$$

$$(n = 38,\ R^2 = 0.73)$$

where R_s refers to the daily mean soil respiration rate in μmol CO_2 m^{-2} s^{-1}; EVI refers to enhanced vegetation index, as a proxy for LAI; and SOC content is the soil organic carbon content (g kg^{-1}) in maize fields of the study area. Eq. (2) was employed to predict the spatial pattern of R_s from spatially distributed EVI and SOC content data during the peak growing season of maize (Figs. 5 and 6). The spatial variation in R_s showed a pattern similar to that in SOC content (Figs. 6 and 7).

Figure 8 shows the accuracy assessment result of the R_s prediction model. The field measured R_s was comparable with the spatial data predicted R_s. Based on the independent test dataset, EVI and SOC content accounted for 69% of the spatial variation in ground-measured R_s, and the RMSE was 0.51 μmol CO_2 m^{-2} s^{-1}. The result of the accuracy assessment suggests that the prediction model, which used EVI and SOC content as the dependent variables, was effective in estimating R_s in maize fields during the peak growing season.

Discussion

Relationships between LAI and three VIs

In this study, in situ measured data were obtained during the peak growing period of maize (corresponding to the tassel stage of maize). The effect of soil background on the spectral reflectance of remote sensing images was negligible during this period because the maize cover was higher with LAI ranging from 2.81 to 4.53. The difference in the capability of spectral vegetation index (VI) responding to LAI variation mainly depended on the sensitivity of VI to the canopy structural variation of maize. Thus, the VI modified the effect of soil reflectance (i.e. MSAVI) did not exhibit a significantly greater advantage than NDVI, which is strongly affected by soil reflectance in sparsely vegetated areas [50]. EVI, which is more sensitive to variation in dense vegetation than

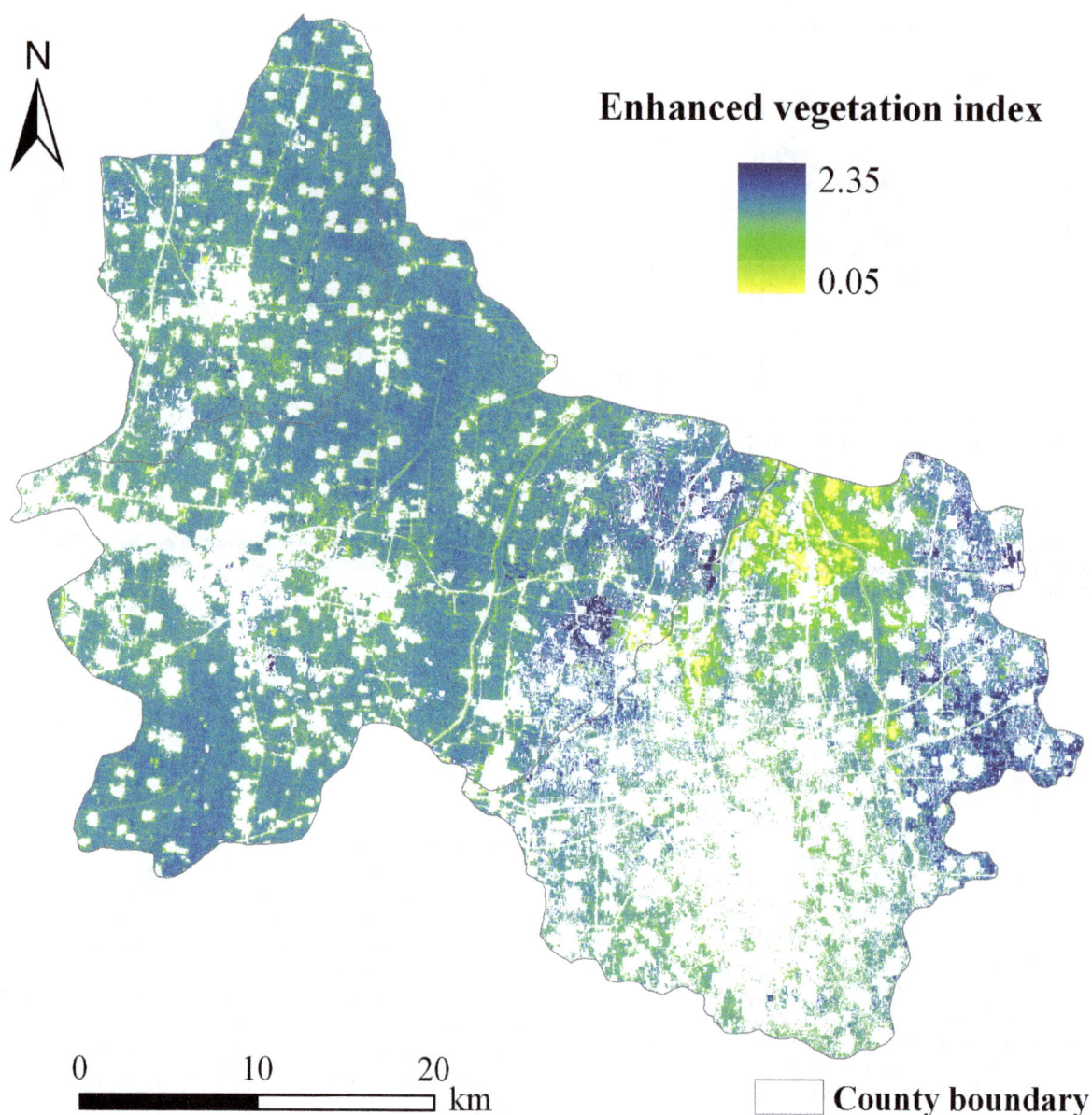

Figure 5. Spatial distribution map of enhanced vegetation index in maize fields in three counties in North China.

NDVI [50], showed the best relationship with the LAI of maize. This result was consistent with our previous study [58] that was conducted in irrigated and rainfed maize fields located at the University of Nebraska, Agricultural and Research Development Center, Mead, Eastern Nebraska, USA.

Measurement accuracy of SOC content

Field measurement data revealed that the SOC content at 0 cm to 20 cm depth in the maize fields ranged from 6.4 g kg^{-1} to 17.3 g kg^{-1}, and the mean value was 12.01 g kg^{-1}. For the mean dry land SOC content in North China, the value appeared to be higher than the previous estimate (0.83 from the average of 268 sample points) [72]. This difference was partly attributed to the fact that only the SOC content in maize fields, not in all dry land types, was considered. Most maize fields in the study site were on a winter wheat/maize rotation, and wheat straw was returned to the soil. The high productivity of maize crops contributed to the

development of a thick A horizon and high SOC content [73,74]. Additionally, only the SOC content in maize fields at 0 cm to 20 cm depth was analyzed, whereas previous studies estimated the SOC content on the basis of organic carbon content to a depth of 1 m [72,75,76]. In agricultural land, soil depth at 0 cm to 20 cm is located in the cultivation layer and has a higher SOC content than the SOC content at the deeper soil layers [34]. This condition contributed to the higher SOC content from the measured soil property data than the previous estimate.

Factors affecting spatial pattern of soil respiration

The spatial differences in R_s at the study site can be mainly attributed to the differences in vegetation productivity and soil property factors among the sample plots, whereas soil temperature and soil moisture served a minor function in regulating the spatial pattern of R_s. A previous study also demonstrated that site variables that reflect site productivity (e.g., LAI or aboveground

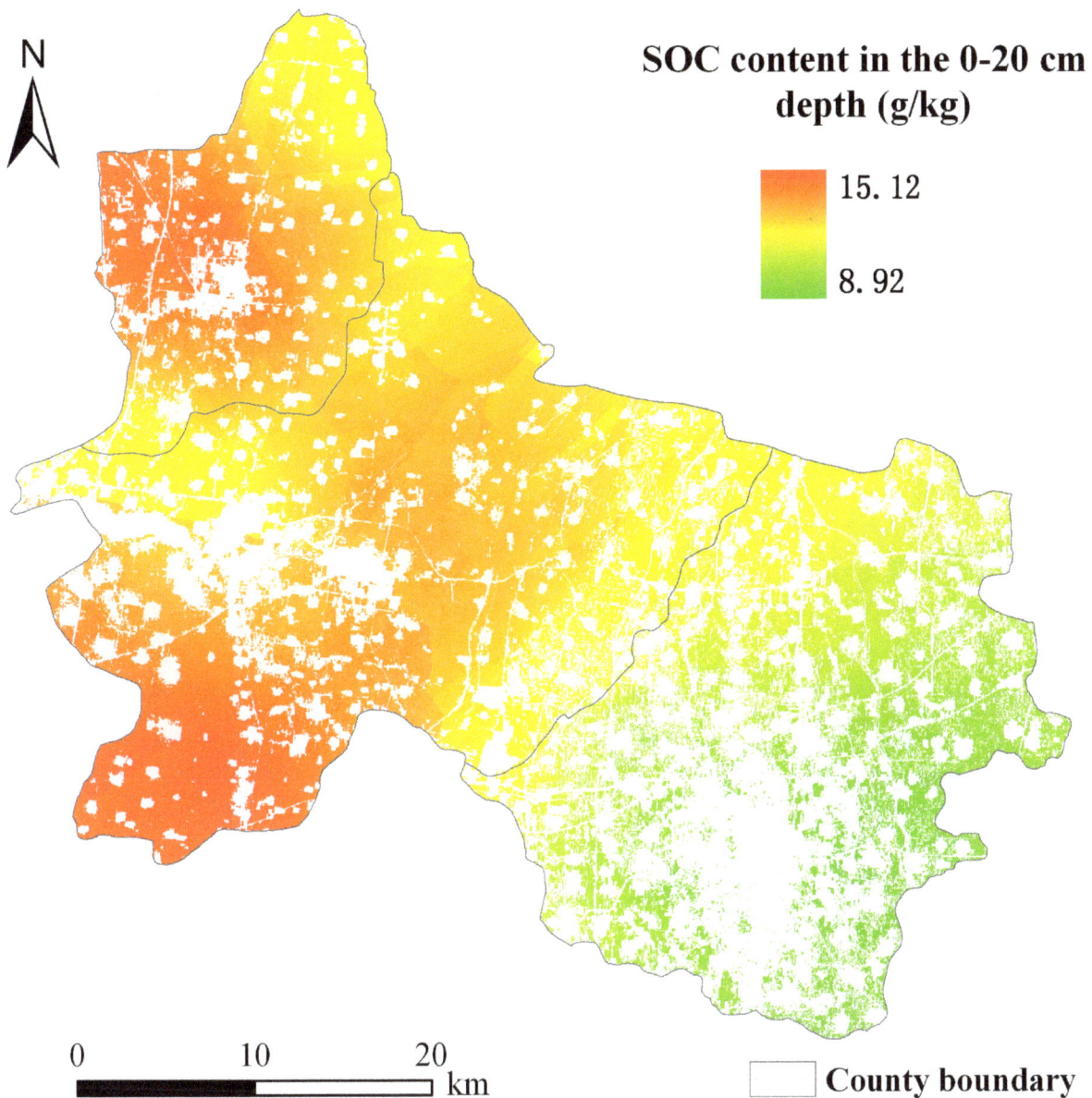

Figure 6. Spatial distribution map of soil organic carbon (SOC) content in the 0–20 cm depth in maize fields in three counties in North China.

net primary productivity) will provide a useful approach for large-scale estimates of regional R_s in terrestrial ecosystems [8]. Soil temperature evidently serves a predominant function in the spatial variations of R_s across sites of climatically contrasting environments [4]. However, at a local scale or under similar climatic conditions, other biological and biophysical factors, such as vegetation productivity and the size of organic carbon pools, may prevail as dominant drivers of R_s [4,77]. At a local scale, the spatial variation in T_{s10} in the study site was small (CV = 4.73%). Thus, soil temperature did not affect the spatial pattern of R_s. Although soil moisture in the maize fields showed a relatively large spatial variation (CV = 12.48%), this variation did not reach a degree that will affect the spatial dynamics of R_s. The soil C quantity and substrate quality factors (i.e., SOC and STN contents) were consistently and strongly correlated with one another and significantly affected the variation in R_s [5,12,13].

However, SEM results showed that the STN content only affected R_s indirectly through the direct effect on the SOC content at the study site.

During the peak growing season of maize, biophysical parameters, such as LAI, Chl_{canopy}, and AGB, were important variables that determined the size of the photosynthetic capacity [56,78]. However, these variables are not truly independent, and a correlation between one of them and R_s may lead to a correlation of the other with R_s. In this study, R_s was strongly correlated with LAI, Chl_{canopy} and AGB of maize fields, whereas LAI was the only variable directly related to R_s during the peak growing season of maize on the basis of SEM analysis.

The direct effect factors of R_s were used to estimate the spatial variability of R_s during the peak growing season of maize in three counties in North China. A simple exponential model, which included the corresponding spatial proxies from remote sensing

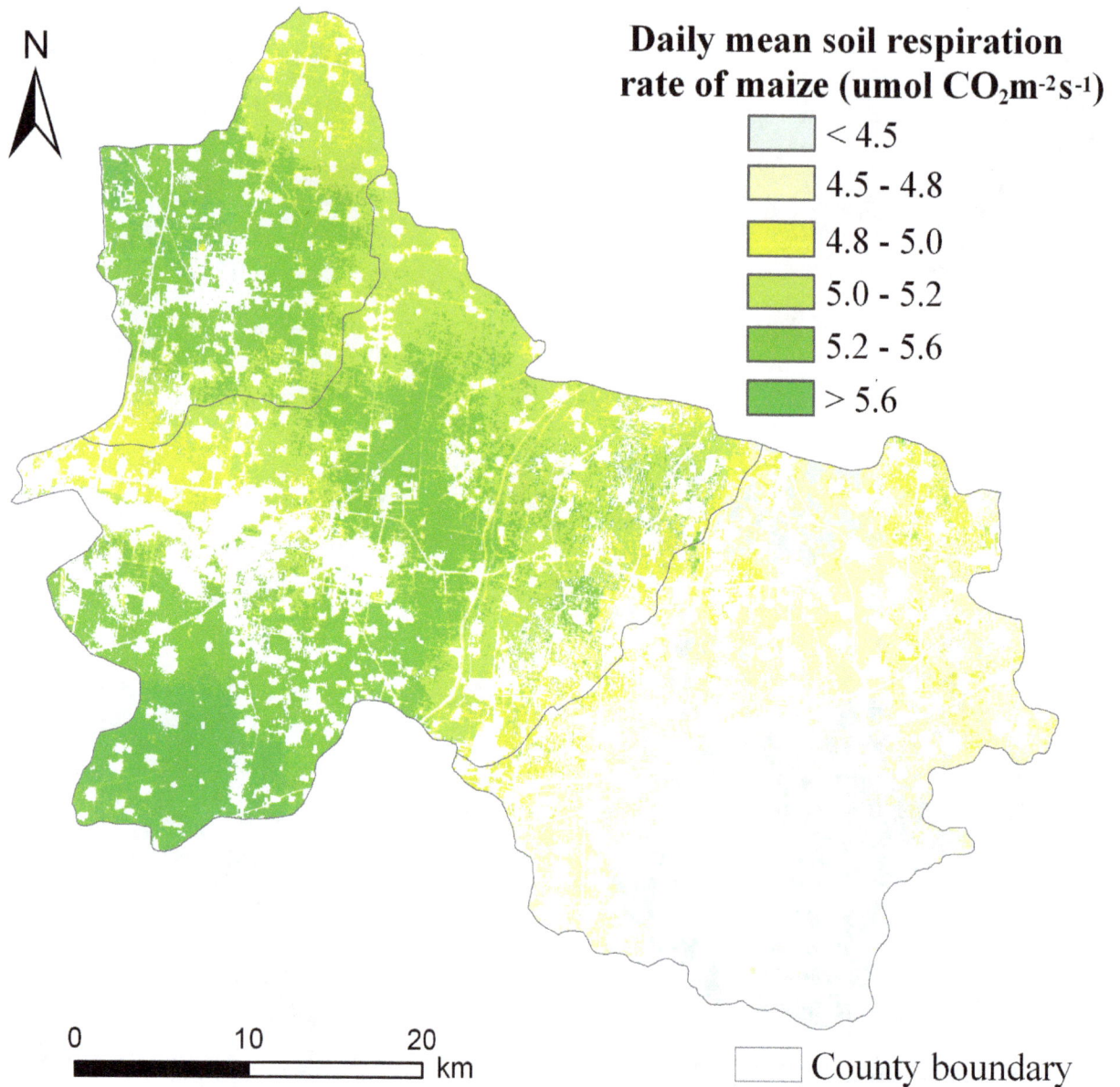

Figure 7. Spatial pattern of daily mean soil respiration rate during the peak growing season of maize in three counties in North China.

and GIS (i.e., EVI and spatially interpolated SOC content), was employed. A similar method was applied to a deciduous broadleaf forest site in the Midwest USA [79]. The independent test data also demonstrated the rationality of this method at the study site to a certain extent (Fig. 8). Regardless of the form of the R_s model, the relationship between LAI and EVI, as well as the kriging interpolation precision of the SOC content, affected the predictive accuracy of the R_s model. A moderate correlation between EVI and LAI (Fig. 4) affected the test accuracy of the exponential model with an R^2 value of 0.69 and an RMSE value of 0.51 μmol CO_2 m^{-2} s^{-1} (Fig. 8). The tendency of kriging to overestimate small values is supported by previous studies [80–82]. This tendency may help explain the bias toward overestimating R_s at low values (Fig. 8). Therefore, improving the accuracy of input parameters from remote sensing or GIS will increase the predictive capability of the R_s model.

Notably, the R_s model developed in this study was applicable to maize fields during the peak growth period in the three counties in North China. However, the model employed in this study does not consider temperature, a main driver of R_s that has high spatial variability. This model may be not used anywhere else or in other stages of the growing season. Furthermore, when spatially distributed data were used in the R_s model, a simple alternative method was employed to estimate the maize LAI by using the remotely sensed EVI, which may be problematic. Verstraeten et al. [83] highlighted that the assimilation of remotely sensed geophysical products into a carbon model is a complex process, and simply exchanging conventional input data for their remotely sensed counterparts is insufficient. Therefore, future research should focus on an integrating spatially distributed R_s datasets and geophysical products from remote sensing and GIS by using the data assimilation method, which has been extensively applied in

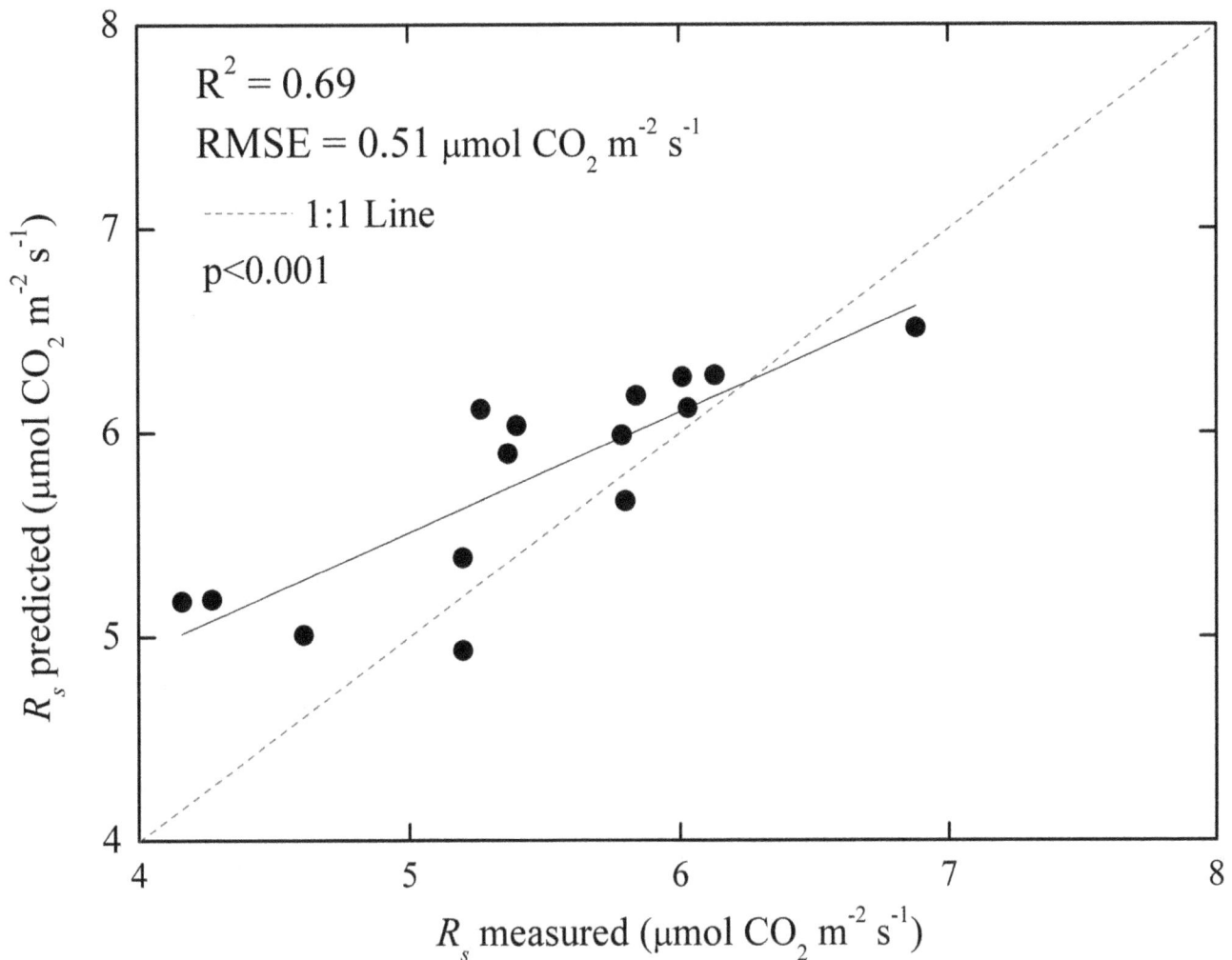

Figure 8. Spatial data predicted soil respiration (R_s) and corresponding ground-based measurements with R^2 and RMSE (µmol CO_2 m^{-2} s^{-1}) during the peak growing season of maize in three counties in North China ($n = 15$). The predicted soil respiration was attained with an exponential model that used EVI and SOC content as dependent variables.

terrestrial carbon cycle research [84–86]. However, this method lack the integration of R_s and spatially distributed data.

Conclusions

This study investigated the potential of spatial data from remote sensing and GIS for estimating the spatial patterns of R_s during the peak growing season of maize in three counties in North China. Based on in situ measurements, plant productivity (i.e., LAI) and soil property (i.e. SOC content) factors were identified as the most important determinants of spatial variability in R_s during the peak growing season of maize, and R_s was weakly related to soil temperature and soil moisture. Spectral VIs calculated from an HJ-1A CCD optical image were used to estimate LAI and EVI was found to be the best proxy for LAI. To derive the spatial pattern of R_s during the peak growing season of maize, a simple

exponential model, which included remotely sensed EVI and GIS spatially interpolated SOC content, was employed. This method was tested by using an independent sample dataset and was shown to be reasonable at the study site.

Acknowledgments

We sincerely thank the anonymous reviewers for their important and constructive revision advices on the manuscript.

Author Contributions

Conceived and designed the experiments: NH LW YQG. Performed the experiments: NH LW PYH. Analyzed the data: NH LW PYH. Contributed reagents/materials/analysis tools: NH LW YQG PYH ZN. Wrote the paper: NH.

References

1. Raich JW, Potter CS (1995) Global patterns of carbon dioxide emissions from soils. Global Biogeochemical Cycles 9: 23–36.

2. Davidson EA, Belk E, Boone RD (1998) Soil water content and temperature as independent or confound factors controlling soil respiration in a temperature mixed hardwood forest. Global Change Biology 4: 217–227.

3. Buchmann N (2000) Biotic and abiotic factors controlling soil respiration rates in Picea abies stands. Soil Biology and Biochemistry 32: 1625–1635.

4. Campbell JL, Sun OJ, Law BE (2004) Supply-side controls on soil respiration among Oregon forests. Global Change Biology 10: 1857–1869.

5. Webster KL, Creed IF, Skowronski MD, Kaheil YH (2009) Comparison of the performance of statistical models that predict soil respiration from forests. Soil Science Society of America Journal 73: 1157–1167.

6. Gaumont-Guay D, Black TA, Griffis TJ, Barr AG, Jassal RS, et al. (2006) Interpreting the dependence of soil respiration on soil temperature and water content in a boreal aspen stand. Agricultural and Forest Meteorology 140: 220–235.

7. Phillips SC, Varner RK, Frolking S, Munger JW, Bubier JL, et al. (2010). Interannual, seasonal, and diel variation in soil respiration relative to ecosystem respiration at a wetland to upland slope at Harvard Forest. Journal of Geophysical Research: Biogeosciences 115: G02019. doi: 10.1029/2008JG000858.

8. Reichstein M, Rey A, Freibauer A, Tenhunen J, Valentini R, et al. (2003) Modeling temporal and large-scale spatial variability of soil respiration from soil water availability, temperature and vegetation productivity indices. Global Biogeochemical Cycles 17: 1104. doi: 10.1029/2003GB002035.

9. Geng Y, Wang Y, Yang K, Wang S, Zeng H, et al. (2012) Soil respiration in Tibetan alpine grasslands: belowground biomass and soil moisture, but not soil temperature, best explain the large-scale patterns. PloS one 7: e34968. doi: 10.1371/journal.pone.0034968.

10. Huang N, Niu Z, Zhan YL, Tappertc MC, Wu CY, et al. (2012) Relationships between soil respiration and photosynthesis-related spectral vegetation indices in two cropland ecosystems. Agricultural and Forest Meteorology 160: 80–89.

11. Chen ST, Huang Y, Zou JW, Shen QR, Hu ZH, et al. (2010) Modeling interannual variability of global soil respiration from climate and soil properties. Agricultural and Forest Meteorology 150: 590–605.

12. Almagro M, Querejeta JI, Boix-Fayos C, Martínez-Mena M (2013) Links between vegetation patterns, soil C and N pools and respiration rate under three different land uses in a dry Mediterranean ecosystem. Journal of Soils and Sediments 13: 641–653.

13. Martin JG, Bolstad PV, Ryu SR, Chen J (2009) Modeling soil respiration based on carbon, nitrogen, and root mass across diverse Great Lake forests. Agricultural and Forest Meteorology 149: 1722–1729.

14. Longley P (Ed.) (2005) Geographic information systems and science. John Wiley & Sons.

15. Weng Q (2001) Modeling urban growth effects on surface runoff with the integration of remote sensing and GIS. Environmental Management 28: 737–748.

16. Lin ML, Chen CW (2011) Using GIS-based spatial geocomputation from remotely sensed data for drought risk-sensitive assessment. International Journal of Innovative Computing, Information and Control 7: 657–668.

17. Shalaby A, Tateishi R (2007) Remote sensing and GIS for mapping and monitoring land cover and land-use changes in the Northwestern coastal zone of Egypt. Applied Geography 27: 28–41.

18. Dewan AM, Yamaguchi Y (2009). Land use and land cover change in Greater Dhaka, Bangladesh: using remote sensing to promote sustainable urbanization. Applied Geography 29: 390–401.

19. Wang XD, Zhong XH, Liu SZ, Liu JG, Wang ZY, et al. (2008). Regional assessment of environmental vulnerability in the Tibetan Plateau: development and application of a new method. Journal of Arid environments 72: 1929–1939.

20. Ceccato P, Connor SJ, Jeanne I, Thomson MC (2005) Application of geographical information systems and remote sensing technologies for assessing and monitoring malaria risk. Parassitologia 47: 81–96.

21. Franklin J (1995) Predictive vegetation mapping: geographic modelling of biospatial patterns in relation to environmental gradients. Progress in Physical Geography 19: 474–499.

22. Raup B, Kääb A, Kargel JS, Bishop MP, Hamilton G, et al. (2007) Remote sensing and GIS technology in the Global Land Ice Measurements from Space (GLIMS) project. Computers & Geosciences 33: 104–125.

23. Keane RE, Burgan R, van Wagtendonk J (2001) Mapping wildland fuels for fire management across multiple scales: integrating remote sensing, GIS, and biophysical modeling. International Journal of Wildland Fire 10: 301–319.

24. He C, Wang S, Xu J, Zhou C (2002) Using remote sensing to estimate the change of carbon storage: a case study in the estuary of Yellow River delta. International Journal of Remote Sensing 23: 1565–1580.

25. Yuan W, Liu S, Yu G, Bonnefond JM, Chen J, et al. (2010) Global estimates of evapotranspiration and gross primary production based on MODIS and global meteorology data. Remote Sensing of Environment 114: 1416–1431.

26. Huang N, He JS, Niu Z (2013) Estimating the spatial pattern of soil respiration in Tibetan alpine grasslands using Landsat TM images and MODIS data. Ecological Indicators 26: 117–125.

27. Davidson EA, Richardson AD, Savage KE, Hollinger DY (2006) A distinct seasonal pattern of the ratio of soil respiration to total ecosystem respiration in a spruce-dominated forest. Global Change Biology 12, 230–239.

28. Vargas R, Baldocchi DD, Allen MF, Bahn M, Black TA, et al. (2010) Looking deeper into the soil: biophysical controls and seasonal lags of soil CO_2 production and efflux. Ecological Applications 20: 1569–1582.

29. Bondeau A, Smith PC, Zaehle S, Schaphoff S, Lucht W, et al. (2007) Modelling the role of agriculture for the 20th century global terrestrial carbon balance. Global Change Biology 13: 679–706.

30. Foley JA, DeFries RS, Asner GP, Barford C, Bonan G, et al. (2005) Global consequences of land use. Science 309: 570–574.

31. Krugh B, Bickham L, Miles D (1994) The solid-state chlorophyll meter: a novel instrument for rapidly and accurately determining the chlorophyll concentrations in seedling leaves. Maize Genetics Cooperation Newsletter 68: 25–27.

32. Wu CY, Wang L, Niu Z, Gao S, Wu MQ (2010) Nondestructive estimation of canopy chlorophyll content using Hyperion and Landsat/TM images. International Journal of Remote Sensing 31: 2159–2167.

33. Gao Y, Xie Y, Jiang H, Wu B, Niu J (2014) Soil water status and root distribution across the rooting zone in maize with plastic film mulching. Field Crops Research 156: 40–47.

34. Kou TJ, Zhu P, Huang S, Peng XX, Song ZW, et al. (2012) Effects of long-term cropping regimes on soil carbon sequestration and aggregate composition in rainfed farmland of Northeast China. Soil and Tillage Research 118: 132–138.

35. Nelson DW, Sommers LE (1982) Total carbon, organic carbon, and organic matter. In: Page AL, Miller RH, Keeney DR. (Eds.), Methods of Soil Analysis. American Society of Agronomy and Soil Science Society of American, Madison. 101–129.

36. Gallaher RN, Weldon CO, Boswell FC (1976) A semiautomated procedure for total nitrogen in plant and soil samples. Soil Science Society of America Journal 40: 887–889.

37. Wardlow BD, Egbert SL, Kastens JH (2007) Analysis of time-series MODIS 250 m vegetation index data for crop classification in the US Central Great Plains. Remote Sensing of Environment 108: 290–310.

38. Wilson EH, Sader SA (2002) Detection of forest harvest type using multiple dates of Landsat TM imagery. Remote Sensing of Environment 80: 385–396.

39. Zhong B, Ma P, Nie A, Yang A, Yao Y, et al. (2014) Land cover mapping using time series HJ-1/CCD data. Science China: Earth Sciences doi: 10.1007/s11430-014-4877-5.

40. Bian JH, Li AN, Jin HA, Lei GB, Huang CQ, et al. (2013) Auto-registration and orthorecification algorithm for the time series HJ-1A/B CCD images. Journal of Mountain Science 10: 754–767.

41. Liu Y, Li M, Mao L, Cheng L, Chen K (2013) Seasonal pattern of tidal-flat topography along the Jiangsu middle coast, China, using HJ-1 optical images. Wetlands 33: 871–886.

42. Wang SD, Miao LL, Peng GX (2012) An Improved Algorithm for Forest Fire Detection using HJ Data. Procedia Environmental Sciences 13: 140–150.

43. Gamon JA, Field CB, Goulden ML, Griffin KL, Hartley AE, et al. (1995) Relationship between NDVI, canopy structure and photosynthesis in three Californian vegetation types. Ecological Applications 5: 28–41.

44. Hansen PM, Schjoerring JK (2003) Reflectance measurement of canopy biomass and nitrogen status in wheat crops using normalized difference vegetation indices and partial least squares regression. Remote Sensing of Environment 86: 542–553.

45. Yu X, Yan Q, Liu Z (2010) Atmospheric correction of HJ-1A multi-spectral and hyper-spectral images. Image and Signal Processing (CISP), 2010 3rd International Congress on. IEEE 5: 2125–2129.

46. Li P, Jiang L, Feng Z (2013) Cross-Comparison of Vegetation Indices Derived from Landsat-7 Enhanced Thematic Mapper Plus (ETM+) and Landsat-8 Operational Land Imager (OLI) Sensors. Remote Sensing 6: 310–329.

47. Rouse JW, Haas RH, Schell JA, Deering DW, Harlan JC (1974) Monitoring the vernal advancements and retrogradation of natural vegetation; In: NASA/GSFC, Final Report, Greenbelt, MD, USA, 1–137.

48. Gamon JA, Field CB, Goulden ML, Griffin KL, Hartley AE, et al. (1995) Relationship between NDVI, canopy structure and photosynthesis in three Californian vegetation types. Ecological Applications 5: 28–41.

49. Qi J, Chehbouni A, Huete AR, Kerr YH, Sorooshian S (1994) A modified soil adjusted vegetation index (MSAVI). Remote Sensing of Environment 48: 119–126.

50. Huete A, Didan K, Miura T, Rodriguez EP, Gao X, et al. (2002) Overview of the radiometric and biophysical performance of the MODIS vegetation indices. Remote Sensing of Environment 83: 195–213.

51. Broge NH, Leblanc E (2001) Comparing prediction power and stability of broadband and hyperspectral vegetation indices for estimation of green leaf area index and canopy chlorophyll density. Remote Sensing of Environment 76: 156–172.

52. Haboudane D, Miller JR, Tremblay N, Zarco-Tejada PJ, Dextraze L (2002) Integrated narrow-band vegetation indices for prediction of crop chlorophyll content for application to precision agriculture. Remote Sensing of Environment 81: 416–426.

53. Gitelson AA, Vina A, Ciganda V, Rundquist DC, Arkebauer TJ (2005) Remote estimation of canopy chlorophyll content in crops. Geophysical Research Letters 32: L08403. doi: 10.1029/2005GL022688.

54. Wu C, Niu Z, Tang Q, Huang W (2008) Estimating chlorophyll content from hyperspectral vegetation indices: Modeling and validation. Agricultural and Forest Meteorology 148: 1230–1241.

55. Hirose T, Ackerly DD, Traw MB, Ramseier D, Bazzaz FA (1997) CO_2 elevation, canopy photosynthesis, and optimal leaf area index. Ecology 78: 2339–2350.

56. Gitelson AA, Vina A, Verma SB, Rundquist DC, Arkebauer TJ, et al. (2006) Relationship between gross primary production and chlorophyll content in crops: Implications for the synoptic monitoring of vegetation productivity. Journal of Geophysical Research-Atmospheres 111: D08S11. doi: 10.1029/2005JD006017.

57. Glenn EP, Huete AR, Nagler PL, Nelson SG (2008) Relationship between remotely-sensed vegetation indices, canopy attributes and plant physiological

processes: what vegetation indices can and cannot tell us about the landscape. Sensors 8: 2136–2160.

58. Huang N, Niu Z (2013) Estimating soil respiration using spectral vegetation indices and abiotic factors in irrigated and rainfed agroecosystems. Plant and Soil 367: 535–550.

59. Chevallier T, Voltz M, Blanchart E, Chotte JL, Eschenbrenner V, et al. (2000) Spatial and temporal changes of soil C after establishment of a pasture on a long-term cultivated vertisol (Martinique). Geoderma 94: 43–58.

60. McGrath D, Zhang C (2003) Spatial distribution of soil organic carbon concentrations in grassland of Ireland. Applied Geochemistry 18: 1629–1639.

61. Liu D, Wang Z, Zhang B, Song K, Li X, et al. (2006) Spatial distribution of soil organic carbon and analysis of related factors in croplands of the black soil region, Northeast China. Agriculture, Ecosystems & Environment 113: 73–81.

62. Matheron G (1963) Principles of geostatistics. Economic geology 58: 1246–1266.

63. Webster R, Oliver MA (2007) Geostatistics for environmental scientists. John Wiley & Sons.

64. Raich JW, Tufekciogul A (2000) Vegetation and soil respiration: correlations and controls. Biogeochemistry 48: 71–90.

65. Schaefer DA, Feng W, Zou X (2009) Plant carbon inputs and environmental factors strongly affect soil respiration in a subtropical forest of southwestern China. Soil Biology and Biochemistry 41: 1000–1007.

66. Curiel Yuste J, Baldocchi DD, Gershenson A, Goldstein A, Misson L, et al. (2007) Microbial soil respiration and its dependency on carbon inputs, soil temperature and moisture. Global Change Biology 13: 2018–2035.

67. Pugesek BH, Tomer A, Von Eye A (Eds.) (2003) Structural equation modeling: applications in ecological and evolutionary biology. Cambridge University Press.

68. Iriondo JM, Albert MJ, Escudero A (2003) Structural equation modelling: an alternative for assessing causal relationships in threatened plant populations. Biological Conservation 113: 367–377.

69. Jonsson M, Wardle DA (2010) Structural equation modelling reveals plant-community drivers of carbon storage in boreal forest ecosystems. Biology Letters 6: 116–119.

70. Kim GS (2010) AMOS 18.0: Structural Equation Modeling. Seoul: Hannarae Publishing Co.

71. National Bureau of statistics of China (2006) China social-economic statistical yearbooks for China's counties and cities. China Statistics Press, Beijing.

72. Wang S, Tian H, Liu J, Pan S (2003) Pattern and change of soil organic carbon storage in China: 1960s–1980s. Tellus B 55: 416–427.

73. West TO, Post WM (2002) Soil organic carbon sequestration rates by tillage and crop rotation. Soil Science Society of America Journal 66: 1930–1946.

74. Wilhelm WW, Johnson JM, Karlen DL, Lightle DT (2007) Corn stover to sustain soil organic carbon further constrains biomass supply. Agronomy journal 99: 1665–1667.

75. Foley JA (1995) An equilibrium model of the terrestrial carbon budget. Tellus B 47: 310–319.

76. Lal R (1999) Soil management and restoration for C sequestration to mitigate the accelerated greenhouse effect. Progress in Environmental Science 1: 307–326.

77. Epron D, Bosc A, Bonal D, Freycon V (2006) Spatial variation of soil respiration across a topographic gradient in a tropical rain forest in French Guiana. Journal of Tropical Ecology 22: 565–574.

78. Suyker AE, Verma SB, Burba GG, Arkebauer TJ (2005) Gross primary production and ecosystem respiration of irrigated maize and irrigated soybean during a growing season. Agricultural and Forest Meteorology 131: 180–190.

79. Huang N, Gu L, Niu Z (2014) Estimating soil respiration using spatial data products: A case study in a deciduous broadleaf forest in the Midwest USA. Journal of Geophysical Research: Atmospheres 119. doi:10.1002/2013JD020515.

80. Hudak AT, Lefsky MA, Cohen WB, Berterretche M (2002) Integration of lidar and Landsat ETM+ data for estimating and mapping forest canopy height. Remote Sensing of Environment 82: 397–416.

81. Meng Q, Cieszewski C, Madden M (2009) Large area forest inventory using Landsat ETM+: a geostatistical approach. ISPRS Journal of Photogrammetry and Remote Sensing 64: 27–36.

82. Tsui OW, Coops NC, Wulder MA, Marshall PL (2013) Integrating airborne LiDAR and space-borne radar via multivariate kriging to estimate above-ground biomass. Remote Sensing of Environment 139: 340–352.

83. Verstraeten WW, Veroustraete F, Wagner W, Van Roey T, Heyns W, et al. (2010) Remotely sensed soil moisture integration in an ecosystem carbon flux model-The spatial implication. Climatic Change 103:117–136.

84. Rayner PJ, Scholze M, Knorr W, Kaminski T, Giering R, et al. (2005) Two decades of terrestrial carbon fluxes from a carbon cycle data assimilation system (CCDAS). Global Biogeochemical Cycles 19.

85. Chevallier F, Bréon FM, Rayner PJ (2007) Contribution of the Orbiting Carbon Observatory to the estimation of CO_2 sources and sinks: Theoretical study in a variational data assimilation framework. Journal of Geophysical Research: Atmospheres 112: D09307. doi: 10.1029/2006JD007375.

86. Knorr W, Kaminski T, Scholze M, Gobron N, Pinty B, et al. (2010) Carbon cycle data assimilation with a generic phenology model. Journal of Geophysical Research: Biogeosciences 115: G04017. doi: 10.1029/2009JG001119.

15

In Situ CO$_2$ Efflux from Leaf Litter Layer Showed Large Temporal Variation Induced by Rapid Wetting and Drying Cycle

Mioko Ataka[1]*, **Yuji Kominami**[2], **Kenichi Yoshimura**[2], **Takafumi Miyama**[2], **Mayuko Jomura**[3], **Makoto Tani**[1]

1 Laboratory of Forest Hydrology, Division of Environmental Science and Technology, Graduate School of Agriculture, Kyoto University, Kyoto, Japan, **2** Kansai Research Center, Forestry and Forest Products Research Institute (FFPRI), Kyoto, Japan, **3** College of Bioresource Sciences, Nihon University, Fujisawa, Kanagawa, Japan

Abstract

We performed continuous and manual in situ measurements of CO$_2$ efflux from the leaf litter layer (R_{LL}) and water content of the leaf litter layer (LWC) in conjunction with measurements of soil respiration (R_S) and soil water content (SWC) in a temperate forest; our objectives were to evaluate the response of R_{LL} to rainfall events and to assess temporal variation in its contribution to R_S. We measured R_{LL} in a treatment area from which all potential sources of CO$_2$ except for the leaf litter layer were removed. Capacitance sensors were used to measure LWC. R_{LL} increased immediately after wetting of the leaf litter layer; peak R_{LL} values were observed during or one day after rainfall events and were up to 8.6-fold larger than R_{LL} prior to rainfall. R_{LL} declined to pre-wetting levels within 2–4 day after rainfall events and corresponded to decreasing LWC, indicating that annual R_{LL} is strongly influenced by precipitation. Temporal variation in the observed contribution of R_{LL} to R_S varied from nearly zero to 51%. Continuous in situ measurements of LWC and CO$_2$ efflux from leaf litter only, combined with measurements of R_S, can provide robust data to clarify the response of R_{LL} to rainfall events and its contribution to total R_S.

Editor: Ben Bond-Lamberty, DOE Pacific Northwest National Laboratory, United States of America

Funding: Funding was provided by the Japan Society for the Promotion of Science (JSPS; grant number 25–2482 (http://www.jsps.go.jp/english/index.html)) Grant-in-Aid for Scientific Research (B) (20380182 (http://www.jsps.go.jp/english/index.html)). The funders had no role in study design, data collection and analysis, decision to publish, or preparation of the manuscript.

Competing Interests: The authors have declared that no competing interests exist.

* Email: teshimamioko@yahoo.co.jp

Introduction

Efflux of CO$_2$ from the soil surface (soil respiration; R_S), which is the sum of respiration by autotrophs and heterotrophs, is an important component of total CO$_2$ efflux from forest ecosystems [1–3]. The R_S: total ecosystem respirations varied from 58% to 76% in a mixed coniferous-deciduous forest [4], depending on interannual and seasonal changes in autotrophic and heterotrophic respiration; variability in R_S can affect the forest carbon balance on daily and seasonal time scales. To explain the cause of variability in R_S, many studies have attempted to separate differing sources of Rs and to examine factors controlling CO$_2$ efflux rate from each source [5–7]. Especially in forest ecosystems, heterotrophic respiration consists of CO$_2$ efflux from various sources (e.g., leaf and root litter, woody debris, soil organic matter) and their rates are controlled by their specific environmental condition such as water content (WC) and temperature [8], physical properties of the substrate (e.g., density and structure) [9,10], and chemical properties (e.g., labile and recalcitrant carbon) [11,12]. Moreover, CO$_2$ efflux from the various heterotrophic sources responds differently to these controlling factors, which illustrates the complexity of R_S. In recent decades, a variety of methods for separating components of heterotrophic respiration

and for determining their contribution to total R_S have been developed [9,13].

Among heterotrophic sources of CO$_2$, the leaf litter layer (L-layer) is a significant reservoir of degradable carbon and a large potential source of CO$_2$ efflux from forest soils [14]. In temperate forests, the contribution of CO$_2$ efflux from the L-layer (leaf litter respiration; R_{LL}) to R_S is reported to range from 23% to 48% [13,15,16]. The L-layer is in direct contact with rainfall, solar radiation, and wind, and environmental conditions (e.g., WC and temperature) can change more dynamically in the L-layer than in lower soil layers. Rapid and transient temporal variation in WC of the L-layer has been observed, especially in warm climates [16,17]. Heterotrophic respiration responds rapidly to changes in moisture status [17,18]; therefore, rapid and transient wetting and drying cycles would produce large temporal variations in R_{LL}. This would significantly affect variation in R_S [17,19], suggesting that R_{LL} is an important controller of temporal (daily and seasonal) patterns in the carbon balance in warm regions [19,20].

Several methods for measuring R_{LL} and for calculating its contribution to R_S have been explored. Cisneros-Dozal et al. [21] used an isotope mass balance method and reported that the contribution of R_{LL} to R_S increased from 5% to 37% in response to water addition after transient drought. Deforest et al. [15]

determined that the annual contribution of R_{LL} to R_S was 48% ±12% by measuring R_S with and without the L-layer, and the ratio was consistent over a range of environmental conditions. However, there is little information about temporal variation in R_{LL} in relation to rainfall events because of the difficulty of continuous and direct measurement of R_{LL} in situ.

To continuously measure CO_2 efflux from the L-layer only, in parallel with measurement of R_S, we developed an approach for measuring R_{LL} using an automated chamber method in a treatment area from which all CO_2 sources except for the L-layer were removed. In parallel with R_{LL} and R_S measurements, we continuously measured water content of the L-layer (LWC) and soil water content (SWC). LWC was measured using a method developed by Ataka et al. [22], in which intact leaf litter was attached to surrounding capacitance sensors. Sensors were also placed on top of the L-layer and at the boundary between the L- and mineral layers. From these continuous in situ measurements, we investigated the response of R_{LL} to rainfall events by comparing R_{LL} with R_S, and examined temporal variation in the contribution of R_{LL} to R_S in a warm temperate forest in Japan.

Materials and Methods

Ethics statement

The study site (Yamashiro Experimental Forest) is maintained by the Forestry and Forest Products Research Institute. All necessary permits were obtained for the field study, and the study did not involve endangered or protected species.

Study site

Our observations of R_{LL} and R_S were conducted at the Yamashiro Experimental Forest in southern Kyoto Prefecture, Japan (34°47′N, 135°50′E). The study site is a 1.7-ha watershed characterized by an annual mean air temperature of 15.5°C (maximum, 34.8°C; minimum, −3.9°C) and annual precipitation of 1449 mm [2]. The rainy season generally occurs from early June to mid-July. Daily rates of evaporation from the forest floor are 0.4–0.8 mm day^{-1} for 1–2 days after precipitation, declining thereafter to 0.2–0.3 mm day^{-1} [23]. The soils are Regosols with sandy loam or loamy sand texture and contain fine gravel (53% by mass) composed of residual quartz crystals from granite parent material [24]. These are immature soils in which the thickness of the A horizon is 2–3 cm. Deciduous broad-leaved, evergreen broad-leaved, and coniferous tree species account for 66%, 28%, and 6% of the living tree biomass, respectively [25]. The forest is dominated by *Quercus serrata* Thunb., which accounts for approximately 33% of the biomass. The L-layer (approximately 3–4 cm thick) consists mainly of fresh *Q. serrata* litter. There is no substantial organic horizon below the L-layer.

Automated chamber method for measuring leaf litter respiration and soil respiration

We measured R_{LL} and R_S using an automated dynamic chamber system with an infrared gas analyzer (IRGA, GMP343; Vaisala Group, Vantaa, Finland) (Fig. 1A). The system consisted of two automated circular chambers for R_{LL} and R_S measurement, four solenoid valves, a pump, mass flow meter, and IRGA. The chambers (surface area 320 cm^2) were made from PVC collars with clear acrylic lids that can be opened and closed automatically using an air cylinder. Air was supplied to the cylinder from a compressor. To ensure a seal between the chamber and the closed lid, a soft rubber gasket was attached to the top edge of the chamber. Opening and closing of the chamber lid and solenoid

valves of each chamber were regulated synchronously by a control unit (ZEN, OMRON, Kyoto, Japan).

The duration of measurement of CO_2 concentration inside each chamber was 6 min and was performed twice per hour. The CO_2 concentration in each chamber was recorded at 1-s intervals using a data logger (GL220, Graphtec, Kanagawa, Japan). We calculated R_{LL} and R_S from the increase in CO_2 concentration (ΔC_{CO2}) using linear regression. Data from the first 2 min were discarded to avoid effects of closing the chamber. R_{LL} and R_S were calculated using the following equation:

$$R = \frac{\Delta C_{CO_2}}{10^6} \times \frac{V}{V_{air}} \frac{273.2}{273.2 + T} \times M_{CO_2} \times \frac{1}{A},\qquad(1)$$

where R is respiration (mg CO_2 m^{-2} s^{-1}), ΔC_{CO2} is the change in CO_2 concentration per unit time (CO_2 ppm s^{-1}), V is the volume of the system (L), V_{air} is the standard gas volume (22.41 L mol^{-1}), T is temperature inside the chamber (°C), M_{CO2} is the molecular weight of CO_2 (44.01 g mol^{-1}), and A is the soil surface area covered by the chamber (m^2).

To continuously measure CO_2 efflux from the L-layer only, we developed an approach for measuring R_{LL} by using an automated chamber method in a treatment area in which all potential CO_2 sources (e.g., organic soil and fine roots) except for the L-layer were replaced with combusted granite soil (Fig. 1B). To prepare the treatment area (1 m^2), we removed surface soil (approximately 5 cm). An acrylic board was placed on the bottom and sides of the treatment area to prevent penetration of roots; a drain tube was located at the bottom of the board to prevent the treatment area from flooding with rainwater. The treatment area was then filled with granite soil combusted in a muffle furnace (500°C for 1 day). For R_{LL} measurement, we placed a PVC collar (320-cm^2 surface area) and acrylic board below the collar. The board was set at a slight incline to drain rainwater from the collar. We added 15 g of newly fallen leaf litter, which represents the average litterfall mass per unit ground surface area at this site, to the collar. We added the leaf litter to each chamber on January 2012. To acquire data on the temporal variation in R_{LL} of fresh leaf litter, we replaced the litter with newly fallen leaf litter in January 2013. The collar for measurement of R_S was placed near the treatment area for R_{LL} measurement and the L-layer inside the collar was removed and leaf litter was supplied similarly as for measurement of R_{LL}. To prevent incorporation of newly fallen litter, we placed a mesh sheet (1×1 mm mesh) on the L-layer inside the chamber, and fallen litter was removed weekly. CO_2 efflux from combusted granite soil was measured 6 months from the start of the R_{LL} measurements. The mean CO_2 flux rate (± standard deviation) was 0.00063±0.00068 mg CO_2 m^{-2} s^{-1} ($n = 16$) when SWC ranged from 0.05 to 0.3 m^3 m^{-3} at temperatures of 24°C. Thus, we assumed that CO_2 efflux from the combusted granite soil was negligible throughout the measurement period.

For continuous in situ measurement of LWC, we used capacitance sensors as described by Ataka et al. [22]. The measurements were performed on the top surface of the L-layer and at the boundary between the L-layer and mineral soil (Fig. 1B), to capture the large vertical distribution of WC within the L-layer. We estimated average LWC from the output voltage (V) of the two sensors using the conversion equation LWC = 12.73 V−3.42 presented by Ataka et al. [22]. LWC at the forest floor shows spatial variability associated with tree canopy conditions. Thus, to reflect the LWC of the L-layer by direct measurement, two capacitance sensors were placed on the L-layer inside the chamber. To check the validity of continuous LWC monitoring, we compared the sensor values with LWC measured

(A)

(B)

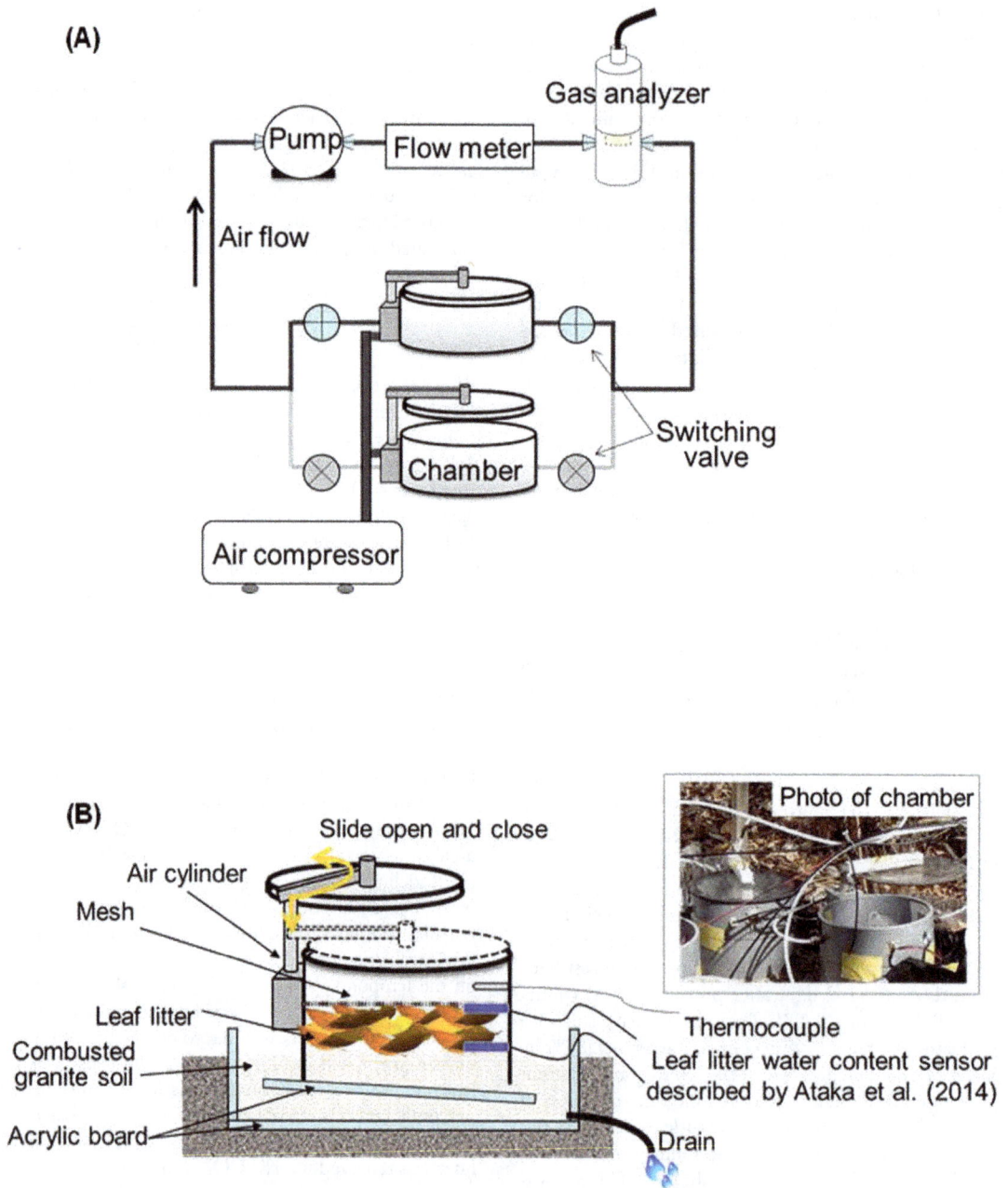

Figure 1. Schematic of the automated chamber system and the experimental design for measurement of CO_2 efflux from the leaf litter layer. A. Schematic of the automated dynamic-closed chamber system for measuring leaf litter respiration and soil respiration. B. The experimental design for continuous measurement of CO_2 efflux from the leaf litter layer only using automated chamber system.

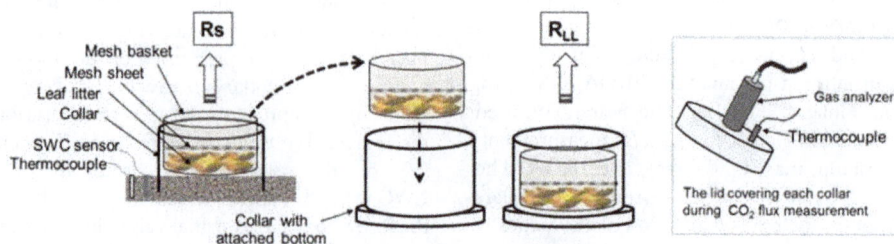

Figure 2. Schematic of the manual chamber system and the experimental design for measurement of CO_2 efflux from the leaf litter layer (R_{LL}) and soil (R_S).

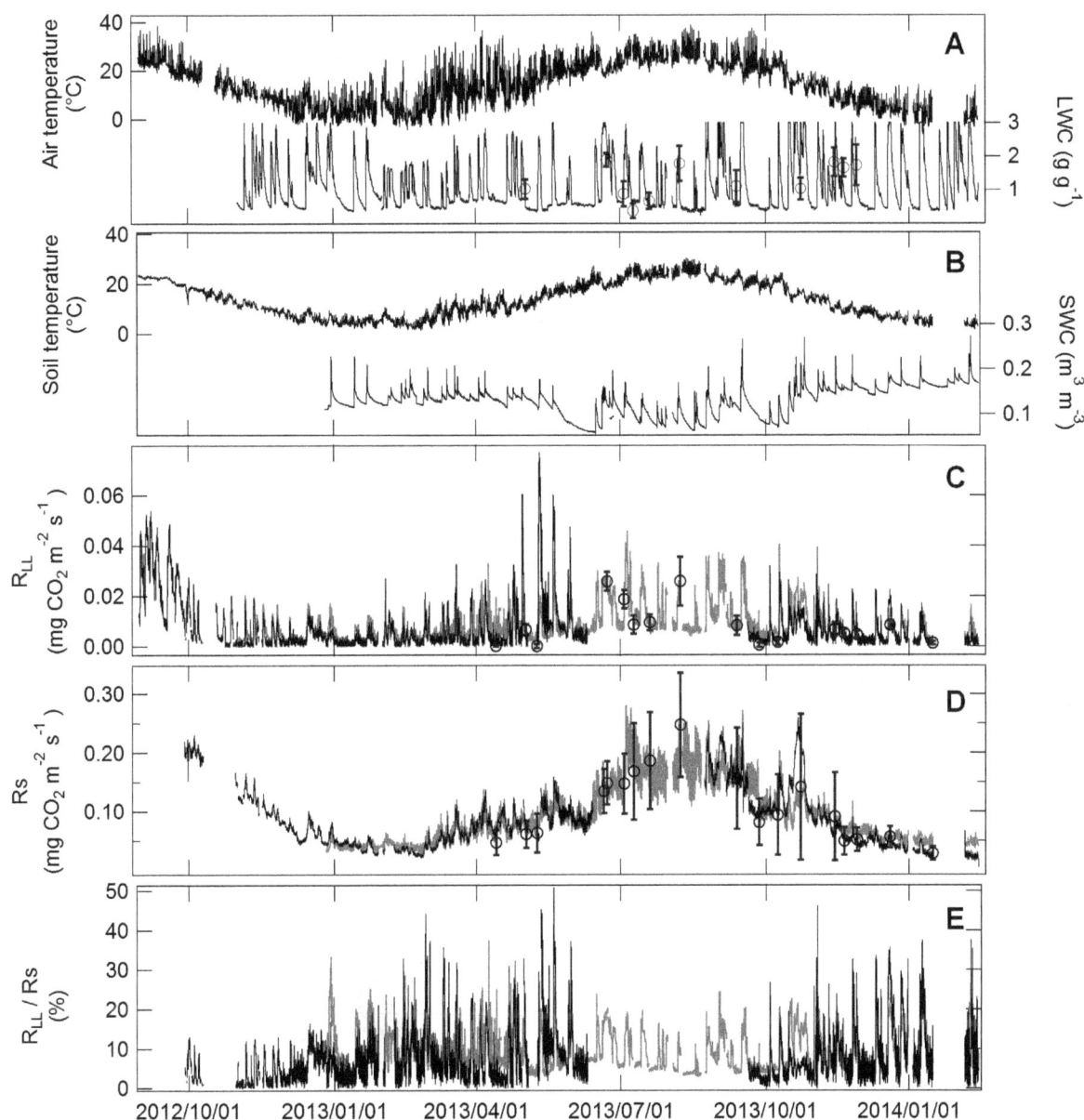

Figure 3. Seasonal variation in environmental factors, CO$_2$ efflux from the leaf litter layer (R_{LL}), and soil respiration (R_S). Data were measured every 30 min between September 2012 and January 2014. **A.** Bold and fine lines show air temperature and water content of the leaf litter layer (LWC), respectively. **B.** Bold and fine lines show soil temperature and soil water content (SWC), respectively. **C.** Black and grey lines show observed and estimated R_{LL}, respectively. **D.** Black and grey lines show observed and estimated R_S, respectively. **E.** Black and grey lines show the ratio of observed and estimated R_{LL} to R_S, respectively. Circles and bars show mean values and standard deviation of manual measurements. Estimated R_{LL} and R_S were calculated from regression equations using temperature (T) and water content (WC): $R_{LL} = 0.29e^{0.059T}[WC/(95.04+WC)]$ and $R_S = 0.031e^{0.10T}[WC/(0.032+WC)]$.

Table 1. Q_{10} of leaf litter respiration (R_{LL}) and soil respiration (R_s) for different water contents of the leaf litter layer (LWC) and soil (SWC).

	R_{LL}			R_s		
	LWC≤1	1<LWC≤2	2<LWC	SWC≤0.1	0.1<SWC≤0.15	0.15<SWC
Q_{10}	1.54	1.88	2.07	1.97	2.12	2.73
a	0.0019	0.0044	0.0064	0.027	0.032	0.025
b	0.043	0.063	0.073	0.068	0.075	0.10

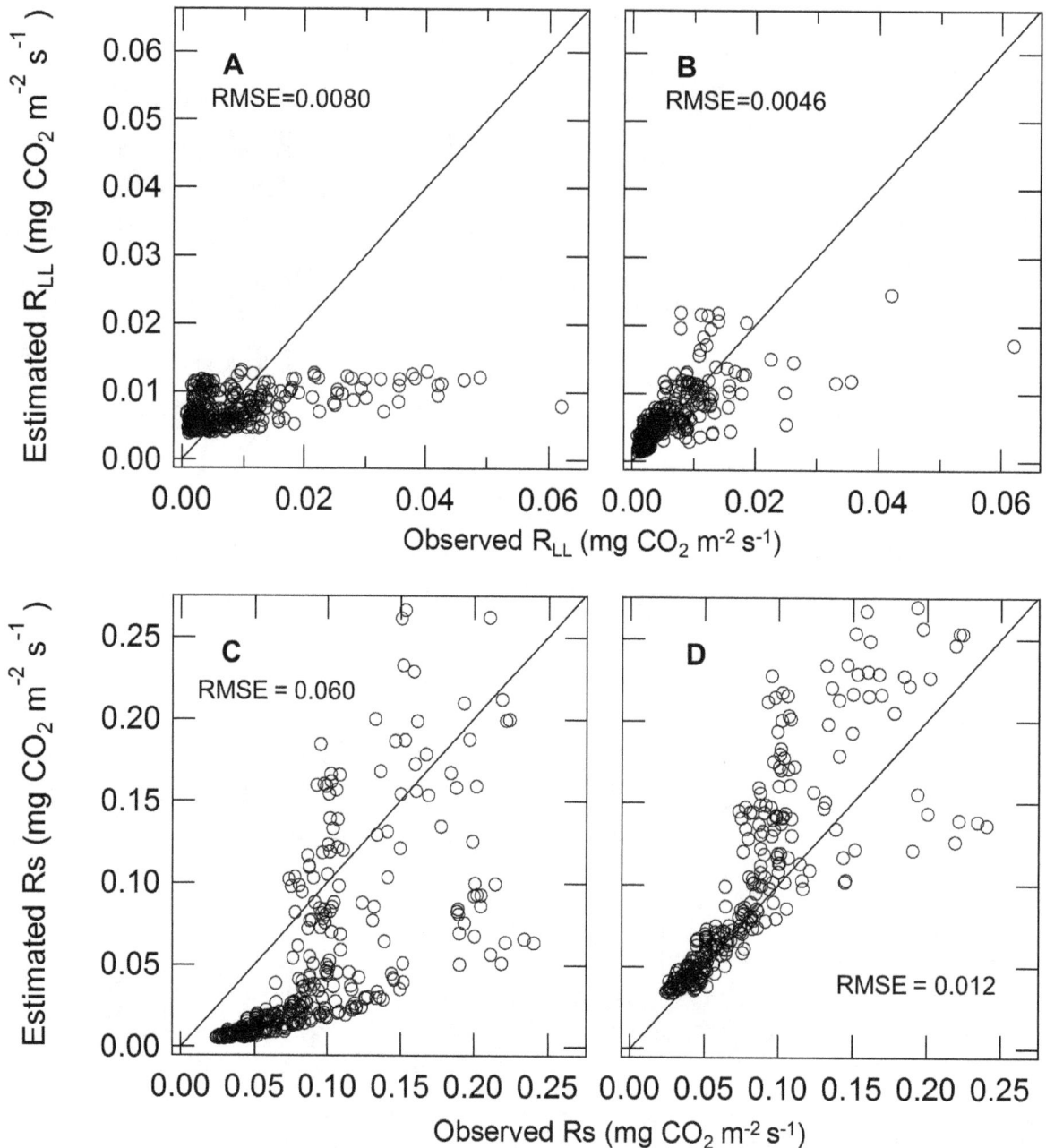

Figure 4. Relationship between observed and estimated CO_2 efflux rate from leaf litter respiration (R_{LL}) and soil respiration (R_S). R_{LL} (A, B) and R_S (C, D) show daily mean values. Estimated respiration rates were calculated using a function of temperature (A, C) from Eq. (5,6) and a function of temperature and water content (B, D) from Eq. (7,8) in the Results. Lines represent the 1:1 ratio. RMSE: root mean square error.

manually as described in the following section. In parallel with LWC measurement, soil temperature (copper-constantan thermo-couple) and soil volumetric water content (ECH₂O EC-5 sensors; Decagon Devices, Pullman, WA, USA) were measured at 5-cm depth near each chamber. The output voltage of all environmental data was recorded every 1 min with a data logger (Datamark LS-3000 PtV; Hakusan, Japan) and average values were computed every 30 min. The environmental data, R_{LL}, and R_S were measured continuously between September 2012 and January 2014. Malfunction of IRGA resulted in a lack of data for R_{LL} and R_S for 31% of the measurements.

Manual chamber method for measuring leaf litter respiration and soil respiration

To determine the validity of R_{LL} and R_S measured using the automated chamber method, respiration was measured using the manual chamber method. We assumed that manual chamber method allow to measure under conditions that were closer to natural than the automated chamber method. We measured R_{LL} and R_S manually using a static chamber system at midday on 18 days between April 2013 and January 2014. Twelve PVC collars (320 cm² surface area) were placed in a 2×4 m area in January 2013. The edges of the collars were inserted approximately 1.5 cm into the soil. To measure R_{LL}, mesh baskets (1×1 mm mesh, the

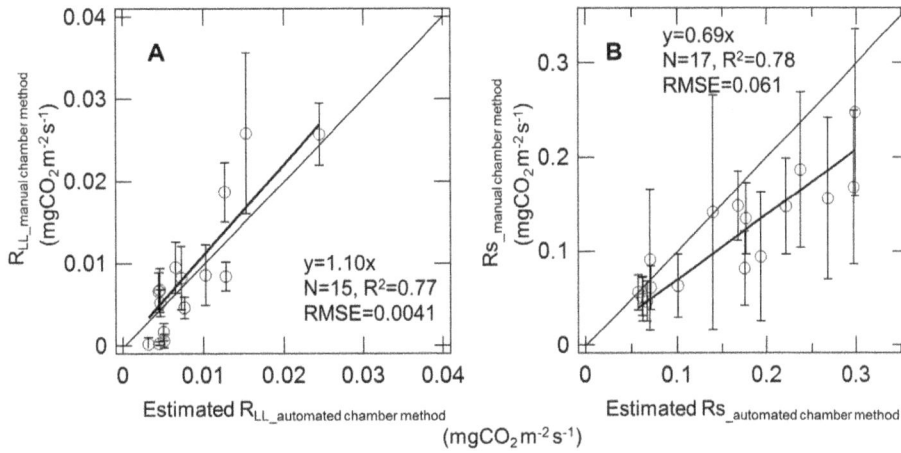

Figure 5. Relationship between respirations measured using a manual chamber method and estimated from automated chamber data. Respiration rate measured with the manual chamber method (R_manual chamber method) show mean value obtained from measurement of 12 collars. Bars show standard deviation. Respiration estimated from automated chamber data (estimated R_automated chamber method) shows daily mean respiration. The estimated R was calculated using a function based on temperature and water content (Eq. 8, 9).

same diameter as the PVC collars; 20 cm) were set into each collar and 15 g (dry weight) of newly fallen leaf litter was placed on the L-layer inside each basket (Fig. 2). To prevent supply of newly fallen litter, we placed a mesh sheet (1×1 mm mesh) on the L-layer inside the chamber, and fallen litter was removed weekly.

For measurement of R_S, the collars were completely covered with lids to which an IRGA and copper-constantan thermocouple were attached. Soil temperature and SWC (5 cm depth) were measured close to the collars when R_S was measured. After completing the measurements of R_S, the mesh baskets were carefully removed from the collars and placed in PVC chambers (20 cm diameter, 7 cm high; Fig. 2). We measured R_{LL} using the same methods as used for R_S measurement. The temperature and CO_2 concentrations in the chamber were recorded at 1-s intervals using a data logger (GL220). Linearity of the CO_2 flux was checked on the data logger monitor at each measurement. The measurement period for each chamber was 10 min and CO_2 data for the middle 5-min intervals were used to determine R_{LL} according to Eq. (1), excluding data from the first 3 min.

For measurement of LWC in the mesh baskets, four or five leaves were removed from each basket and immediately placed in sealed plastic bags. Fresh weight of the leaf litter was measured in the laboratory within 24 h of sampling. Leaf litter samples were oven dried at 65°C for 48 h, and water content (WC; g g^{-1}) was calculated using Eq. 2 as follows:

$$WC = \frac{(FW - DW)}{DW}, \quad (2)$$

where FW is the fresh mass of the sample (g), and DW is the dry mass of the sample (g). Samples were returned to each mesh basket within 1 week after sampling.

Leaf litter respiration and soil respiration rates as a function of environmental factors

Respiration models are fundamentally described by nonlinear functions. We used the following function to investigate the response of respiration to temperature:

$$R = a \ \exp(bT), \quad (3)$$

where T is temperature (leaf litter temperature for R_{LL} measurement or soil temperature for R_S measurement) and a and b are constants. Leaf litter temperature was assumed to be same as air temprature. b is related to the Q_{10} parameter ($Q_{10} = e^{10b}$). To determine the effects of temperature and water content on R_{LL} and R_S, we used a function that was previously applied to estimate soil respiration by Subke and Schlesinger [26]:

$$R = a \ \exp(bT) \left(\frac{WC}{c + WC}\right), \quad (4)$$

where a, b, and c are constants. LWC or SWC was used as WC in this equation. These nonlinear regressions were performed using a modified Levenberg–Marquardt method with Igor Pro 6.0 software (WaveMetrics, Lake Oswego, OR, USA). The estimated respiration values presented in this manuscript were calculated using Eq. 4.

Short-term changes in R_{LL} and LWC on wetting and drying cycle

To evaluate short-term changes in R_{LL} and LWC after rainfall events, we chose eight typical periods that included one wetting and drying cycle and had consecutive no rainfall days for at least 3 days. We used daily mean R_{LL} and LWC before the day on which precipitation occurred as the pre-wetting condition, and these values after precipitation as the post-wetting condition. Daily mean R_{LL} was calculated from R_{LL} values observed using the automated chamber method.

Effect of wetting and drying cycle of the L-layer on R_{LL} and Rs on the annual time scale

To investigate the effects of wetting and drying of the L-layer on R_{LL} on the annual time scale, we separated the estimated daily mean R_{LL} in 2013 into 'Dry' and 'Wet' periods based on daily mean LWC as a threshold value. The threshold LWC value that separated 'Dry' and 'Wet' periods for R_{LL} was estimated by the abovementioned short-term analyses. Daily mean R_{LL} was calculated from the estimated R_{LL} values because there were gaps in the continuous R_{LL} data observed using the automated

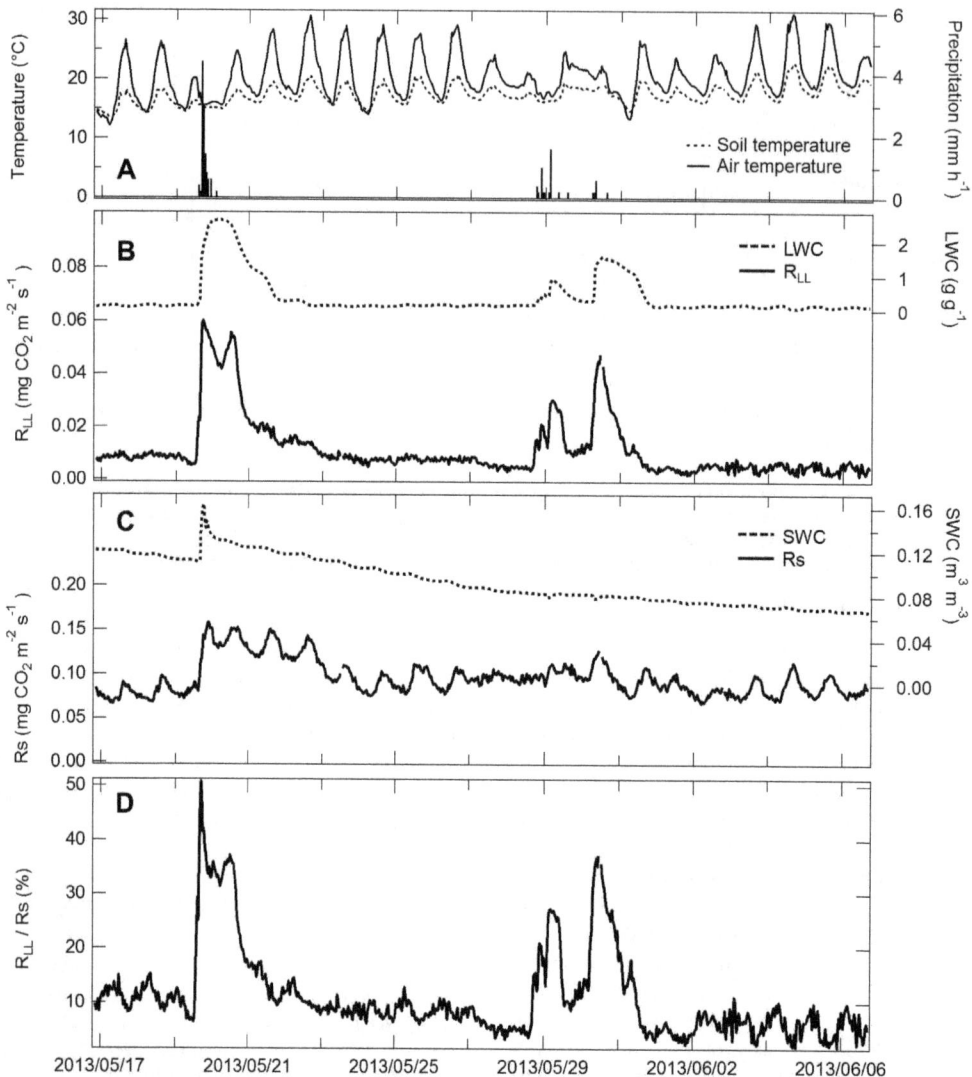

Figure 6. Temporal variation in environmental factors, CO$_2$ efflux from the leaf litter layer (R_{LL}), soil respiration (R_S), and the ratio of R_{LL} to R_S. Data was measured at one collar every 30 min between May 17 and June 6, 2013. **A.** Soil and air temperature. Spikes on the x-axis indicate precipitation events (mm h^{-1}). **B.** R_{LL} and water content of the leaf litter layer (LWC). **C.** R_S and soil water content (SWC). **D.** The ratio of R_{LL} to R_S (%).

chambers. We estimated the contribution of R_{LL} accumulated during the wet and dry period to total R_S.

Results

Seasonal variation in R_{LL} and R_S

The magnitude of the peak in the observed R_{LL} pulse was higher in summer than in winter (Fig. 3C). R_{LL} values were low when LWC was low (Fig. 3A, C). R_S changed substantially according to temperature (Fig. 3B, D), with higher values in summer than in winter. The relationships between respiration and temperature were described by the following functions:

$$R_{LL}(mg\,CO_2\,m^{-2}\,s^{-1}) = 0.0038\exp(0.065 \times T_{LL}),\quad (5)$$

$$Rs\,(mg\,CO_2\,m^{-2}\,s^{-1}) = 0.0031\exp(0.19 \times Ts),\quad (6)$$

where T_{LL} is leaf litter temperature and Ts is soil temperature (°C).

To evaluate effect of WC on the temperature sensitivity of respiration, the measured respiration data was separated into three groups based on WC (Table 1). More than 14% of total respiration data was included in each WC group. R_{LL} showed low values when WC values were low in spite of high temperature. Consequently, calculated Q$_{10}$ values for not only R_{LL} but also R_S decreased with decreasing WC. The relationships between respiration and temperature and WC were described by the following functions:

$$R_{LL}\,(mg\,CO_2\,m^{-2}\,s^{-1}) =$$
$$0.29\exp(0.059 \times T_{LL})\left(\frac{LWC}{95.04 + LWC}\right),\quad (7)$$

$$Rs\,(mg\,CO_2\,m^{-2}\,s^{-1}) =$$
$$0.031\exp(0.10 \times Ts)\left(\frac{SWC}{0.032 + SWC}\right),\quad (8)$$

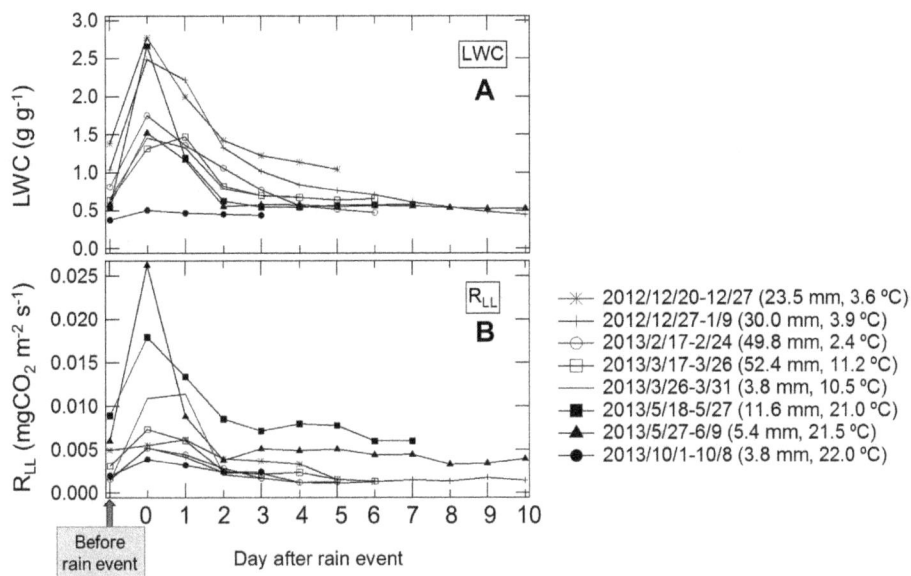

Figure 7. Temporal variation in water content of the leaf litter layer (LWC) and CO$_2$ efflux from the leaf litter layer (R_{LL}) after rainfall events. LWC (A) and R_{LL} (B) show the daily mean values. The rainfall intensity of each precipitation event was 23.5 mm in 2 days (2012/12/20–12/27, mean air temperature; 3.6°C); 30.0 mm in 3 days (2012/12/27–1/9, 3.9°C); 49.8 mm in 2 days (2013/2/17–2/24, 2.4°C); 52.4 mm in 3 days (2013/3/17–3/26, 11.2°C); 3.8 mm in 2 days (2013/3/26–3/31, 10.5°C); 11.6 mm in 2 days (2013/5/18–5/27, 21.0°C); 5.4 mm in 3 days (2013/5/27–6/9, 21.5°C); and 3.8 mm in 4 days (2013/10/1–10/8, 22.0°C).

where LWC (g g^{-1}) and SWC (m^3 m^{-3}) are water content of leaf litter and soil, respectively. The RMSE between observed and estimated daily mean respiration based on temperature (R_{LL}, 0.0080 mg CO$_2$ m^{-2} s^{-1}; R_S, 0.060 mg CO$_2$ m^{-2} s^{-1}) was larger than that based on temperature and WC (R_{LL}, 0.0046 mg CO$_2$ m^{-2} s^{-1}; R_S, 0.012 mg CO$_2$ m^{-2} s^{-1}) (Fig. 4). Estimated respiration was calculated using the equation based on temperature and WC because of the lower RMSE. Throughout the measurement period, the contribution of observed R_{LL} to variation in R_S changed from nearly zero to 51% following a rainfall event (Fig. 3E).

To consider the validity of R_{LL} and R_S estimated from continuous measurement, we compared these values with respiration rates measured using the manual chamber method (Fig. 5). Estimated respiration was very similar to that observed using manual measurements. The RMSE between estimated and observed respiration were 0.0041 and 0.061 mg CO$_2$ m^{-2} s^{-1} for R_{LL} and R_S, respectively.

Temporal changes in R_{LL} and R_S on the short-term scale

To show clear temporal variation in R_{LL} and R_S, the period between May 17 and June 6, 2013 (Fig. 6) was chosen because this

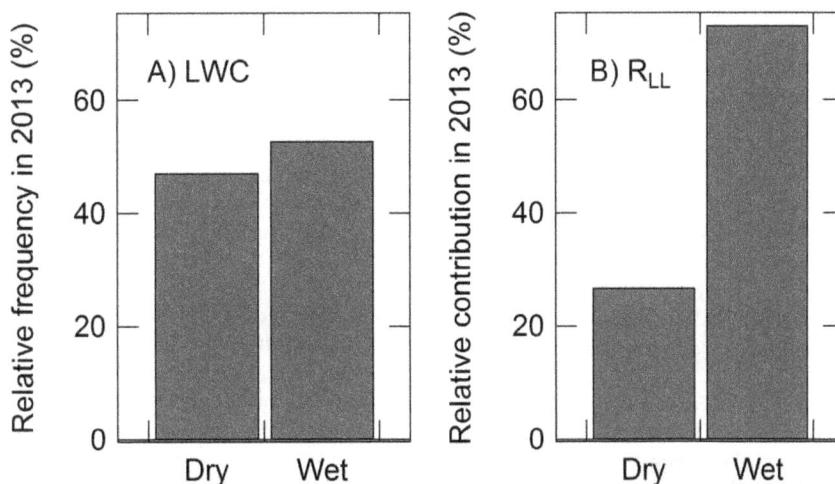

Figure 8. Histograms of the relative frequency of "Dry" and "Wet" periods in relation to water content of the leaf litter layer (LWC), and the relative contribution of estimated leaf litter respiration (R_{LL}) in 2013. The daily mean LWC (A) and R_{LL} (B) were used to present histograms. Estimated respiration rates were calculated using a function based on temperature (T) and water content (WC). $R_{LL} = 0.29e^{0.059T}$[WC/(95.04+WC)]. The daily mean LWC and R_{LL} were defined as Dry or Wet based on LWC. Days in which daily mean LWC <0.75 g g^{-1} were defined as Dry periods, while days in which daily mean LWC ≥0.75 g g^{-1} were defined as Wet periods.

period included two characteristic rainfall events. The rainfall intensity was 11.6 mm over 13 h during the first event and 5.4 mm over 46 h during the second event. LWC and SWC increased from 0.11 to 2.64 g g^{-1} and from 0.11 to 0.16 m^3 m^{-3}, respectively, following the first rainfall event (Fig. 6B, C). LWC increased from 0.16 to 1.58 g g^{-1} but SWC did not increase after the second rainfall event.

Temporal variation in R_{LL} measured using the automated chamber system changed according to wetting and drying of the L-layer (Fig. 6B), reaching a maximum of 0.060 and 0.047 mg CO_2 m^{-2} s^{-1} during first and second rainfall events, respectively. R_S increased following the increase in SWC and subsequently decreased gradually with diurnal variation according to temperature (Fig. 6C). Between May 17 and June 6, 2013, the contribution of R_{LL} to R_S increased from 6.5% to 51%, with a peak value of 51% during the first rainfall event and 37% during the second rainfall event (Fig. 6D).

Both R_{LL} and LWC reached a peak during or one day after rainfall events (Fig. 7). The peak of R_{LL} and LWC varied from 0.0020 to 0.026 mg CO_2 m^{-2} s^{-1} and from 0.50 to 2.66 g g^{-1}, respectively. Peak value of each rainfall event highly depended on air temperature. High peaks of R_{LL} were observed in the warm season (0.017 mg CO_2 m^{-2} s^{-1}; 2013/5/18–5/27, 0.026 mg CO_2 m^{-2} s^{-1}; 2013/5/27–6/9 in Fig. 7). Also, the peak value was related to LWC: low peak of R_{LL} was observed when LWC was low (0.004 mg CO_2 m^{-2} s^{-1}; 2013/10/1–10/8 in Fig. 7). The relationship between LWC and amout of precipitation was not clear. In the cold season, peak values of R_{LL} were relatively low (e.g., 0.005 mg CO_2 m^{-2} s^{-1}; 2013/2/17–2/24, 0.006 mg CO_2 m^{-2} s^{-1}; 2012/12/20–12/27 in Fig. 7) even when the L-layer was wet enough (LWC more than 1.5 g g^{-1}). The peak values of R_{LL} were 1.2- to 8.6-fold higher than the R_{LL} values before rainfall events, and R_{LL} fell to pre-wetting levels within 2–4 days after rainfall events and peak LWC values were 1.3- to five-fold higher than LWC before rainfall, and LWC also dropped to pre-wetting levels within 2–4 days after rainfall events. We defined R_{LL} from the period just after rainfall events through 2–4 days later as the "R_{LL} pulse".

Effects of wetting and drying of the L-layer on R_{LL} and R_S on the annual time scale

Estimated daily mean R_{LL} in 2013 was separated into 'Dry' and 'Wet' periods based on daily mean LWC. Days for which mean LWC was <0.75 g g^{-1} were categorized as Dry, while days for which mean LWC ≥0.75 g g^{-1} were categorized as Wet. The threshold value (0.75 g g^{-1}) was obtained from mean LWC 3 days after a rainfall event (Fig. 7A). The relative frequency of Dry and Wet periods in 2013 were 47.2% and 52.8%, respectively, while the relative contributions of daily mean R_{LL} during the Dry and Wet periods in 2013 were 26.9% and 73.2%, respectively (Fig. 8). Annual R_{LL} and R_S in 2013 were estimated to be 0.69 and 7.94 t C ha^{-1} y^{-1}, respectively. The RMSE between continuous respiration measured and estimated based on temperature and WC was 0.011 and 0.029 t C ha^{-1} y^{-1}, respectively.

The contribution of annual R_{LL} to R_S was 8.6%. The relative frequency of LWC was similar during Dry and Wet periods, while the contribution of R_{LL} during the Wet period was approximately three-fold higher than that during the Dry period (Fig. 8).

Discussion

As seen in Fig. 6, R_{LL} immediately increased with wetting of the L-layer and decreased to pre-wetting levels within 2–4 days after rainfall events, which was consistent with observations made in previous studies [17,19]. R_{LL} showed no diurnal variation despite a diurnal temperate range >10°C. Consequently, the Q_{10} of R_{LL} increased with increasing LWC (Table. 1). The variation in Q_{10} would be directly related to water stress experienced by microorganism. This indicated that LWC can reach to adequate low value, suspected as water stress for microorganism, within several days after rainfall. On the one hand, R_S increased during rainfall and subsequently decreased, showing diurnal variation. The Q_{10} of R_S also increased with increasing SWC. Dannoura et al. [27] reported that root respiration showed little change with variation in SWC compared with changes in R_S. Therefore, the increased Q_{10} of R_S with increasing SWC might be highly affected by not only R_{LL} but also by respiration from other heterotrophic sources.

Although the relative frequency of LWC was similar during Dry and Wet periods, the contribution of annual R_{LL} during the Wet period was approximately three-fold higher than that during the Dry period (Fig. 8), indicating strong effect of rainfall on R_{LL}. Although the R_{LL} pulse can last for only 3–4 days after a rainfall event, this pulse would determine a large part of annual R_{LL}. This suggests that the magnitude of total R_{LL} may be influenced by the frequency of rainfall events, especially in summertime, rather than the intensity of rainfall. Still, the cumulative R_{LL} in the Dry period contributed 26.9% of annual R_{LL} in 2013, even though instantaneous R_{LL} was very low. There may be large vertical variability in WC and R_{LL} within the L-layer, indicating that higher WC and R_{LL} occur in lower parts of the L-layer during the drying process because the upper L-layer dries more rapidly [28]. In that case, although the mean WC of the L-layer was very low, local wetting in lower sections would produce small CO_2 fluxes. Despite low instantaneous R_{LL}, the accumulation of R_{LL} over a long time period (approximately 6 mo) resulted in a substantial contribution (27%) of Dry-period respiration to annual R_S.

Raindrops first reach the L-layer and then percolate to the soil layers below. Small amounts of precipitation caused no change in SWC or R_S, but R_{LL} increased rapidly with increasing LWC (Fig. 6). In semi-arid and arid ecosystems, wetting of the L-layer and surface soil by small fog-drop pulses during the dry season can contribute up to 35% of R_S [29]. Although such small water inputs (e.g., brief rain showers and fog), which mainly affect the surface of the forest floor, can be significant drivers of temporal variation in R_S, the soil water content sensors (generally inserted at depths > 5 cm) could not capture these inputs. Continuous measurement of LWC allowed for realistic modeling of the effects of rapid changes in LWC on R_{LL}.

Although the annual contribution of R_{LL} to R_S was relatively small (8.6%), this contribution showed large temporal variation according to rainfall, ranging from nearly zero to 51%. Several other studies have described similar results [17,21]. For example, Borken et al. [17] reported that peaks in R_{LL} during addition of water ranged from 0.031 to 0.071 mg CO_2 m^{-2} s^{-1} in vitro, which represented 11–26% of maximum in situ R_S in the Harvard forest, although R_{LL} before addition of water was nearly zero. These findings indicate that R_{LL} is a significant component of rapid and transient temporal variation in R_S in relation to rainfall events. Although numerous studies have examined CO_2 efflux from mineral soils in relation to the intensity, duration, and frequency of rainfall [30,31], few studies have focused on R_{LL} because of the difficulty in measuring this dynamic. Here, R_{LL} pulses were observed only during and several days after rainfall events. Thus, periodic sampling (e.g., twice per week) might be insufficient to capture the contribution of the R_{LL} pulse to R_S. Moreover, manual flux measurements are usually not performed during precipitation events because of difficulties that can occur with electronic instruments and sampling methods. In our view, conducting in situ measurements of CO_2 efflux from the L-layer only over short time intervals (e.g., up to 1 h) produces robust data for understanding the response of R_{LL} to rainfall events and its contribution to R_S.

The contribution of R_{LL} to annual R_S was 8.6% in our site. In an oak forest, the contribution of R_{LL} to R_S was 23%, according to model simulation based on temperature and LWC by Hanson et al. [13]. Ngao et al. [32] reported a lower contribution (8%) in a beech forest, estimated using an isotope mass balance approach, which was close to the value observed at our site (8.6%). However, simple quantitative comparisons between studies are difficult because of the use of different methods. In addition, some technical problems remain at our site. First, we performed R_{LL} measurements in the treatment area in which the mineral soil below the L-layer was replaced with combusted granite soil. This treatment may have affected the microbial community and environmental conditions in the L-layer. Secondly, each continuous measurement of R_{LL} and R_S was performed with single chambers, so spatial heterogeneity in R_{LL} and R_S were not considered. Automated chamber methods allowed high-interval measurements of temporal variation in respiration but had poorer spatial distribution compared with the manual chamber method. The balance of trade-offs between automated and manual chamber method is subject to the relative importance of characterizing temporal and spatial variability of individual CO_2 sources. The number of chambers used can enhance the accuracy of measured mean values. Loescher et al. [33] reported that the number of chambers needs to be >100 to adequately represent spatial variability. However, this is not a feasible experimental design because of practical limitations to sampling efforts. To improve estimation of R_{LL} and R_S at the forest stand level, and to better understand the soil carbon budget, a comprehensive comparison of the diverse C pools and fluxes in forest soils is required.

Conclusions

In our study, the rapid and transient variation in R_{LL} induced by rainfall; the peak R_{LL} was observed during or one day after rainfall, and R_{LL} subsequently decreased to pre-wetting levels within 2–4 days after rainfall events, following the decrease in LWC. On the one hand, CO_2 efflux from coarse woody debris found in our site decreased during rainfall events, and subsequently, a gradual increase in CO_2 efflux continued for at least 14 days until next rainfall [34]. Therefore, coarse woody debris was a CO_2 efflux source over longer time scales, while R_{LL} approached nearly zero within a few days after rainfall events, even at high temperatures. Such specific temporal CO_2 efflux patterns for each heterotrophic source when subjected to wetting and drying cycles would be a result of substrate properties (e.g., specific surface area). In our view, continuous and direct measurements of CO_2 efflux and environmental conditions characterized by substrate properties of individual CO_2 sources could improve understanding of the processes that regulate variation in heterotrophic respiration and R_S and enable progress beyond empirical models that are primarily based on simple temperature and SWC relationships.

Moreover, the magnitude of heterotrophic respiration under wetting and drying cycles is strongly related to microbial physiology and community composition. For example, Schnurer et al. [35] showed that longer-duration wetting could promote microbial biomass, causing an increase in basal respiration. Fierer et al. [36] showed the influence of drying and rewetting frequency on microbial (fungi and bacteria) community composition. To improve understanding of heterotrophic respiration associated with response and adaptation of microorganisms under climatic changes, collected continuous in situ data for CO_2 efflux and environmental conditions (e.g., temperature and WC) of individual CO_2 sources should be combined with analyses of microbial physiology and community composition.

Acknowledgments

We greatly thank Dr. Yoshiko Kosugi and the staff of the Forest Hydrology Laboratory of Kyoto University for assistance in the field and for helpful advice.

Author Contributions

Conceived and designed the experiments: MA YK TM. Performed the experiments: MA KY MT. Analyzed the data: MA YK MT. Contributed reagents/materials/analysis tools: MA MJ. Contributed to the writing of the manuscript: MA.

References

1. Curtis PS, Hanson PJ, Bolstad P, Barford C, Randolph JC, et al. (2002) Biometric and eddy-covariance based estimates of annual carbon storage in five eastern North American deciduous forests. Agricultural and Forest Meteorology 113: 3–19.
2. Kominami Y, Jomura M, Dannoura M, Goto Y, Tamai K, et al. (2008) Biometric and eddy-covariance-based estimates of carbon balance for a warm-temperate mixed forest in Japan. Agricultural and Forest Meteorology 148: 723–737.
3. Keith H, Leuning R, Jacobsen KL, Cleugh HA, van Gorsel E, et al. (2009) Multiple measurements constrain estimates of net carbon exchange by a Eucalyptus forest. Agricultural and Forest Meteorology 149: 535–558.
4. Yuste JC, Nagy M, Janssens IA, Carrara A, Ceulemans R (2005) Soil respiration in a mixed temperate forest and its contribution to total ecosystem respiration. Tree Physiology 25: 609–619.
5. Hanson PJ, Edwards NT, Garten CT, Anderson JA (2000) Separating root and soil microbial contributions to soil respiration: A review of methods and observations. Biogeochemistry 48: 115–146.
6. Kuzyakov Y (2006) Sources of CO_2 efflux from soil and review of partitioning methods. Soil Biology & Biochemistry 38: 425–448.
7. Moyano FE, Kutsch WL, Rebmann C (2006) Soil respiration fluxes in relation to photosynthetic activity in broad-leaf and needle-leaf forest stands. Agricultural and Forest Meteorology 148: 135–143.
8. Suseela V, Conant RT, Wallenstein MD, Dukes JS (2012) Effects of soil moisture on the temperature sensitivity of heterotrophic respiration vary seasonally in an old-field climate change experiment. Global Change Biology 18: 336–348.
9. Jomura M, Kominami Y, Tamai K, Miyama T, Goto Y, et al. (2007) The carbon budget of coarse woody debris in a temperate broad-leaved secondary forest in Japan. Tellus B 59: 211–222.
10. Matsumoto A, Kominami Y, Ishii H (2010) Field measurement of heterotrophic respiration of root litter using a small chamber system. Journal of Forest Research 92(5): 269–272.
11. Tewary CK, Pandey U, Singh JS (1982) Soil and litter respiration rates in different microhabitats of a mixed oak-conifer forest and their control by edaphic conditions and substrate quality. Plant and Soil 65: 233–238.
12. Kirschbaum MUF (2013) Seasonal variations in the availability of labile substrate confound the temperature dependence of organic matter decomposition. Soil Biology & Biochemistry 57: 568–576.
13. Hanson PJ, O'Neill EG, Chambers MLS, Riggs JS, Joslin JD, et al. (2003) Soil respiration and litter decomposition. In: North America Temperate Deciduous Forest Responses to Changing Precipitation Regimes (eds Hanson PJ and Wullschleger SD). Springer, New York.
14. Andersson M, Kjoller A, Struwe S (2004) Microbial enzyme activities in leaf litter, humus and mineral soil layers of European forests. Soil Biology & Biochemistry 36: 1527–1537.
15. DeForest JL, Chen J, McNulty SG (2009) Leaf litter is an important mediator of soil respiration in an oak-dominated forest. International Journal of Biometeorology 53(2): 1432–1254.
16. Wilson TB, Kochendorfer J, Meyers TP, Heuer M, Sloop K, et al. (2014) Soil respiration in a mixed temperate forest and its contribution to total ecosystem respiration. Agricultural and Forest Meteorology 192–193: 42–50.

17. Borken W, Davidsona EA, Savagea K, Gaudinskib J, Trumborec SE (2003) Drying and wetting effects on carbon dioxide release from organic horizons. Soil Science Society of American Journal 67: 1888–1896.

18. Orchard VA, Cook FJ (1983) Relationship between soil respiration and soil moisture. Soil Biology & Biochemistry 15(4): 447–453.

19. Lee X, Wu HJ, Sigler J, Oishi C, Siccama T (2004) Rapid and transient response of soil respiration to rain. Global Change Biology 10: 1017–1026.

20. Goulden ML, Miller SD, da Rocha HR, Menton MC, de Freitas HC, et al. (2004) Diel and seasonal patterns of tropical forest CO_2 exchange. Ecological Application 14: 42–54.

21. Cisneros Dozal LM, Trumbore S, Hanson PJ (2007) Effect of moisture on leaf litter decomposition and its contribution to soil respiration in a temperate forest. Journal of Geophysical Research 112: 148–227.

22. Ataka M, Kominami, Miyama T, Yoshimura K, Jomura M, et al. (2014) Using capacitance sensors for the continuous measurement of the water content in the litter layer of forest soil. Applied and Environmental Soil Science 2014:

23. Tamai K, Hattori S (1994) Modeling of evaporation from forest floor in a deciduous broad-leaved forest and its application to basin. Journal of the Japanese Forestry Society 76: 233–241 (in Japanese with English summary).

24. Kaneko S, Akieda N, Naito F, Tamai K, Hirano Y (2007) Nitrogen budget of a rehabilitated forest on a degraded granitic hill. Journal of Forest Research 12: 38–44.

25. Goto Y, kominami Y, Miyama T, Tamai K, Kanazawa Y (2003) Aboveground biomass and net primary production of a broad-leaved secondary forest in the southern part of Kyoto prefecture, central Japan. Bulletin of FFPRI 387: 115–147 (in Japanese with English summary).

26. Subke JA, Reichstein M, Tenhunen JD (2003) Explaining temporal variation in soil CO_2 efflux in a mature spruce forest in Southern Germany. Soil Biology & Biochemistry 35: 1467–1483.

27. Dannoura M, Kominami Y, Tamai K, Jomura M, Miyama T, et al. (2006) Development of an automatic chamber system for long-term measurements of CO_2 flux from roots. Tellus 58B: 502–512.

28. Ataka M, Kominami Y, Jomura M, Yoshimura K, Uematsu C (2014) CO_2 efflux from leaf litter focused on spatial and temporal heterogeneity of moisture. Journal of Forest Research 19: 295–300.

29. Carbone MS, Still CJ, Ambrose AR, Dawson TE, Williams AP, et al. (2011) Seasonal and episodic moisture controls on plant and microbial contributions to soil respiration. Oecologia 167: 265–278.

30. Borken W, Matzner E (2009) Reappraisal of drying and wetting effects on C and N mineralization and fluxes in soils. Global Change Biology 15: 808–824.

31. Birch HF (1958) The effect of soil drying on humus decomposition and nitrogen availability. Plant and Soil 10: 9–31.

32. Ngao J (2005) Estimating the contribution of leaf litter decomposition to soil CO_2 efflux in a beech forest using 13C-depleted litter. Global Change Biology 11(10): 1768–1776.

33. Loescher HW, Law BE, Mahrt L, Hollinger DY, Campbell J, et al. (2006) Uncertainties in, and interpretation of, carbon flux estimates using the eddy covariance technique. Journal of Geophysical Research 111: D21S90.

34. Jomura M, Kominami Y, Kanazawa Y (2005) Long-term measurements of the CO_2 flux from coarse woody debris using an automated chamber system. Journal of the Japanese Forest Society 87(2): 138–144 (in Japanese with English summary).

35. Schnurer J, Clarholm M, Bostrom S, Rosswall T (1986) Effects of moisture on soil microorganisms and nematodes: A field experiment. Microbial Ecology 12(2): 217–230.

36. Fierer N, Schimel JP, Holden PA. (2003) Influence of drying-rewetting frequency on soil bacterial community structure. Microbial Ecology 45(1): 63–71.

16

Reciprocal Effects of Litter from Exotic and Congeneric Native Plant Species via Soil Nutrients

Annelein Meisner[1]*, Wietse de Boer[2], Johannes H. C. Cornelissen[3], Wim H. van der Putten[1,4]

1 Department of Terrestrial Ecology, Netherlands Institute of Ecology (NIOO-KNAW), Wageningen, The Netherlands, **2** Department of Microbial Ecology, Netherlands Institute of Ecology (NIOO-KNAW), Wageningen, The Netherlands, **3** Systems Ecology, Department of Ecological Science, Faculty of Earth and Life Sciences, Vrije Universiteit (VU) Amsterdam, Amsterdam, The Netherlands, **4** Laboratory of Nematology, Wageningen University, Wageningen, The Netherlands

Abstract

Invasive exotic plant species are often expected to benefit exclusively from legacy effects of their litter inputs on soil processes and nutrient availability. However, there are relatively few experimental tests determining how litter of exotic plants affects their own growth conditions compared to congeneric native plant species. Here, we test how the legacy of litter from three exotic plant species affects their own performance in comparison to their congeneric natives that co-occur in the invaded habitat. We also analyzed litter effects on soil processes. In all three comparisons, soil with litter from exotic plant species had the highest respiration rates. In two out of the three exotic-native species comparisons, soil with litter from exotic plant species had higher inorganic nitrogen concentrations than their native congener, which was likely due to higher initial litter quality of the exotics. When litter from an exotic plant species had a positive effect on itself, it also had a positive effect on its native congener. We conclude that exotic plant species develop a legacy effect in soil from the invaded range through their litter inputs. This litter legacy effect results in altered soil processes that can promote both the exotic plant species and their native congener.

Editor: Justin Wright, Duke University, United States of America

Funding: This study was funded by the Dutch Research Council ALW-Vici project (number 865.05.002) to WHV. The funders had no role in study design, data collection and analysis, decision to publish, or preparation of the manuscript.

Competing Interests: The authors have declared that no competing interests exist.

* E-mail: A.Meisner@nioo.knaw.nl; AnneleinMeisner@gmail.com

Introduction

Plant species can be introduced into new ecosystems by humans via transport, tourism, trade [1,2] or changes in climate [3,4,5]. Some of these introductions result in biological invasions, which can have profound effects on the invaded habitats and the biodiversity therein [6,7]. One of the strongest impacts of exotic plant species on ecosystem processes operates via altered quality of litter inputs, which can alter the cycling of nutrients [8,9,10]. These altered soil processes have been hypothesized to provide a positive feedback to the exotic plant species through changes in litter inputs [9,11,12,13], but there are very few experimental tests showing that exotic plants indeed influence the legacy of the soil to their own benefit [10]. Here, we present results of an experimental study on litter effects of exotic and congeneric plant species, which are native in the invaded habitat, on soil processes and individual performance of exotic and native congener.

Differences in initial litter chemistry between exotic and native plant species are important for soil processes involved in litter decomposition [14,15] and are mediated indirectly by the soil decomposer subsystem [16,17,18]. For example, a higher lignin content can slow down the phased processes of litter breakdown [19], because this recalcitrant component needs specialist lignolytic fungi for degradation and can shield the more easily available components (e.g. cellulose) from decomposers during the earliest phases of litter breakdown [20,21]. Therefore, litter inputs of exotic plant species that differ in litter quality from native

species have been shown to increase or decrease soil processes [22,23,24], which may remain in the soil as a legacy.

These litter legacies can affect the performance of exotic or native plant species [25,26]. When litter deposition increases the soil nutrient status, this may create a positive legacy effect to the subsequent plant species, either native or exotic (Fig. 3.11c in [27]). For example, litter addition from an exotic grass has been observed to increase biomass of the exotic grass itself and of a native shrub [28]. In contrast, litter can create a negative legacy effect when litter releases compounds into the soil during litter decomposition that inhibit plant growth [29,30]. A variety of long-term soil legacy effects of exotic plant species has been reported, including positive as well as negative legacy effects to native plant species [31,32].

Altered cycling of nutrients by exotic plant species is often hypothesized to promote exotic plant species exclusively (e.g. [33,34,35]). A relatively large number of studies have analyzed exotic litter effects in a context of plant community interactions. However, less is known about individual effects of exotic plant litter on exotic and native plant species [10]. Here, we study if the legacy of litter from exotics and congeneric natives reciprocally affect their performance when grown in monocultures via changes in soil processes. When litter of exotic plant species is of higher quality than of native plant species, this may increase soil nutrient mineralization [33,36] and nutrient availability [37,38]. Recently established exotic plant species in the Netherlands may have higher litter quality than congeneric native species [39].

Therefore, we test the hypothesis that litter from these exotic plant species provides a positive feedback to itself and inhibits natives through soil legacy effects. In order to avoid confounding effects due to major differences in plant chemistry and other traits that might differ between species [40], we compared exotic plant species with congeneric natives that co-occur in the invaded habitat.

Our hypothesis was tested by three experiments. In the first two experiments, we tested how soil mixed with litter from exotic plant species influenced soil respiration, soil mineralization and soil availability of nitrogen compared to soil mixed with litter from native plants species. In the third experiment, we tested how decomposing litter from exotic and native plant species affected germination rates and plant biomass of both exotic and native plant species. We performed the experiments with three genera of exotic and congeneric native plants that all co-occur in the same invaded habitat (Table 1).

Results

Experiment 1: Soil respiration

Exotic litter-inoculated soils showed (or in the case of *Rorippa* tended to show) a larger increase in cumulative respiration over time (Figure 1) as indicated by the Time by Origin interactions (Table 2).

Experiment 2: litter effects on soil N, enzyme activities and fungal biomass

Soil with litter from exotic *Artemisia* and *Senecio* accumulated more inorganic N than soil with litter from their congeneric native species (Figure 2A and 2C), as indicated by the origin by time interaction (Table 3). There was also an origin by time interaction for *Rorippa* (Table 2), because soil with litter from exotic *R. austriaca* had lower N concentration than soil with litter from native *R. sylvestris* only after 2 weeks of incubation (Figure 2B). These differences in inorganic N accumulation between soils with litter from exotic and native plant species corresponds with the initial litter N concentrations (Table 1). Soil with litter from exotic plant species had less fungal biomass than soil with litter from native plant species in the case of *Rorippa* and *Senecio*, but not in the case of *Artemisia* (Table 3, Figure 2D, E, F). The highest activity of cellulase was observed after 9 weeks of incubation (Figure 2G, H, I, Table 3). Significant differences at peak activity were observed in the case of *Artemisia* (Table 3), where litter from exotic *A. biennis* induced the highest cellulase activity (Figure 2G). Mn-peroxidase

activity in soil with litter was relatively low and did not show significant differences between soil with litter from exotics and natives (Table 3, see Figure S1A, B, C). Soil pH showed some significant, but minor differences (Table 3, see Figure S1D, E, F).

Experiment 3: Litter effects on seedling germination and plant biomass

Seed germination and root sprouting of natives were not inhibited by litter from their congeneric exotic. In contrary, we observed a positive trend that litter from the exotic *R. austriaca* increased the rate of sprouting of both *R. sylvestris* and *R. austriaca* (Table 4, Figure 3). The rates of germination (and sprouting) of exotic plant species were lower than of natives for *Artemisia* and *Rorippa*, whereas the reverse was observed for *Senecio* (Figure 3A, B, C, Table 4).

Litter from exotics did not reduce biomass production of congeneric natives (Figure 3). Instead, *A. biennis* and *A. vulgaris* produced more biomass in soil with litter from the exotic *A. biennis* than from the native *A. vulgaris* (Table 4, Figure 3D). There was a similar trend for *Senecio* (Table 4, Figure 3F). *Rorippa austriaca* produced more biomass than *R. sylvestris*, whereas biomass was not different between exotic and native species in the case of *Artemisia* and *Senecio* (Table 4, Figure 3).

Discussion

Our results reject the hypothesis that litter from exotic plant species inhibits native plant species while promoting themselves. Instead, we observed that if litter from an exotic plant species increased its own biomass production or germination rate, this litter also promoted biomass and germination of its native congener. Moreover, negative litter effects by litter from exotic plant species were not observed in our study. Our comparison was made within plant genera, but our results are in agreement with two other studies on litter effects of exotic species on natives. *Senecio jacobaea*, an exotic species introduced in New Zealand, increased biomass production of native plant species from New Zealand [41]. In addition, litter of an exotic grass in the USA favored not only its own biomass production, but also biomass production of a native shrub [28]. These studies and our results suggest that not only exotic plant species exclusively, but also native plant species may benefit from the litter of exotic plant species.

The positive effect of litter from exotic plant species may have been due to differences in initial litter quality, because litter from exotics contained less lignin and lower lignin: N ratios than litter of

Table 1. Plant species used in experiments.

Plant name[1]	Plant origin[2]	Time of introduction[2]	Litter chemistry		
			% C	% N	Lignin (mg C/g litter)
Artemisia biennis	North-Asia	1950–1975	44	2.5	121
Artemisia vulgaris	Native[3]		46	1.7	205
Rorippa austriaca	East Europe	1900–1925	35	1.3	43
Rorippa sylvestris	Native[3]		39	2.2	84
Senecio inaequidens	South-Africa	1925–1950	46	2.3	113
Senecio jacobaea[4]	Native[3]		44	1.8	130

[1]Nomenclature according to Van der Meijden [80].
[2][69].
[3]Native to the Netherlands.
[4]recently *Senecio jacobaea* has been renamed as *Jacobaea vulgaris* [81].

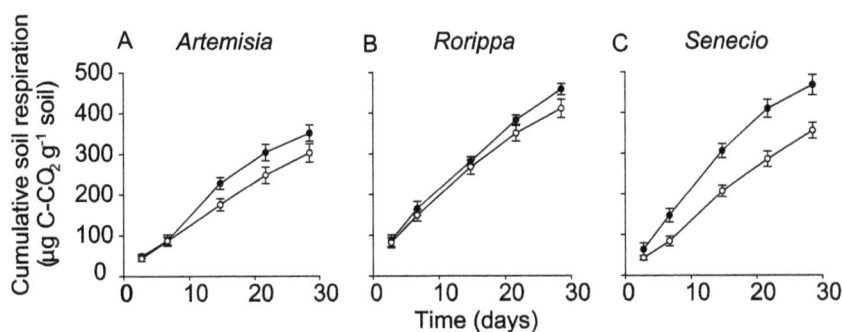

Figure 1. Mean cumulative soil respiration. ($\pm$ SE). Measured in flasks with litter from exotic (filled circles) and native plant species (open circles) for *Artemisia* (a), *Rorippa* (b), and *Senecio* (c).

the congeneric natives (Table 1). The higher litter quality of exotic species may have increased microbial activity as shown by higher cumulative respiration rates, because the degradable carbon pool in litter from exotics was likely better accessible to decomposers than in litter from natives [42]. Based on cellulase-activities it seems that cellulose was only more available in litter from the exotic *A. biennis*. Soil available N concentrations reflected initial litter N concentrations, which were highest in litter from exotic *Artemisia* and *Senecio* species. In the case of *Rorippa*, there was no such an effect. The increased cumulative respiration rates and mineral N concentration in soil incubated with litter from exotic plant species could be the result of degradation of litter itself as well as from stimulation of degradation of soil organic matter (priming) [43]. This priming-induced increase of soil organic matter mineralization has also been proposed to be an important consequence of exotic grass invasion into hardwood forest [44]. Fungal biomass was more often lower in soil with litter from exotics than litter from natives, which is likely due to the lower initial lignin concentration of exotics [21,45]. Therefore, litter from exotic species may change the soil food-web to a more bacterial dominated one if this litter is of higher quality than litter from native plant species [46,47].

Other studies showed that differences in litter decomposition rates between exotic and native plant species strongly depend on initial litter quality (e.g. [23,33], but see [48]). Our results indicate that these differences in litter decomposition rates between exotic and native plant species can result in altered soil processes and

nutrient availability. Moreover, differences in initial litter quality between native and exotic plant species may explain the site-dependent differences in nutrient concentrations, litter decomposition and carbon mineralization between invaded and uninvaded sites in Europe [49,50,51].

The native plant species used in our study are also invasive in other parts of the world. It has been proposed that comparisons between exotic plant species and native plant species that are invasive elsewhere, may be complicated, as the natives have traits that can promote their invasiveness [52]. In that case, a congeneric comparison of exotics and natives should not result in differences, whereas our study showed that litter from exotics clearly promoted soil respiration and nitrogen availability compared with litter from natives. Species that are introduced into other regions often pass through environmental filters, which can result in rapid evolution of these plant species [53,54]. As a result, invasive and native populations of the same species do not necessarily have the same traits [55,56]. Our congeneric comparisons made it less likely that differences in litter effect may be due to secondary defense compounds exclusively produced by exotic plants [57]. Nevertheless, in cases of differences in secondary defense compounds, or when slow growing native plant species with poor litter quality are being replaced by fast growing exotics with high litter quality [58], it is possible that exotic species benefit disproportionally from their own litter.

Litter legacy effects are important for the dominance of individual plant species in plant communities in the next growing season [25,26]. Litter legacies that increase soil nutrient concentrations may increase the dominance of exotic plant species when they take more advantage of these nutrients than the competing natives. Therefore, interactions with other mechanisms that increase the performance of exotics more than natives should be considered when explaining exotic plant dominance in ecosystems [59,60]. For example, a modeling study showed that an exotic invasive wetland plant has likely evolved a mechanism to produce litter of lower quality that decomposes slower, which reduces the dominance of the native plant species due to competition for light [61]. Another mechanism that could interact with a positive litter legacy effect on soil processes is the release from belowground enemies when an exotic plant species invades a new range (e.g. [62,63,64]). Indeed, two exotics in our study have been shown to experience a less negative effect from their rhizosphere biota [65]. In that case, litter of exotic plants may cause a legacy effect favoring the exotic over natives when they are released from soil-borne enemies. Therefore, future experiments may be needed to untangle these interacting mechanisms, for example by growing exotic and native species in competition.

Table 2. Repeated-measure ANOVA for soil respiration.

Factors	Plant genera								
	Artemisia			**Rorippa**			**Senecio**		
	d.f.	**F**	**P**	**d.f.**	**F**	**P**	**d.f.**	**F**	**P**
Between subject									
Origin (O)	1	2.77	0.13	1	0.96	0.36	1	13.9	0.004
Error	10			8			10		
Within subject									
Time (T)	1.4	361	<0.001	1.2	1141	<0.001	1.6	635	<0.001
T×O	1.4	5.47	0.027	1.2	4.50	0.054	1.6	13.9	<0.001
Error	14			9.9			16		

Litter from exotic versus native plant species (named Origin) of three genera (*Artemisia*, *Rorippa* and *Senecio*) were compared.

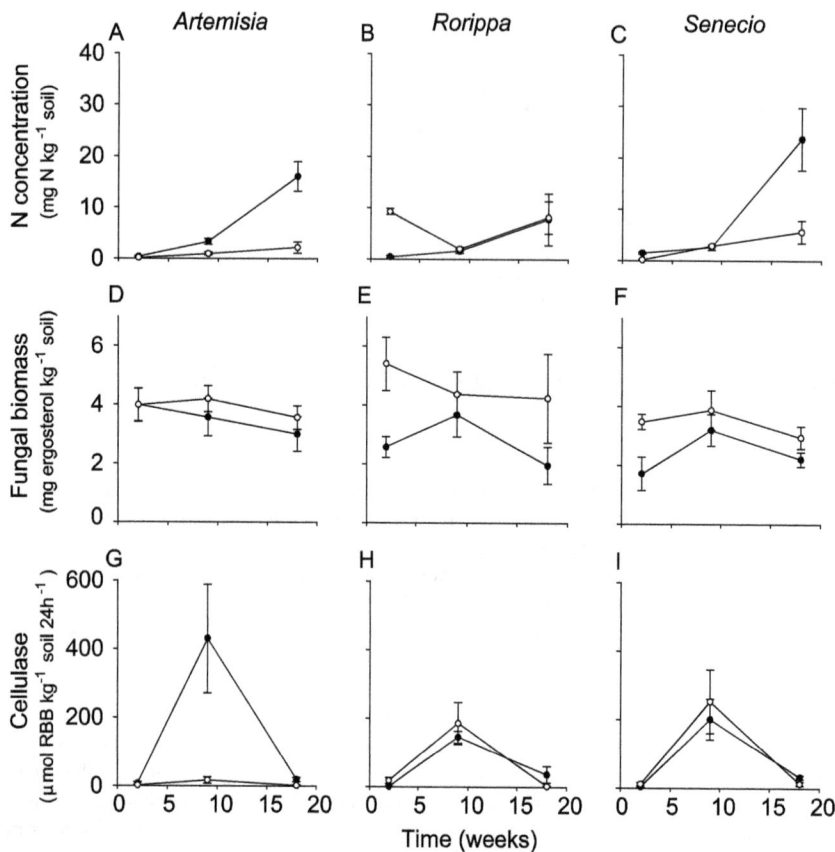

Figure 2. Effects of litter on nitrogen, fungal biomass and cellulase activity. Soil available inorganic nitrogen (N) (A, B, C), fungal biomass (D, E, F) and cellulase activity (G, H, I) in soil mixed with litter from exotic plant species (filled circles) and litter from native plant species (open circles). Means (± SE) are presented for *Artemisia* (A, D, G), *Rorippa* (B, E, H) and *Senecio* (C, F, I).

We conclude that monocultures of the exotic plant species and their congeneric native can benefit from increased soil nutrient availability through the legacy of exotic litter. Litter legacy effects on soil processes alone may, therefore, disproportionally benefit exotic over native plant species only in interaction with other mechanisms [66].

Materials and Methods

Ethics Statement

All necessary permits to collect soil and plant material from the Gelderse Poort region were obtained from Staatsbosbeheer regio Oost, the Netherlands.

Plant selection

We made a phylogenetically controlled comparison of exotics and congeneric natives (e.g. [23,37,67]), to ensure that differences in litter effects would not be influenced by differences in major classes of plant chemistry within a plant pair. The three plant pairs all co-occurred in the same riverine habitat and the exotic and native congeners occurred in mixed stands [68]. Therefore, species interactions through litter are realistically occurring in the field. Three exotic and their congeneric native plant species were selected using the national standard list of the Dutch flora [39,65,69]. We chose exotic plant species that are recent invaders and have increased in frequency in the second half of the 20th century in order to include exotic species with invasive potential [5]. Finally, a practical point was that sufficient amounts of litter,

and seeds or root fragments had to be available to conduct the experiment. All plants co-occurred in the Gelderse Poort region, which is where the River Rhine enters the Netherlands. Three species pairs that could be selected according to the above-mentioned criteria were: *Artemisia biennis* and *A. vulgaris*; *Rorippa austriaca* and *R. sylvestris*; *Senecio inaequidens* and *S. jacobaea* (Table 1). The three native species are all invasive in other parts of the world [70,71,72].

Collection of plant and soil material

Soil, litter, seeds and root fragments were all collected from the Gelderse Poort region. Root fragments were collected for *Rorippa*, because this genus and especially the exotics has very difficult seeds to collect [73]. Soil was collected from 5 locations in Milli-ngerwaard, a nature reserve within this region (51°52′N; 5°59′E). After sampling, soil was homogenized and sieved through a 10 mm mesh to remove coarse fragments and plant material. The homogenized soil had a pH of 7.8 and a moisture content of 14.7% (w/w) [39].

In autumn 2008, litter was collected from the Gelderse Poort region by selecting senesced leaves from standing plants [74]. Litter was collected from at least 10 individuals per plant species at multiple locations within the Gelderse Poort region. Litter was air-dried, stored in paper bags until use, chopped into 0.5×0.5 cm pieces and mixed for subsequent use in the experiment. Initial chemical composition of litter was determined on dried (at 70°C) and then ground litter (see Table 1). Total carbon (C) and nitrogen (N) were determined using a NC analyzer (Thermo flash EA

Table 3. ANOVA for effects of litter on soil properties.

Factors	Plant genera					
	Artemisia[1]		Rorippa[1]		Senecio[1]	
	F	P	F	P	F	P
Soil Inorganic N						
Origin (O)	51.7	<0.001	12.0	0.005	18.6	<0.001
Time (T)	55.0	<0.001	6.82	0.01	34.7	<0.001
OxT	13.1	<0.001	10.6	0.002	5.91	0.008
Fungal biomass						
Origin (O)	0.80	0.38	5.10	0.043	7.57	0.01
Time (T)	1.00	0.38	0.49	0.63	2.70	0.087
OxT	0.20	0.82	0.54	0.59	0.85	0.44
Cellulase activity						
Origin (O)	28.1	<0.001	0.02	0.89	0.05	0.83
Time (T)	16.7	<0.001	16.5	<0.001	30.7	<0.001
OxT	2.97	0.07	5.77	0.018	3.03	0.07
Mn-peroxidase activity						
Origin (O)	0.89	0.35	0.44	0.42	0.18	0.67
Time (T)	14.2	<0.001	0.36	0.67	6.29	0.006
OxT	0.57	0.57	0.44	0.34	0.29	0.75
pH						
Origin (O)	4.40	0.046	11.9	0.005	4.00	0.057
Time (T)	43.9	<0.001	23.7	<0.001	36.8	<0.001
OxT	1.90	0.17	0.78	0.78	4.30	0.026

Litter from exotic or native species (Origin) were compared for three plant genera (Artemisia, Rorippa and Senecio) at three destructive sampling points (Time).
[1]Numerator d.f. is 2 for time, 1 for origin and 2 for Time×Origin. Denominator d.f. is 24 for Artemisia and Senecio and 12 for Rorippa pair.

Table 4. ANOVA for effects of litter effects on plant performance.

Factors	Plant genera					
	Artemisia[1]		Rorippa[1]		Senecio[1]	
	F	P	F	P	F	P
Germination/sprouting						
Litter (L)	1.78	0.20	4.13	0.06	1.86	0.19
Plant (P)	23.7	<0.001	17.7	<0.001	13.9	0.002
LxP	0.02	0.88	0.06	0.81	0.79	0.39
Plant biomass						
Litter (L)	9.54	0.007	1.23	0.29	3.56	0.078
Plant (P)	1.04	0.32	7.47	0.016	0.03	0.87
LxP	0.02	0.89	1.52	0.24	0.86	0.37

Litter effects from exotic versus native plant species (Litter) on germination or (in the case of Rorippa) sprouting rates and plant biomass production as well as the differences between exotic and native plant species (Plant) within three genera (Artemisia, Rorippa, and Senecio).
[1]Numerator d.f. is 1 for all factors. Denominator d.f. is 16, except for Rorippa-pair where denominator d.f. is 14.

1112). Lignin content was determined according to Poorter and Villar [75]. Briefly, the litter material was subjected to polar, non-polar and acid extraction steps. The mass of the remaining residue was corrected for ash and the ash-adjusted C and N content of the residue was used to calculate lignin concentrations. This lignin fraction has been used successfully as litter quality index, but may contain small amounts of other recalcitrant C compounds besides lignin [29].

Seeds were collected in autumn 2008. Root fragments were collected for Rorippa-pair in spring 2009. Root fragments and seeds were surface-sterilised in a 0.5% sodium hypochlorite solution to kill potential root and seed pathogens. Root fragments of R. sylvestris were also rinsed with 70% ethanol, because a pilot showed higher root sprouting.

Experiment 1: litter effects on soil respiration

In order to determine the effects of litter on soil respiration, each litter was mixed with field soil and placed in flasks. Per plant species, six flasks of 315 ml were used (four flasks for R. austriaca due to limited amount of available litter). Each flask received an amount of field-moist soil equivalent to 40 gram dry weight and on top of this soil a 29.6 gram mixture of soil and litter (71.6:1) was placed, representing an average yearly amount of litter per unit of soil in temperate systems [76]. Six flasks without litter in the top layer were included as control. Soil was kept at 50% water holding capacity (WHC), which equals 17.7% w/w. Flasks were closed

with a rubber septum, placed in randomized order in an incubation chamber and incubated at 10°C, which is the yearly average temperature of the Netherlands (www.knmi.nl). At days 3, 7, 15, 22 and 29, gas samples were collected from the headspace using a gastight syringe and stored in an Exetainer® vial until analysis. After each sampling, flasks were opened to allow ventilation for an hour to prevent high CO_2 levels in the flasks and to adjust the moisture if needed by adding demineralized water. CO_2-concentrations were measured against a reference line on a Thermo FOCUS GC equipped with a RT-QPLOT column from Restek (30 m long and 0.53 mm diameter). The average CO_2 concentration in control pots was subtracted from the CO_2 concentration in the pots that contained litter. Cumulative CO_2 production was calculated for each litter type.

Experiment 2: litter effects on soil N, enzyme activities and fungal biomass

In order to determine how litter influenced soil N availability, enzyme activities and fungal biomass, litter of each plant species was mixed with field soil and placed in cubic microcosms of 0.5 L with a surface area of 81 cm^2. There were 15 replicates for each litter (8 replicates for R. austriaca and 10 for R. sylvestris due to limited availability of litter). Each microcosm received an amount of field-moist soil equivalent to 450 gram dry soil and on top of this soil 83 gram of the same litter-soil mixture as used in experiment 1 was added. The microcosms were incubated in a climate room at 10°C, 83% humidity and soil was kept at 50% WHC (= 17.7% w/ w). Five random microcosms were harvested after 2, 9 and 18 weeks of incubation, after which the top layer of soil was analyzed.

Available mineral N was extracted by shaking moist soil (equivalent to 10 g dry weight) in 50 ml 1 M KCl for 2 h. N-NH_4^+ and N-NO_3^- concentrations were measured on a Technicon TrAAcs 800 auto-analyzer. pH_{water} was measured in a 1: 2.5 soil to water ratio. Ergosterol, a specific fungal biomarker in the cell wall, was used to measure fungal biomass. This biomarker is not present in arbuscular mycorrhizal fungi (AMF) [77]. Ergosterol was extracted from soil using an alkaline-extraction method and measured on a Dionex HPLC equipped

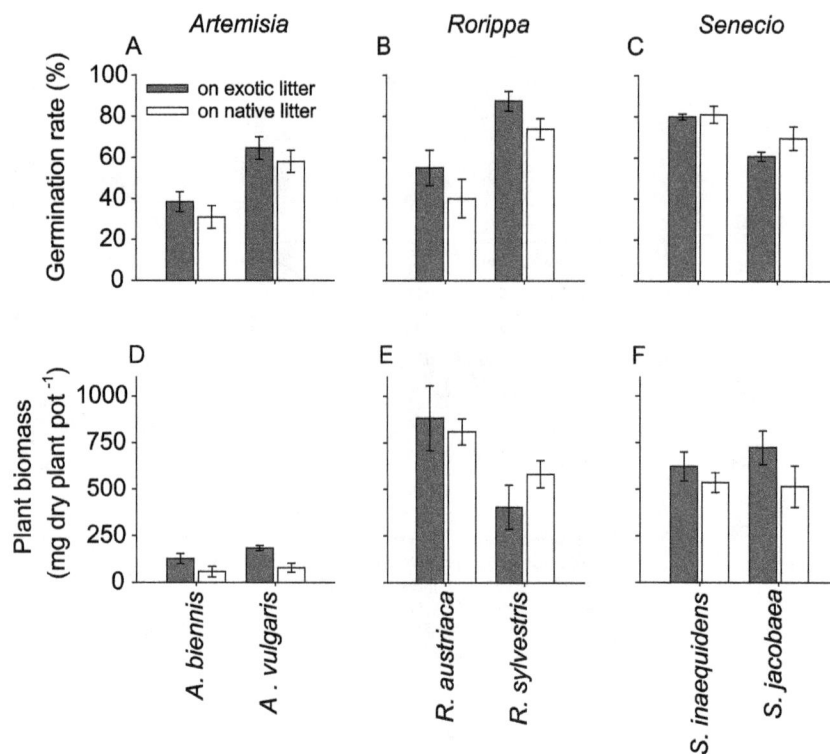

Figure 3. Effects of litter on germination rates and plant biomass production. Mean (± SE) for germination or (in the case of *Rorippa*) sprouting rate (A, B, C) and plant biomass (D, E, F) production of exotic and native plant species in litter from exotic (grey bars) or native plant species (white bars) belonging to three genera. Exotic plant species are: *A. biennis*, *R. austriaca* and *S. inaequidens*. Native plant species are: *A. vulgaris*, *R. sylvestris* and *S. jacobaea*. Significances of litter effects and plant effects are given in Table 4.

with a C 18 reverse-phase column and a UV-detector set at 282 nm [78]. Lignin degrading enzyme activity (Mn-peroxidase) and cellulose degrading enzyme activity (endo-1,4-β-glucanase) were measured according to Van der Wal et al. [79], modified by extracting 6 gram of soil with 9 ml of milli-q water. Endo-1,4-β-glucanase is an indicator of cellulase activity and is therefore called cellulase in the main text.

Experiment 3: litter effects on seedling germination and plant biomass production

In order to determine how litter influenced seedling germination and plant biomass production, seeds of exotic and native plant species were placed on soil that had been incubated with their own litter, as well as on soil that had been incubated with the litter of the congener. We created a series of 10 microcosms (8 for *R. austriaca*) per litter origin, which were pre-incubated for 18 weeks as in experiment 2 in order to mimic litter decomposition in winter prior to plant growth in spring. For *Artemisia* and *Senecio*, 50 seeds of exotic or native plant species were placed on half of the microcosm within the genera to create five microcosms per litter origin for each plant origin within genera. For *Rorippa*, 10 root fragments of exotic or native species were placed in the soil of half of the microcosm. Germination or sprouting rates were registered after 17 days for *Senecio*, after 22 days for *Rorippa*, and after 36 days for *Artemisia*, because the time of germination or sprouting differed between genera. After germination, seedlings or cuttings were thinned so that one seedling with median length was left. Microcosms were harvested after 9.5 weeks of incubation. All harvested plants were dried to constant weight at 70°C and weighed. Microcosms were placed in a climate chamber at 19°C/

10°C and 83% humidity (average May–September growing conditions for plant species in the Netherlands, www.knmi.nl) with daylight for 16 h per 24 h.

Data analysis

The results were analyzed with Statistica version 9.0 (StatSoft, Inc. (2009), Tulsa, USA) by considering the three genera separately. Repeated measures ANOVAs were performed per genus-pair for soil respiration with origin (litter from exotic or native plant species) as the between-subject factor. As the sphericity assumption was violated for all genus-pairs, Greenhouse-Geisser adjusted P values and degrees of freedom were calculated (Table 2). An ANOVA was performed for the effects of litter on soil per genus-pair with origin (litter from exotic or native plant species) and time (2, 9 and 18 weeks of incubation) as fixed factors. Cellulase was log-transformed to meet assumptions of ANOVA. Inorganic N concentration was log-transformed for the genera *Artemisia* and *Rorippa* and fourth-root transformed for *Senecio* to meet assumptions of ANOVA. Effects of litter origin on germination rates and plant biomass production were analyzed per genus-pair by ANOVA with litter (litter from exotic or native plant species) and plant (exotic or native plant species) as fixed factors. Germination rates were arcsine transformed and biomass was log transformed to meet assumptions of ANOVA.

Supporting Information

Figure S1   Effects of litter on Mn-peroxidase activity and pH. Mn-peroxidase activity (A, B, C) and pH (D, E, F) in soil incubated with litter from exotic plant species (filled circles) or with litter from native plant species (open circles). Means (± SE) are presented for *Artemisia* (A, D), *Rorippa* (B, E) and *Senecio* (C, F).

Acknowledgments

We thank Staatsbosbeheer regio Oost for giving us permission to collect plant and soil material in the Gelderse Poort region. We thank Ciska Raaijmakers, Wiecher Smant, Gera Hol, Henk Duyts, Paulien Klein Gunnewiek and Richard van Logtestijn for their help and advice during the experiment; Heike Schmitt for advice on respiration measurements; Daan Blok for help with setting up the experiment and for helpful comments on the manuscript; Fernando Monroy Martinez, Remy Hillekens, Mirka Macel and Tim Engelkes for discussion about the experiment; Koen Verhoeven, Martijn Bezemer and Arjen Biere for discussions about statistics; two anonymous referees for comments on previous versions of this manuscript. This is NIOO publication 5204.

Author Contributions

Conceived and designed the experiments: AM WD WHV JHCC. Performed the experiments: AM. Analyzed the data: AM. Contributed reagents/materials/analysis tools: WHV JHCC. Wrote the paper: AM WD JHCC WHV.

References

1. Hodkinson DJ, Thompson K (1997) Plant dispersal: the role of man. J Appl Ecol 34: 1484–1496.
2. Mack RN, Simberloff D, Lonsdale WM, Evans H, Clout M, et al. (2000) Biotic invasions: Causes, epidemiology, global consequences, and control. Ecol App 10: 689–710.
3. Parmesan C, Yohe G (2003) A globally coherent fingerprint of climate change impacts across natural systems. Nature 421: 37–42.
4. Walther GR, Roques A, Hulme PE, Sykes MT, Pysek P, et al. (2009) Alien species in a warmer world: risks and opportunities. Trends Ecol Evol 24: 686–693.
5. Tamis WLM, Van't Zelfde M, Van der Meijden R, De Haes HAU (2005) Changes in vascular plant biodiversity in the Netherlands in the 20th century explained by their climatic and other environmental characteristics. Climatic Change 72: 37–56.
6. Chapin FSI, Zavaleta ES, Eviner VT, Naylor RL, Vitousek PM, et al. (2000) Consequences of changing biodiversity. Nature 405: 234–242.
7. Vitousek PM, Dantonio CM, Loope LL, Rejmanek M, Westbrooks R (1997) Introduced species: A significant component of human-caused global change. N Z J Ecol 21: 1–16.
8. Ehrenfeld JG (2010) Ecosystem consequences of biological invasions. Annu Rev Ecol Evol Syst 41: 59–80.
9. Liao C, Peng R, Luo Y, Zhou X, Wu X, et al. (2008) Altered ecosystem carbon and nitrogen cycles by plant invasion: a meta-analysis. New Phyt 177: 706–714.
10. Levine JM, Vila M, D'Antonio CM, Dukes JS, Grigulis K, et al. (2003) Mechanisms underlying the impacts of exotic plant invasions. P Roy Soc Lond B Bio 270: 775–781.
11. Farrer EC, Goldberg DE (2009) Litter drives ecosystem and plant community changes in cattail invasion. Ecol App 19: 398–412.
12. Raizada P, Raghubanshi AS, Singh JS (2008) Impact of invasive alien plant species on soil processes: A review. Proc Nat Acad Sci India Sect B 78: 288–298.
13. Ehrenfeld JG (2004) Implications of invasive species for belowground community and nutrient. Weed Technology 18: 1232–1235.
14. Wardle DA, Barker GM, Bonner KI, Nicholson KS (1998) Can comparative approaches based on plant ecophysiological traits predict the nature of biotic interactions and individual plant species effects in ecosystems? J Ecol 86: 405–420.
15. Meier CL, Bowman WD (2008) Links between plant litter chemistry, species diversity, and below-ground ecosystem function. Proc Natl Acad Sci U S A 105: 19780–19785.
16. Wardle DA, Bardgett RD, Klironomos JN, Setälä H, Van der Putten WH, et al. (2004) Ecological linkages between aboveground and belowground biota. Science 304: 1629–1633.
17. Aerts R, Chapin FSI (2000) The mineral nutrition of wild plants revisited: A re-evaluation of processes and patterns. Adv Ecol Res 30: 1–67.
18. Hobbie SE (1992) Effects of plant-species on nutrient cycling. Trends Ecol Evol 7: 336–339.
19. Cornwell WK, Cornelissen JHC, Amatangelo K, Dorrepaal E, Eviner VT, et al. (2008) Plant species traits are the predominant control on litter decomposition rates within biomes worldwide. Ecol Lett 11: 1065–1071.
20. De Boer W, Folman LB, Summerbell RC, Boddy L (2005) Living in a fungal world: impact of fungi on soil bacterial niche development. FEMS Microbiol Rev 29: 795–811.
21. Osono T (2007) Ecology of ligninolytic fungi associated with leaf litter decomposition. Ecol Res 22: 955–974.
22. Rothstein DE, Vitousek PM, Simmons BL (2004) An exotic tree alters decomposition and nutrient cycling in a Hawaiian montane forest. Ecosystems 7: 805–814.
23. Godoy O, Castro-Diez P, Van Logtestijn RSP, Cornelissen JHC, Valladares F (2010) Leaf litter traits of invasive species slow down decomposition compared to Spanish natives: a broad phylogenetic comparison. Oecologia 162: 781–790.
24. Drenovsky RE, Batten KM (2007) Invasion by *aegilops triuncialis* (barb goatgrass) slows carbon and nutrient cycling in a serpentine grassland. Biol Invasions 9: 107–116.
25. Facelli JM, Facelli E (1993) Interactions after death - plant litter controls priority effects in a successional plant community. Oecologia 95: 277–282.
26. Berendse F (1994) Litter decomposability - a neglected component of plant fitness. J Ecol 82: 187–190.
27. Bardgett RD, Wardle DA (2010) Aboveground- belowground linkages: biotic interactions, ecosystem processes and global change. New York, USA: Oxford University Press.
28. Wolkovich EM, Bolger DT, Cottingham KL (2009) Invasive grass litter facilitates native shrubs through abiotic effects. J Veg Sci 20: 1121–1132.
29. Dorrepaal E, Cornelissen JHC, Aerts R (2007) Changing leaf litter feedbacks on plant production across contrasting sub-arctic peatland species and growth forms. Oecologia 151: 251–261.
30. Callaway RM, Ridenour WM (2004) Novel weapons: invasive success and the evolution of increased competitive ability. Frontiers in Ecology and the Environment 2: 436–443.
31. Scharfy D, Funk A, Olde Venterink H, Güsewell S (2011) Invasive forbs differ functionally from native graminoids, but are similar to native forbs. New Phyt 189: 818–828.
32. Yelenik SG, Levine JM (2011) The role of plant–soil feedbacks in driving native-species recovery. Ecology 92: 66–74.
33. Allison SD, Vitousek PM (2004) Rapid nutrient cycling in leaf litter from invasive plants in Hawai'i. Oecologia 141: 612–619.
34. Hawkes CV, Wren IF, Herman DJ, Firestone MK (2005) Plant invasion alters nitrogen cycling by modifying the soil nitrifying community. Ecol Lett 8: 976–985.
35. Sperry LJ, Belnap J, Evans RD (2006) *Bromus tectorum* invasion alters nitrogen dynamics in an undisturbed arid grassland ecosystem. Ecology 87: 603–615.
36. Petsikos C, Dalias P, Troumbis AY (2007) Effects of *Oxalis pes-caprae* L. invasion in olive groves. Agricult Ecosys Environ 120: 325–329.
37. Ashton IW, Hyatt LA, Howe KM, Gurevitch J, Lerdau MT (2005) Invasive species accelerate decomposition and litter nitrogen loss in a mixed deciduous forest. Ecol App 15: 1263–1272.
38. Ehrenfeld JG (2003) Effects of exotic plant invasions on soil nutrient cycling processes. Ecosystems 6: 503–523.
39. Meisner A, de Boer W, Verhoeven KJF, Boschker HTS, van der Putten WH (2011) Comparison of nutrient acquisition in exotic plant species and congeneric natives. J Ecol 99: 1308–1315.
40. Pyšek P, Richardson DM (2007) Traits Associated with Invasiveness in Alien Plants: Where Do we Stand? In: Nentwig W, ed. Biological Invasions: Springer Berlin Heidelberg. pp 97–125.
41. Wardle DA, Nicholson KS, Rahman A (1995) Ecological effects of the invasive weed species Senecio jacobaea L. (ragwort) in a New Zealand pasture. Agricult Ecosys Environ 56: 19–28.
42. Berg B, McClaugherty C (2008) Plant litter: decomposition, humus formation, carbon sequestration. Berlin-Heidelberg: Springer.
43. Kuzyakov Y, Friedel JK, Stahr K (2000) Review of mechanisms and quantification of priming effects. Soil Biol Biochem 32: 1485–1498.
44. Strickland MS, Devore JL, Maerz JC, Bradford MA (2010) Grass invasion of a hardwood forest is associated with declines in belowground carbon pools. Global Change Biol 16: 1338–1350.
45. Cadisch G, Giller KE (1997) Driven by nature: plant litter quality and decomposition. Wallingford, UK: CAB International.
46. Bardgett RD (2005) The biology of soil: A community and ecosystem approach. Oxford: Oxford University Press.
47. Coleman DC, Reid CPP, Cole CV (1983) Biological strategies of nutrient cycling in soil systems. Adv Ecol Res 13: 1–55.
48. Kurokawa H, Peltzer DA, Wardle DA (2010) Plant traits, leaf palatability and litter decomposability for co-occurring woody species differing in invasion status and nitrogen fixation ability. Funct Ecol 24: 513–523.
49. Koutika LS, Vanderhoeven S, Chapuis-Lardy L, Dassonville N, Meerts P (2007) Assessment of changes in soil organic matter after invasion by exotic plant species. Biol Fertil Soils 44: 331–341.
50. Dassonville N, Vanderhoeven S, Vanparys V, Hayez M, Gruber W, et al. (2008) Impacts of alien invasive plants on soil nutrients are correlated with initial site conditions in NW Europe. Oecologia 157: 131–140.
51. Vila M, Tessier M, Suehs CM, Brundu G, Carta L, et al. (2006) Local and regional assessments of the impacts of plant invaders on vegetation structure and soil properties of Mediterranean islands. J Biogeography 33: 853–861.

52. Van Kleunen M, Dawson W, Schlaepfer D, Jeschke JM, Fischer M (2010) Are invaders different? A conceptual framework of comparative approaches for assessing determinants of invasiveness. Ecol lett 13: 947–958.

53. Müller-Schärer H, Schaffner U, Steinger T (2004) Evolution in invasive plants: implications for biological control. Trends Ecol Evol 19: 417–422.

54. Lachmuth S, Durka W, Schurr FM (2011) Differentiation of reproductive and competitive ability in the invaded range of Senecio inaequidens: the role of genetic Allee effects, adaptive and nonadaptive evolution. New Phyt 192: 529–541.

55. Güsewell S, Jakobs G, Weber E (2006) Native and introduced populations of Solidago gigantea differ in shoot production but not in leaf traits or litter decomposition. Funct Ecol 20: 575–584.

56. Feng YL, Lei YB, Wang RF, Callaway RM, Valiente-Banuet A, et al. (2009) Evolutionary tradeoffs for nitrogen allocation to photosynthesis versus cell walls in an invasive plant. Proc Natl Acad Sci U S A 106: 1853–1856.

57. Inderjit, Evans H, Crocoll C, Bajpai D, Kaur R, et al. (2011) Volatile chemicals from leaf litter are associated with invasiveness of a Neotropical weed in Asia. Ecology 92: 316–324.

58. Walker LR, Vitousek PM (1991) An invader alters germination and growth of a native dominant tree in Hawaii. Ecology 72: 1449–1455.

59. Catford JA, Jansson R, Nilsson C (2009) Reducing redundancy in invasion ecology by integrating hypotheses into a single theoretical framework. Divers Distrib 15: 22–40.

60. Blumenthal D, Mitchell CE, Pysek P, Jarosik V (2009) Synergy between pathogen release and resource availability in plant invasion. Proc Natl Acad Sci U S A 106: 7899–7904.

61. Eppinga MB, Kaproth MA, Collins AR, Molofsky J (2011) Litter feedbacks, evolutionary change and exotic plant invasion. J Ecol 99: 503–514.

62. Van Grunsven RHA, Van der Putten WH, Bezemer TM, Berendse F, Veenendaal EM (2010) Plant- soil interactions in the expansion and native range of a poleward shifting plant species. Global Change Biol 16: 380–385.

63. Callaway RM, Thelen GC, Rodriguez A, Holben WE (2004) Soil biota and exotic plant invasion. Nature 427: 731–733.

64. Reinhart KO, Packer A, Van der Putten WH, Clay K (2003) Plant-soil biota interactions and spatial distribution of black cherry in its native and invasive ranges. Ecol lett 6: 1046–1050.

65. Engelkes T, Morriën E, Verhoeven KJF, Bezemer MT, Biere A, et al. (2008) Successful range expanding plants have less aboveground and belowground enemy impact. Nature 456: 946–948.

66. Inderjit, van der Putten WH (2010) Impacts of soil microbial communities on exotic plant invasions. Trends Ecol Evolut 25: 512–519.

67. Agrawal AA, Kotanen PM, Mitchell CE, Power AG, Godsoe W, et al. (2005) Enemy release? An experiment with congeneric plant pairs and diverse above- and belowground enemies. Ecology 86: 2979–2989.

68. Dirkse GM, Hochstenbach SMH, Reijerse AI (2007) Flora van Nijmegen en Kleef 1800–2006/Flora von Nimwegen und Kleve 1800–2006. Mook, The Netherlands: KNNV, printed at Zevendal.

69. Tamis WLM, Van der Meijden R, Runhaar J, Bekker RM, Ozinga WA, et al. (2005) Anex: standaardlijst van de Nederlandse flora 2003. Gorteria supplement 6: 135–229.

70. Barney JN (2006) North American history of two invasive plant species: phytogeographic distribution, dispersal vectors, and multiple introductions. Biol Invasions 8: 703–717.

71. Wardle DA (1987) The ecology of ragwort (Senecio-jacobaea l) - a review. N Z J Ecol 10: 67–76.

72. Stuckey RL (1966) The distribution of Rorippa sylvestris (Cruciferae) in North America. Sida 2: 361–376.

73. Dietz H, Köhler A, Ullmann I (2002) Regeneration Growth of the Invasive Clonal Forb Rorippa austriaca (Brassicaceae) in Relation to Fertilization and Interspecific Competition. Plant Ecol 158: 171–182.

74. Cornelissen JHC (1996) An experimental comparison of leaf decomposition rates in a wide range of temperate plant species and types. J Ecol 84: 573–582.

75. Poorter H, Villar R (1997) The Fate of Acquired Carbon in Plants: Chemical Composition and Construction Costs. In: Bazzaz FA, Grace J, eds. Plant Resource Allocation. San Diego: Academic Press. pp 39–72.

76. Penuelas J, Prieto P, Beier C, Cesaraccio C, de Angelis P, et al. (2007) Response of plant species richness and primary productivity in shrublands along a north-south gradient in Europe to seven years of experimental warming and drought: reductions in primary productivity in the heat and drought year of 2003. Global CHange Biol 13: 2563–2581.

77. Olsson PA, Larsson L, Bago B, Wallander H, van Aarle IM (2003) Ergosterol and fatty acids for biomass estimation of mycorrhizal fungi. New Phyt 159: 7–10.

78. De Ridder-Duine AS, Smant W, Van der Wal A, Van Veen JA, De Boer W (2006) Evaluation of a simple, non-alkaline extraction protocol to quantify soil ergosterol. Pedobiologia 50: 293–300.

79. Van der Wal A, De Boer W, Smant W, Van Veen JA (2007) Initial decay of woody fragments in soil is influenced by size, vertical position, nitrogen availability and soil origin. Plant Soil 301: 189–201.

80. Van der Meijden R (2005) Heukels' flora van Nederland. , The Netherlands: Wolters-Noordhoff bv.

81. Pelser PB, Veldkamp J-F, Van der Meijden R (2006) New combinations in Jacobaea Mill. (Asteraceae - Senecioneae). Compositae Newsletter 44: 1–11.

17

Effects of Precipitation Increase on Soil Respiration: A Three-Year Field Experiment in Subtropical Forests in China

Qi Deng[1,2], Dafeng Hui[3]*, Deqiang Zhang[1], Guoyi Zhou[1], Juxiu Liu[1], Shizhong Liu[1], Guowei Chu[1], Jiong Li[1]

1 South China Botanical Garden, Chinese Academy of Sciences, Guangzhou, China, 2 Wuhan Botanical Garden, Chinese Academy of Sciences, Wuhan, China, 3 Department of Biological Sciences, Tennessee State University, Nashville, Tennessee, United States of America

Abstract

Background: The aim of this study was to determine response patterns and mechanisms of soil respiration to precipitation increases in subtropical regions.

Methodology/Principal Findings: Field plots in three typical forests [i.e. pine forest (PF), broadleaf forest (BF), and pine and broadleaf mixed forest (MF)] in subtropical China were exposed under either Double Precipitation (DP) treatment or Ambient Precipitation (AP). Soil respiration, soil temperature, soil moisture, soil microbial biomass and fine root biomass were measured over three years. We tested whether precipitation treatments influenced the relationship of soil respiration rate (R) with soil temperature (T) and soil moisture (M) using $R = (a+cM)\exp(bT)$, where a is a parameter related to basal soil respiration; b and c are parameters related to the soil temperature and moisture sensitivities of soil respiration, respectively. We found that the DP treatment only slightly increased mean annual soil respiration in the PF (15.4%) and did not significantly change soil respiration in the MF and the BF. In the BF, the increase in soil respiration was related to the enhancements of both soil fine root biomass and microbial biomass. The DP treatment did not change model parameters, but increased soil moisture, resulting in a slight increase in soil respiration. In the MF and the BF, the DP treatment decreased soil temperature sensitivity b but increased basal soil respiration a, resulting in no significant change in soil respiration.

Conclusion/Significance: Our results indicate that precipitation increasing in subtropical regions in China may have limited effects on soil respiration.

Editor: Ben Bond-Lamberty, DOE Pacific Northwest National Laboratory, United States of America

Funding: This work was financially supported by National Basic Research Program of China (2009CB42110×), Strategic Priority Research Program-Climate Change: Carbon Budget and Relevant Issues of Chinese Academy of Sciences (XDA05050205), Dinghushan Forest Ecosystem Research Station, and the National Science Foundation (0933958). The funders had no role in study design, data collection and analysis, decision to publish, or preparation of the manuscript.

Competing Interests: The authors have declared that no competing interests exist.

* E-mail: dhui@tnstate.edu

Introduction

Soil respiration in terrestrial ecosystems plays an important role in global carbon cycling and climate change [1–4]. However, our understanding of precipitation impacts on soil respiration is still very limited, particularly in tropical and subtropical forests [5]. As greater intensity of precipitation and more severe droughts and floods are predicted in the future [6–7], such changes in precipitation may have significant influences on soil moisture and soil respiration in terrestrial ecosystems. Compared to drought, few studies have been done on the influence of heavy precipitation on soil respiration [8–14]. Considering that tropical and subtropical forests contain more than 25% of the carbon in the terrestrial biosphere, it is imperative to improve our mechanistic understanding of soil respiration responses to pre-cipitation and soil moisture changes [15,16].

Soil respiration includes both respiration of living roots and microbial respiration resulted from microbial decomposition of litter and soil organic matter [3,5,10,12]. Root activity and microbial decomposition are often subject to both environmental factors and substrate changes related to phenological processes [17–20]. Any changes in root biomass, soil organic matter, root and microbial activities due to precipitation change could influence soil respiration. Like many biological processes, soil respiration is also influenced by soil temperature and moisture in many different ecosystems [3,21–23]. While it is generally accepted that global warming could influence the relationship of soil respiration and temperature, how precipitation treatments would influence soil respiration and its relationship to soil moisture has not been well investigated. When treatments such as warming, precipitation, or CO_2 concentration changes are applied, response variables may respond directly to changes in environmental factors as well as alter their relationships with environmental factors. Thus, soil respiration responses to precipitation treatments could be caused by either changes in environmental factors such as soil temperature and moisture, or functional changes – which are defined as changes in model parameters of soil respiration with soil temperature and moisture, or both [23]. For example, functional change due to a change in soil temperature sensitivity may increase or decrease soil respiration even when soil temperature is not influenced by precipitation treatments. Functional change could be attributed to the changes in phenological process, substrate or microbial activity in an ecosystem [2,5,16,23].

Changes in soil moisture under different precipitation treatments could influence the responses of soil respiration to precipitation. There is no doubt that precipitation is usually the driving factor of the dynamics in soil moisture. However, soil water storage after precipitation events depends on vegetation types and covers, soil characteristics (e.g., infiltration rates, slopes, textures, depths, impermeable layers), and losses to deep drainage, lateral flow, and evaporation [24]. Thus, the response of soil moisture to precipitation treatments often varies in different ecosystems. For example, drought treatments using automated retractable curtains reduced soil moisture by 32–48%, 15–61%, and 19–25% at three heathlands [25], and double precipitation increased soil moisture by only 10% in Oklahoma grassland [10]. How precipitation changes influence soil moisture in subtropical forests may have significant impacts on soil respiration.

Functional changes (i.e. changes in model parameters of soil respiration with soil temperature and moisture) reflect underlying biological changes in the response of soil respiration to precipitation changes. Many empirical models of soil respiration and soil moisture have been developed [26–29]. Response of soil respiration to soil moisture is usually nonlinear, with soil respiration increases with soil moisture increases, levels off at high soil moisture, and even decreases when soil moisture is too high [3,28,29]. However, linear regression seems to work well in many different ecosystems, including boreal forests, sub-Antarctic island ecosystems, temperate grasslands, temperate forests, Mediterranean ecosystems, and particularly, tropical and subtropical forests [22,30–34]. The slope of the linear regression model can be considered as soil moisture sensitivity, as it reflects an average change in soil respiration due to one unit change of soil moisture. While many precipitation manipulation experiments have been performed [9–15,21,22], only a few studies have attempted to study the soil moisture sensitivity change under climate change, particularly precipitation [3,35–37].

Another important functional relationship is the response of soil respiration to soil temperature [2,38,39]. Soil temperature is the major control of soil respiration due to its influences on the kinetics of microbial decomposition, root respiration and diffusion of enzymes and substrates [32,40]. Numerous studies have focused on the responses of soil respiration to soil temperature. The most widely used model is an exponential equation ($R = R_0\exp(bT)$) where R is soil respiration, T is soil temperature, and parameter R_0 is basal soil respiration, and b is related to soil temperature sensitivity ($Q_{10} = \exp(10b)$) [41,42]. Many studies reported that soil temperature sensitivity may decrease under high temperature treatments [2,30,38,39] and increase under low temperature [38,40–43]. Several studies also indicated that soil water stress or excess may decrease soil temperature sensitivity of soil respiration [27,42,44]. Since soil temperature and soil moisture may interactively regulate soil respiration in field conditions, relationships of soil respiration with both soil temperature and moisture have also been proposed [20,36,45]. Whether and how soil moisture and temperature sensitivities vary with precipitation increase have not been well investigated [5,14].

We conducted a precipitation manipulation field experiment in subtropical forests in Southern China with an overall aim to understand the responses of soil respiration to precipitation increase. We selected three common forests at the study site, established two precipitation treatments in each forest, and measured soil respiration over three years. Double precipitation was realized through automatic interception-redistribution systems that delivering intercepted precipitation from nearby plots of the same size [10]. Adjacent control plots received ambient precipitation (AP). We addressed the following three questions in this study: 1) what are the response patterns of soil respiration to precipitation increase in the subtropical forests? 2) Do different forest sites respond differently to precipitation increase? 3) Does precipitation increase influence soil temperature and moisture sensitivities? The conclusions obtained in this study will enrich our knowledge of soil respiration responses to precipitation changes in subtropical forests in China and may have potentially significant implications for terrestrial ecosystem carbon cycling.

Materials and Methods

Ethics Statement

The study site is maintained by the South China Botanical Garden, Chinese Academy of Sciences. The location is within the Dinghushan Forest Ecosystem Research Station, Chinese Ecosystem Research Network (CERN). All necessary permits were obtained for the described field study. The field study did not involve endangered or protected species. Data will be made available upon request.

Site Description

The study site is located in the center of Guangdong Province in southern China (112°13′39″–112°33′41″ E, 23°09′21″–23°11′30″ N). Climate in the region is typical south subtropical monsoon climate, with mean annual temperature of 21.4°C, and mean annual precipitation of 1956 mm [33], of which nearly 80% falls in the hot-humid wet/rainy season (April-September) and 20% in the dry season (October-March). The bedrock is sandstone and shale. Three common subtropical forests (at elevations ranging from 200 to 300 m, less than 500 m from one another and facing the same slope direction) were selected including a coniferous Masson pine forest (PF), a conifer and broadleaf mixed forest (MF), and an evergreen broadleaf forest (BF). The three forests also represent forests in early-, middle-, and advanced-successional stages in the region [46,47]. Soil properties and major stand information are listed in Table 1. The PF (approximately 22 ha), originally planted by local people in the 1950 s, was dominated by *Pinus massoniana* in the tree layer and *Baeckea frutescens*, *Rhodomyrtus tomenosa*, and *Dicranopteris linearis* in the shrub and herb layers. The MF (approximately 557 ha) was developed from artificial pine forest with a gradual invasion of some pioneer broadleaf species through natural succession. The upper canopy of the community is dominated by *Schima superba*, *Castanopsis chinensis*, and *Craibiodendron scleranthum var. kwangtungense*. Artificial disturbances have not occurred in the MF for about 100 years. The BF (approximately 218 ha) located in the central area of the reserve was dominated by *Castanopsis chinensis*, *Cryptocarya concinna*, *Schima superba*, *Machilus chinensis* without any *Pinus massoniana*. No disturbance was recorded for the past 400 years in the BF [37–38].

Experimental Design

We used a two-factor experimental design considering forest ecosystem type and precipitation treatment. At each forest site, a randomized block design was used with three blocks. In each block, one double precipitation (DP) treatment plot and one

Table 1. Stand characteristics of the pine forest (PF), the mixed forest (MF) and the broadleaf forest (BF) at the Dinghushan Forest Ecosystem Research Station.

Forests	PF	MF	BF
Elevation (m)	200–300	220–300	220–300
Stand age (year)	50–60	About 110	About 400
Successional stage	Early	Middle	Advanced
Biomass (Mg C ha^{-1})[a]	61.3	82.1	145.2
Standing litter (g m^{-2})[b]	436±146	497±103	328±71
Abovegroud litter input (g m^{-2} yr^{-1})[b]	699±76	801±142	631±105
LAI[c]	4.3±0.4	6.5±0.7	7.8±0.5
SOM (0–10 cm) (g kg^{-1} soil)[c]	23.3±1.1	26.8±1.3	38.9±1.6
Bulk density (0–10 cm) (g cm^{-3})[c]	1.32±0.04	1.10±0.08	0.86±0.06
SOC (0–60 cm) (Mg C ha^{-1})[d]	105.2	111.3	164.1
Gravel (%)[e]	34.7	19.8	12.7
Sand (%)[e]	48.8	48.0	38.1
Silt (%)[e]	26.3	22.1	26.7
Clay (%)[e]	23.9	29.9	35.2
Soil pH value[f]	3.79±0.05	3.86±0.03	3.92±0.03

[a]From Liu et al. (2007) [57].
[b]Unpublished data from the Dinghushan Forest Ecosystem Research Station (2007-2009).
[c]From Zhang et al. (2006) [33]. SOM and LAI represent soil organic matter in the top 10 cm depth and leaf area index, respectively.
[d]From Fang et al. (2003) [60]. SOC represents soil organic carbon in the top 60 cm depth.
[e]From He and others (1982) [61].
[f]From Yan et al. (2009) [56].

control plot were arranged. For the DP plot, precipitation was intercepted in a nearby plot with same size as the treatment plot using transparent polyvinyl chloride (PVC) sheer roof and was redistributed to the DP plot using pipes similar to those used in [10]. The control plot that received ambient precipitation (AP) was built next to the treatment plot. Each plot was 3×3 m^2 and the distance between the DP and AP plots was more than one meter.

Soil Respiration Measurements

Five PVC soil collars (80 cm^2 in area and 5 cm in height) were permanently installed 3 cm into the soil in each plot in November 2006. The distance between adjacent collars was more than 50 cm. Soil respiration was measured three times a month from January 2007 to December 2008 and two times a month in 2009 using a Li-6400 infrared gas analyzer (Li-COR, Inc., Lincoln, Nebraska, USA) connected to a Li-6400-09 soil respiration chamber (9.55 cm diameter) (Li-COR, Inc., Lincoln, Nebraska, USA). The measurements were made between 9:00 am and 12:00 pm local time. Previous work at this study site has demonstrated that soil respiration in forests measured during this period was close to daily mean [30,48]. Soil respiration was measured three cycles for each soil collar and the CO_2 concentration change in the chamber to complete one cycle was set as 10 ppm above the set point. Soil respiration in a treatment plot was calculated as the mean of five collar measurements (the measurement at five collars in a plot mostly differed by less than 5% at any measurement period). Soil temperature at 5 cm below the soil surface was also monitored with a thermocouple sensor

attached to the respiration chamber during the soil respiration measurement. Volumetric soil moisture of the top 5 cm soil layer was measured on five random locations within a treatment plot using a PMKit [34] at the same time when the soil respiration measurements were being taken.

Soil Microbial Biomass and Fine Root Biomass Measurements

Soil samples were collected in February 2008 to determine soil microbial biomass C content, and three more times in May 2008, August 2008 and November 2008. Two samples of six cores (2.5 cm diameter) were randomly collected from each plot in the three forests. After removing roots and plant residues, the composited samples were immediately sieved through a 2-mm mesh sieve. The soil microbial biomass carbon was determined by the fumigation-extraction technique. The soil microbial biomass carbon was extracted with potassium sulfate on both fumigated and unfumigated soil [49,50]. The carbon content of the extract was tested and the biomass was calculated based on the difference between the carbon content of fumigated vs. the unfumigated soil [49,50].

To measure fine root biomass (diameter≤3 mm), we collected soil cores (0–20 cm depth) in February 2008 using a 10 cm diameter stainless-steel corer, and three more times in April 2008, August 2008 and October 2008. Each sample was randomly collected from each plot in each forest. Fine roots were separated by washing and sieving, dried at 60°C for 48 h and weighed.

Statistical Analysis

Soil respiration rate and soil temperature in a plot were calculated as the means of five collar measurements. Soil moisture was calculated as the mean of five measurements at random locations in a plot. We used repeated measure Analysis of Variance (ANOVA) to test the differences in soil respiration rate, soil temperature and soil moisture among forests, precipitation treatments, and years. Each treatment was replicated three times (three blocks). Multiple comparisons (Least Significant Difference, LSD method) were conducted if significant effects of forest ecosystem types, precipitation treatments or years were found. Previous work at study sites demonstrated that soil respiration increases exponentially with soil temperature and linearly with soil moisture [30,33,34]. Thus, we first developed the relationship between soil respiration and soil temperature with an exponential function and the relationship between soil respiration and soil moisture with a linear regression mode. Considering that soil temperature and moisture may interactively regulate soil respiration, we also fit soil respiration (R) with soil temperature (T) and soil moisture (M) together using $R = (a+cM)\exp(bT)$, where a is parameter related to basal soil respiration when both $T = 0$ and $M = 0$; b and c are parameters related to the soil temperature and moisture sensitivities of soil respiration, respectively. Like most studies, we used measurements of soil respiration, soil temperature and moisture of whole years here. One caveat of this approach was that seasonal variations of tree roots growth, carbon substrate in the soils, and soil microbial community would influence soil respiration, but were difficult to quantify. Non-linear least square method was used to derive the model parameters using SAS NLIN procedure [51]. Soil temperature and moisture sensitivities were derived for different precipitation treatments in the three forests. All data analyses were carried out using SAS software Version 9.1 [51] (SAS Institute Inc., Cary, NC, USA).

Results

Effects of Precipitation Treatments on Soil Temperature and Moisture

There were strong seasonal variations of precipitation in all three years, with intensive precipitation occurring from April through September (i.e., wet season) (Figure 1). The annual precipitation amount was 1341.6, 2925.8, and 1864.4 mm in 2007, 2008, and 2009, respectively. The very high precipitation in 2008 was mostly attributed to two heavy precipitation months (May and June) which had 50% of the total annual precipitation (Figure 1). The high precipitation intensity and large interannual variability in precipitation throughout the three years were typical in subtropical China. Mean annual air temperature did not vary much and was 22.77, 22.08, 22.71°C in 2007, 2008, and 2009, respectively. The monthly mean air temperature ranged from 11.35°C (February 2008) to 30.11°C (July 2007).

The seasonal patterns of soil temperature in three forests were similar to the pattern of air temperature (Figure 2a). Among the three forests, soil in the PF was significantly warmer (22.42°C) than that in the MF (20.20°C) and the BF (20.32°C) (Tables 2 and 3). No significant difference in annual mean soil temperature was found between the MF and the BF. Precipitation treatments did not change soil temperature in all three forests.

Soil moisture was significantly influenced by precipitation treatments and varied among forest ecosystem types and years (Table 2). Soil moisture in both the DP and AP treatments showed strong variations in all three forests (Figure 2b). Soil moisture was maintained at about 29% vol. in the BF and the MF, but only 20% vol. in the PF over the observation period (Table 2; Figure 2b). The DP treatment slightly increased annual mean soil moisture by approximately 11.4% compared to the AP treatment.

Effects of Precipitation Treatments on Soil Respiration, Soil Microbial Biomass and Fine Root Biomass

The soil respiration rate was significantly influenced by forest ecosystems and precipitation treatments, and the effects of precipitation treatments varied among the three forest ecosystems (Table 2). Soil respiration was significantly lower in the PF (2.37 $\mu mol\ CO_2\ m^{-2}\ s^{-1}$), compared to that in the BF (3.07 $\mu mol\ CO_2\ m^{-2}\ s^{-1}$) and MF (3.15 $\mu mol\ CO_2\ m^{-2}\ s^{-1}$), averaged over three years of the experiment. The DP treatment increased mean annual soil respiration in the PF (15.4%), and did not show significant change in the BF or the MF.

The responses of soil microbial biomass and fine root biomass to precipitation treatment also varied among forest ecosystems (Figure 3). Soil microbial biomass in the DP treatment increased by 19.0% and 24.0% in the MF and the PF, respectively, compared to the AP treatment (Figure 3), but did not change in the BF. The DP treatment enhanced soil microbial biomass in both the wet and dry seasons in the PF, but only in the wet season in the MF. The DP treatment increased fine root biomass by 31.2% in the PF, but not in the MF and the BF (Figure 3). Fine root biomass in the PF was enhanced in the dry season by the DP treatment.

Effects of Precipitation Treatments on the Functional Relationships of Soil Respiration with Soil Temperature and Moisture

Under both precipitation treatments and in all three forests, soil respiration responded exponentially to soil temperature and linearly to soil moisture (Figure 4). The DP treatment reduced soil temperature sensitivity in the BF and the MF, but not in the PF. Soil moisture sensitivity was not influenced by the DP treatment. Since soil temperature and soil moisture interactively regulate soil respiration, we considered both soil temperature and soil moisture and fit a combination model [30]. The best regression models explained 75–93% of soil respiration variations under two precipitation treatments in three forests (Table 4). The DP treatment decreased soil temperature sensitivities in the BF and the MF, but did not change soil moisture sensitivity. Basal soil respiration was enhanced under the DP treatment in both the BF and the MF. Under high temperature and heavy precipitation conditions, soil respiration under the DP treatment was lower than that under the AP treatment (Table 4), but in the PF, the DP treatment did not change the functional relationship of soil respiration with soil temperature and moisture developed under the AP control.

Discussion

The findings from our three-year precipitation manipulation experiment provide insights into the effects of precipitation increase on forest ecosystem soil respiration in subtropical monsoon areas and may have significant implications in modeling soil respiration. First, we found that unlike in arid and semi-arid ecosystems, soil respiration in the subtropical forests showed little response to precipitation increase, even when the precipitation was doubled. Second, we proposed to differentiate two reasons of soil respiration changes in response to precipitation increase (i.e., changes due to climate factor change and/or functional change) and demonstrated that different mechanisms may lead to different responses of soil respiration to precipitation treatments in different forest sites. The DP treatment increased soil moisture, enhanced basal soil respiration, but decreased soil temperature sensitivity in the BF and MF, resulting in no change in soil respiration. The increase in soil respiration in the PF under the DP treatment was solely caused by an increase in soil moisture, as no functional change was detected. Third, the slight increase in soil respiration under the DP treatment in the PF was supported by increases in soil microbial biomass and fine root biomass. As no changes in soil microbial biomass and fine root biomass were observed in the BF treatment and only slight change in soil microbial biomass in the MF, little change in soil respiration was observed in the MF and the BF. Our findings indicate that total soil respiration might not change much in the subtropical forests if precipitation increases in the future.

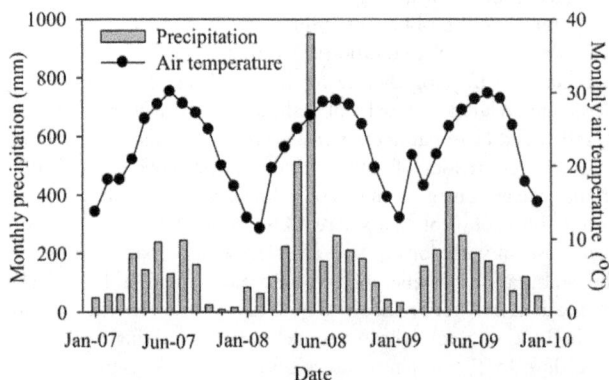

Figure 1. Monthly rainfall and mean air temperature at the Dinghushan Forest Ecosystem Research Station in Southern China during the experimental period from 2007 to 2009.

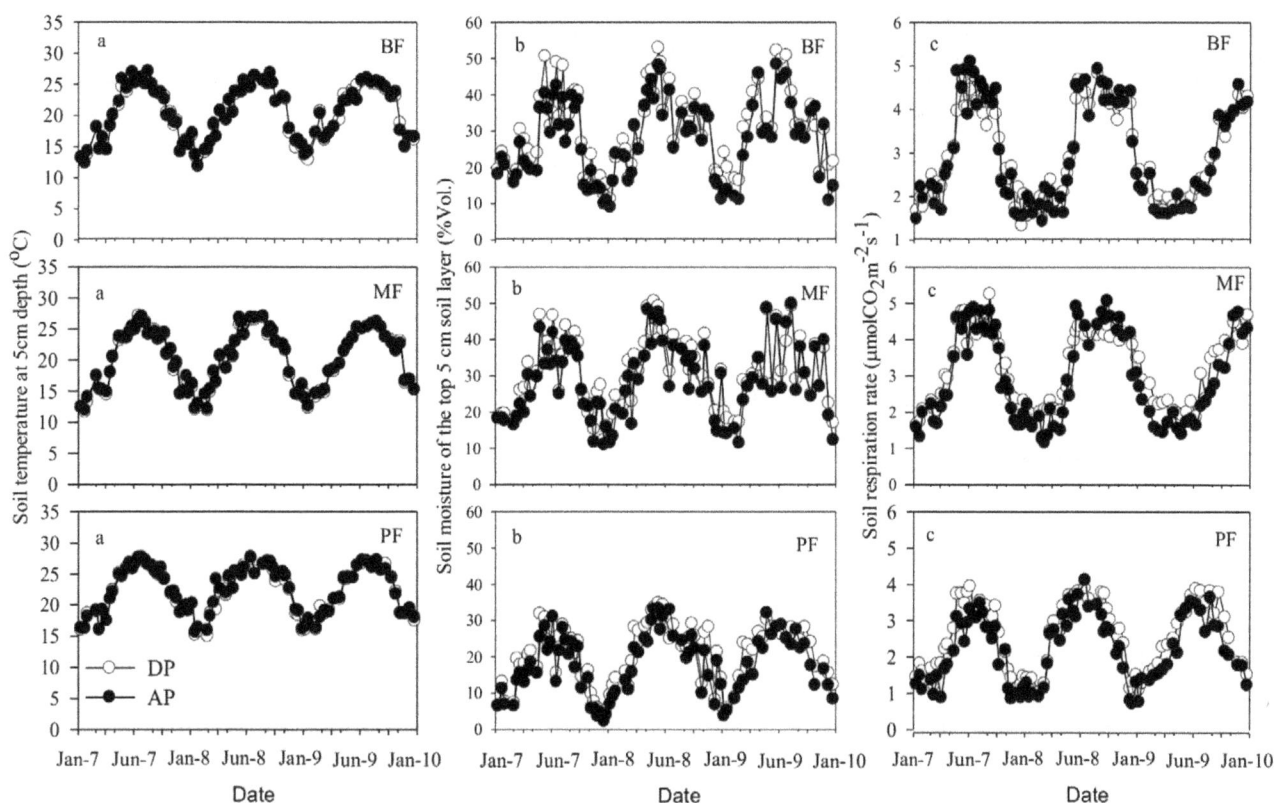

Figure 2. Seasonal dynamics of soil temperature at 5 cm depth, soil moisture of the top 5 cm soil layer, and soil respiration under ambient precipitation (AP) and double precipitation (DP) treatments in the broadleaf forest (BF), the mixed forest (MF) and pine forest (BF).

Responses of Soil Respiration to Precipitation Treatments

Previous studies have indicated that the water status of an ecosystem may influence the direction of soil respiration to either reduction or increase in precipitation treatments [25]. In this study, we found 15.4% annual increase in soil respiration in the PF and no change of soil respiration in the BF and the MF (Table 3). Different responses might be attributed to differences in soil condition and vegetation at these study sites. Soil in the PF contains more sand, less clay, and more gravel, and had lower ambient soil moisture content than those in the BF and the MF (Table 1). Trees in the PF were younger and smaller in biomass and LAI [29]. As a result, we found that soil

respiration in the PF was low, but showed a significant influence by precipitation increase. Responses of soil respiration to precipitation increase also varied among different studies. For example, the DP treatment resulted in an increase of 9.0% in soil respiration in a tallgrass prairie [10]. But a large increase of 31% in soil respiration was reported in arid and semiarid grassland with 30% increase in annual precipitation [36]. Results from a recent study indicated that soil respiration may be decreased under precipitation increase in a humid tropical forest [13].

Table 2. Significance test using Analysis of Variance (ANOVA).

Source	df	Soil respiration	Soil temperature	Soil moisture
Forest	2	79.97**	41.88**	158.98**
Precipitation	1	11.56**	0.57	30.58**
Forest×Precipitation	2	2.97*	0.01	0.08
Year	2	0.18	0.39	25.24**
Forest×Year	4	1.45	0.10	0.75
Precipitation×Year	2	0.10	0.03	0.05
Forest×Precipitation×Year	4	0.08	0.02	0.25

Significance of the effects of forest type, precipitation treatment, year and their interactions on soil respiration rate, soil temperature, and soil moisture at the Dinghushan Forest Ecosystem Research Station are tested using ANOVA. Numbers are F-values. Stars indicate the level of significance (*p<0.05, **p<0.01).

Table 3. Mean value and significance of soil temperature, moisture and soil respiration from 2007 to 2009 between precipitation treatments in the pine forest (PF), the mixed forest (MF) and the broadleaf forest (BF), respectively.

Variable	Broadleaf forest (BF)		Mixed forest (MF)		Pine forest (PF)	
	DP	AP	DP	AP	DP	AP
Soil temperature	20.25[a]	20.39[a]	20.12[a]	20.29[a]	22.32[a]	22.52[a]
(°C)	±0.48	±0.48	±0.52	±0.51	±0.42	±0.41
Soil moisture	30.36[a]	27.40[b]	30.98[a]	28.46[b]	21.11[a]	18.13[b]
(% Vol.)	±1.21	±1.20	±1.10	±1.14	±0.97	±0.94
Soil respiration	3.08[a]	3.06[a]	3.25[a]	3.04[a]	2.54[a]	2.20[b]
(μmol CO_2 m^{-2} s^{-1})	±0.11	±0.13	±0.12	±0.13	±0.11	±0.11

Table shows means and standard errors of soil temperature at 5 cm depth, soil moisture of the top 5 cm soil layer, and soil respiration rate under ambient precipitation (AP) and double precipitation (DP) treatments from the broadleaf forest, the mixed forest and the pine forest.
Mean values in each forest within a row with different letter have significant differences at $\alpha = 0.05$ level.

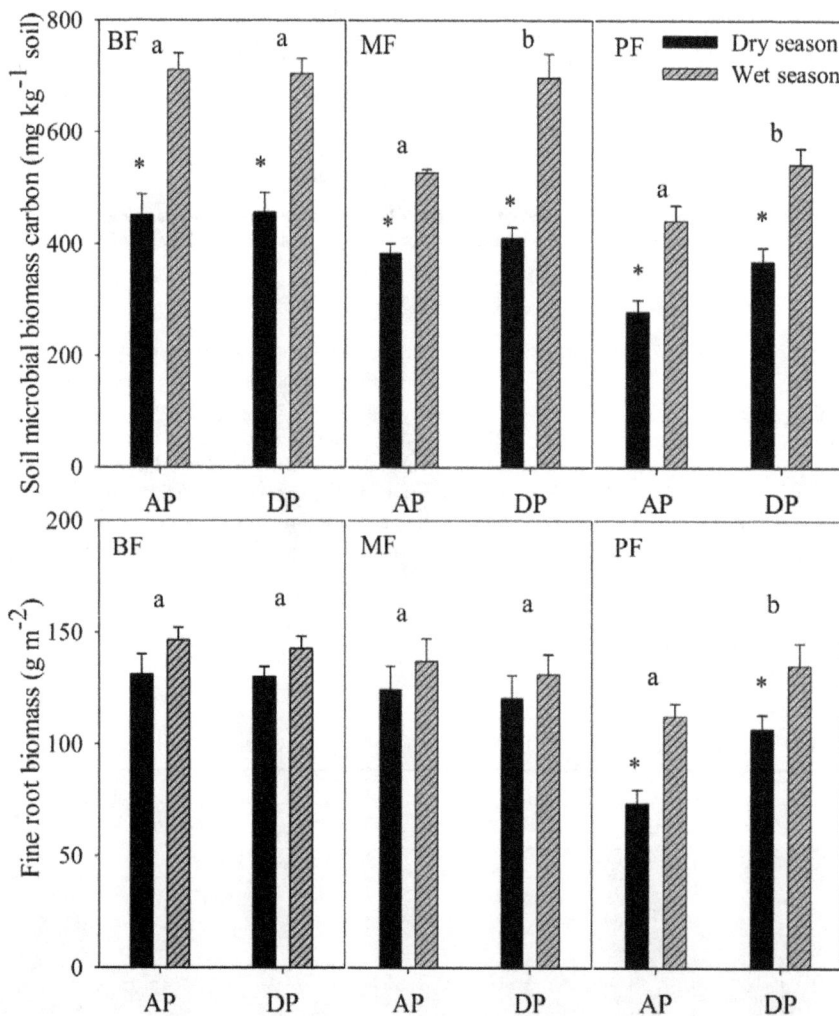

Figure 3. Soil microbial biomass carbon content and fine root biomass (diameter≤3 mm) under ambient precipitation (AP) and double precipitation (DP) treatments in the broadleaf forest (BF), the mixed forest (MF) and the pine forest (PF). Error bars are standard errors, sample size n = 6 for soil microbial biomass carbon content, sample size n = 3 for fine root biomass. Different letters in each forest denote significant difference (p<0.05) among precipitation treatments. *indicates significant difference between wet and dry seasons.

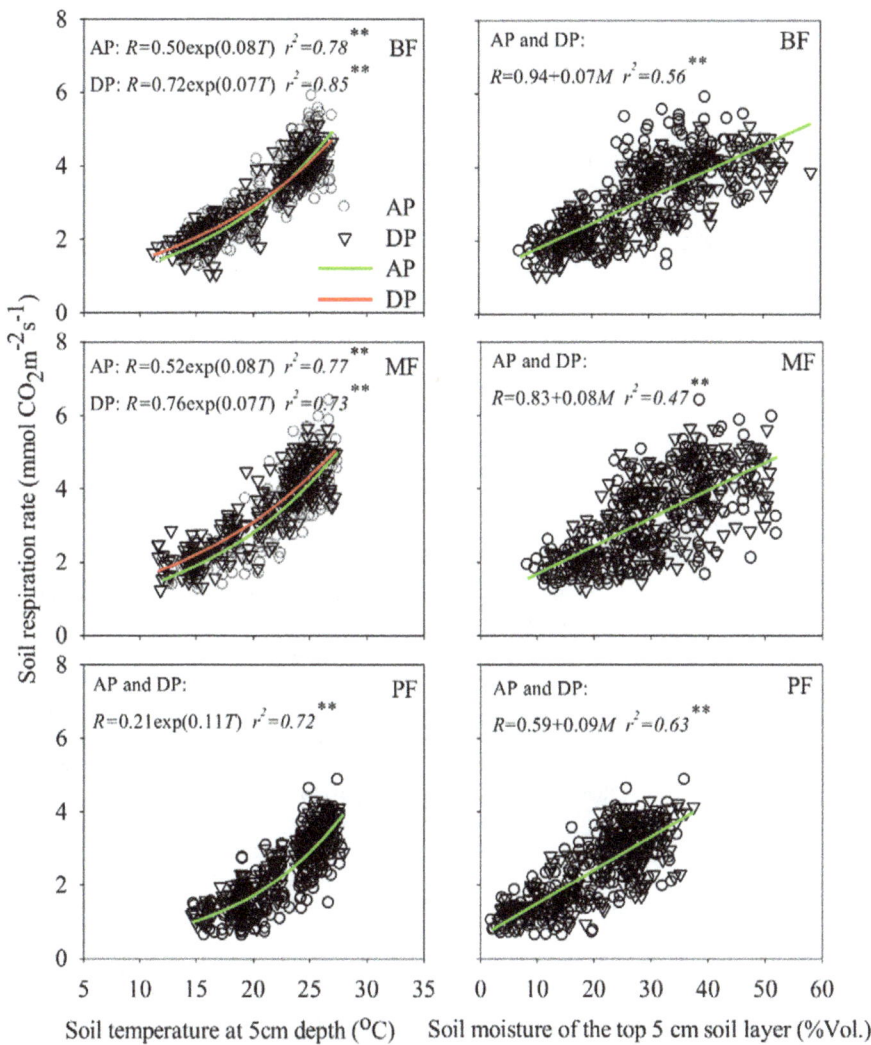

Figure 4. Relationship of soil respiration with soil temperature or soil moisture under ambient precipitation (AP) and double precipitation (DP) treatments in the broadleaf forest (BF), the mixed forest (MF) and the pine forest (PF). If fitted models aren't significantly different between in the AP and DP treatments, one single model is fitted for all data. **indicates significant relationship at $\alpha = 0.01$ levels.

Table 4. Functional relationship and significant test of model parameters.

Forest	Treatment	a	c	b	Q_{10}	R^2
Broadleaf forest	DP	0.7158 ± 0.0267^a	0.0088 ± 0.0009^a	0.0544 ± 0.0020^a	1.72	0.91**
	AP	0.5486 ± 0.0358^b	0.0082 ± 0.0013^a	0.0646 ± 0.0033^b	1.91	0.83**
Mixed forest	DP	0.8077 ± 0.0458^a	0.0068 ± 0.0015^a	0.0556 ± 0.0029^a	1.74	0.79**
	AP	0.5698 ± 0.0395^b	0.0064 ± 0.0013^a	0.0660 ± 0.0035^b	1.93	0.80**
Pine forest	DP	0.2541 ± 0.0145^a	0.0074 ± 0.0008^a	0.0786 ± 0.0026^a	2.19	0.93**
	AP	0.2691 ± 0.0333^b	0.0111 ± 0.0026^a	0.0657 ± 0.0064^a	1.93	0.75**

Relationship of soil respiration rate (R, $\mu mol\ CO_2\ m^{-2}\ s^{-1}$) with soil temperature at 5 cm below the soil surface (T,°C) and soil moisture of the top 5 cm soil layer (M, % vol.) is developed using $R = (a + cM)exp(bT)$ (parameter estimate $\pm$ standard error). R^2 in the table is the determination of coefficient, $Q_{10} = exp(10b)$ is temperature sensitivity coefficient, and slope c is soil moisture sensitivity. The treatments are: AP = ambient precipitation, DP = double precipitation. Different letters in each forest within a column denote significant difference ($p < 0.05$) between the two precipitation treatments. **$p < 0.01$. Numbers in bold indicates significant differences with the AP treatment.

Functional Changes of Soil Respiration to Precipitation Treatments

Functional change of soil respiration to soil temperature/ moisture under climate change is common and contributes to the responses of soil respiration in different ecosystems. A study in grasslands found that soil respiration was more sensitive to soil moisture than to soil temperature during prolonged drying cycles [52]. Ecosystems in xeric regions often possess lower soil respiration and higher soil moisture sensitivity than those in mesic regions [30,53]. But the response of soil respiration to soil moisture change may be different in wet subtropical forests. We found that the DP treatment did not change soil moisture sensitivity, but decreased soil temperature sensitivity significantly in the BF and the MF ecosystems (Table 4). Many other studies have also found that soil respiration is insensitive to soil moisture unless that soil moisture is below levels at which metabolic activity decreases [20,26,54].

The lower temperature sensitivities under the DP treatment here may be due to the following two reasons. 1) Enhanced soil moisture under the DP treatment might decrease soil aeration and soil oxygen concentration [14], thus, more activation energy was needed to stimulate enzymatic rates [20]. Due to the subtropical monsoon climate, forests in the study site receive abundant heat, light, and water [55,56]. Therefore, soils in these wet forests are often limited by soil oxygen concentration and nutrients, especially during the hot-humid season (April-September) [47]. 2) Greater leaching of dissolved organic carbon and nutrients under the DP treatment may reduce substrate availability [15,57], and result in a decline in the Q_{10} values of soil respiration [8]. Previous work in this experiment has also shown that the active organic carbon, in particular particulate and light fraction organic carbon, often infiltrated to deeper soil layers with precipitation increase in the MF and BF [58,59]. In the PF where soil was relative drier, the DP treatment stimulated fine root biomass and microbial activity (Figure 3). The greater soil microbial activity could release more nutrients from soil organic matter for fine root uptake, and increase soil respiration. The DP treatment in the BF and the MF did not stimulate soil microbe or fine root biomass, and caused little change in soil respiration in these forests.

Environmental Factor Changes Alone may Contribute to Soil Respiration Changes Under Precipitation Treatments

Environmental factor changes induced by climate change alone could have significant influences on ecosystem responses. In this study, we found that the functional response of soil respiration to soil temperature and moisture in the PF under the DP treatment was not changed compared to the AP treatment (Table 4). However, increases in soil moisture under the DP treatment slightly enhanced soil respiration. A similar result was reported recently in a Mediterranean evergreen forest [37]. They found

that when 27% of throughfall was excluded over three years, soil moisture was reduced by 7–10%. While the three-year throughfall exclusion did not change functional properties of the response of soil respiration to soil water content and soil temperature, soil respiration decreased by 11% due to the environmental factor change.

Limitation of the Study

In this study, we selected three typical forest ecosystems in the south of China and tested the effects of precipitation increase on soil respiration. One shortcoming of the experimental design was unreplicated forest ecosystem types. While three replicated plots were employed for each precipitation treatment (i.e. DP and AP) at each forest ecosystem site, the forest types were not replicated. Thus, the inferences regarding the response differences among forest ecosystems should be read with caution. Further studies are needed to draw rigorous conclusions regarding forest ecosystem responses using replicated forest types.

Conclusions

Using a three-year field experiment in subtropical forests in China, we demonstrated that soil respiration under the DP treatment was not changed in the BF and the MF, but slightly increased in the PF. The lower response of soil respiration was consistent with small or no change of fine root biomass and microbial biomass under the DP treatment. The different responses in the three forests were associated with both functional change and environmental factor change induced by the precipitation treatments. Changes in soil temperature sensitivity and basal soil respiration together with change in soil moisture help us understand soil respiration responses at different forest sites. The shift of soil temperature sensitivity and basal soil respiration under different precipitation regimes may have potentially significant implications for terrestrial ecosystem carbon cycling, and should be considered in terrestrial ecosystem models. Whether soil moisture sensitivity of soil respiration is changed by precipitation treatments, particularly drought, may warrant further study.

Acknowledgments

The authors are grateful to Dr. Ben Bond-Lamberty for his insightful comments and valuable suggestions. We also thank Drs. Yiqi Luo, Robert B. Jackson and Phillip Ganter for their constructive comments on an early version of this manuscript. Ms. Jennifer Cartwright provided critical editing of the manuscript.

Author Contributions

Conceived and designed the experiments: DZ DH GZ. Performed the experiments: QD J. Liu SL GC J. Li. Analyzed the data: QD DH DZ GZ. Wrote the paper: DH QD DZ GZ.

References

1. Valentini R, Matteucci G, Dolman AJ, Schulze E-D, Rebmann C, et al. (2000) Respiration as the main determinant of carbon balance in European forests. Nature 404: 861–865.
2. Luo Y, Wan S, Hui D, Wallace LL (2001) Acclimatization of soil respiration to warming in a tall grass prairie. Nature 413: 622–625.
3. Hui D, Luo Y (2004) Evaluation of soil CO_2 production and transport in Duke Forest using a process-based modeling approach. Global Biogeochem Cycles 18: GB4029, doi:10.1029/2004GB002297.
4. Jackson RB, Cook CW, Pippen JS, Palmer SM (2009) Increased belowground biomass and soil CO_2 fluxes after a decade of carbon dioxide enrichment in a warm-temperate forest. Ecology 90: 3352–3366.
5. Wu Z, Dijkstra P, Koch GW, Penuelas J, Hungate BA (2011) Responses of terrestrial ecosystems to temperature and precipitation change: a meta-analysis of experimental manipulation. Global Change Biol 2(17): 927–942.
6. Intergovernmental Panel on Climate Change (IPCC) (2007) Climate Change 2007: The Physical Science Basis. Solomon, S., Qin, D., Manning, M, et al. eds. Contribution of Working Group I to the Fourth Assessment Report of the Intergovernmental Panel on Climate Change. Cambridge, UK and New York, USA: Cambridge University Press, 996pp.
7. Allan RP, Soden BJ (2008) Atmospheric warming and the amplification of precipitation extremes. Science 321: 1481–1484.
8. Harper CW, Blair JM, Fay PA, Knapp AK, Carlisle JD (2005) Increased rainfall variability and reduced rainfall amount decreases soil CO_2 flux in a grassland ecosystem. Global Change Biol 11: 322–334.
9. Borken W, Savage K, Davidson EA, Trumbore SE (2006) Effects of experimental drought on soil respiration and radiocarbon efflux from a temperate forest soil. Global Change Biol 12: 177–193.

10. Zhou XH, Sherry RA, An Y, Wallace LL, Luo Y (2006) Main and interactive effects of warming, clipping, and doubled precipitation on soil CO_2 efflux in a grassland ecosystem. Global Biogeochem Cycles 20: GB1003.

11. Davidson EA, Nepstad DC, Ishida FY, Brando PM (2008) Effects of an experimental drought and recovery on soil emissions of carbon dioxide, methane, nitrous oxide, and nitric oxide in a moist tropical forest. Global Change Biol 14: 582–2590.

12. Jenerette GD, Scott RL, Huxman TE (2008) Whole ecosystem metabolic pulses following precipitation events. Funct Ecol 22: 924–930.

13. Cleveland CC, Wieder WR, Reed SC, Townsend AR (2010) Experimental drought in a tropical rain forest increases soil carbon dioxide losses to the atmosphere. Ecology 91(8): 2313–2323.

14. van Straaten O, Veldkamp E, Köhler M, Anas I (2010) Spatial and temporal effects of drought on soil CO_2 efflux in a cacao agroforestry system in Sulawesi, Indonesia. Biogeosciences 7 : 1223–1235.

15. Knapp AK, Beier C, Briske DD, Classen AT, Luo Y, et al. (2008) Consequences of more extreme precipitation regimes for terrestrial ecosystems. Bioscience 58: 811–821.

16. Beier C, Beierkuhnlein C, Wohlgemuth T, Penuelas J, Emmett B, et al. (2012). Precipitation manipulation experiments - challenges and recommendations for the future. Ecology Letters, doi: 10.1111/j.1461-0248.2012.01793.x.

17. Rustad LE, Huntington TG, Boone RD (2000) Controls on soil respiration: implications for climate change. Biogeochemistry 48: 1–6.

18. Bond-Lamberty B, Thomson A (2010) Temperature-associated increases in the global soil respiration record. Nature, 464, 579–582.

19. Hartley IP, Armstrong AF, Murthy R, Barron-Gafford, Ineson P, et al. (2006) The dependence of respiration on photosynthetic substrate supply and temperature: integrating leaf, soil and ecosystem measurements. Global Change Biol 12: 1954–1968.

20. Fang C, Moncrieff JB (2001) The dependence of soil CO_2 efflux on temperature. Soil Biol Biochem 33: 155–165.

21. Curriel Yuste J, Janssens IA, Carrara A, Meiresonne L, Ceulemans R (2003) Interactive effects of temperature and precipitation on soil respiration in a temperate maritime pine forest. Tree Physiol 23 : 1263–1270.

22. Scotta ED, Veldkamp E, Schwendenmann, Guimarães BR, Paixão RK, et al. (2007) Effects of an induced drought on soil CO_2 efflux and soil CO_2 production in an eastern Amazonian rainforest, Brazil. Global Change Biol 13: 546–560.

23. Hui D, Luo Y, Katul G (2003) Partitioning interannual variability in net ecosystem exchange into climatic variability and functional change. Tree Physiol 23: 433–442.

24. Brady NC, Weil RR (2002) The Nature and Properties of Soils. 13th ed. Upper Saddle River (NJ): Prentice Hall.

25. Sowerby A, Emmett BA, Tietema A, Beier C (2008) Contrasting effects of repeated summer drought on soil efflux in hydric and mesic heathland soils. Global Change Biol 14: 2388–2404.

26. Davidson EA, Belk E, Boone RD (1998) Soil water content and temperature as independent or confounded factors controlling soil respiration in a temperate mixed hardwood forest. Global Change Biol 4: 217–227.

27. Reichstein M, Rey A, Freibauer A, Tenhunen J, Valentini R, et al. (2003) Modeling temporal and large-scale spatial variability of soil respiration from soil water availability, temperature and vegetation productivity indices. Global Biogeochem Cycles 17: 1–15.

28. Ilstedt U, Nordgren A, Malmer A (2000) Optimum soil water for soil respiration before and after amendment with glucose in humid tropical acrisols and a boreal mor layer. Soil Biol Biochem 32: 1591–1599.

29. Deng Q, Zhou GY, Liu SZ, Chu GW, Zhang DQ (2011) Responses of soil CO_2 efflux to precipitation pulses in two subtropical forests in southern China, Environ manage 48: 1182–1188.

30. Tang XL, Liu SG, Zhou GY, Zhang DQ, Zhou CY (2006) Soil-atmospheric exchange of CO_2, CH_4, and N_2O in three subtropical forest ecosystems in southern China. Global Change Biol 12: 546–560.

31. Schwendenmann L, Veldkamp E (2005) The role of dissolved organic carbon, dissolved organic nitrogen, and dissolved inorganic nitrogen in a tropical wet forest ecosystem. Ecosystems 8: 339–351.

32. Luo Y, Zhou X (2006) Soil and Respiration Environment. Academic Press, San Diego.

33. Zhang DQ, Sun XM, Zhou GY, Yan JH, Wang YS, et al. (2006) Seasonal dynamics of soil CO_2 effluxes with responses to environmental factors in lower subtropical forest of China. Science in China Ser, D-Earth Sciences 49 (S1): 139–149.

34. Deng Q, Zhou GY, Liu JX, Liu S, Duan H, et al. (2010) Responses of soil respiration to elevated carbon dioxide and nitrogen addition in young subtropical forest ecosystems in China. Biogeosciences 7: 315–328.

35. Noormets A, Desai A, Cook BD, Ricciuto DM, Euskirchen ES, et al. (2008) Moisture sensitivity of ecosystem respiration: comparison of 14 forest ecosystems in the Upper Great Lakes Region, USA. Agric For Meteorol 148: 216–230.

36. Liu W, Zhang Z, Wan S (2009) Predominant role of water in regulating soil and microbial respiration and their responses to climate change in a semiarid grassland. Global Change Biol 15: 184–195.

37. Misson L, Rocheteau A, Rambal S, Ourcival JM, Limousin JM, et al. (2010) Functional changes in the control of carbon fluxes after 3 years of increased drought in a Mediterranean evergreen forest? Global Change Ecol 16: 2461–2475.

38. Davidson EA, Janssens IA (2006) Temperature sensitivity of soil carbon decomposition and feedbacks to climate change. Nature 440: 165–173.

39. Davidson EA, Janssens I, Luo YQ (2006) On the variability of respiration in terrestrial ecosystems: moving beyond Q_{10}. Global Change Biol 12: 154–164.

40. Kirschbaum MUF (2010) The temperature dependence of organic matter decomposition: seasonal temperature variations turn a sharp short-term temperature response into a more moderate annually averaged response. Global Change Biol 16: 2117–2129.

41. Cox PM, Betts RA, Jones CD, Spall SA, Totterdell IJ (2000). Acceleration of global warming due to carbon-cycle feedbacks in a coupled climate model. Nature 408 : 184–187.

42. Xu M, Qi Y (2001) Spatial and seasonal variations of Q_{10} determined by soil respiration measurements at a Sierra Nevadan forest. Global Biogeochem Cycles 15: 687–697.

43. Zhou T, Shi PJ, Hui D, Luo Y (2009) Global pattern of temperature sensitivity of soil heterogeneous respiration (Q_{10}) and its implications for carbon-climate feedback. J Geophy Res - Biogeosciences 114: G02016.

44. Jassal RS, Black TA, Novak MD, Gaumont-Guay Nesic Z (2008) Effect of soil water stress on soil respiration and its temperature sensitivity in an 18-year-old temperate Douglas-fir stand. Global Change Biol 14: 1–14.

45. Schaefer K, Denning AS, Suits N, Kaduk J, Baker I, et al. (2002) Effect of climate on interannual variability of terrestrial CO_2 fluxes. Global Biogeochem Cycles 16: GB1102.

46. Peng SL, Wang BS (1995) Forest succession at Dinghushan, Guangdong, China. Chinese Journal of Botany 7: 75–80.

47. Wang BS, Ma MJ (1982) The successions of the forest community in Dinghushan. Tropical and Subtropical Forest Ecosystem Research 1: 142–156 (in Chinese with English abstract).

48. Shen CD, Yi WX, Sun YM, Xing CP, Yang Y, et al. (2001) Distribution of ^{14}C and ^{13}C in forest soils of the Dinghushan Biosphere Reserve. Radiocarbon 43(2B): 671–678.

49. Jenkinson DS (1987) Determination of microbial biomass carbon and nitrogen in soil. In: Wilson JR, eds. Advances in N Cycling in Agricultural Ecosystem. Wallingford, UK: Commonwealth Agricultural Bureau International press, 368–386.

50. Vance ED, Brookes SA, Jenkinson DS (1987) An extraction method for measuring soil microbial biomass C. Soil Biol Biochem 19: 703–707.

51. Hui D, Jiang C (1996) Practical SAS Usage, Beijing University of Aeronautics & Astronautics Press, Beijing, China.

52. Fay PA, Kaufman DM, Nippert JB, Carlisle JD, Harper CW (2008) Changes in grassland ecosystem function due to extreme rainfall events: implications for responses to climate change. Global Change Biol 14: 1600–1608.

53. Shen W, Jenerette GD, Hui D, Phillips RP, Ren H (2008) Effects of changing precipitation regimes on dryland soil respiration and C pool dynamics at rainfall event, seasonal and interannual scales. J Geophys Res – Biogeosciences 113: G03024, doi:10.1029/2008JG000685.

54. Palmroth S, Maier CA, McCarthy HR, Oishi AC, Kim HS, et al. (2005) Contrasting responses to drought of forest floor CO_2 efflux in a Loblolly pine plantation and a nearby Oak–Hickory forest. Global Change Biol 11: 421–434.

55. Ding MM, Brown S, Lugo AE (2001) A continental subtropical forest in China compared with an insular subtropical forest in the Caribbean. General Technical Report IITF-17, USDA (United States Department of Agriculture) Forest Service, International Institute of Tropical Forestry, Rio Piedras, Puerto Rico.

56. Yan JH, Zhang DQ, Zhou GY, Liu JX (2009) Soil respiration associated with forest succession in subtropical forests in Dinghushan Biosphere Reserve. Soil Biol Biochem 9: 1–9.

57. Liu JX, Zhang DQ, Zhou GY, Faivre-Vuillin B, Deng Q, et al. (2008) CO_2 enrichment increases cation and anion loss in leaching water of model forest ecosystems in southern China. Biogeosciences 5: 1783–1795.

58. Chen XM, Liu JX, Deng Q, Chu GW, Zhou GY, et al. (2010) Effects of precipitation intensity on soil organic carbon fractions and their distribution under subtropical forests of South China. Chinese Journal of Applied Ecology, 21: 1210–1216 (in Chinese with English abstract).

59. Liu S, Luo Y, Huang YH, Zhou GY (2007) Studies on the community biomass and its allocations of five forest types in Dinghushan Nature Reserve. Ecol Science 26(5), 387–393.

60. Fang YT, Mo JM, Peng SL, Li DJ (2003) Role of forest succession on carbon sequestration of forest ecosystems in lower subtropical China. Acta Ecologica Sinica, 23, 1685–1694.

61. He C, Chen S, Liang Y (1982) The soils of Dinghushan Biosphere Reserve. Tropical and Subtropical Forest Ecosystem, 1, 25–38.

18

Soil Respiration and Organic Carbon Dynamics with Grassland Conversions to Woodlands in Temperate China

Wei Wang[1]*, **Wenjing Zeng**[1,2], **Weile Chen**[1,2], **Hui Zeng**[2], **Jingyun Fang**[1]

1 Department of Ecology, College of Urban and Environmental Sciences, and Key Laboratory for Earth Surface Processes of the Ministry of Education, Peking University, Beijing, China, **2** Key Laboratory for Urban Habitat Environmental Science and Technology, Peking University Shenzhen Graduate School Shenzhen, China

Abstract

Soils are the largest terrestrial carbon store and soil respiration is the second-largest flux in ecosystem carbon cycling. Across China's temperate region, climatic changes and human activities have frequently caused the transformation of grasslands to woodlands. However, the effect of this transition on soil respiration and soil organic carbon (SOC) dynamics remains uncertain in this area. In this study, we measured *in situ* soil respiration and SOC storage over a two-year period (Jan. 2007–Dec. 2008) from five characteristic vegetation types in a forest-steppe ecotone of temperate China, including grassland (GR), shrubland (SH), as well as in evergreen coniferous (EC), deciduous coniferous (DC) and deciduous broadleaved forest (DB), to evaluate the changes of soil respiration and SOC storage with grassland conversions to diverse types of woodlands. Annual soil respiration increased by 3%, 6%, 14%, and 22% after the conversion from GR to EC, SH, DC, and DB, respectively. The variation in soil respiration among different vegetation types could be well explained by SOC and soil total nitrogen content. Despite higher soil respiration in woodlands, SOC storage and residence time increased in the upper 20 cm of soil. Our results suggest that the differences in soil environmental conditions, especially soil substrate availability, influenced the level of annual soil respiration produced by different vegetation types. Moreover, shifts from grassland to woody plant dominance resulted in increased SOC storage. Given the widespread increase in woody plant abundance caused by climate change and large-scale afforestation programs, the soils are expected to accumulate and store increased amounts of organic carbon in temperate areas of China.

Editor: Han Y.H. Chen, Lakehead University, Canada

Funding: This research was supported by the National Basic Research Program of China (No. 2012CB956303 and 2010CB950600), projects of the National Natural Science Foundation of China (31222011, 31270363 and 31070428), and projects supported by the Foundation for Innovative Research Groups of the National Natural Science Foundation of China (No. 31021001). The funders had no role in study design, data collection and analysis, decision to publish, or preparation of the manuscript.

Competing Interests: The authors have declared that no competing interests exist.

* E-mail: wangw@urban.pku.edu.cn

Introduction

Soils are the largest store of carbon in the biosphere [1], so small changes in soil organic carbon (SOC) storage will profoundly influence atmospheric CO_2 concentrations and potentially influence the global climate [2]. Moreover, soil respiration is the second largest flux of carbon between terrestrial ecosystems and the atmosphere [3]. Global changes have substantially impacted soil respiration and, in turn, SOC dynamics [4,5]. However, soils are the largest source of uncertainty in the terrestrial carbon balance [6].

Natural and anthropogenic-induced vegetation-type conversions are among the most important components of global changes [7]. The shifts between grasslands and plant communities dominated by woody vegetation are one of the most frequent occurring vegetation transition types [8,9,10,11,12]. For instance, deforestation is believed to be a major anthropogenic source of CO_2 to the atmosphere [13,14,15,16,17]. In contrast, large scale forest expansion and re-growth may be important sources for the missing carbon sink [18,19]. Vegetation-type conversions influence the balance of organic carbon in soil and hence may cause changes in soil respiration [20,21]. Changes in vegetation-type are expected to have major effects on the terrestrial carbon balance [22].

Shifts in vegetation types may profoundly affect the dynamics of soil respiration and SOC by influencing soil microclimate and the production and transfer of aboveground photosynthate to belowground [23,24,25,26,27]. However, the direction of changes in the soil respiration and the consequent changes in organic carbon storage in soil within adjacent grass-woody vegetative transition is still controversial [28,29,30]. The inconsistencies may, to a large degree, be caused by the differences in the various locations and the types of transition occurring [31].

Because regional aspects of the global carbon cycle are drawing increasing scientific and political interest, there is a strong impetus to better understand how land use change effects China's carbon balance [32,33,34]. However, few reports on soil respiration and SOC dynamics are available. Furthermore, the currently available studies were mainly conducted in China's southern tropical and sub-tropical areas [35,36,37,38,39,40]. Nevertheless, the temperate areas of northern China are also experiencing frequent, diverse and continuous transitions in the vegetation types, which should substantially affect SOC dynamics and soil respiration in this area. Since the 1970s, the Chinese government has implemented several ecological restoration projects, including the Three-North Shelterbelt Program covering 41% area of the country, across the temperate regions of China that receive less than 400 mm of precipitation annually. These reforestation and afforestation activities were believed to influence carbon cycling and carbon storage in this area [41,42]. In addition, the study of dynamics of organic carbon in soil shows the level of organic carbon in soil is relatively sensitive to increasing temperatures in the temperate climatic zone [43]. Therefore, evaluating how large-scale transitions of vegetation types influence soil respiration and consequent SOC storage is critical to calculating temperate China's carbon budget under the scenario of global change.

In this study, we quantify soil respiration and SOC dynamics from five adjacent grass-woody vegetation types in the temperate areas of northern China. We aimed to 1) measure annual soil respiration as well as SOC storage and residence time, and 2) explore the major drivers for the variations in soil respiration among different vegetation types. We hypothesized that 1) soil respiration as well as SOC storage and residence time were higher in woody vegetation types than in grasslands, and 2) vegetation-mediated change in soil microenvironments was a major driver for the variation of soil respiration.

Materials and Methods

Ethics Statement

The administration of the Saihanba Forestry Center gave permission for this research at each study site. We confirm that the field studies did not involve endangered or protected species.

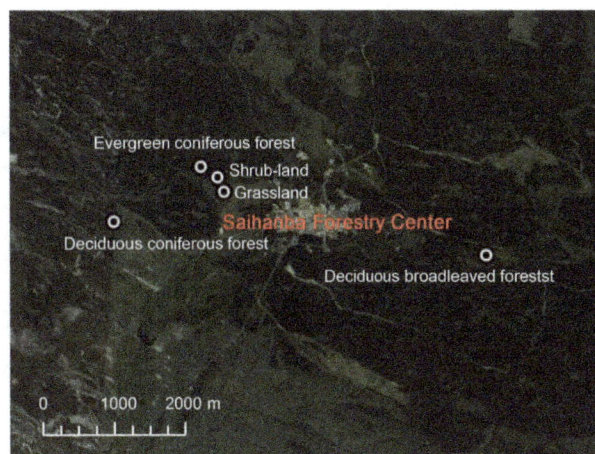

Figure 2. The location of five vegetation types in Saihanba Forestry Center.

Site description and land-use history

The study was conducted at the Saihanba Forestry Center in Hebei Province and Inner Mongolia Autonomous Area, northern China (117°12′–117°30′E, 42°10′–42°50′N, 1,400 m a.s.l.). The study area has a semi-arid and semi-humid temperate climate and lies in a typical forest-steppe ecotone on predominately sandy soils with long and cold winters (November to March), and short springs and summers. Annual mean air temperature and precipitation from 1964 to 2004 were $-1.4°C$ and 450.1 mm, respectively.

This area contains the largest of plantation forests in China, with evergreen *Pinus sylvestris* L. var. *mongolica* Litv. (Mongolia pine) and deciduous *Larix principis-rupprechtii* Mayr (larch) as dominant

Figure 1. Current land-use patterns at the study site. From a 1:1,000,000 scale map of the vegetation types of China [72].

Table 1. Site characteristics and physical and chemical properties of topsoil (0–20 cm).

Site	Vegetation types	Domain species	NDVI	ST (°C)	SWC (%)	SOC (g m^{-2})	STN (g m^{-2})	Soil pH	SBD* (g cm^{-3})
GR	grassland	*Leymus chinensis*	0.44[a]	3.8[a]	8.4[a]	1476.7[a]	136.5[a]	6.28[a]	0.92[a]
SH	shrubland	*Rosa bella* Rehd. et Wils & *Malus baccata**	0.38[b]	3.4[a]	12.6[b]	1841.5[a]	189.4[b]	6.30[a]	0.71[b]
EC	~15 yr old evergreen coniferous plantation	*Pinus sylvestris* var. *mongolica*	0.55[c]	3.2[a]	7.3[a]	2022.7[a]	146.2[a]	6.45[a]	0.98[a]
DC	~15 yr old deciduous coniferous plantation	*Larix principis-rupprechtii*	0.36[b]	3.6[a]	7.8[a]	2993.9[b]	172.8[a]	6.30[a]	1.06[a]
DB	~45 yr old deciduous broadleaved forest	*Betula platyphylla*	0.51[d]	1.7[b]	19.7[c]	2830.7[b]	287.7[c]	5.92[b]	0.69[b]

*SH contained two different dominant species in separate plots and the results of those plots were averaged in our study.
NDVI = Normalized Difference Vegetation Index, soil temperature = ST, SWC = soil water content, SOC = soil organic carbon, STN = soil total nitrogen, SBD = soil bulk density. Different lowercase letters indicated significant differences ($P<0.05$).

species; secondary deciduous forests mainly consist of *Betula platyphylla* Sukaczev (birch). In addition, shrublands dominated by *Rosa bella* Rehd. et Wils. (Solitary rose) and *Malus baccata* (L.) Borkh. (Siberian crabapple) and meadow grasslands are also very common (Fig. 1). The herbaceous layers of Mongolia pine and larch are similar, and are composed of *Sanguisorba officinalis* L. (Radix Sanguisorbae), *Thalictrum aquilegifolium* L., *Agrimonia pilosa* Ledeb. and *Carex stenophylla* Wahlenb., while the herbaceous layer of birch is made up of *Agrimonia pilosa* Ledeb. and Radix Sanguisorbae. The herbaceous layer of Siberian crabapple is dominated by *Veronica linariifolia* Pall. ex Link, *Galium verum* L., *Heteropappus hispidus* (Thunb.) Less., *Trollius chinensis* Bunge, and *Bupleurum chinense* DC. The herbaceous layer of Solitary rose

consists of *Leymus chinensis* (Trin.) Tzvelev. The meadow grassland is zonal vegetation dominated by *L. chinensis*.

The current land-cover pattern resulted from both natural and human-induced vegetation type transitions: from ~5900 to ~2900 ^{14}C years BP, the original deciduous broadleaf forest (DB) were gradually replaced by evergreen coniferous forest (EC) and deciduous coniferous forest (DC) in those places when climate changed from humid to arid; after ~2900 ^{14}C years BP, EC and DC shifted to grassland (GR) in some drier places [44]. In the late 1900 s, the remaining primary forests were harvested by large scale industrial logging and initially became grasslands, but more recently the grasslands have been replaced by secondary SH, DB and plantations of EC and DC. Furthermore, based on the trends for increasing temperature and precipitation in this area [45], together with the large-scale reforestation and afforestation policy of the Chinese government [42], the cover area of woody vegetation types is predicted to increase in the future.

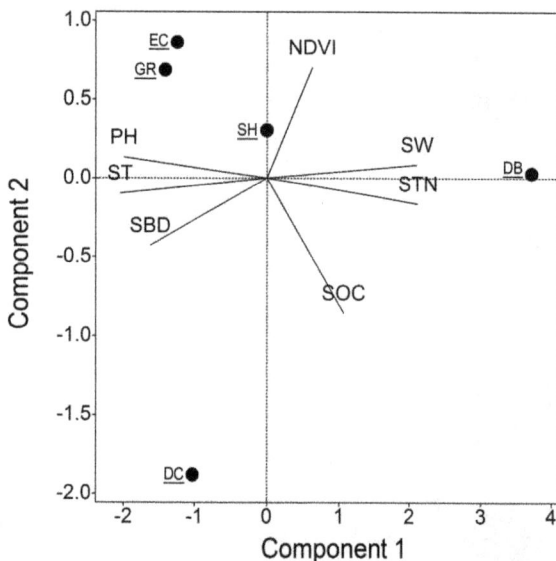

Experimental plot design

The abundant vegetation types co-occurring in our study area provide an excellent opportunity to examine how ecological processes respond to changes in vegetation type. We selected five adjacent grass-woody vegetation types (Fig. 2) to study the influences of vegetation type transitions on biogeochemical processes. All sites for these vegetation types were less than 5 km apart to ensure each site had the same climatic and edaphic condition. Table 1 summarizes the characteristics and species composition of each vegetation type. Three replicates were designed for each of five vegetation types including: GR (*L. chinensis*), SH (*R. bella* & *M. baccata*), EC (~15 year old *P. sylvestris* var. *mongolica*), DC (~15 year old *L. principis-rupprechtii*), and DB

Figure 3. Principal component analysis of site properties. Score plot of five vegetation types during Principal Component Analysis of site properties, including normalized difference vegetation index (NDVI), soil temperature (ST), soil water content (SWC), soil organic carbon (SOC), soil total nitrogen (STN), soil pH, and soil bulk density (SBD). Habitats were grassland (GR), shrubland (SH), evergreen coniferous forest (EC), deciduous coniferous forest (DC), and deciduous broadleaved forest (DB).Seasonal dynamics of soil respiration

Table 2. The proportion of variation explained from principal component analysis on the seven environmental variables.

Component	Eigenvalue	Proportion	Cumulative
1	4.61	0.66	0.66
2	1.21	0.17	0.83
3	1.01	0.15	0.98
4	0.16	0.02	1.00

Table 3. The loading scores of traits on each component from principal component analysis on the seven environmental variables.

Variable	Component 1	Component 2	Component 3
NDVI	0.14	0.58	−0.71
ST	−0.44	−0.08	0.28
SWC	0.46	0.07	0.18
SOC	0.23	−0.71	−0.37
STN	0.46	−0.14	0.02
pH	−0.43	0.11	−0.10
SBD	−0.35	−0.35	−0.49

NDVI = normalized difference vegetation index, ST = soil temperature, SWC = soil water content, SOC = soil organic carbon, STN = soil total nitrogen, pH = soil pH, SBD = soil bulk density.

(~45 year old *B. platyphylla*). Each 20 m×20 m plot was sampled with five subsamples (i.e. soil respiration measurement collars).

Soil respiration, soil temperature and moisture

Soil respiration (SR) was measured using a Li-8100 soil CO_2 flux system (LI-COR Inc. Lincoln, NE, USA) from Jan 2007 to Dec 2008. During the growing season (April to October), five polyvinyl chloride (PVC) collars (10 cm inside diameter, 6 cm height above the soil surface) were inserted 3 cm into the soil in each plot and were left in the same locations throughout the study period. These five PVC collars were placed in each plot, one in the center and one in each corner. Living plants inside the collars were clipped at the soil surface 1 day before each measurement to exclude the effect of aboveground vegetation. SR was measured every 10–15 days. Measurements were made between 08:00 and 11:00 am (based on our measurements of diurnal changes in SR, data not shown) to minimize the daily variation in SR and obtain mean daily SR. For each measurement, respiration rates were calculated as means of three plots for each stand. During winter (November to March), longer soil collars (determined by snow

depth, less than 30 cm) were inserted into the soil surface and stabilized for 24 h before measurement of the winter SR [46,47]. The Li-8100 soil CO_2 flux system was kept in an isolated and heated container to keep its temperature above freezing point.

Soil temperature (ST) was recorded during respiration measurements near each collar at 5 cm soil depth with the LI-COR 8100 temperature probe. Continuous measurements of ST at 5 cm depth were recorded at 30-min intervals with StowAway loggers (Onset Comp. Corp., Bourne, MA, USA) inserted in the soil near one collar at each study site. Soil volumetric water content (SWC) at a depth of 0–10 cm was measured using time domain reflectometry (Soil Moisture Equipment Corp., Santa Barbara, CA, USA). SWC was only obtained during the growing season because the probe could not be fully inserted into the frozen soil in winter.

Soil sampling and measurements

Because the carbon stored in topsoil is the carbon pool that is most sensitive to land management practices [48,49], we sampled mineral soils at depths of 0–10 and 10–20 cm from five random locations per plot using 5.8-cm diameter soil cores during the summers of 2007 and 2008. Soil bulk density (SBD) of the two soil horizons was quantified in all soil surveys from the mass of the oven-dry soil (105°C) divided by the volume of the soil cores. Next, all plant materials were removed from fresh soil samples, and soil was passed through a 2-mm sieve. *In situ* root biomass per unit area was determined by the entire root biomass in the soil core divided by the cross section area of the core. Soil pH was determined from air-dried soil samples in distilled H_2O solution, with a pH meter (Model PHS-2, INESA Instrument, Shanghai, China). The SOC content and soil total nitrogen (STN) were determined from oven-dried soil samples with an elemental analyzer (Vario EL III Universal CHNOS Elemental Analyzer, Elementar, Hanau, Germany). The mass-based SOC and STN were converted into area-based with soil bulk density of each horizon (0–10 and 10–20 cm depth).

NDVI (Normalized Difference Vegetation Index) data

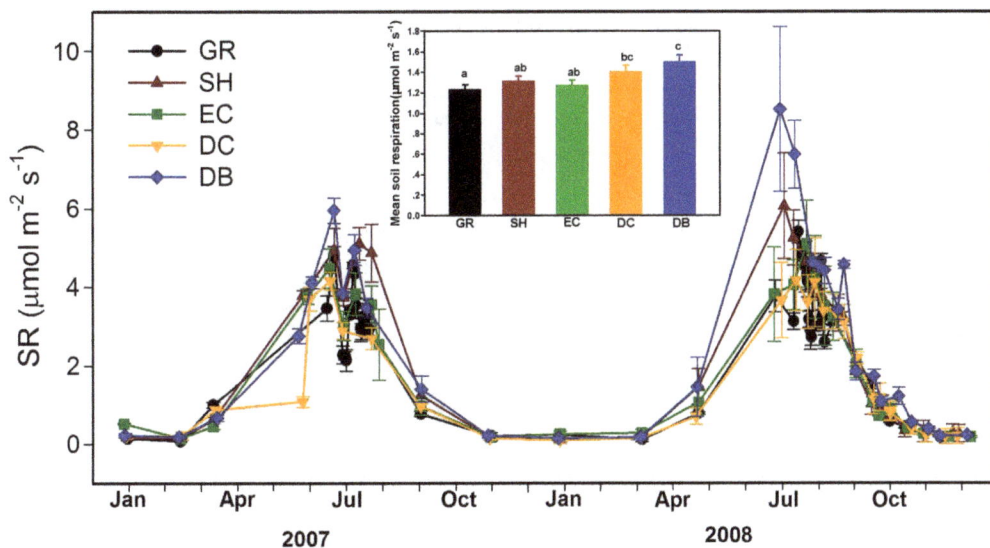

Figure 4. Seasonal dynamics of soil respiration (SR) among five adjacent vegetation types. Habitats are as listed in Fig. 2.

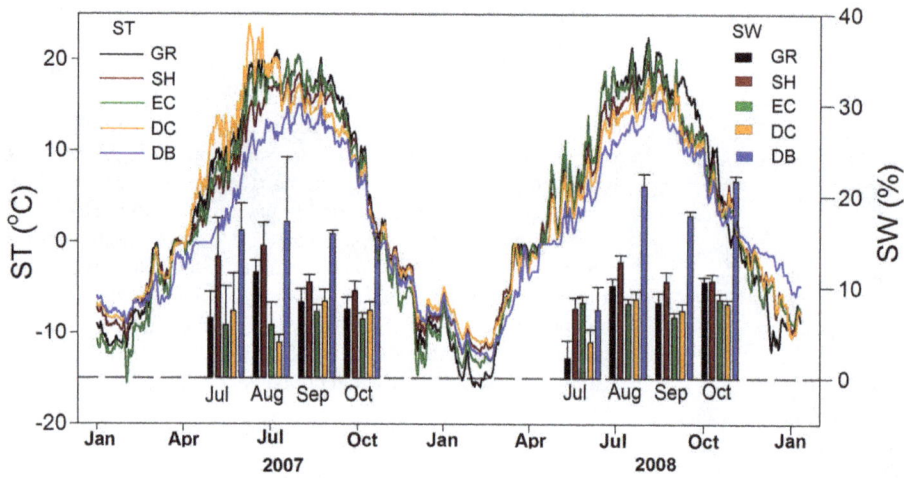

Figure 5. Seasonal dynamics of soil temperature and soil water content across five adjacent vegetation types. ST = soil temperature at 5 cm depth, SWC = soil water content at 10 cm depth. Habitats are as listed in Fig. 2.

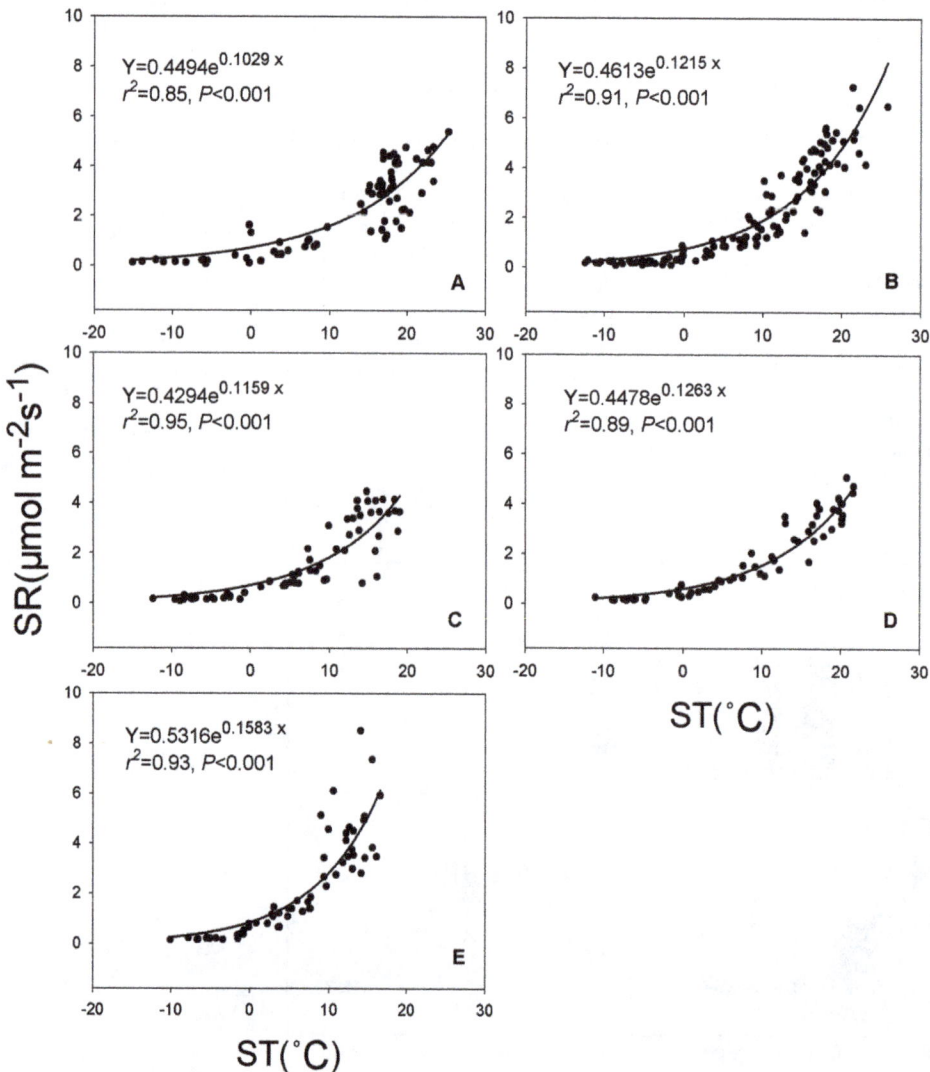

Figure 6. Relationships between the seasonal dynamics of soil respiration and soil temperature at 5 cm depth across habitats as listed in Fig. 2.

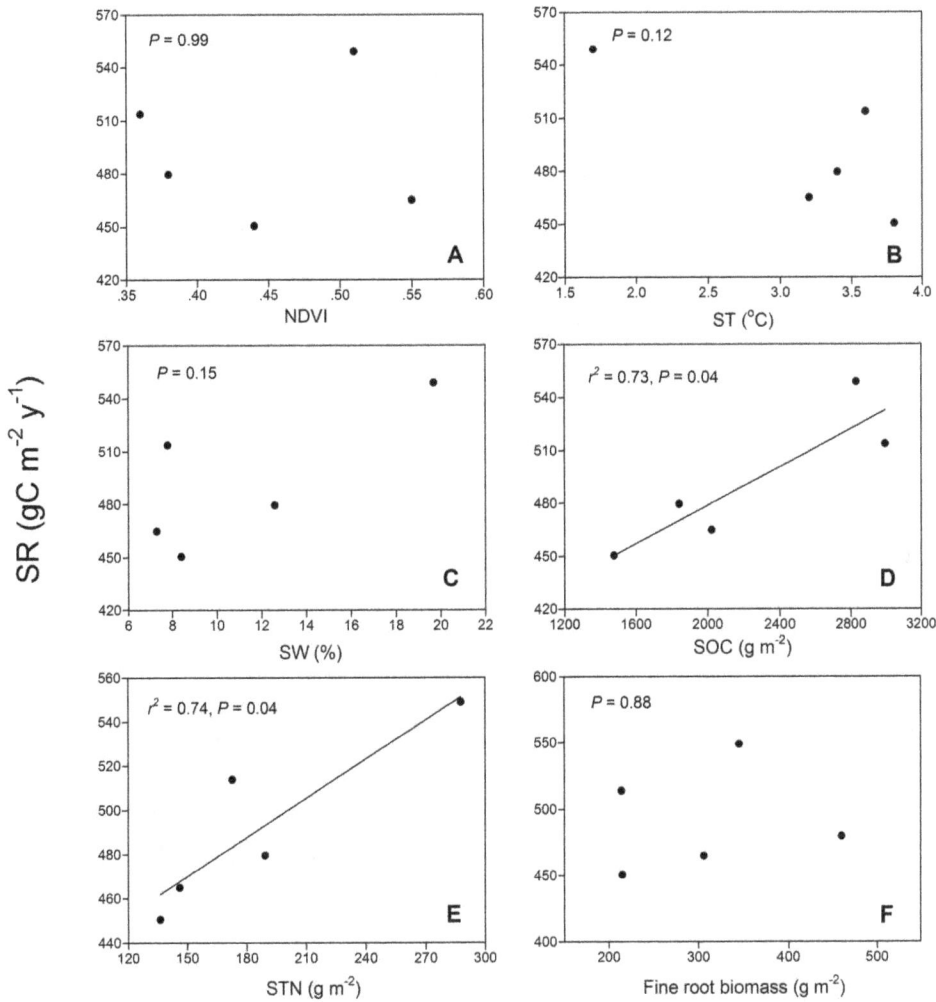

Figure 7. Relationships between annual soil respiration and NDVI (A), soil temperature at 5 cm depth (ST) (B), soil water content at 10 cm depth (SWC) (C), soil organic carbon (SOC) (D), soil total nitrogen (STN) (E) from the top 20 cm depth and fine root biomass at 0–30 cm depth (F) across the five adjacent vegetation types.

NDVI is derived from the red: near-infrared reflectance ratio:

$$NDVI = \frac{(NIR - VIS)}{(NIR + VIS)}, \qquad (1)$$

where NIR and VIS stand for the spectral reflectance measurements acquired in the near-infrared and visible (red) regions of the spectrum, respectively [50]. The NDVI depends on photosynthesis and is closely correlated to GPP [51,52]. We calculated the mean annual NDVI of each vegetation type from 2007–2008 to analyze the influence of aboveground carbon input on SR. Because the five vegetation types studied here covered large areas and had uniform distributions and sparse understories, NDVI of 16-Day L3Global 250 m product (MOD13 Q1) could well represent our plot measurements. We downloaded satellite data from our study period from https://wist.echo.nasa.gov/api. Harmonic (Fourier) analysis was used to remove the considerable noise remaining in the NDVI time series from satellite data to obtain reasonably smooth continuous data [53].

Statistical analysis

We examined the relationships between SR and ST by fitting exponential functions to the data from each vegetation type using the following equation:

$$SR = a \times e^{\beta \times ST}, \qquad (2)$$

where SR is observed soil respiration (plot-wide averages measured periodically throughout the year), ST is the concurrently measured soil temperature (5 cm depth), with a and β being the fitted parameters obtained using least squares nonlinear regression with SigmaPlot V. 8.02.

Annual SR was estimated with the yearly period continuously measured ST and the exponential function between SR and ST for each vegetation type. The mean residence time of SOC was estimated for each vegetation type by dividing the mass of SOC in the top 20 cm of the soil profile by the heterotrophic respiration flux, which equals the total SR minus root respiration. Root respiration of each vegetation type was obtained based on the estimates from Wang et al. [54], who measured excised root respiration for all the five vegetation types we studied at the same sites during the same period.

Table 4. Annual soil respiration, contribution of root respiration to total soil respiration, and soil organic carbon residence time in the top 20 cm of soil.

Site	SR (g C m^{-2} yr^{-1})	Root respiration contribution* (%)	HR (g C m^{-2} yr^{-1})	SOC (g m^{-2})	Residence time (yr)
GR	450.4[a]	4.7	429.0	1476.7[a]	3.4
SH	479.5[b]	23.5	366.8	1841.5[a]	5.0
EC	464.8[c]	18.3	380.0	2022.7[a]	5.3
DC	513.7[d]	24.0	390.2	2993.9[b]	7.7
DB	548.9[e]	17.9	450.5	2830.7[b]	6.3

*The estimates were derived from those of Wang et al. [54].
SR = annual soil respiration, HR = annual heterotrophic respiration, SOC = soil organic carbon, GR = grassland, SH = shrubland, EC = evergreen coniferous forest, DC = deciduous coniferous forest, DB = deciduous broadleaved forest. Different lowercase letters in mean SR and SOC indicated significant differences ($P<0.05$).

To analyze the environmental differences among vegetation types, principal component analysis (PCA) was performed on correlations among ST, SWC, NDVI, SOC, STN, soil pH and SBD, and the five vegetation types were ordered by their scores on the first two principal components. The relationships between annual SR and ST, SWC, NDVI, SOC, STN and live fine root biomass were examined by linear regression. The differences of SR among vegetation types were tested using one-way ANOVA. We used the averaged value of the five subsamples in each plot to

conduct statistical analysis. All statistical analyses were performed with a significance level of 0.05, using SPSS software (2009, ver. 18.0, SPSS Inc., Chicago, IL, USA).

Results

Microenvironment of different vegetation types

The annual average ST at 5 cm deep was not significantly different between grassland and woody vegetation types, with one exception (Table 1). The DB habitat had significantly lower ST at 5 cm deep and higher SWC at 10 cm deep. No significant differences in SWC occurred among GR, EC and DC habitats. The SH and DB habitats had lower soil bulk density than other vegetation types, but the DC forest and DB forest showed higher SOC content (Table 1, $P<0.05$). Furthermore, STN content was highest in DB forest, and its pattern was consistent with that of SWC across different vegetation types. There is no significant difference in soil pH among all vegetation types (Table 1).

PCA identified three significant principle components (eigenvalue >1) of variations (Fig. 3). The first two principle components explained 83% of the total variance in the dataset (Table 2, Table 3). The first principal component was mainly associated with the differences in ST, SWC, STN and soil pH across different vegetation types. The second and third principal components were correlated to SOC and NDVI, respectively. Among the four woody vegetation types, EC is the most similar environmentally to GR, followed by SH. However, DB and DC were the different from GR along the first and second principal component, respectively.

As expected, SR (Fig. 4) and ST (Fig. 5) were higher in summer and lower in winter (Fig. 4). The mean SR showed significant differences among five vegetation types, in the order of GR < EC < SH < DC < DB (Fig. 4). The seasonal dynamics of SR were exponentially related to ST across different vegetation types, which explained 85%, 91%, 95%, 89% and 93% of the variation in SR for GR, SH, EC, DC and DB, respectively (Fig. 6).

Annual soil respiration and soil organic carbon turnover

Annual SR was lower in GR than in the woody vegetation types (Table 4). The annual SR increased by 3% following the conversion from GR to EC, 6% to SH, 14% to DC, and 22% to DB. The variations in the annual SR among the five vegetation types were significantly correlated with SOC and STN, but not with ST, SWC, NDVI and fine root biomass (Fig. 7).

In contrast with the pattern of total SR, GR had higher annual heterotrophic respiration than SH, EC and DC, but lower than

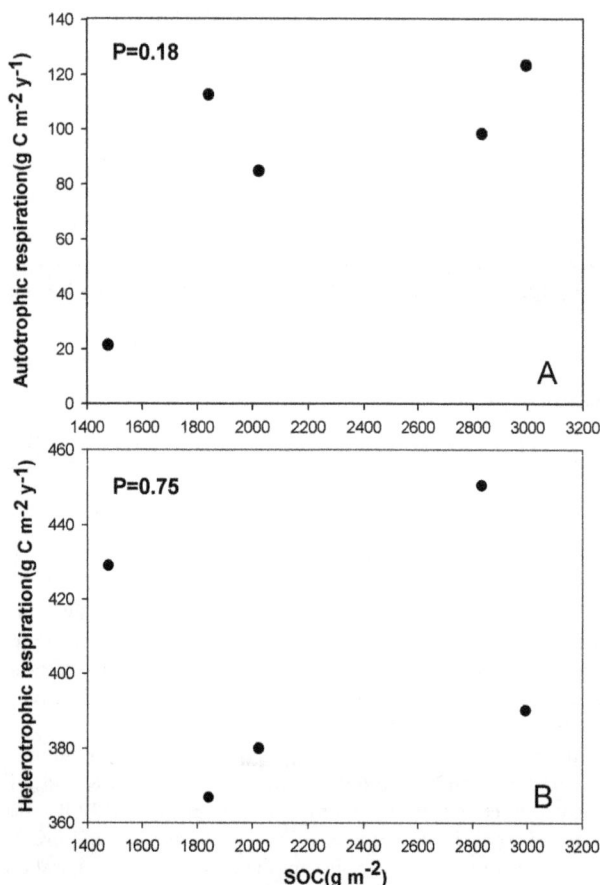

Figure 8. Relationships between heterotrophic respiration (A), autotrophic respiration (B) and soil organic carbon (SOC) across the five adjacent vegetation types.

DB (Table 4). Because GR had a lower SOC content and higher heterotrophic respiration, the residence time of SOC in GR is shorter (3.4 yr), in comparison with the relatively longer time in woody vegetation types (5.0 yr in SH, 5.3 yr in EC, 6.3 yr in DB and 7.7 yr in DC, Table 4).

Discussion

Effects of vegetation types on soil respiration

Our estimates of annual SR ranged from 450.4 to 548.9 g C m^{-2} yr^{-1}(Table 4), which fell into the range reported in temperate areas [55,56]. Among the five vegetation types, GRs showed lower SR than woodlands, which contradicted with the conclusions made using a synthesis of global data that reported SR from various types of GRs was averaged about 20% greater than various types of forests [57]. Additionally, in a juniper woodland-grassland pair in Kansas (USA), SR from GRs was 38% higher than from woodlands [58]. Our findings therefore did not support earlier generalizations that state GRs tended to allocate large proportions of their photosynthate belowground, and this results in higher SR than occurs woodlands [58]. However, in a subalpine Australian ecosystem rates of respiration in woodland soils were twice more than those in nearby grassland soils, which is similar to our results [27].

Three factors, microclimate (ST and moisture), aboveground photosynthetic supply to roots and substrate availability have been known to be important controls on SR [57]. Our results suggested that SOC and STN were major contributing factors for the variations of SR among different vegetation types (Fig. 7). Heterotrophic respiration was the dominant component of total SR, ranging from 76% to 95% (Table 4). Therefore, we expected the correlation between SR and SOC was derived from the component of heterotrophic respiration, because heterotrophic respiration is a result of the mineralization of SOC that is stored in large stocks [59,60,61] while autotrophic respiration depends on fresh photosynthates [62]. However, we did not observe a correlation between HR and SOC (Fig. 8A). Moreover, we observed an increasing trend of root respiration with an increase in SOC (Fig. 8B) although statistical test was not significant. Therefore, root respiration may be the main driver of the differences in SR between the different vegetation types. The correlation between root respiration and SOC may be attributable to the fact that higher root respiration is connected to higher photosynthetic activity [63,64,65], which will increase the carbon input to soil and therefore SOC content. Therefore, accurate discrimination of root respiration and heterotrophic respiration from the total SR was critical for gaining an improved understanding of the driving factors of SR among different vegetation types. Our results thus indicate that when modeling SR across a temperate heterogeneous landscape, we should pay more attention to the differences in heterotrophic respiration and root respiration components among different vegetation types.

Effects of vegetation types on soil carbon storage and turnover

Reported changes of SOC storage varied widely with reported increases [28,30,66], no change [58] and decreases [67,68,69]

after grasslands were converted to woodlands. Our results showed that SOC storage was 92% higher in DB and 103% higher in DC compared with GR (Table 4), suggesting possible larger amounts of organic carbon are input to soil through litterfall and root turnover in woodlands than grassland. Moreover, grasslands have been reported to have stronger wind erosion than woodlands in our study area, which reduced the soil clay and silt content, possibly explaining the potential reductions of surface organic matter in our study area [70,71]. However, no significant increase was observed when EC and SH (Table 4) were compared with GR, possibly because of the small sample size and large spatial heterogeneity seen in SOC in this study.

In our study area, mean residence time of the near-surface SOC pool in woody communities exceeded that of GR (Table 4). Similarly, McCulley *et al.* [66] also observed that both SR and mean residence time of the near-surface SOC pool in wooded communities (11 years) exceeded that of GRs (6 years) in a subtropical ecosystem. However, in paired juniper woodland and C$_4$-dominated grassland sites, longer woodland topsoil residence time (33 years) was observed than in GRs (18 years) [58]. In addition, our estimates of SOC residence time were shorter than those of McCulley *et al.* [66] and Smith & Johnson [58], partly because of the differences in the estimates of heterotrophic respiration. In our study, heterotrophic respiration comprised a large portion of total SR (76% to 95%, Table 4), whereas, Smith & Johnson [58] assumed root respiration is 50% of SR and McCulley *et al.* [66] used three scenarios in which they assumed root respiration comprised 30%, 50%, or 70% of total SR.

Conclusion

In conclusion, by determining SR and SOC dynamics in five adjacent vegetation types (GR, SH, EC, DC and DB) in the temperate area of northern China, we identified an increase in both annual SR and residence time of SOC from grassland to woody vegetation types. The increase in annual SR was coupled with changes in soil substrate availability (SOC and STN). The increases in SR suggest an increase in landscape-scale carbon emissions occurred during both natural and anthropogenic transitions occurred from grassland to plant communities dominated by woody vegetation. However, the SOC pool storage and its residence time also increased, suggesting a larger increase in carbon input than in carbon loss from the surface soil layer, thus implying an accumulation of SOC during grassland conversion into woodlands in temperate China.

Acknowledgments

We thanked two anonymous reviewers for their very constructive and valuable comments for its early version of our manuscript.

Author Contributions

Conceived and designed the experiments: WW. Performed the experiments: WW. Analyzed the data: WW WC WZ. Contributed reagents/materials/analysis tools: HZ JF. Wrote the paper: WW WZ WC.

References

1. Bond-Lamberty B, Thomson A (2010) Temperature-associated increases in the global soil respiration record. Nature 464: 579–582.
2. IPCC (2007) Climate Change 2007: The Physical Science Basis. Cambridge, UK: Contributions of Working Group I to the 4th Assessment Report of the Intergovernmental Panel on Climate Change.
3. Bahn M, Rodeghiero M, Anderson-Dunn M, Dore S, Gimeno C, et al. (2008) Soil respiration in European grasslands in relation to climate and assimilate supply. Ecosystems 11: 1352–1367.
4. Bond-Lamberty B, Bunn AG, Thomson AM (2012) Multi-year lags between forest browning and soil respiration at high northern latitudes. PloS one 7: e50441.

5. Davidson EA, Janssens IA (2006) Temperature sensitivity of soil carbon decomposition and feedbacks to climate change. Nature 440: 165–173.

6. Piao SL, Fang JY, Ciais P, Peylin P, Huang Y, et al. (2009) The carbon balance of terrestrial ecosystems in China. Nature 458: 1009–1013.

7. Houghton RA (2010) How well do we know the flux of CO_2 from land-use change? Tellus Series B-Chemical and Physical Meteorology 62: 337–351.

8. Creamer CA, Filley TR, Boutton TW, Oleynik S, Kantola IB (2011) Controls on soil carbon accumulation during woody plant encroachment: evidence from physical fractionation, soil respiration, and delta C-13 of respired CO_2. Soil Biology & Biochemistry 43: 1678–1687.

9. Eclesia RP, Jobbagy EG, Jackson RB, Biganzoli F, Pineiro G (2012) Shifts in soil organic carbon for plantation and pasture establishment in native forests and grasslands of South America. Global Change Biology 18: 3237–3251.

10. Eldridge DJ, Bowker MA, Maestre FT, Roger E, Reynolds JF, et al. (2011) Impacts of shrub encroachment on ecosystem structure and functioning: towards a global synthesis. Ecology Letters 14: 709–722.

11. Livesley SJ, Kiese R, Miehle P, Weston CJ, Butterbach-Bahl K, et al. (2009) Soil-atmosphere exchange of greenhouse gases in a *Eucalyptus marginata* woodland, a clover-grass pasture, and *Pinus radiata* and *Eucalyptus globulus* plantations. Global Change Biology 15: 425–440.

12. Wheeler CW, Archer SR, Asner GP, McMurtry CR (2007) Climatic/edaphic controls on soil carbon/nitrogen response to shrub encroachment in desert grassland. Ecological Applications 17: 1911–1928.

13. Don A, Schumacher J, Freibauer A (2011) Impact of tropical land-use change on soil organic carbon stocks – a meta-analysis. Global Change Biology 17: 1658–1670.

14. Dutra Aguiar AP, Ometto JP, Nobre C, Lapola DM, Almeida C, et al. (2012) Modeling the spatial and temporal heterogeneity of deforestation-driven carbon emissions: the INPE-EM framework applied to the Brazilian Amazon. Global Change Biology 18: 3346–3366.

15. Saner P, Loh YY, Ong RC, Hector A (2012) Carbon stocks and fluxes in tropical lowland dipterocarp rainforests in Sabah, Malaysian Borneo. PloS one 7: e29642.

16. Throop HL, Archer SR (2008) Shrub (*Prosopis velutina*) encroachment in a semidesert grassland: spatial-temporal changes in soil organic carbon and nitrogen pools. Global Change Biology 14: 2420–2431.

17. van der Werf GR, Morton DC, DeFries RS, Olivier JGJ, Kasibhatla PS, et al. (2009) CO_2 emissions from forest loss. Nature Geoscience 2: 829–829.

18. Paul KI, Polglase PJ, Nyakuengama JG, Khanna PK (2002) Change in soil carbon following afforestation. Forest Ecology and Management 168: 241–257.

19. Perez-Quezada JF, Bown HE, Fuentes JP, Alfaro FA, Franck N (2012) Effects of afforestation on soil respiration in an arid shrubland in Chile. Journal of Arid Environments 83: 45–53.

20. Poeplau C, Don A (2013) Sensitivity of soil organic carbon stocks and fractions to different land-use changes across Europe. Geoderma 192: 189–201.

21. Wiesmeier M, Sporlein P, Geuss U, Hangen E, Haug S, et al. (2012) Soil organic carbon stocks in southeast Germany (Bavaria) as affected by land use, soil type and sampling depth. Global Change Biology 18: 2233–2245.

22. Lai LM, Zhao XC, Jiang LH, Wang YJ, Luo LG, et al. (2012) Soil respiration in different agricultural and natural ecosystems in an arid region. PloS one 7: e48011.

23. Arora VK, Boer GJ (2010) Uncertainties in the 20th century carbon budget associated with land use change. Global Change Biology 16: 3327–3348.

24. Browning DM, Archer SR, Asner GP, McClaran MP, Wessman CA (2008) Woody plants in grasslands: post-encroachment stand dynamics. Ecological Applications 18: 928–944.

25. Carbone MS, Winston GC, Trumbore SE (2008) Soil respiration in perennial grass and shrub ecosystems: linking environmental controls with plant and microbial sources on seasonal and diel timescales. Journal of Geophysical Research-Biogeosciences 113: G02022.

26. Gimeno TE, Escudero A, Delgado A, Valladares F (2012) Previous land use alters the effect of climate change and facilitation on expanding woodlands of Spanish Juniper. Ecosystems 15: 564–579.

27. Jenkins M, Adams MA (2010) Vegetation type determines heterotrophic respiration in subalpine Australian ecosystems. Global Change Biology 16: 209–219.

28. Boutton TW, Liao JD, Filley TR, Archer SR (2009) Belowground carbon storage and dynamics accompanying woody plant encroachment in a subtropical savanna. In: Lal R, Follett R, editors. Soil Carbon Sequestration and the Greenhouse Effect: Soil Science Society of America, Madison, WI. 181–205.

29. Marin-Spiotta E, Sharma S (2013) Carbon storage in successional and plantation forest soils: a tropical analysis. Global Ecology and Biogeography 22: 105–117.

30. McKinley DC, Blair JM (2008) Woody plant encroachment by Juniperus virginiana in a mesic native grassland promotes rapid carbon and nitrogen accrual. Ecosystems 11: 454–468.

31. Barger NN, Archer SR, Campbell JL, Huang CY, Morton JA, et al. (2011) Woody plant proliferation in North American drylands: a synthesis of impacts on ecosystem carbon balance. Journal of Geophysical Research-Biogeosciences 116: G00K07.

32. Houghton RA, Hackler JL (2003) Sources and sinks of carbon from land-use change in China. Global Biogeochemical Cycles doi:10.1029/2002GB001970.

33. Schimel DS, House JI, Hibbard KA, Bousquet P, Ciais P, et al. (2001) Recent patterns and mechanisms of carbon exchange by terrestrial ecosystems. Nature 414: 169–172.

34. Smith P, Davies CA, Ogle S, Zanchi G, Bellarby J, et al. (2012) Towards an integrated global framework to assess the impacts of land use and management change on soil carbon: current capability and future vision. Global Change Biology 18: 2089–2101.

35. Iqbal J, Hu R, Du L, Lan L, Shan L, et al. (2008) Differences in soil CO_2 flux between different land use types in mid-subtropical China. Soil Biology & Biochemistry 40: 2324–2333.

36. Li HM, Ma YX, Aide TM, Liu WJ (2008) Past, present and future land-use in Xishuangbanna, China and the implications for carbon dynamics. Forest Ecology and Management 255: 16–24.

37. Liu F, Wu XB, Bai E, Boutton TW, Archer SR (2011) Quantifying soil organic carbon in complex landscapes: an example of grassland undergoing encroachment of woody plants. Global Change Biology 17: 1119–1129.

38. Liu H, Zhao P, Lu P, Wang YS, Lin YB, et al. (2008) Greenhouse gas fluxes from soils of different land-use types in a hilly area of South China. Agriculture Ecosystems & Environment 124: 125–135.

39. Sheng H, Yang YS, Yang ZJ, Chen GS, Xie JS, et al. (2010) The dynamic response of soil respiration to land-use changes in subtropical China. Global Change Biology 16: 1107–1121.

40. Tang XL, Liu SG, Zhou GY, Zhang DQ, Zhou CY (2006) Soil-atmospheric exchange of CO_2, CH_4, and N_2O in three subtropical forest ecosystems in southern China. Global Change Biology 12: 546–560.

41. Berthrong ST, Jobbagy EG, Jackson RB (2009) A global meta-analysis of soil exchangeable cations, pH, carbon, and nitrogen with afforestation. Ecological Applications 19: 2228–2241.

42. Fang JY, Chen AP, Peng CH, Zhao SQ, Ci L (2001) Changes in forest biomass carbon storage in China between 1949 and 1998. Science 292: 2320–2322.

43. Henry HAL (2008) Climate change and soil freezing dynamics: historical trends and projected changes. Climatic Change 87: 421–434.

44. Zhang Y, Liu H (2010) How did climate drying reduce ecosystem carbon storage in the forest-steppe ecotone? A case study in Inner Mongolia, China. Journal of Plant Research 123: 543–549.

45. You L, Shen JG, Pei H (2002) Climatic changes in recent 50 years and forecast for the next 10~25 years in Inner Mongolia. Meteorology of Inner Mongolia 4: 14–18 (in Chinese).

46. Elberling B (2007) Annual soil CO_2 effluxes in the High Arctic: the role of snow thickness and vegetation type. Soil Biology & Biochemistry 39: 646–654.

47. Kurganova I, De Gerenyu VL, Rozanova L, Sapronov D, Myakshina T, et al. (2003) Annual and seasonal CO_2 fluxes from Russian southern taiga soils. Tellus Series B-Chemical and Physical Meteorology 55: 338–344.

48. Kirschbaum MUF (2004) Soil respiration under prolonged soil warming: are rate reductions caused by acclimation or substrate loss? Global Change Biology 10: 1870–1877.

49. Leifeld J, Kogel-Knabner I (2005) Soil organic matter fractions as early indicators for carbon stock changes under different land-use? Geoderma 124: 143–155.

50. Gamon JA, Field CB, Goulden ML, Griffin KL, Hartley AE, et al. (1995) Relationships between NDVI, canopy structure, and photosynthesis in 3 Californian vegetation types. Ecological Applications 5: 28–41.

51. Ahrends HE, Etzold S, Kutsch WL, Stoeckli R, Bruegger R, et al. (2009) Tree phenology and carbon dioxide fluxes: use of digital photography at for process-based interpretation the ecosystem scale. Climate Research 39: 261–274.

52. Richardson AD, Braswell BH, Hollinger DY, Jenkins JP, Ollinger SV (2009) Near-surface remote sensing of spatial and temporal variation in canopy phenology. Ecological Applications 19: 1417–1428.

53. Jakubauskas ME, Legates DR, Kastens JH (2001) Harmonic analysis of time-series AVHRR NDVI data. Photogrammetric Engineering and Remote Sensing 67: 461–470.

54. Wang W, Peng SS, Fang JY (2010) Root respiration and its relation to nutrient contents in soil and root and EVI among 8 ecosystems, northern China. Plant and Soil 333: 391–401.

55. Raich JW, Schlesinger WH (1992) The global carbon-dioxide flux in soil respiration and its relationship to vegetation and climate. Tellus Series B-Chemical and Physical Meteorology 44: 81–99.

56. Wang W, Chen WL, Wang SP (2010) Forest soil respiration and its heterotrophic and autotrophic components: global patterns and responses to temperature and precipitation. Soil Biology & Biochemistry 42: 1236–1244.

57. Raich JW, Tufekcioglu A (2000) Vegetation and soil respiration: correlations and controls. Biogeochemistry 48: 71–90.

58. Smith DL, Johnson L (2004) Vegetation-mediated changes in microclimate reduce soil respiration as woodlands expand into grasslands. Ecology 85: 3348–3361.

59. Fang CM, Smith P, Moncrieff JB, Smith JU (2005) Similar response of labile and resistant soil organic matter pools to changes in temperature. Nature 433: 57–59.

60. Knorr W, Prentice IC, House JI, Holland EA (2005) Long-term sensitivity of soil carbon turnover to warming. Nature 433: 298–301.

61. Reichstein M, Falge E, Baldocchi D, Papale D, Aubinet M, et al. (2005) On the separation of net ecosystem exchange into assimilation and ecosystem respiration: review and improved algorithm. Global Change Biology 11: 1424–1439.

62. Högberg P, Nordgren A, Buchmann N, Taylor AFS, Ekblad A, et al. (2001) Large-scale forest girdling shows that current photosynthesis drives soil respiration. Nature 411: 789–792.
63. Bahn M, Schmitt M, Siegwolf R, Richrer A, Brüggemann N (2009) Does photosynthesis affect grassland soil-respired CO_2 and its carbon isotope composition on a diurnal timescale? New Phytologist 182: 451–460.
64. Högberg P, Singh B, Löfvenius MO, Nordgren A (2009) Partitioning of soil respiration into its autotrophic and heterotrophic components by means of tree-girdling in old boreal spruce forest. Forest Ecology & Management 257: 1764–1767.
65. Schindlbacher A, Zechmeister-Boltenstern S, Jandl R (2009) Carbon losses due to soil warming: Do autotrophic and heterotrophic soil respiration respond equally? Global Change Biology 15: 901–913.
66. McCulley RL, Archer SR, Boutton TW, Hons FM, Zuberer DA (2004) Soil respiration and nutrient cycling in wooded communities developing in grassland. Ecology 85: 2804–2817.

67. Gill RA, Burke IC (1999) Ecosystem consequences of plant life form changes at three sites in the semiarid United States. Oecologia 121: 551–563.
68. Jackson RB, Banner JL, Jobbagy EG, Pockman WT, Wall DH (2002) Ecosystem carbon loss with woody plant invasion of grasslands. Nature 418: 623–626.
69. Wei XR, Shao MG, Fu XL, Horton R, Li Y, et al. (2009) Distribution of soil organic C, N and P in three adjacent land use patterns in the northern Loess Plateau, China. Biogeochemistry 96: 149–162.
70. Zhou RL, Li YQ, Zhao HL, Drake S (2008) Desertification effects on C and N content of sandy soils under grassland in Horqin, northern China. Geoderma 145: 370–375.
71. Zhang YK, Liu HY (2010) How did climate drying reduce ecosystem carbon storage in the forest-steppe ecotone? A case study in Inner Mongolia, China. Journal of Plant Research 123: 543–549.
72. Hou XY (2001) Vegetation Atlas of China (1: 1,000,000). Beijing, China: Science Press.

19

Temperature Sensitivity and Basal Rate of Soil Respiration and Their Determinants in Temperate Forests of North China

Zhiyong Zhou[1,2]*, Chao Guo[1,2], He Meng[3]

1 Ministry of Education Key Laboratory for Silviculture and Conservation, Beijing Forestry University, Beijing, China, **2** The Institute of Forestry and Climate Change Research, Beijing Forestry University, Beijing, China, **3** College of Forestry, Inner Mongolia Agriculture University, Hohhot, China

Abstract

The basal respiration rate at 10°C (R_{10}) and the temperature sensitivity of soil respiration (Q_{10}) are two premier parameters in predicting the instantaneous rate of soil respiration at a given temperature. However, the mechanisms underlying the spatial variations in R_{10} and Q_{10} are not quite clear. R_{10} and Q_{10} were calculated using an exponential function with measured soil respiration and soil temperature for 11 mixed conifer-broadleaved forest stands and nine broadleaved forest stands at a catchment scale. The mean values of R_{10} were 1.83 μmol CO_2 m^{-2} s^{-1} and 2.01 μmol CO_2 m^{-2} s^{-1}, the mean values of Q_{10} were 3.40 and 3.79, respectively, for mixed and broadleaved forest types. Forest type did not influence the two model parameters, but determinants of R_{10} and Q_{10} varied between the two forest types. In mixed forest stands, R_{10} decreased greatly with the ratio of coniferous to broadleaved tree species; whereas it sharply increased with the soil temperature range and the variations in soil organic carbon (SOC), and soil total nitrogen (TN). Q_{10} was positively correlated with the spatial variances of herb-layer carbon stock and soil bulk density, and negatively with soil C/N ratio. In broadleaved forest stands, R_{10} was markedly affected by basal area and the variations in shrub carbon stock and soil phosphorus (P) content; the value of Q_{10} largely depended on soil pH and the variations of SOC and TN. 51% of variations in both R_{10} and Q_{10} can be accounted for jointly by five biophysical variables, of which the variation in soil bulk density played an overwhelming role in determining the amplitude of variations in soil basal respiration rates in temperate forests. Overall, it was concluded that soil respiration of temperate forests was largely dependent on soil physical properties when temperature kept quite low.

Editor: Dafeng Hui, Tennessee State University, United States of America

Funding: This study was jointly funded by the National Foundation of Natural Science of China (Grant No. 41003029) and by the Special Research Program for Public-Welfare Forestry of the State Forestry Administration of China (Grant No. 201104008). The funders had no role in study design, data collection and analysis, decision to publish, or preparation of the manuscript.

Competing Interests: The authors have declared that no competing interests exist.

* E-mail: zhiyong@bjfu.edu.cn

Introduction

CO_2 emission from soil and plants to the atmosphere determines the amplitude of feedbacks of forest ecosystems to global climate change. Accurate prediction of the amount of CO_2 respired by forest soil is of great importance in evaluating the carbon balance of forest ecosystems. In most cases, soil respiration rate at a given temperature can be estimated by the empirical functions using soil basal respiration rate (R_{10}, soil respiration rate at 10°C) and the temperature sensitivity of soil respiration (Q_{10}, a proportional change in soil respiration with a 10°C increase in temperature) [1,2,3]. Therefore, it seems vital to identify the biophysical variables driving these two parameters to advance the research on soil carbon turnover.

Soil respiration is mostly controlled by soil temperature [3,4], secondarily by soil moisture, nutrients [5], vegetation type [6], tree species composition [7], topography, and climate [8]. To increase the comparability of soil respiration rate under different environmental conditions, a standardized parameter (e.g. R_{10}) is proposed when emphasizing the effects of biophysical factors other than temperature. Although soil basal respiration may also be influenced by the similar variables mentioned above [9], it is still of importance to make clear the relationship of soil basal respiration with biophysical variables in improving the precision of simulation models. This is because, for a specific forest ecosystem, some biophysical factors can be considered as additional predictive variables when estimating soil respiration rate using empirical methods [3,10].

Great effort has been exerted to the response of soil respiration to a change in temperature in recent decades [2,11], which is denoted in most studies to be the temperature sensitivity of soil respiration, and is theoretically represented by an invariant coefficient (Q_{10}) of ~2, especially in coupled climate-carbon cycle models [12,13]. The extensive use of a fixed Q_{10} has brought large convenience in calculating the amount of CO_2 respired from soil, but it has also evoked a controversy between theoretical studies and incubation experiments or field measurements [14]. It is demonstrated that the temperature sensitivity of soil respiration (Q_{10}) can be influenced in ecosystems by many biophysical or physicochemical factors, including the forest floor conditions [15], soil physical properties [16], soil nutrients [17], and vegetation type [18]. Therefore, the Q_{10} originated from the temperature dependence equation shows distinct intersite difference or temporal variation [16,17,18]. Obviously, the application of a constant Q_{10} can not lead to an unbiased estimation of soil respiration rate for the studying ecosystem type any more.

Being illustrated by the calculation process of the common empirical function, an inherent correlation apparently exists between basal soil respiration and the temperature sensitivity [1,19]. Mathematically, Q_{10} is dependent on, and acts as a multiplier of R_{10} [19]. Any effort paid on the single parameter has limited use in improving the estimating precision of the extensively applied empirical functions.

Temperate forests in northern China mainly extend along the mountain ridge with heterogeneous growing conditions, which provide a natural experimental place for continuing similar research work on model parameters of soil respiration. In this study, we investigated the instantaneous rate of soil respiration and environmental variables at a representative catchment of the temperate forests in China, and calculated the two model parameters using the temperature dependent function. Herein hypotheses were proposed that the apparent temperature sensitivity of soil respiration could display detectable variations among forest types with different micro-environmental properties, and biophysical variables other than soil temperature could play an important role in determining soil basal respiration rate when the temperature decreased to a comparatively low level. Accordingly, our main objectives of this paper were to: 1) quantify the changing magnitude of model parameters of soil respiration within or between forest types; 2) identify the predominant variables controlling the spatial heterogeneity of the two parameters on the catchment scale in temperate forests.

Materials and Methods

Ethics statement

This research was conducted on field sites with the permission of the Taiyueshan Long-Term Forest Ecosystem Research Station. We declare that no privately owned land was used in this study, and that the field investigation did not involve any protected or endangered plant and animal species, and that no human or animal subjects were used in this study. The research has adhered to the legal requirements of China during the field study period.

Study site and experimental layout

This study was carried out at the catchment named after Xiaoshegou near the Taiyueshan Long-Term Forest Ecosystem Research Station (latitude 36°04′N, longitude 112°06′E; elevation 600 – 2600 m a.s.l), which is about 190 km southwest of Taiyuan in Shanxi province of China. Annual mean temperature varies between 10°C and 11°C, with 26°C in the warmest month of July and −23°C in coldest month of January; whilst mean annual precipitation ranges from 500 mm to 600 mm [20]. The hill in the study area is at an elevation of 1800 m with its bottom of 1200 m a.s.l. The soil type of the hill slope belongs to a Eutric Cambisols (FAO classification) or a Cinnamon soil (Chinese classification) with the mean soil depth of 30 cm to 50 cm, soil organic carbon content (SOC) of 0.77% to 5.47%, total nitrogen content (TN) of 0.036% to 0.232%, and soil pH from 6.9 to 7.6. The proportion of <0.01 mm and <0.001 mm soil fraction varies within the range of 46.54% to 63.10% and of 18.88% to 41.45%, respectively [21]. The dominant tree species in the forests are *Pinus tabuliformis*, *Quercus wutaishanica*, *Betula dahurica*, *Larix gmelinii var. principis-rupprechtii*, *Tilia mongolica*. The understory shrub community mainly consists of *Corylus mandshurica*, *Corylus heterophylla*, *Acer ginnala*, *Lespedeza bicolor*, *Philadelphus incanus*, *Rosa bella*, *Lonicera chrysantha*.

The herbaceous community is commonly composed of *Carex lanceolata*, *Spodiopogon sibiricus*, *Rubia chinensis*, *Thalictrum petaloideum*, *Melica pappiana*.

Twenty 20×20 m plots spread along four hill ridges with different topography at the small catchment, including 9 broadleaved forest stands and 11 mixed conifer-broadleaved forest stands. The forest type was classified by the basal area ratio of coniferous to broadleaved tree species. The forest community was classified as the mixed forest type when its ratio fell within the range of 20% to 80%. Forest community structure was investigated in later Aug-2009. Each plant with diameter at breast height (DBH) >5 cm was measured for values of DBH and height respectively basing on tree species for these 20 plots. On each plot, five 5×5 m subplots were established for the investigation of shrub community, and five 1×1 m subplots for herbaceous community.

Measurements of soil respiration

Soil respiration rate was measured once per month for each forest stand during the growing season of May to November in 2008 and 2009, using a Li-Cor infrared gas analyzer (LI-8100, Li-Cor Inc., Lincoln, NE, U.S.A.) equipped with a portable chamber. The chamber was put on the top of installed collars for 2 minutes before measurements. In early April, nine polyvinyl chloride (PVC) collars were evenly placed on each plot with eight collars arranged in a circle at 5 m to the plot center and one right at the center. The PVC collar of 10 cm in diameter and 5 cm in height was permanently inserted 3 cm into the soil with 2 cm remaining above the surface of the forest floor. The live herbs or seedlings were carefully removed out the collars to avoid bias due to its respiratory activity just after plant growth occurred. Concurrently, soil temperature at 10 cm depth adjacent to each PVC collar was monitored using a thermocouple probe attached to LI-8100 system. The averaged data of soil respiration and soil temperature across the nine PVC collars per month were fitted to the following exponential model [1,10] to calculate basal parameters of soil respiration for each forest plot.

$$R_s = \alpha \times e^{\beta T} \tag{1}$$

where R_s is in situ soil respiration rate measured in the field, α and β are model parameters, T is the measured soil temperature. According to equation (1), the temperature sensitivity of soil respiration was calculated by:

$$Q_{10} = e^{10\beta} \tag{2}$$

Soil basal respiration was calculated by:

$$R_{10} = \alpha \times e^{10\beta} \tag{3}$$

Measurements of environmental variables

Shrub community was investigated by species for plant density and biomass of a representative sampling plant. The sampling plants were harvested and brought back to laboratory, and oven dried at 75°C to constant weight. The biomass of each shrub species within the community was estimated basing on plant density and its mean weight. The herbaceous plants in the 1×1 m subplot were all harvested for aboveground components. Additionally, litter on the forest floor was also collected in five 30×30 cm subplots on each plot. The herbaceous plant samples and litter were separately placed in envelope, transported to laboratory, and oven dried at 75°C for at least 48 h before weighing.

Soil cores of 4 cm in diameter and 20 cm in depth were sampled at five measurement points on each plot in later growing season of 2009. The air dried soil samples were mound to pass a 0.2 mm sieve for nutrient analysis after visible litter segments were picked out by hand. SOC and TN were determined separately following the modified Mebius method [22] and the Kjeldahl digestion procedure [23]. Soil phosphorus (P) was measured using the colorimetric determination method described by John [24]. Soil pH was measured in deionized H_2O using Sartorius AG (PB-10, Sartorius, Germany), after equilibration for 1 h in a water: soil ratio of 2.5:1. Soil cores were additionally excavated by a cylindrical sampler of 100 cm^3 at five sampling positions on each plot, and oven dried at 110°C for at least 48 h in laboratory to measure the soil bulk density.

Data analyses

Soil physicochemical properties were also averaged for each plot when their effects on R_{10} and Q_{10} were analyzed. A two-tailed t-test was applied to detect the differences of R_{10} and Q_{10} between these two forest types at $\alpha = 0.05$. The spatial variability was expressed using the coefficient of variation (CV) calculated as the following.

$$CV = \frac{SD}{M} \times 100\% \qquad (4)$$

where SD means standard deviation, and M represents mean value.

All these data analyses were carried out using the software of SPSS 15.0. Figures were made using the software of SigmaPlot in version 10.0.

In order to test the combined contribution of biophysical variables to the variability of R_{10} and Q_{10}, redundancy analysis (RDA) [25] was conducted with R_{10} and Q_{10} as dependent variables and with selected biophysical variables, i.e. DBH, soil pH, variances in soil bulk density, soil TN and soil pH, as explanatory variables. RDA was performed using the software of Canoco for Windows 4.5.

Results

Inter- and intra-forest-type variations in basal parameters of soil respiration

R_{10} and Q_{10} were on average 10% and 11% higher in the broadleaved forest stands than in the mixed forest stands, although no statistically significant difference was detected between these two forest types ($P = 0.25$ for R_{10} and 0.91 for Q_{10}). There existed large spatial heterogeneity in temperature sensitivity and basal rate of soil respiration among forest stands. The CV of R_{10} ranged from 11% in the broadleaved forest to 19% in the mixed forest, and the CV of Q_{10} varied from 24% in the broadleaved forest to 29% in the mixed forests (Table 1).

Particularly, in the mixed forest stands, R_{10} was significantly affected by the basal area ratio between coniferous and broadleaved tree species, and greatly declined with the percentage of coniferous tree species. No significant correlation was found between Q_{10} and the basal area ratio in mixed forest stands (Fig. 1).

Determinant variables of soil basal respiration rate

R_{10} was mainly influenced by soil nutrient content and rose linearly with CV of SOC and CV of TN; soil temperature range during which soil respiration was monitored was significantly correlated with R_{10} in the mixed forest (Fig. 2A, 2B, and 3). Contrarily, in the broadleaved forest stands, R_{10} was largely determined by the basal area and the spatial variations of shrub carbon stock and soil phosphorus content (Fig. 2C and 4). There was a linearly inverse relationship between R_{10} and the basal area, whereas R_{10} increased differentially with increasing variations in shrub carbon stock and soil P in the broadleaved forest stands.

Determinant variables of the temperature sensitivity of soil respiration

In the mixed forest, Q_{10} was positively correlated with CV of herbaceous carbon stock and CV of soil bulk density (Fig. 5 and 6B), and negatively with soil C/N ratio (Fig. 7C). In the broadleaved forest, Q_{10} notably decreased with soil pH (Fig. 6A), but significant positive correlations were found between Q_{10} and CV of SOC and TN (Fig. 7A and B).

Combined relationships among R_{10}, Q_{10} and biophysical factors

Although many environmental factors were found in this study to independently exert significant effects on individual parameter of R_{10} or Q_{10}, 51% of the variations in both R_{10} and Q_{10} on the spatial scale were explained jointly by five biophysical variables, i.e., CV of soil bulk density, DBH, CV of soil TN, soil pH, and CV of soil pH, after forward selection of environmental variables. Particularly, most of the variations in R_{10} and Q_{10} were mainly ascribed to the variance of soil bulk density (Table 2). In addition, the importance of these selected factors was also highlighted by the result of Redundancy analysis, which showed that Axis 1 and Axis 2 accounted for 86.3% and 13.7% of the total variance in basal parameters of soil respiration, respectively (Fig. 8).

Discussion

Variation in R_{10} and its determining variables

Soil respiration rate at 10°C has received little attention in contrast to the instantaneous rate of soil respiration in the study of soil carbon cycle. Moreover, the comparability of R_{10} under changing circumstances is more reasonable than that of normally measured soil respiration rate. Even at the same temperature of 10°C, soil basal respiration still exhibits a large variation within or across forest types. On the scale of the catchment, R_{10} varies in a range of 1.25 μmol CO_2 m^{-2} s^{-1} to 2.30 μmol CO_2 m^{-2} s^{-1} in the mixed forest with a mean value of 1.83 μmol CO_2 m^{-2} s^{-1}. R_{10} changes from 1.59 μmol CO_2 m^{-2} s^{-1} to 2.46 μmol CO_2 m^{-2} s^{-1} in the broadleaved forest with a mean rate of 2.01 μmol CO_2 m^{-2} s^{-1}. These values of R_{10} just fall well in the range of R_{10} of different forests in the same region [3], but they are slightly higher than those for pine and oak forests in Brasschaat [26]. Given that the similar empirical function has been applied in calculating the basal rate of soil respiration, the variation of R_{10} is greatly induced by environmental factors other than soil temperature.

Stand structure has been indicated to be a dominant factor accounting for the spatial variation in soil respiration in beech and mixed-dipterocarp forests. Basal area and DBH exert a significant positive effect on soil respiration [4,27]. But, as to the specific results of this study, significant negative correlations are found between stand structure parameters and R_{10} for both forest types in the temperate region of North China. R_{10} declines differentially with the percentage of coniferous tree species in mixed forest community, and with the basal area across broadleaved forest stands. This intriguing scenario may be ascribed to the complexity

Table 1. Variation in basal parameters of soil respiration within or between forest types.

	R_{10} (μmol CO_2 m^{-2} s^{-1})	CV of R_{10} (%)	Q_{10}	CV of Q_{10} (%)
Broadleaved forest	2 a	11	4 a	24
Mixed forest	1 a	19	3 a	29

The significance of differences of basal parameters between forest types were separately tested by independent t - test (two - tailed) at $\alpha = 0.05$ (n = 9 in the broadleaved forest, and 11 in the mixed forest). Same lowercase letter means no significant difference is detected at $\alpha = 0.05$ within 95% confidence interval between the two forest types.

of CO_2 production in forest soils. Soil respiration consists of autotrophic respiration from roots and rhizosphere and heterotrophic respiration from microbial decomposition. Total rate and basal rate of soil respiration have been found to be slightly higher in the pure broadleaved forest stands than in the pure coniferous forest stands [3]. R_{10} is apparently depressed by the increasing admixed proportion of needle leaf tree species in the mixed forests. Perhaps, it is ascribed to the physiological differences between coniferous and broadleaved tree species. R_{10} is indicated to be modulated by plant photosynthesis (i.e. gross primary productivity) [9] via determining the activity of rhizosphere respiration [28]. Autotrophic respiration accounts for ~50% of total soil respiration, which may even be higher in growing season for temperate forests [29]. In cold weather with temperature at ~10°C, photosynthetic activity of the mixed forest stands with higher basal area can be heavily impeded, subsequently resulting in a lower R_{10}.

Conversely, R_{10} significantly increases with the heterogeneous properties of shrub carbon stock and soil P content in the broadleaved forest stands and by the spatial variations in soil organic carbon content and TN content in the mixed forest stands. This may be due to the dominance of microbial respiratory fraction in total soil respiration at lower temperature. It is the microbial community composition and climatic factors that control forest soil respiration in cold seasons [30]. Additionally, soil microbial biomass and respiration have been eventually influenced by soil biophysical properties [31] and by environmental biochemical processes [32] through substrate availability, which

indicates that soil respiration is essentially an enzymatic controlled process [9,19].

Variation in Q_{10} and its determinants

The temperature sensitivity of soil respiration, Q_{10}, shows evident intra- and inter-forest-type variations on the catchment scale in temperate forests of northern China. The average Q_{10} values across forest stands for each forest type are larger than those of young plantations and a secondary *Populus davidiana* stand in the semiarid Loess Plateau [3]. However, the average Q_{10} is comparable to the reported values by Peng et al. [18] and by Zheng et al. [17] through synthesizing a great number of studies about temperature sensitivity of soil respiration on the spatial scales from region to country. In essence, variant Q_{10} values demonstrate the deficiency of the temperature dependent functions in describing the sensitivity of soil respiration to temperature.

The apparent Q_{10} derived from field experiment is actually a combined temperature sensitivity of different fractions of soil CO_2 flux [10]. Particularly, the enzymatic reactivity of substrate decomposition to temperature is considered as the intrinsic Q_{10} [19]. Although the Q_{10} value of experimental study is suggested to be influenced by a wide range of ecological variables from molecular structure to climatic factors [18,19], the direct determinant of the temperature sensitivity of soil respiration is still dependent on the substrate availability [19]. In this study, we find that Q_{10} could be markedly influenced by soil C/N ratio, soil pH, and the spatial heterogeneous properties of herbaceous carbon stock, SOC, soil TN, and soil bulk density. It is also worth mentioning that the contributors to the variations of Q_{10} differ

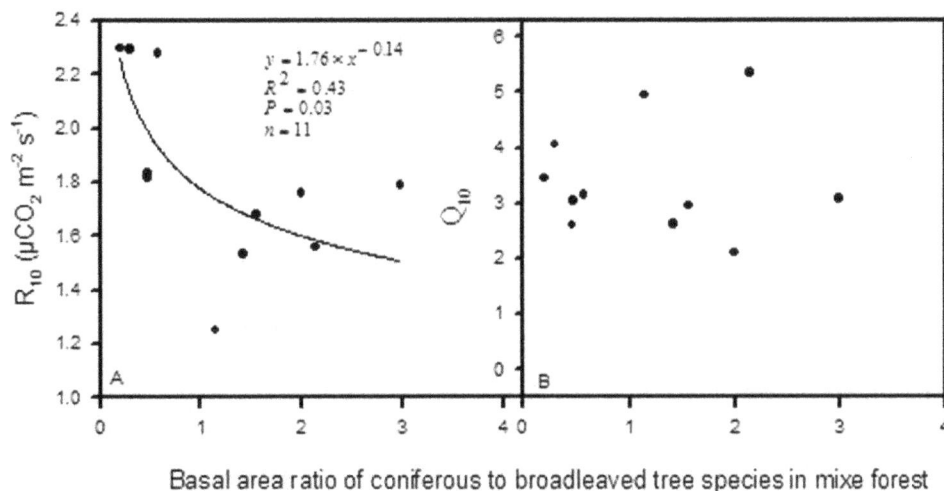

Basal area ratio of coniferous to broadleaved tree species in mixe forest

Figure 1. Trends of R_{10} and Q_{10} with basal area ratio of coniferous to broadleaved tree species.

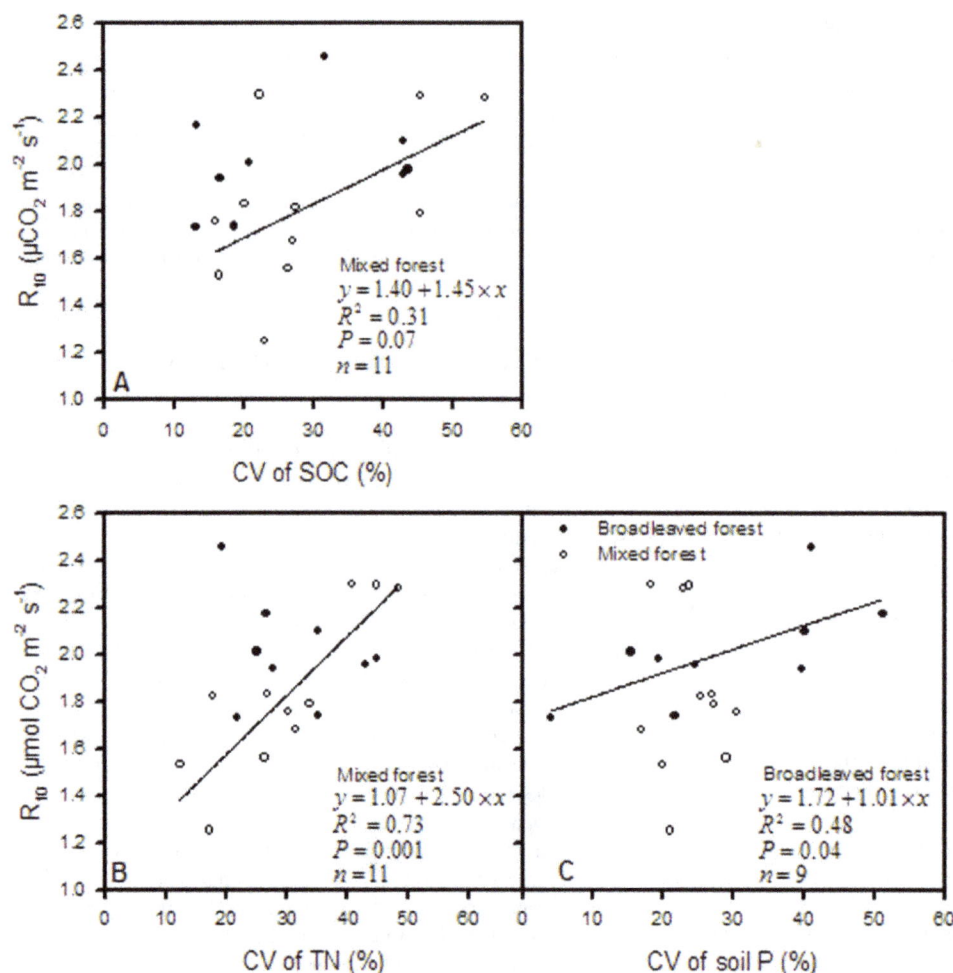

Figure 2. Correlations of R_{10} with the variations of SOC, soil TN, and soil P in broadleaved and mixed forests.

considerably between these two forest types. Similar results have been reported that the Q_{10} values changed with the alteration of ecosystems and vegetation types [17,18]. Indeed, the extrinsic

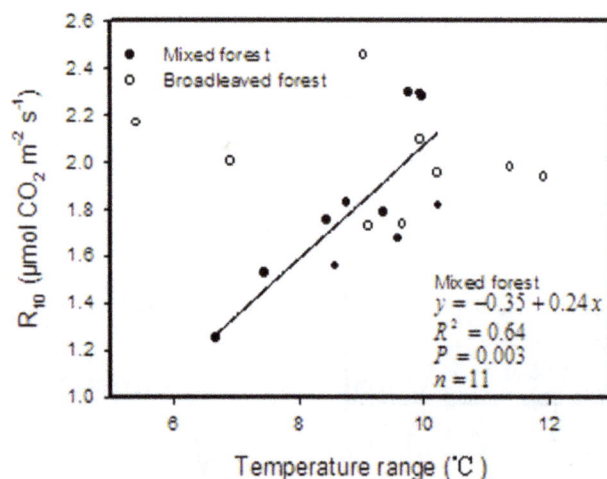

Figure 3. Correlations of R_{10} with soil temperature range for both forest types.

factors pose the effect on temperature sensitivity mainly via the primary control of substrate availability.

It has been recognized that the assumption of constant Q_{10} of soil respiration is incorrect [9,16], because the sensitivity of soil respiration to temperature is a complex reactivity; also it is more than being described by a simple parameter of temperature-dependent models. To date, a consensus has not been reached to clarify the mechanism underpinning the temperature sensitivity of soil respiration, but the study of easily monitored variables, such as soil physicochemical properties, SOC, soil TN, and forest type, etc., will help add extra predictive factors other than temperature in interpreting the variability of the apparent Q_{10}.

The effects of forest types on the correlations between biophysical variables and R_{10} and Q_{10}

R_{10} and Q_{10} have been demonstrated by our results to be influenced by biophysical factors and their spatial variation in forest stands. Although similar intrinsic mechanisms account for the variations of R_{10} and Q_{10} with forest microenvironments, the specific determining factors of soil basal respiration still vary with forest type. This is because forest type consisting of different tree species displays great distinctions in biotic and abiotic variables, which ultimately manipulate the changing gradient and direction of R_{10} and Q_{10}.

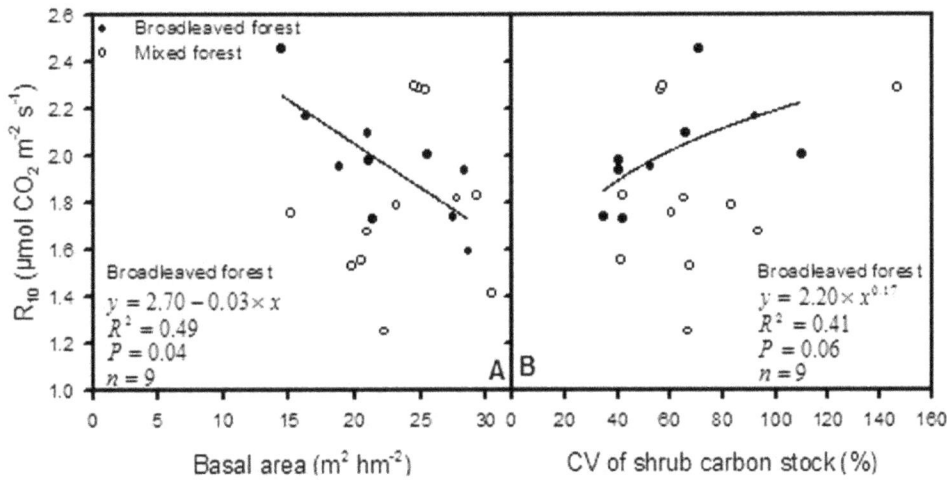

Figure 4. Correlations of R_{10} with basal area and CV of shrub carbon stock separately for both forest types.

At our study site, the mixed forest type is mainly composed of the coniferous tree species *P. tabuliformis* and the broadleaved tree species *Q. wutaishanica*. *P. tabuliformis* in forest ecosystem perhaps takes responsibility for the distinct correlations of R_{10} and Q_{10} with biophysical variables between broadleaved and mixed forest types, because tree species determines not only the microbial community structure but also the decomposition dynamics of forest litter [33,34]. Differences in the mycorrhizospheres and hyphospheres are the substantial way through which tree species affect soil microbial community including bacteria, archaea, fungi, and both free-living and symbiotic organisms [34]. Greater catabolic diversity and different bacterial and fungal communities have been found in the surface soil layer beneath mixed species plantations [35], and ectomycorrhizal fungi has been indicated to correlate with the presence of pine trees [36]. In addition, labile or soluble organic matter could also be affected by forest type with quantity and quality differences in litter and root exudates. This may induce the variations of soil microbial and enzyme activities between broadleaved and mixed forest types [37,38]. Obviously, the anisotropic response of heterotrophic respiration derived from microbial activity to biophysical factors may account for the variant correlations of Q_{10} and R_{10} with measured variables between the two forest types.

In general, Q_{10} has a mathematical interrelationship with R_{10} and they also can be expressed by each other [10]. Furthermore, both Q_{10} and R_{10} display the confounding reactions of the complex process of soil respiration to the changes in exterior environmental factors. Therefore it can improve the overall understanding of the underlying mechanism driving soil respiration to concurrently analyze the variances of R_{10} and Q_{10} and their determinants. Although a single variable can explain larger variance of R_{10} or Q_{10}, the comparatively lower attribution of the

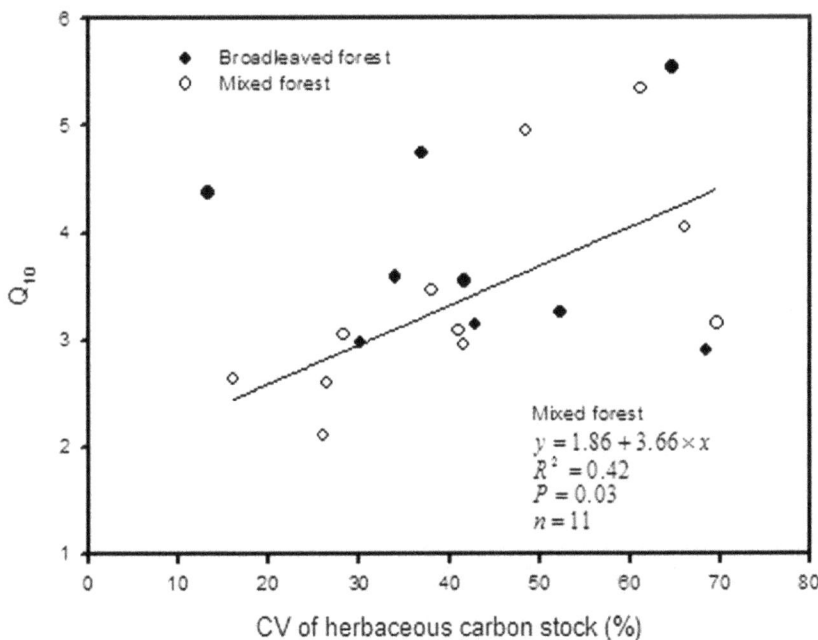

Figure 5. Relationships of Q_{10} with variation of herbaceous carbon stock in both forest types.

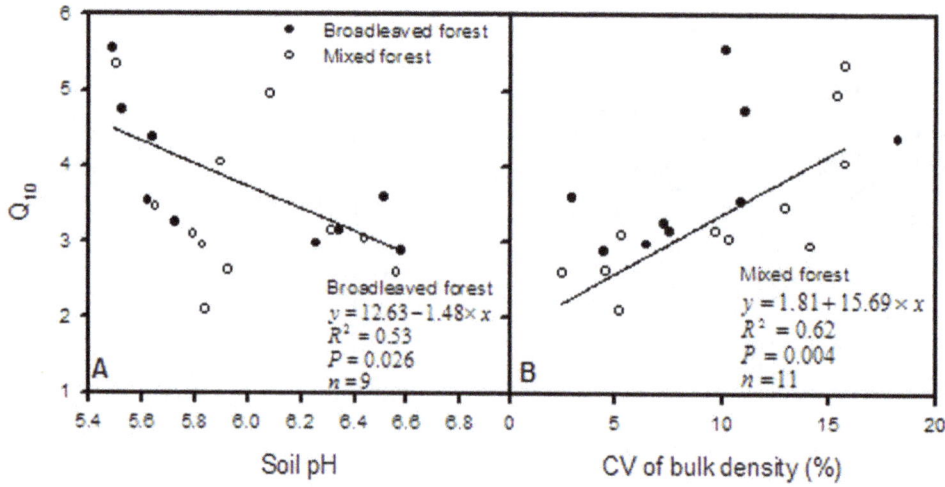

Figure 6. Relationships of Q_{10} with soil physical factors separately in broadleaved and mixed forest types.

variation of Q_{10} and R_{10} to the five selected variables demonstrates the inherent interactions existing among biotic and abiotic variables. It is the internal interaction that determines the amplitudes of soil basal respiration rates with varying environmental conditions across temporal or spatial scales [10]. This viewpoint is also supported by the study of SØE and Buchmann [4]. Therefore, a more accurate estimation of soil CO_2 efflux cannot be achieved for a specific forest ecosystem until the changes in Q_{10} and R_{10} are concurrently taken into account with alterations of microenvironment.

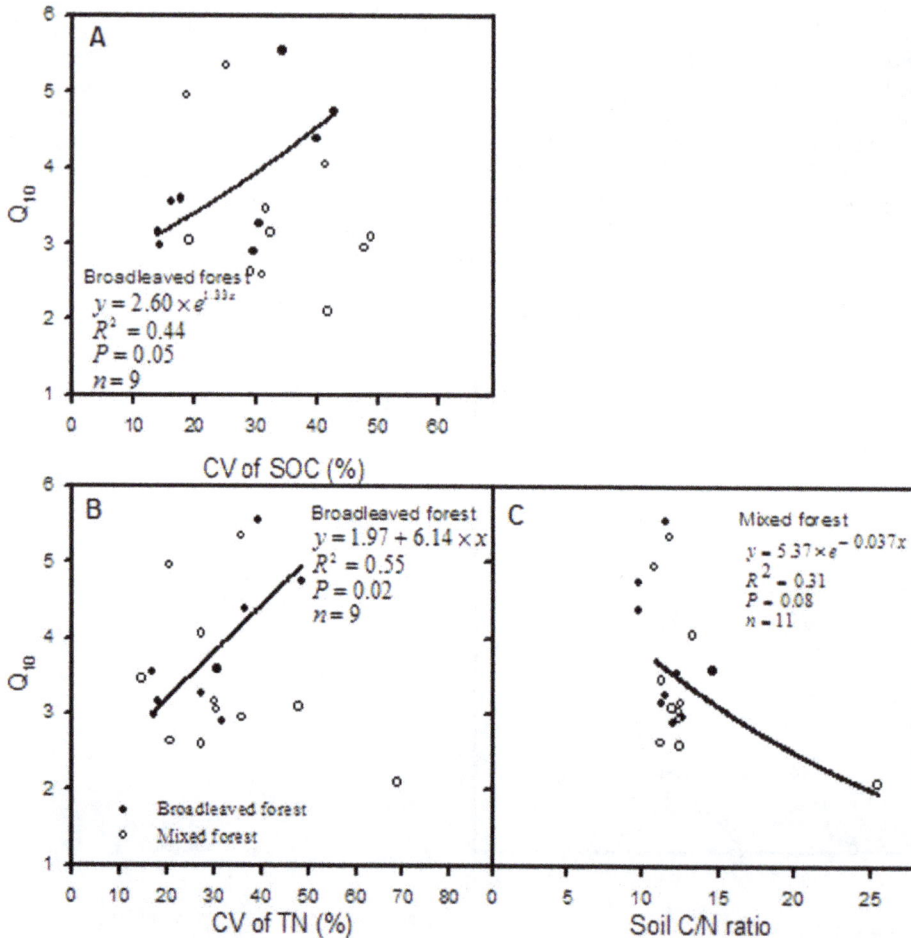

Figure 7. Relationships between soil chemical properties and Q_{10} respectively in broadleaved and mixed forest types.

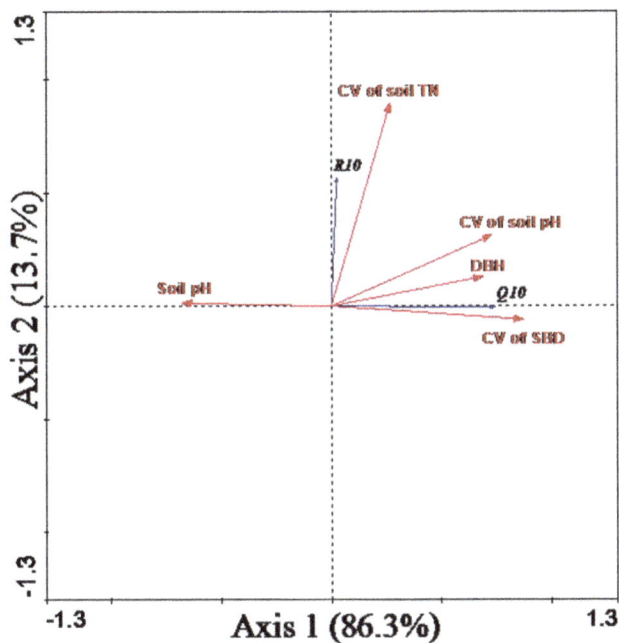

Figure 8. Redundancy analysis (RDA) among Q_{10}, R_{10} and the biophysical variables. DBH means diameter at breast height; SBD means soil bulk density.

Table 2. The effects of biophysical variables on R_{10} and Q_{10} analyzed by the method of Redundancy Analysis (RDA).

Variables	Lambda-1[a]	Lambda-A[b]	P[c]	F[d]
CV of SBD	0.340	0.34	0.004	9.28
DBH	0.244	0.07	0.152	2.01
CV of soil TN	0.213	0.06	0.178	1.93
Soil pH	0.204	0.03	0.454	0.81
CV of pH	0.088	0.01	0.788	0.25

SBD: soil bulk density; DBH: diameter at breast height.
[a]Describe marginal effects, which shows the variance when the variable is used as the only factor.
[b]Describe conditional effects, which shows the additional variance each variable explains when it is included in the model.
[c]The level of significance corresponding to Lambda-A when performing Monte Carlo test (with 499 random permutations) at the 0.05 significance level.
[d]The Monte Carlo test statistics corresponding to Lambda-A at the 0.05 significance level.

figure edition. The Taiyueshan Long-Term Forest Ecosystem Research Station is also appreciated for the access to conduct field work on its experimental sites. Fengfeng Kang is especially acknowledged by us for editing an earlier version of the manuscript. We also thank two anonymous reviewers for their valuable comments and suggestions in improving the manuscript.

Acknowledgments

We are grateful to Zhongkui Luo for his assistance in field investigation and to Yeming You and Hua Su for their advices on Redundancy analysis and

Author Contributions

Conceived and designed the experiments: ZZ HM. Performed the experiments: ZZ CG. Analyzed the data: ZZ. Contributed reagents/materials/analysis tools: ZZ CG. Wrote the paper: ZZ.

References

1. van't Hoff, JH (1898) Lectures on theoretical and physical chemistry. In: Lehfeldt, RA. (Ed.), Chemical Dynamics: Part 1. Edwart Arnold, London, UK, pp. 224–229 (translated from German).

2. Boone RD, Nadelhoffer KJ, Canary JD, Kaye JP (1998) Roots exert a strong influence on the temperature sensitivity of soil respiration. Nature 396: 570–572.

3. Zhou Z, Zhang Z, Zha T, Luo Z, Zheng J, et al. (2013) Predicting soil respiration using carbon stock in roots, litter and organic matter in forests of Loess Plateau in China. Soil Biol Biochem 57: 135–143. doi:10.1016/j.soilbio.2012.08.010

4. SØE ARB, Buchmann N (2005) Spatial and temporal variations in soil respiration in relation to stand structure and soil parameters in an unmanaged beech forest. Tree Physiology 25: 1427–1436.

5. Rodeghiero M, Cescatti A (2005) Main determinants of forest soil respiration along an elevation/temperate gradient in the Italian Alps. Global Change Biol 11: 1024–1041. doi:10.1111/j.1365-2486.2005.00963.x

6. Jenkins M, Adams MA (2010) Vegetation type determines heterotrophic respiration in subalpine Australian ecosystems. Global Change Biol 16: 209–219. doi:10.1111/j.1365-2486.2009.01954.x

7. Bréchet L, Ponton S, Roy J, Freycon V, Coûteaux M-M, et al. (2009) Do tree species characteristics influence soil respiration in tropical forests? A test based on 16 tree species planted in monospecific plots. Plant Soil 319: 235–246. doi: 10.1007/s11104-008-9866-z

8. Kang SY, Doh S, Lee D, Lee D, Jin VL, et al. (2003) Topographic and climatic controls on soil respiration in six temperate mixed-hardwood forest slopes, Korea. Global Change Biol 9: 1427–1437.

9. Sampson DA, Janssens IA, Curiel Yuste J, Ceulemans R (2007) Basal rates of soil respiration are correlated with photosynthesis in a mixed temperate forest. Global Change Biol 13: 2008–2017. doi: 10.1111/j.1365-2486.2007.01414.x

10. Davidson EA, Janssens IA, Luo YQ (2006) On the variability of respiration in terrestrial ecosystems: moving beyond Q_{10}. Global Change Biol 12: 154–164.

11. Lloyd J, Taylor JA (1994) On the temperature dependence of soil respiration. Funct Ecol 8: 315–323.

12. Tjoelker MG, Oleksyn J, Reich PB (2001) Modelling respiration of vegetation: evidence for a general temperature-dependent Q_{10}. Global Change Biol 7: 223–230.

13. Friedlingstein P, Cox P, Betts R, Bopp L, von Bloh W, et al. (2006) Climate-carbon cycle feedback analysis: Results from the C^{4}MIP model intercomparison. J Climate 16: 3337–3353.

14. Knorr W, Prentice IC, House JI, Holland EA (2005) Long-term sensitivity of soil carbon turnover to warming. Nature 433: 298–301. doi:10.1038/nature03226

15. Malcolm GM, López-Gutiérrez JC, Koide RT (2009) Temperature sensitivity of respiration differs among forest floor layers in a Pinus resinosa plantation. Soil Biol Biochem 41: 1075–1079. doi:10.1016/j.soilbio.2009.02.011

16. Janssens IA, Pilegaard K (2003) Large seasonal changes in Q_{10} of soil respiration in a beech forest. Global Change Biol 9: 911–917.

17. Zheng ZM, Yu GR, Fu YL, Wang YS, Sun XM, et al. (2009) Temperature sensitivity of soil respiration is affected by prevailing climatic conditions and soil organic carbon content: A trans-China based case study. Soil Biol Biochem 41: 1531–1540. doi:10.1016/j.soilbio.2009.04.013

18. Peng S, Piao S, Wang T, Sun J, Shen Z (2009) Temperature sensitivity of soil respiration in different ecosystems in China. Soil Biol Biochem 41: 1008–1014. doi:10.1016/j.soilbio.2008.10.023

19. Davidson EA, Janssens IA (2006) Temperature sensitivity of soil carbon decomposition and feedbacks to climate change. Nature 440: 165–173. doi:10.1038/nature04514

20. Ma QY (1983) Determination of the biomass of individual trees in stands of Pinus Tabulaeformis plantation in north China. J Beijing Forestry College 4: 1–18. (in Chinese with English abstract)

21. Guo JT, Wang WX, Cao XF, Wang ZQ, Lu BL (1992) A study on forest soil groups of Taiyue mountain. Journal of Beijing Forestry University 14: 134–142. (in Chinese with English abstract)

22. Nelson DW, Sommers LE (1982) Total carbon, organic carbon, and organic matter. In: Page AL, Miller RH, Keeney DR (eds) Methods of soil analysis. American Society of Agronomy and Soil Science Society of American, Madison, pp 101–129.

23. Gallaher RN, Weldon CO, Boswell FC (1976) A semiautomated procedure for total nitrogen in plant and soil samples. Soil Sci Soc Am J 40: 887–889.

24. John MK (1970) Colorimetric determination of phosphorus in soil and plant materials with ascorbic acid. Soil Sci 109: 214–220.

25. Lep J, milauer P (2003) Multivariate analysis of ecological data using Canoco, 1st edn. Cambridge University Press, New York.

26. Curiel Yuste J, Janssens IA, Carrara A, Ceulemans R (2004) Annual Q_{10} of soil respiration reflects plant phenological patterns as well as temperature sensitivity. Global Change Biol 10: 161–169. doi:10.1111/j.1529-8817.2003.00727.x

27. Katayama A, Kume T, Komatsu H, Ohashi M, Nakagawa M, et al. (2009) Effect of forest structure on the spatial variation in soil respiration in a Bornean

tropical rainforest. Agr Forest Meteorol 149: 1666–1673. doi: 10.1016/j.agroformet.2009.05.007

28. Levy-Varon JH, Schuster WSF, Griffin KL (2012) The autotrophic contribution to soil respiration in a northern temperate deciduous forest and its response to stand disturbance. Oecologia 169: 211–220. doi:10.1007/s00442-011-2182-y

29. Hanson PJ, Edwards NT, Garten CT, Andrews JA (2000) Separating root and soil microbial contribution to soil respiration: A review of methods and observations. Biogeochemistry 48: 115–146.

30. Monson RK, Lipson DL, Burns SP, Turnipseed AA, Delany AC, et al. (2006) Winter forest soil respiration controlled by climate and microbial community composition. Nature 439: 711–714. doi:10.1038/nature04555

31. Dupuis EM, Whalen JK (2007) Soil properties related to the spatial pattern of microbial biomass and respiration in agroecosystems. Can J Soil Sci 87: 479–484.

32. Resat H, Bailey V, McCue LA, Konopka A (2012) Modeling microbial dynamics in heterogeneous environments: growth on soil carbon sources. Microb Ecol 63: 883–897. doi:10.1007/s00248-011-9965-x

33. Vivanco L, Austin AT (2008) Tree species identity alters forest litter decomposition through long-term plant and soil interactions in Patagonia, Argentina. J Ecol 96: 727–736. doi: 10.1111/j.1365-2745.2008.01393.x

34. Prescott CE, Grayston SJ (2013) Tree species influence on microbial communities in litter and soil: Current knowledge and research needs. Forest Ecol Manag 309: 19–27.

35. Jiang Y, Chen C, Xu Z, Liu Y (2012) Effects of single and mixed species forest ecosystems on diversity and function of soil microbial community in subtropical China. J Soil Sediment 12: 228–240.

36. Hackl E, Pfeffer M, Donat C, Bachmann G, Zechmeister-Boltenstern S (2005) Compostion of the microbial communities in the mineral soil under different types of natural forest. Soil Biol Biochem 37: 661–671.

37. Xing SH, Chen CR, Zhou BQ, Zhang H, Nang ZM, et al. (2010) Soil soluble organic nitrogen and active microbial characteristics under adjacent coniferous and broadleaf plantation forests. J Soil Sediment 10: 748–757.

38. Cheng F, Peng XB, Zhao P, Yuan J, Zhong CG, et al. (2013) Soil microbial biomass, basal respiration and enzyme activity of main forest types in the Qinling Mountains. Plos One 8: e67353. doi:10.1371/journal.pone.0067353.

20

A New Estimation of Global Soil Greenhouse Gas Fluxes Using a Simple Data-Oriented Model

Shoji Hashimoto*

Department of Forest Site Environment, Forestry and Forest Products Research Institute (FFPRI), 1 Matsunosato, Tsukuba, Ibaraki, Japan

Abstract

Soil greenhouse gas fluxes (particularly CO_2, CH_4, and N_2O) play important roles in climate change. However, despite the importance of these soil greenhouse gases, the number of reports on global soil greenhouse gas fluxes is limited. Here, new estimates are presented for global soil CO_2 emission (total soil respiration), CH_4 uptake, and N_2O emission fluxes, using a simple data-oriented model. The estimated global fluxes for CO_2 emission, CH_4 uptake, and N_2O emission were 78 Pg C yr^{-1} (Monte Carlo 95% confidence interval, 64–95 Pg C yr^{-1}), 18 Tg C yr^{-1} (11–23 Tg C yr^{-1}), and 4.4 Tg N yr^{-1} (1.4–11.1 Tg N yr^{-1}), respectively. Tropical regions were the largest contributor of all of the gases, particularly the CO_2 and N_2O fluxes. The soil CO_2 and N_2O fluxes had more pronounced seasonal patterns than the soil CH_4 flux. The collected estimates, including both the previous and the present estimates, demonstrate that the means of the best estimates from each study were 79 Pg C yr^{-1} (291 Pg CO_2 yr^{-1}; coefficient of variation, CV = 13%, $N=6$) for CO_2, 21 Tg C yr^{-1} (29 Tg CH_4 yr^{-1}; CV = 24%, $N=24$) for CH_4, and 7.8 Tg N yr^{-1} (12.2 Tg N_2O yr^{-1}; CV = 38%, $N=11$) for N_2O. For N_2O, the mean of the estimates that was calculated by excluding the earliest two estimates was 6.6 Tg N yr^{-1} (10.4 Tg N_2O yr^{-1}; CV = 22%, $N=9$). The reported estimates vary and have large degrees of uncertainty but their overall magnitudes are in general agreement. To further minimize the uncertainty of soil greenhouse gas flux estimates, it is necessary to build global databases and identify key processes in describing global soil greenhouse gas fluxes.

Editor: Carl J. Bernacchi, University of Illinois, United States of America

Funding: This research was supported by Forestry and Forest Products Research Institute, Japan. Also, this work was supported by a grant from the Japan Society for the Promotion of Science KAKENHI (grant number 24510025). The funders had no role in study design, data collection and analysis, decision to publish, or preparation of the manuscript.

Competing Interests: The author has declared that no competing interests exist.

* E-mail: shojih@ffpri.affrc.go.jp

Introduction

Soil greenhouse gas (GHG; particularly CO_2, CH_4, and N_2O) fluxes are a key component to understanding climate change. CO_2 is produced by mostly heterotrophic organisms and plant root respiration and is emitted from the soil surface to the atmosphere [1–2]. Soil is generally a sink of atmospheric CH_4 through oxidation in the soil [3–4], but the soil in wetlands is a strong source of CH_4. In general, N_2O is released from the soil surface to the atmosphere [5–6] and is the result of N_2O production and consumption processes in soil [7]. The soil CO_2 flux is the largest component of the soil GHG fluxes, and it nearly counterbalances the plant carbon fixation. However, considering their global warming potentials, CH_4 and N_2O fluxes are also important components. Moreover, it is reported that recent changes in the climate may increase these soil GHG fluxes both globally and regionally [2] [8].

Despite the importance of these soil GHG fluxes, the number of reports on global soil GHG fluxes remains limited. In general, these estimations have been performed using detailed process-oriented models [6] [9] or simple data-oriented models [2] that entail data synthesis, and these two approaches compensate for the disadvantages of each. For example, simple data-oriented models cannot trace detailed processes and may not be suitable for long-term predictions, but they can provide more data-oriented estimates. Also, simple data-oriented models provide benchmarks

against results from more detailed, process-oriented models [1] [10].

The objective of this paper is to report new global estimates of soil CO_2 emission (total soil respiration), CH_4 uptake, and N_2O emission fluxes. First, I report new global estimates that were estimated using a simple data-oriented model [8] [11]. The soil GHG flux submodels describe each gas flux simply in terms of three functions: the soil physiochemical properties, water-filled pore space, and soil temperature. The total fluxes, spatial distribution, and seasonality of each flux were estimated. Here, the average fluxes between 1980 and 2009 are provided. Second, the global estimates reported in previous studies were compiled, and I report the means of the best estimates from each study.

Results

The estimated global fluxes of CO_2 emission, CH_4 uptake, and N_2O emission were 78 Pg C yr^{-1} (Monte Carlo 95% confidence interval, 64–95 Pg C yr^{-1}), 18 Tg C yr^{-1} (11–23 Tg C yr^{-1}), and 4.4 Tg N yr^{-1} (1.4–11.1 Tg N yr^{-1}), respectively. The uncertainty was the largest for the N_2O flux and smallest for the CO_2 flux. Respectively, the boreal (mean annual temperature, T<2.0°C), temperate (2.0≤T≤17.0°C), and tropical (T>17.0°C) ecosystems contribute 10%, 19%, and 70% to the total global CO_2 flux, 18%, 26%, and 56% to the total global CH_4 flux, and 5%, 18%, and 77% to the total global N_2O flux. The contribution of the tropical ecosystems was the highest for all of the gases, especially for CO_2 and N_2O.

Figure 1 shows the estimated spatial distributions of the soil CO_2 emission, CH_4 uptake, and N_2O emission fluxes; the relationships between each gas flux are shown in Figure 2. The CO_2 and N_2O fluxes showed clear spatial patterns that were controlled mainly by temperature. The fluxes were higher in the tropical regions, and they decreased at higher latitudes, yet the two gas fluxes do not always co-occur (Figure 1AC and Figure 2C). The fluxes from the $+30°$ to $-30°$ latitude belt were high for CO_2 and N_2O, but the belt seems to be wider for CO_2 than N_2O. For N_2O at the latitude regions of approximately $+30°$ and $-30°$, only the fluxes from east of North America and East Asia, east of South America, and east of Australia were high. In contrast, the CH_4 flux did not show clear temperature-induced spatial patterns. Hot spots of CH_4 uptake were observed in North and South America, Kamchatka, Japan, and New Zealand, corresponding to the distribution of highly porous soils (Andosols). The distribution patterns of the frequencies differed among the three gases (Figure 3). The CO_2 flux showed a wider and flatter range than the CH_4 flux and exhibited a relatively low peak value (300–450 g C m^{-2} yr^{-1}). The CH_4 flux has a single peak in the middle of the range. The N_2O flux had a long, right-skewed distribution, which is often observed in field studies [12]. The distributions in the histograms correspond to the spatial distribution of each gas flux. The distinct spatial distribution patterns for CO_2 and N_2O (Figure 1) resulted in the broad distributions of the CO_2 and N_2O fluxes in the histograms (Figure 3); the wide spatial distribution of the smaller flux resulted in peaks in the low values in CO_2 and N_2O flux histograms (Figure 3).

Seasonal changes in the CO_2 emission, CH_4 uptake, and N_2O emission fluxes are shown in Figure 4. Except in low-latitude regions, CO_2 and N_2O showed clear seasonality, being high during the summer and low during the winter. As observed in Figure 1, the belt of large flux around the tropical regions was narrower for N_2O than CO_2, and a north-south asymmetry can be observed for N_2O. The seasonal changes in the CH_4 flux were not as large as the other two gases. The CH_4 uptake flux was relatively higher in the middle latitudes and was high during the summer and low during the winter. The seasonality seemed to be the opposite at low latitudes ($+20°$ and $-20°$).

Discussion

I compiled reports on global soil CO_2 emission, CH_4 uptake, and N_2O emission fluxes [1–6] [9–10] [13–41], and the estimates

in this study were comparable to those of previous studies (Figure 5). The estimate for CO_2 was within the range of previous studies but was relatively smaller than the latest estimate derived from the synthesis of global data [2]. For the CH_4 uptake, the estimate in this study was intermediate among the previous estimates, and the CH_4 estimates had greater variance when compared with the CO_2 estimates. In my literature survey, the number of estimates for CH_4 was the largest among the three gases. The estimate for N_2O was of the same magnitude as the previous estimate but was relatively smaller than those of previous studies. When evaluating the uncertainty of each study, the uncertainties for the N_2O and CH_4 estimates were quite large. The uncertainty for the CO_2 flux appears to be smallest; however, it should be emphasized that the uncertainty for the CO_2 estimate would still have the highest impact on the uncertainty in terms of the global GHG budget because, among the three gases, the soil CO_2 efflux is the largest component in global warming potentials. The means of the best estimates from each study were 79 Pg C yr^{-1} (291 Pg CO_2 yr^{-1}; coefficient of variation, CV = 13%, $N = 6$) for CO_2, 21 Tg C yr^{-1} (29 Tg CH_4 yr^{-1}; CV = 24%, $N = 24$) for CH_4, and 7.8 Tg N yr^{-1} (12.2 Tg N_2O yr^{-1}; CV = 38%, $N = 11$) for N_2O. For N_2O, the earliest two estimates (the estimate of *Banin et al.* (1984) [38] and *Banin* (1986) [39], and that of *Bowden* (1986) [37]) are markedly higher than the others values. Accordingly, the mean calculated without these two estimates was 6.6 Tg N yr^{-1} (10.4 Tg N_2O yr^{-1}; CV = 22%, $N = 9$). The base years of the estimates compiled in Figure 5 vary among the estimates. Moreover, it was found that the base year of each estimate is not always stated in each reference. Because the climate is changing, and interannual climate variation should not be regarded as being negligible, the difference in the selected base years should be an important consideration. In addition, the vegetation, land cover, or soil type that was masked out in each simulation varies among these studies, which is one of the sources of variations in the estimates. The compilation presented here provides approximate overall estimates based on historic reports; however, the consideration of the different calculation conditions used in various studies is one of the important process for lessening the variation of estimates among studies. Another issue is that the source of uncertainty and the definition of uncertainty differ among studies, which hinders the comparison of uncertainty in published estimates.

More distinct spatial distribution patterns and seasonality were found for CO_2 and N_2O than for CH_4. This difference is mostly attributable to the high temperature sensitivity of CO_2 and N_2O

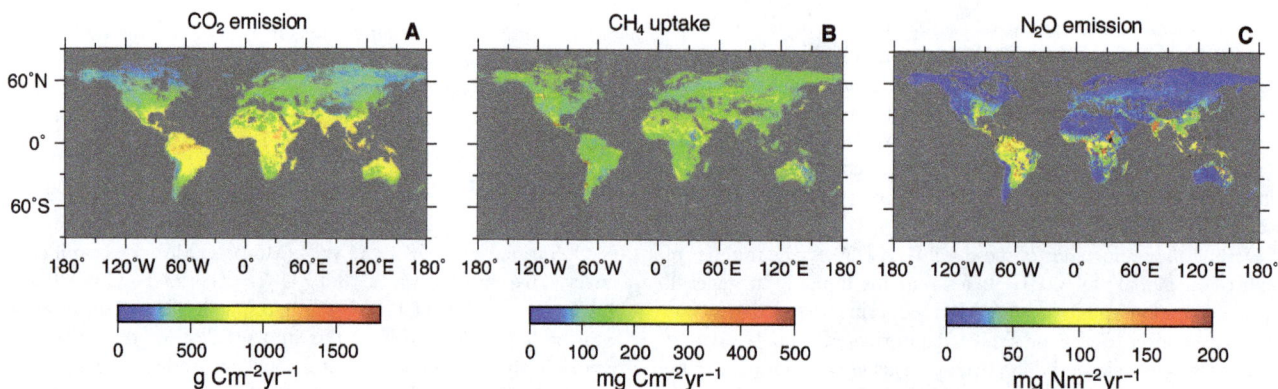

Figure 1. Global maps of the estimated rates of fluxes. (**A**) CO_2 emission flux, (**B**) CH_4 uptake flux, and (**C**) N_2O emission flux. The values are the averages between 1980 and 2009.

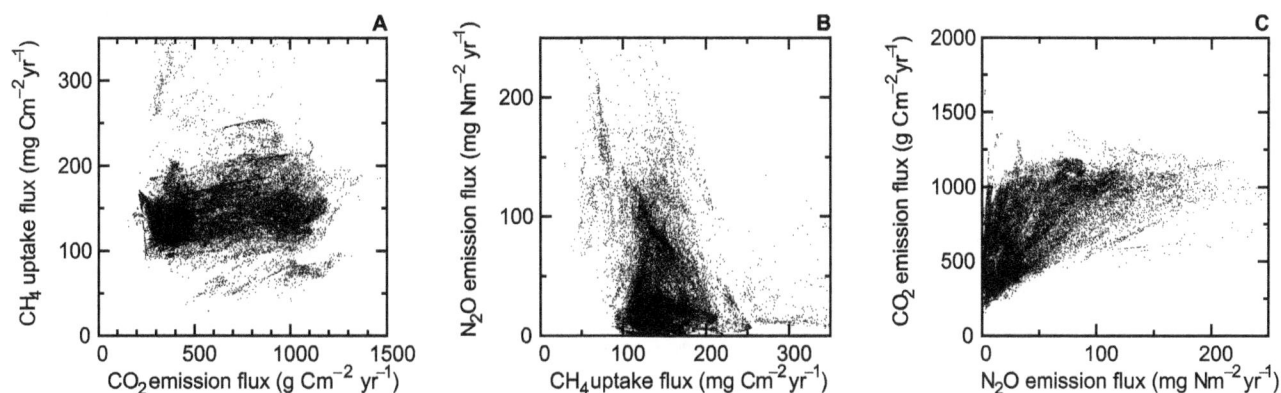

Figure 2. Relationships between each flux. (A) CO_2 emission flux and CH_4 uptake flux, (B) CH_4 uptake flux and N_2O emission flux, and (C) N_2O emission flux and CO_2 emission flux.

and the low temperature sensitivity of CH_4 in the model structure. Similar spatial distribution patterns were found in previous studies. For example, it is reported that the contribution of CO_2 flux from tropical ecosystems was 67% [2]. Also it is estimated that more than 60% of the global N_2O flux occurred via tropical forest and savanna ecosystems [6]. For the CH_4 uptake flux, the global distribution pattern still appears to vary among models [4] [21]; some studies estimated distinct spatial distribution patterns, whereas others did not. For example, four schemes for CH_4 uptake (the algorithms of *Potter et al.* (1996) [9], *Ridgewell et al.* (1999) [23], *Del Grosso et al.* (2000) [42], and *Curry* (2007) [22]) were used for global CH_4 uptake flux estimates [21]; the comparison demonstrated that the total CH_4 uptake fluxes estimated by the four schemes were comparable, but the fluxes showed the different spatial distribution patters.

One of the limitations of the model used in this study could be the simple exponential function that is used to estimate the temperature response of GHG fluxes, especially for the CO_2 flux. It has been reported that the temperature sensitivity of soil CO_2 fluxes changes depending on the temperature; in particular, it has been noted that soil CO_2 flux shows a greater temperature sensitivity at low temperatures [43]. For simplicity, the gas flux submodels used here adopts a simple exponential temperature response. This simplification may lead to errors in the estimation of soil CO_2 fluxes in cooler regions, although this limitation likely has a small effect on the global estimates because the contributions of temperate and tropical regions dominate the global soil CO_2 flux. Another limitation is that our simulations did not distinguish

between forested and agricultural areas. The gas flux submodels were parameterized using data observed in forested areas and do not include the effects of agricultural activity (e.g., N fertilizer sources). The N_2O flux, in particular, substantially differs between forested areas and agricultural areas. Therefore, the estimates reported in this study only account for so-called background emissions from agricultural areas.

An advantage of the present study is that the estimates are based on the simple data-oriented models that were data-assimilated with multi-site data using Bayesian calibration; therefore, the model estimates were well constrained by the observed data and are shown with uncertainty. To obtain more data-constrained estimates of global soil GHG fluxes, however, it would be important to constrain models with the global dataset via the data-assimilation process. One of the key factors is the development of global datasets [44]. Another key is to include necessary, though not too many, processes in the model. Simpler models are easy to data-assimilate and can provide more data-constrained estimates, but they may not be good for long-term estimations because a variety of potential feedback processes should affect the fluxes. However, too many detailed-process-oriented models can provide possible feedback processes but are not easily data-constrained with global datasets, and they increase uncertainty. Therefore, to identify essential processes in describing global soil GHG fluxes, closer collaborations between modelers and experimenters/observers and inter-model comparisons are vital.

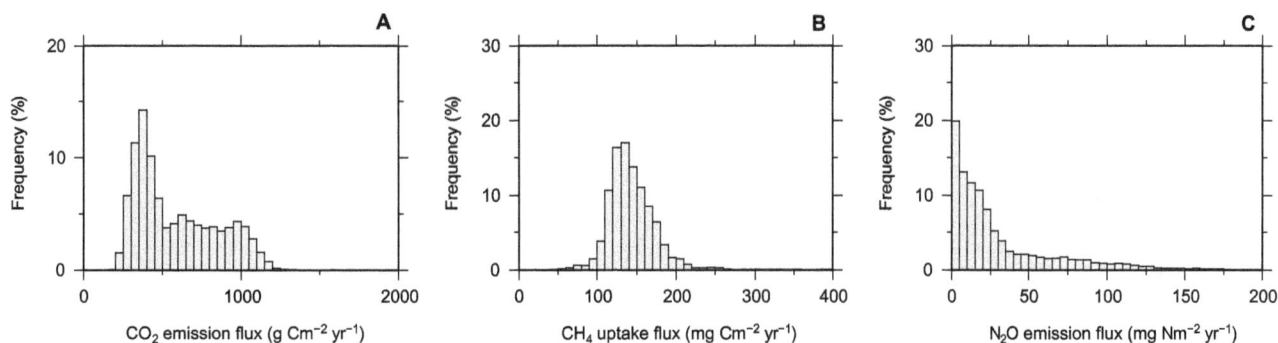

Figure 3. Histograms of modeled soil GHG fluxes by gridded cells. (A) CO_2 emission flux, (B) CH_4 uptake flux, and (C) N_2O emission flux.

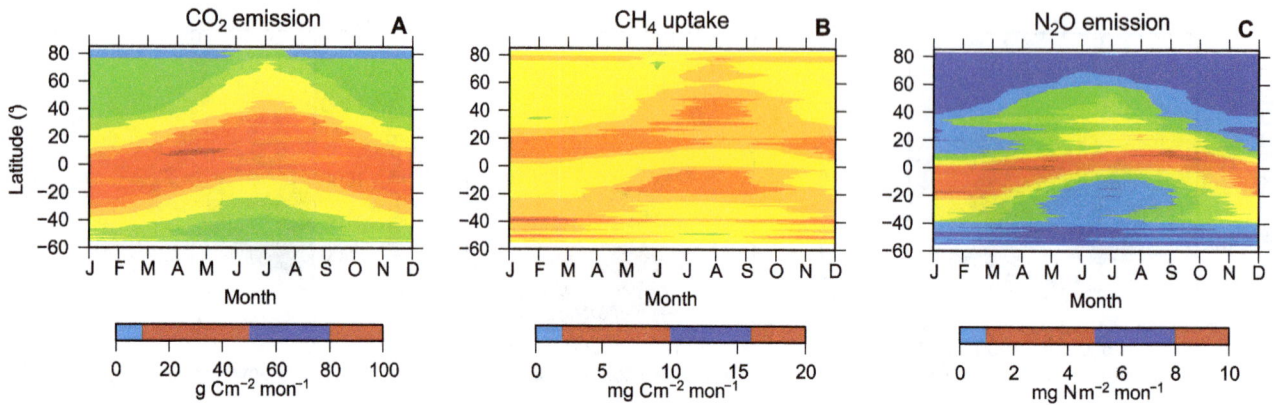

Figure 4. Seasonal and latitudinal distributions of the fluxes. (**A**) CO_2 emission flux, (**B**) CH_4 uptake flux, and (**C**) N_2O emission flux.

This study reported new global estimates of soil CO_2 emission, CH_4 uptake, and N_2O emission fluxes, which were estimated using a simple data-oriented model. The estimates were comparable to the previous estimates for all of the gases evaluated. The simulation results clearly demonstrated differences and similarities in spatial distribution patterns and in the seasonality of the three gas fluxes. The results, including both previous and the present estimates, revealed that the reported estimates vary and have large uncertainties but that the overall magnitudes are in general agreement. To lessen the uncertainty in soil GHG flux estimates

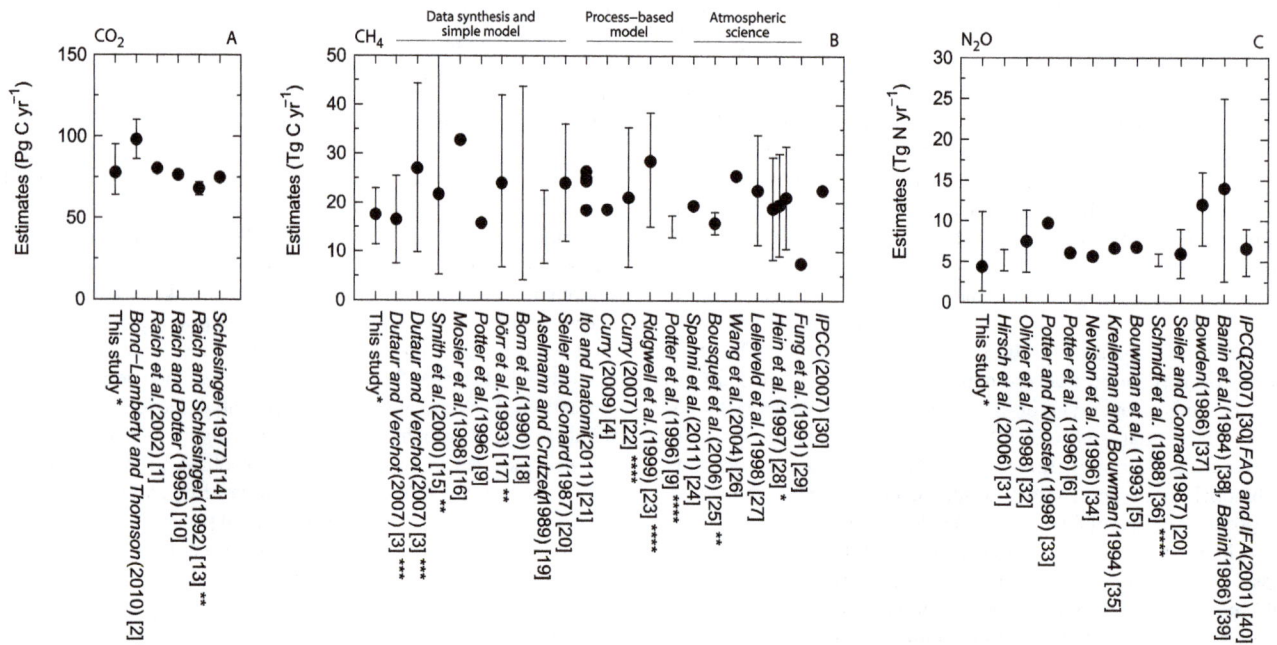

Figure 5. Comparison of the global estimates for each flux. (**A**) CO_2 emission flux, (**B**) CH_4 uptake flux, and (**C**) N_2O emission flux. The estimates are in reverse chronological order. For the CH_4 flux, the studies were divided according to the methodologies because the number of studies was large. The values in "data synthesis and simple model" include estimates from data synthesis and extrapolations. For the CO_2 flux, all estimates are from data synthesis and simple modeling. For the N_2O flux, only the estimate in *Hirsch et al.* (2006) [31] is from atmospheric inversion, and the estimates from *Potter and Klooster* (1998) [33] to *Bouwman et al.* (1993) [5] are from process-based model. Other estimates are from data synthesis. The definitions of the bars differ (*95% confidence interval; **standard deviation; ***standard error; ****based on two different model assumptions or parameters; no-mark: no uncertainty was reported or the definition of the bar could not be explicitly identified.). The higher end of the bar of *Smith et al.* (2000) [15] is 90 Tg C yr^{-1} (**B**). The values in *Ito and Inatomi* (2011) [21] are the results from four models (**B**). The values in *Hein et al.* (1997) [28] are the results from three different assumptions (**B**). The value in *Hirsch et al.* (2006) [31] is the preindustrial flux (i.e., the anthropogenic terrestrial flux enhancement was removed), and the value in *Olivier et al.* (1998) [32] is the sum of the soil microbial production, grasslands, and background emissions arable land sources (**C**). For *Banin et al.* (1984) [38] and *Banin* (1986) [39], the estimate without cultivated land is plotted (**C**). When cultivated land is include, the estimate ranges from 4 to 29 Tg N yr^{-1}. For the estimates of IPCC, only the latest estimates were included (*IPCC*, 2007) [30] (**B,C**). In this synthesis, I did not include estimates that appeared to be the citation of the estimates in IPCC reports. *Bouwman et al.* (1995) [41] reported two estimates of N_2O emission flux that were calculated by overlaying the emission inventories from *Bouwman et al.* (1993) [5] and *Kreileman and Bouwman* (1994) [35] with a new land cover database. The estimates (7.0 and 6.6 Tg N yr^{-1}) were slightly different from original estimates (6.8 and 6.7 Tg N yr^{-1}), but were approximately the same as the originals; therefore, these estimates of *Bouwman et al.* (1995) [41] were not included in this compilation.

further, it is necessary to build global databases and identify key processes in describing global soil GHG fluxes.

Materials and Methods

The SGR, a regional, simple soil greenhouse gas flux model, was used [8] [11]; the SGR model consists of submodels of soil temperature, water, and GHG fluxes (Figure S1). A monthly time step was adopted, and the inputs for the model were the monthly mean air temperature and the monthly precipitation. The soil physical and chemical properties were also required. The soil temperature submodel calculates the soil temperature using the mean air temperature and the snow cover, and the soil water submodel calculates the water-filled pore space (WFPS) using the air temperature, the potential evapotranspiration [45], and the precipitation. The soil water characteristic was estimated using the generalized soil-water relationship [46]. The Bayesian calibration scheme was used to parameterize the snow, soil temperature, WFPS, and soil gas submodels. The scheme is an optimization scheme that uses Monte Carlo sampling and a model-data synthesis scheme. In each grid, the snow cover and potential evapotranspiration were calculated using monthly air temperature and precipitation data, and the soil temperature and WFPS were subsequently simulated. Using the soil physiochemical property, WFPS, and soil temperature, the flux model for each gas yields a monthly flux. The model is described in detail elsewhere [8] [11–12], and all parameters are shown in Table S1.

Gas Flux Submodel

The SG models were used for the soil GHG fluxes [11]. In these models, each gas flux (CO_2, μg C m^{-2} s^{-1}; CH_4, μg C m^{-2} h^{-1}; and N_2O, μg N m^{-2} h^{-1}) is described by the same three factors: soil physiochemical properties, soil water, and soil temperature:

$$\text{Gas flux} = f(SP)g(WFPS)h(T), \qquad (1)$$

where $f(SP)$ is the function for the soil physiochemical properties (SP, 0–5-cm soil layer), $g(WFPS)$ is the function for the WFPS (5-cm depth), and $h(T)$ is the function for the soil temperature (5-cm depth).

The $f(SP)$ is defined as follows: the function for the CO_2 flux was defined to increase with increasing C/N ratios (CNR, 0–5-cm soil layer):

$$f(CNR) = me^{nCNR} \qquad (2)$$

The function for the CH_4 flux was defined to decrease with increasing bulk density (BD, Mg m^{-3}, 0–5-cm soil layer):

$$f(BD) = me^{-nBD} \qquad (3)$$

For the N_2O flux, the function was defined to decrease with decreasing CNR:

$$f(CNR) = me^{-nCNR} \qquad (4)$$

The function for the WFPS (5 cm) was defined by the following equation and was used for every gas model:

$$g(WFPS) = \left(\frac{WFPS - a}{b - a}\right)^d \left(\frac{WFPS - c}{b - c}\right)^{-d\frac{b-c}{b-a}}, \qquad (5)$$

where the parameters a and c are the minimum and maximum values of the WFPS, respectively (i.e., $g(a) = g(c) = 0$). Parameter b, which ranges between a and c, is the optimum parameter (i.e., $g(b) = 1$). Parameter d controls the curvature of the function, but the three other parameters also affect the shape. The function has a convex shape, and the values range from 0 to 1.

The exponential function was used for the soil temperature for every gas flux as follows:

$$h(T) = e^{pT}, \qquad (6)$$

where p is the parameter and T is the soil temperature (°C, 5 cm). The value of $h(T)$ is 1 when the soil temperature is 0°C.

The gas flux submodels were calibrated using multi-site data, which were gathered monthly in Japanese forests between 2002 and 2004 (36 sites, $N = 768$ in total for each gas flux) [11]. After parameterisation, the values of the root mean square errors (RMSE) for the CO_2, CH_4, and N_2O fluxes were 10.25 μg C m^{-2} s^{-1}, 29.29 μg C m^{-2} h^{-1}, and 5.65 μg N m^{-2} h^{-1} ($N = 768$ for each gas), respectively.

Snow Submodel

I adopted a simple snow model that calculates the snow accumulation and snowmelt based on the air temperature and the precipitation [47].

$$\text{If } T_{\text{air}} \leq T_{\text{snow}} \text{ then Snowfall} = PRE \qquad (7)$$

$$\text{If } T_{\text{air}} \geq T_{\text{melt}} \text{ then Snowmelt} = S_{\text{melt}}(T_{\text{air}} - T_{\text{melt}}) \qquad (8)$$

where T_{air} is the monthly air temperature (°C), T_{snow} is the maximum temperature at which precipitation becomes snow (°C), T_{melt} is the minimum temperature at which snowmelt occurs (°C), S_{melt} is the snow melting rate (mm°C^{-1}), and PRE is the precipitation (mm). This simple snow model was used to estimate whether soil is covered with snow. In this model, the amount/depth of snow accumulation does not affect the simulation. Instead, the model output is affected by whether the soil is covered with snow via the soil temperature submodel.

Soil Water Submodel

Because the gas flux models require the WFPS, the WFPS was calculated in the soil water submodel. First, an index of wetness was defined as follows:

$$r_i = \frac{R_{pre}PRE_i + (1 - R_{pre})PRE_{i-1}}{R_{pet}PET_i + (1 - R_{pet})PET_{i-1}} \qquad (9)$$

where r_i is the wetness index of the month (ratio). PRE_i and PRE_{i-1} are the precipitation for the month and the last month (mm), respectively, and PET_i and PET_{i-1} are the potential evapotranspiration of the month and the last month (mm), respectively. R_{pre} and R_{pet} are constants (ratio) that indicate the weights of the precipitation and potential evapotranspiration of the month, respectively. The function indicates that the wetness of the site, r_i, is affected by not only the precipitation and potential evapotranspiration of the month but also those of the last month.

Second, the WFPS was calculated using the following functions:

$$\text{If } r_i \geq 1 \text{ then } WFPS_i = S_{WW} \ln(r_i) + WS_0 \quad (10)$$

$$\text{If } r_i < 1 \text{ then } WFPS_i = S_{WD} \ln(r_i) + WS_0 \quad (11)$$

$$\text{If } T_{air} < T_W \text{ then } WFPS_i = WFPS_{i-1} \quad (12)$$

where WS_0 is a WFPS when r is 1 (or $\ln(r) = 0$) and is defined as the WFPS of a 30-kPa soil water potential and $WFPS_{i-1}$ is the WFPS of the last month. It is assumed that the WFPS does not change when the air temperature is low (lower than T_W °C) because of the low evapotranspiration and the minor amount of snowmelt.

The potential evapotranspiration was estimated using the Thornthwaite method [45], which calculates the potential evapotranspiration using the air temperature and the longitude. The generalized soil−water characteristics model [46] was used to calculate the soil water characteristics (WS_0) from the soil texture. The default parameters were used for the potential evapotranspiration submodel [45] and the soil water characteristics submodel [46].

Soil Temperature Submodel

A linear model was used for soil temperature (T_{soil}, °C): when the soil is not covered with snow, the soil temperature is calculated with a linear function of air temperature (T_{air}, °C); when soil is covered with snow, a constant temperature was assumed.

$$\text{If Snow} = 0 \text{ then } T_{soil} = T_{air} - (S_{st} T_{air} + I_{st}) \quad (13)$$

$$\text{If Snow} > 0 \text{ then } T_{soil} = T_{snowsoil} \quad (14)$$

where S_{st}, I_{st}, and $T_{snowsoil}$ are constant (°C).

Effect of Atmospheric CH₄ Concentration on CH₄ Uptake

Although uncertain feedbacks between soil nitrogen and CH_4 oxidation in soil have been suggested [48], the CH_4 uptake is generally expected to increase with the atmospheric CH_4 concentration [4]. The effect of atmospheric CH_4 was therefore included by multiplying the factor of CH_4 concentration, $j([CH_4])$, which was calculated using the relative concentration of atmospheric CH_4.

$$CH_4 \text{ flux} = f(SP)g(WFPS)h(T)j([CH_4]). \quad (15)$$

Driving Data and Simulations

The gas fluxes were evaluated with a spatial resolution of $0.5° \times 0.5°$. The air temperature and precipitation were derived from the CRU 3.1 (Climate Research Unit) climate data [49], and the global grid area data in the EOS-WEBSTER were used. The ISRIC-WISE global dataset of soil properties was used for the distribution of the soil physiochemical properties [50]. The soil physiochemical properties in the ISRIC-WISE dataset were converted to those of the 0–5-cm soil layer using ISRIC-WISE global soil profile data [51]. Soils with distinctively small bulk density (≤ 0.28 Mg m^{-3} in ISRIC-WISE) were excluded because they were presumed to be peat soils. The data of atmospheric CH_4

concentrations observed at the Ryori BAPMon station, from the GLOBALVIEW-CH4 database [52], were used to calculate $j([CH_4])$.

A Monte Carlo approach was used to evaluate the uncertainty of the estimates. For each simulation, new parameters were chosen from the uncertainty for each parameter, as determined through the Bayesian calibration. A normal distribution with a 10% coefficient of variance was assumed for each parameter that did not undergo Bayesian calibration. The model was run 1000 times, and the results were analyzed using the R statistical computing software (version 2.11.1). The codes for the SGR and Bayesian calibration were written in C.

Here, the average CO_2 emission flux, CH_4 uptake flux, and N_2O emission flux between 1980 and 2009 are shown. The SGR models do not include CH_4 emissions; therefore, this study focuses on the soil CH_4 uptake. Areas of ice, permanent water, mangrove, and peat soils (see above) were masked out. The cultivated area was included in this study.

Comparison with Data from a Global Database of Soil CO₂ Flux (Soil Respiration)

A global database of soil CO_2 flux (soil respiration) was released recently (https://code.google.com/p/srdb/) [44]. Although the mismatch in scale between site-scale measurements and the coarse resolution of the simulation ($0.5° \times 0.5°$) should be an issue, the results of the simulation were compared with the data in the database (version 20100517a). For the comparison, the data from non-agricultural ecosystems without experimental manipulation measured using infrared gas analyzer or gas chromatography were extracted. The data with quality check flags, except for Q01, Q02, and Q03, were excluded (please see the database). A total of 1464 data points met the above conditions, and 1246 data points where the measurement locations (latitude and longitude) corresponded to the simulated area were included. The comparison showed that the two agreed in their magnitude and were positively correlated ($R = 0.43$) (Figure S2). However there was some mismatches: the variation in the simulated values was less than that of observed data points. In particular, the simulation did not produce large fluxes (e.g. >1500 g C m^{-2} yr^{-1}). This difference is partly due to the different scale in the field measurements and the simulation. The second difference is that the fluxes generated by the simulation were smaller than those of the database. This difference resulted in the gap between the estimate in this study and the global estimate reported by Bond-Lamberty (2010) [2], which is based on the global database (Figure 5). This gap would suggest that the global estimate substantially varies depending on the data used to constrain the model, although the differences in model structures and the scale mismatch between measurements and simulations should be taken into account.

Supporting Information

Figure S1 Schematic diagram of the modeling approach.

Figure S2 Comparison between data in a global dataset [44] and those of the simulations. The data from non-agricultural ecosystems without experimental manipulation measured using infrared gas analyzer or gas chromatography were extracted. The data with quality check flags, except for Q01, Q02, and Q03, were excluded (please see the database). A total of 1464 data points met the above conditions, and 1246 data points where the measurement locations (latitude and longitude) corresponded to the simulated area were included. The broken line is y = 0.17x+418 ($P<0.0001$). The Pearson's correlation coefficient was 0.43.

Acknowledgments

I thank Shigehiro Ishizuka, Tadashi Sakata, and Tomoaki Morishita for their discussions. I am also grateful for the constructive comments of Carl J. Bernacchi, Ben Bond-Lamberty, and one anonymous reviewer.

References

1. Raich JW, Potter CS, Bhagawati D (2002) Interannual variability in global soil respiration, 1980−94. Glob Change Biol 8: 800–812.
2. Bond-Lamberty B, Thomson A (2010) Temperature-associated increases in the global soil respiration record. Nature 464: 579–582.
3. Dutaur L, Verchot LV (2007) A global inventory of the soil CH_4 sink. Global Biogeochem Cy 21: GB4013, doi: 10.1029/2006GB002734.
4. Curry CL (2009) The consumption of atmospheric methane by soil in a simulated future climate. Biogeosciences 6: 2355–2367.
5. Bouwman AF, Fung I, Matthews E, John J (1993) Global analysis of the potential for N_2O production in natural soils. Global Biogeochem Cy 7: 557–597.
6. Potter CS, Matson PA, Vitousek PM, Davidson EA (1996) Process modeling of controls on nitrogen trace gas emissions from soils worldwide. J Geophys Res 101: 1361–1377.
7. Chapuis-Lardy L, Wrage N, Metay A, Chotte J-L, Bernoux M (2007) Soils, a sink for N_2O? A review. Glob Change Biol 13: 1–17.
8. Hashimoto S, Morishita T, Sakata T, Ishizuka S (2011) Increasing trends of soil greenhouse gas fluxes in Japanese forests from 1980 to 2009. doi: 10.1038/srep00116.
9. Potter CS, Davidson EA, Verchot LV (1996) Estimation of global biogeochemical controls and seasonality in soil methane consumption. Chemosphere 32(11): 2219–2246.
10. Raich JW, Potter CS (1995) Global patterns of carbon dioxide emissions from soils. Global Biogeochem Cy 9: 23–36.
11. Hashimoto S, Morishita T, Sakata T, Ishizuka S, Kaneko S, et al. (2011) Simple models for soil CO_2, CH_4, and N_2O fluxes calibrated using a Bayesian approach and multi-site data. Ecol Modell 222: 1283–1292.
12. Morishita T, Sakata T, Takahashi M, Ishizuka S, Mizoguchi T, et al. (2007) Methane uptake and nitrous oxide emission in Japanese forest soils and their relationship to soil and vegetation types. Soil Sci Plant Nutr 53(5): 678–691.
13. Raich JW, Schlesinger WH (1992) The global carbon dioxide flux in soil respiration and its relationship to vegetation and climate. Tellus Ser B 44: 81–99.
14. Schlesinger WH (1977) Carbon balance in terrestrial detritus. Annu Rev Ecol Syst 8: 51–81.
15. Smith KA, Dobbie KE, Ball BC, Bakken LR, Sitaula BK, et al. (2000) Oxidation of atmospheric methane in Northern European soils, comparison with other ecosystems, and uncertainties in the global terrestrial sink. Glob Change Biol 6: 791–803.
16. Mosier AR, Duxbury JM, Freney JR, Heinemeyer O, Minami K, et al. (1998) Mitigating agricultural emissions of methane. Clim Change 40: 39–80.
17. Dörr H, Katruff L, Levin I (1993) Soil texture parameterization of the methane uptake in aerated soils. Chemosphere 26: 697–713.
18. Born M, Dörr H, Levin I (1990) Methane consumption in aerated soils of the temperate zone. Tellus Ser B 42: 2–8.
19. Aselmann I, Crutzen PJ (1989) Global distribution of natural freshwater wetlands and rice paddies, their net primary productivity, seasonality and possible methane emissions. J Atm Chem 8: 307–358.
20. Seiler W, Conrad R (1987) Contribution of tropical ecosystems to the global budgets of trace gases, especially CH_4, H_2, CO, and N_2O. In: Dickinson RE editor. Geophysiology of Amazonia: Vegetation and climate interactions. New York: Wiley and Sons. 133–160.
21. Ito A, Inatomi M (2012) Use of a process-based model for assessing the methane budgets of global terrestrial ecosystems and evaluation of uncertainties. Biogeosciences 9: 759–773.
22. Curry CL (2007) Modeling the soil consumption of atmospheric methane at the global scale. Global Biogeochem Cy 21: GB4012, doi: 10.1029/2006GB002818.
23. Ridgwell AJ, Marshall SJ, Gregson K (1999) Consumption of atmospheric methane by soils: a process-based model. Global Biogeochem Cy 13: 59–70.
24. Spahni R, Wania R, Neef L, van Weele M, Pison I, et al. (2011) Constraining global methane emissions and uptake by ecosystems. Biogeosciences 8: 1643–1665.
25. Bousquet P, Ciais P, Miller JB, Dlugokencky EJ, Hauglustaine DA, et al. (2006) Contribution of anthropogenic and natural sources to atmospheric methane variability. Nature 443: 439–443.
26. Wang JS, Logan JA, McElroy MB, Duncan BN, Megretskaia IA, et al. (2004) A 3-D model analysis of the slowdown and interannual variability in the methane growth rate from 1988 to 1997. Global Biogeochem Cy 18: GB3011, doi: 10.1029/2003GB002180.
27. Lelieveld J, Crutzen PJ, Dentener FJ (1998) Changing concentration, lifetime and climate forcing of atmospheric methane. Tellus Ser B 50: 128–150.
28. Hein R, Crutzen PJ, Heimann M (1997) An inverse modeling approach to investigate the global atmospheric methane cycle. Global Biogeochem Cy 11: 43–76.
29. Fung I, John J, Lerner J, Matthews E, Prather M, et al. (1991) Three-dimensional model synthesis of the global methane cycle. J Geophys Res 96: 13033–13065.
30. IPCC (2007) Climate change 2007: The physical science basis. Cambridge: Cambridge University Press, 996 p.
31. Hirsch AI, Michalak AM, Bruhwiler LM, Peters W, Dlugokencky EJ, et al. (2006) Inverse modeling estimates of the global nitrous oxide surface flux from 1998−2001. Global Biogeochem Cy 20: GB1008, doi: 10.1029/2004GB002443.
32. Olivier JGJ, Bouwman AF, Van der Hoek KW, Berdowski JJM (1998) Global air emission inventories for anthropogenic sources of NO_x, NH_3 and N_2O in 1990. Environ Pollut 102: 135–148.
33. Potter CS, Klooster SA (1998) Interannual variability in soil trace gas (CO_2, N_2O, NO) fluxes and analysis of controllers on regional to global scales. Global Biogeochem Cy 12: 621–635.
34. Nevison CD, Esser G, Holland EA (1996) A global model of changing N_2O emissions from natural and perturbed soils. Clim Change 32: 327–378.
35. Kreileman GJJ, Bouwman AF (1994) Computing land use emissions of greenhouse gases. Water Air Soil Poll 76: 231–258.
36. Schmidt J, Seiler W, Conrad R (1988) Emission of nitrous oxide from temperate forest soils into the atmosphere. J Atm Chem 6: 95–115.
37. Bowden WB (1986) Gaseous nitrogen emissions from undisturbed terrestrial ecosystems: an assessment of their impacts on local and global nitrogen budgets. Biogeochemistry 2: 249–279.
38. Banin A, Lawless JG, Whitten RC (1984) Global N_2O cycles–Terrestrial emissions, atmospheric accumulation and biospheric effects. Adv Space Res 4: 207–216.
39. Banin A (1986) Global budget of N_2O: The role of soils and their change. Sci Total Env 55: 27–38.
40. FAO and IFA (2001) Global estimates of gaseous emissions of NH_4, NO and N_2O from agricultural land. Rome, 106 p.
41. Bouwman AF, Van der Hoek KW, Olivier JGJ (1995) Uncertainties in the global source distribution of nitrous oxide. J Geophys Res 100: 2785–2800.
42. Del Grosso SJ, Parton WJ, Mosier AR, Ojima DS, Potter CS, et al. (2000) General CH_4 oxidation model and comparisons of CH_4 oxidation in natural and managed systems. Global Biogeochem Cy 14: 999–1019.
43. Kirschbaum MUF (1995) The temperature dependence of soil organic matter decomposition, and the effect of global warming on soil organic C storage. Soil Biol Biochem 27: 753–760.
44. Bond-Lamberty B, Thomson A (2010) A global database of soil respiration data. Biogeosciences 7: 1915–1926.
45. Thornthwaite CW (1948) An approach toward a rational classification of climate. Geogr Rev 38: 55–94.
46. Saxton KE, Rawls WJ, Romberger JS, Papendick RI (1986) Estimating generalized soil-water characteristics from texture. Soil Sci Soc Am J 50: 1031–1036.
47. Thornton PE, Hasenauer H, White MA (2000) Simultaneous estimation of daily solar radiation and humidity from observed temperature and precipitation: an application over complex terrain in Austria. Agr Forest Meteorol 104: 255–271.
48. King GM, Schnell S (1994) Effect of increasing atmospheric methane concentration on ammonium inhibition of soil methane consumption. Nature 370: 282–284.
49. Mitchell TD, Jones PD (2005) An improved method of constructing a database of monthly climate observations and associated high-resolution grids. Int J Clim 25: 693–712.
50. Batjes NH (2005) ISRIC-WISE global data set of derived soil properties on a 0.5 by 0.5 degree grid (Version 3.0). Wageningen: ISRIC-World Soil Information.
51. Batjes NH (1995) A homogenized soil data file for global environmental research: a subset of FAO, ISRIC and NRCS profiles (Version 1.0). Wageningen: International Soil Reference and Information Centre.
52. GLOBALVIEW-CH4 (2009) Cooperative Atmospheric Data Integration Project - Methane. CD-ROM, NOAA ESRL, Boulder, Colorado [Also available on Internet via anonymous FTP to ftp.cmdl.noaa.gov, Path: ccg/ch4/GLOBALVIEW].

Author Contributions

Conceived and designed the study: SH. Performed modeling and analyses: SH. Wrote the manuscript: SH.

PERMISSIONS

LIST OF CONTRIBUTORS

Philippe Saner and Andy Hector
Institute of Evolutionary Biology and Environmental Studies, University of Zurich, Zurich, Switzerland

Yen Yee Loh
School of International Tropical Forestry, Universiti Malaysia Sabah, Sabah, Malaysia

Robert C. Ong
Forest Research Center Institute, Sabah, Malaysia

Yan Geng and Jin-Sheng He
Department of Ecology, College of Urban and Environmental Sciences, and Key Laboratory for Earth Surface Processes of the Ministry of Education, Peking University, Beijing, China
Key Laboratory of Adaptation and Evolution of Plateau Biota, Northwest Institute of Plateau Biology, Chinese Academy of Sciences, Xining, China

Yonghui Wang, Kuo Yang and Shaopeng Wang
Department of Ecology, College of Urban and Environmental Sciences, and Key Laboratory for Earth Surface Processes of the Ministry of Education, Peking University, Beijing, China

Hui Zeng
Department of Ecology, College of Urban and Environmental Sciences, and Key Laboratory for Earth Surface Processes of the Ministry of Education, Peking University, Beijing, China
Shenzhen Key Laboratory of Circular Economy, Shenzhen Graduate School, Peking University, Shenzhen, China

Frank Baumann, Peter Kuehn and Thomas Scholten
Department of Geoscience, Physical Geography and Soil Science, University of Tuebingen, Tuebingen, Germany

Bing Gao, Xiaotang Ju, Fang Su, Fengbin Gao, Qingsen Cao, Peter Christie, Xinping Chen and Fusuo Zhang
College of Resources and Environmental Sciences, China Agricultural University, Beijing, China

Oene Oenema
Wageningen University and Research Center, Alterra, Wageningen, The Netherlands

Shuli Niu and Linghao Li
State Key Laboratory of Vegetation and Environmental Change, Institute of Botany, Chinese Academy of Sciences, Xiangshan, Beijing, China

Bing Song, Zhe Zhang and Haijun Yang
State Key Laboratory of Vegetation and Environmental Change, Institute of Botany, Chinese Academy of Sciences, Xiangshan, Beijing, China
Graduate School of Chinese Academy of Sciences, Yuquanlu, Beijing, China

Shiqiang Wan
Key Laboratory of Plant Stress Biology, College of Life Sciences, Henan University, Kaifeng, Henan, China

Wei Feng, Yuqing Zhang, Xin Jia, Bin Wu, Tianshan Zha, Shugao Qin, Ben Wang, Chenxi Shao, Jiabin Liu and Keyu Fa
Yanchi Research Station, College of Soil and Water Conservation, Beijing Forestry University, Beijing, China

Evgenia Blagodatskaya
Soil Science of Temperate Ecosystems, Büsgen-Institute, University of Göttingen, Göttingen, Germany
Institute of Physicochemical and Biological Problems in Soil Science, Russian Academy of Sciences, Pushchino, Russia
Agricultural Soil Science, Büsgen-Institute, University of Göttingen, Göttingen, Germany

Sergey Blagodatsky
Institute of Physicochemical and Biological Problems in Soil Science, Russian Academy of Sciences, Pushchino, Russia
Institute for Plant Production and Agroecology in the Tropics and Subtropics, University of Hohenheim, Stuttgart, Germany

Traute-Heidi Anderson
Thünen-Institute of Climate-Smart Agriculture (vTI), Braunschweig, Germany

Yakov Kuzyakov
Soil Science of Temperate Ecosystems, Büsgen-Institute, University of Göttingen, Göttingen, Germany
Agricultural Soil Science, Büsgen-Institute, University of Göttingen, Göttingen, Germany

Tana E. Wood
International Institute of Tropical Forestry, USDA
Forest Service, Río Piedras, Puerto Rico, United States
of America
Fundación Puertorriqueña de Conservación, San Juan,
Puerto Rico, United States of America

Matteo Detto
Smithsonian Tropical Research Institute, Apartado
Balboa, Republic of Panama

Whendee L. Silver
Department of Environmental Science, Policy and
Management, University of California, Berkeley,
California, United States of America

Wei Wang, Wenjing Zeng and Weile Chen
Department of Ecology, College of Urban and
Environmental Sciences, and Key Laboratory for Earth
Surface Processes of the Ministry of Education, Peking
University, Beijing, China

Yuanhe Yang
Institute of Botany, The Chinese Academy of Sciences,
Beijing, China

Hui Zeng
Shenzhen Graduate School, Key Laboratory for Urban
Habitat Environmental Science and Technology,
Peking University, Shenzhen, China

Liming Yan
State Key Laboratory of Vegetation and Environmental
Change, Institute of Botany, Chinese Academy of
Sciences, Beijing, China
School of Life Sciences, Fudan University, Shanghai,
China

Shiping Chen
State Key Laboratory of Vegetation and Environmental
Change, Institute of Botany, Chinese Academy of
Sciences, Beijing, China

Jianyang Xia and Yiqi Luo
Department of Microbiology and Botany, University
of Oklahoma, Norman, Oklahoma, United States of
America

Zhijian Mu
Chongqing Key Laboratory of Soil Multi-scale
Interfacial Processes, College of Resources &
Environment, Southwest University, Chongqing,
China

Aiying Huang
College of Agronomy & Biotechnology, Southwest
University, Chongqing, China

Jiupai Ni and Deti Xie
Chongqing Engineering Research Center for
Agricultural Non-point Source Pollution Control
in Three -Gorges Region, College of Resources &
Environment, Southwest University, Chongqing,
China

Jianjian Wang
Key Laboratory of Resource Plants, Beijing Botanical
garden, West China Subalpine Botanical Garden,
Institute of Botany, Chinese Academy of Sciences,
Beijing, China
University of Chinese Academy of Sciences, Beijing,
China

Yuan Tian and Xi Chen
Xinjiang Institute of Ecology and Geography, Chinese
Academy of Sciences, Urumqi, Xinjiang, China

**Xuechun Zhao, Lianhe Jiang, Yuanrun Zheng and
Liming Lai**
Key Laboratory of Resource Plants, Beijing
Botanical garden, West China Subalpine
Botanical Garden, Institute of Botany,
Chinese Academy of Sciences, Beijing,
China

Yong Gao
Inner Mongolia Agricultural University, Hohhot, Inner
Mongolia, China

Shaoming Wang
Key Laboratory of Oasis Ecological Agriculture of
Xinjiang Bingtuan, Shihezi, Xinjiang, China

**Fei Peng, Quangang You, Manhou Xu, Jian Guo, Tao
Wang and Xian Xue**
Key Laboratory of Desert and Desertification, Chinese
Academy of Sciences, Cold and Arid Regions
Environmental and Engineering Research Institute,
Chinese Academy of Sciences, Lanzhou, China

**Chelsea Arnold, Teamrat A. Ghezzehei and Asmeret
Asefaw Berhe**
School of Natural Sciences, University of California
Merced, Atwater, California, United States of America

Ni Huang, Li Wang, Pengyu Hao and Zheng Niu
The State Key Laboratory of Remote Sensing Science,
Institute of Remote Sensing and Digital Earth, Chinese
Academy of Sciences, Beijing, China

Yiqiang Guo
Land Consolidation and Rehabilitation Center,
Ministry of Land and Resources, Beijing, China

Mioko Ataka and Makoto Tani
Laboratory of Forest Hydrology, Division of Environmental Science and Technology, Graduate School of Agriculture, Kyoto University, Kyoto, Japan,

Yuji Kominami, Kenichi Yoshimura and Takafumi Miyama
Kansai Research Center, Forestry and Forest Products Research Institute (FFPRI), Kyoto, Japan,

Mayuko Jomura
College of Bioresource Sciences, Nihon University, Fujisawa, Kanagawa, Japan

Annelein Meisner
Department of Terrestrial Ecology, Netherlands Institute of Ecology (NIOO-KNAW), Wageningen, The Netherlands

Wietse de Boer
Department of Microbial Ecology, Netherlands Institute of Ecology (NIOO-KNAW), Wageningen, The Netherlands,

Johannes H. C. Cornelissen
Systems Ecology, Department of Ecological Science, Faculty of Earth and Life Sciences, Vrije Universiteit (VU) Amsterdam, Amsterdam, The Netherlands

Wim H. van der Putten
Department of Terrestrial Ecology, Netherlands Institute of Ecology (NIOO-KNAW), Wageningen, The Netherlands
Laboratory of Nematology, Wageningen University, Wageningen, The Netherlands

Qi Deng
South China Botanical Garden, Chinese Academy of Sciences, Guangzhou, China
Wuhan Botanical Garden, Chinese Academy of Sciences, Wuhan, China

Dafeng Hui
Department of Biological Sciences, Tennessee State University, Nashville, Tennessee, United States of America

Deqiang Zhang, Guoyi Zhou, Juxiu Liu, Shizhong Liu, Guowei Chu and Jiong Li
South China Botanical Garden, Chinese Academy of Sciences, Guangzhou, China

Wei Wang and Jingyun Fang
Department of Ecology, College of Urban and Environmental Sciences, and Key Laboratory for Earth Surface Processes of the Ministry of Education, Peking University, Beijing, China

Wenjing Zeng and Weile Chen
Department of Ecology, College of Urban and Environmental Sciences, and Key Laboratory for Earth Surface Processes of the Ministry of Education, Peking University, Beijing, China
Key Laboratory for Urban Habitat Environmental Science and Technology, Peking University Shenzhen Graduate School Shenzhen, China

Hui Zeng
Key Laboratory for Urban Habitat Environmental Science and Technology, Peking University Shenzhen Graduate School Shenzhen, China

Zhiyong Zhou and Chao Guo
Ministry of Education Key Laboratory for Silviculture and Conservation, Beijing Forestry University, Beijing, China
The Institute of Forestry and Climate Change Research, Beijing Forestry University, Beijing, China

He Meng
College of Forestry, Inner Mongolia Agriculture University, Hohhot, China

Shoji Hashimoto
Department of Forest Site Environment, Forestry and Forest Products Research Institute (FFPRI), 1 Matsunosato, Tsukuba, Ibaraki, Japan

Index